MOLECULAR THERMODYNAMICS

PHYSICAL CHEMISTRY MONOGRAPH SERIES

Edited by Walter Kauzmann, Princeton University

MOLECULAR THERMODYNAMICS

RICHARD E. DICKERSON, *California Institute of Technology*

RATES AND MECHANISMS OF CHEMICAL REACTIONS

W. C. GARDINER, JR., *University of Texas*

QUANTUM MECHANICS IN CHEMISTRY, Second Edition

MELVIN W. HANNA, *University of Colorado*

THERMAL PROPERTIES OF MATTER

WALTER KAUZMANN, *Princeton University*

VOLUME I: KINETIC THEORY OF GASES

VOLUME II: THERMODYNAMICS AND STATISTICS: WITH APPLICATIONS TO GASES

INTRODUCTION TO CRYSTALLOGRAPHY

DONALD E. SANDS, *University of Kentucky*

MOLECULAR THERMODYNAMICS

Richard E. Dickerson

California Institute of Technology

W. A. BENJAMIN, INC.
Menlo Park, California • Reading, Massachusetts
London • Amsterdam • Don Mills, Ontario • Sydney

MOLECULAR THERMODYNAMICS

Library of Congress Catalog Card Number 70–794801

MANUFACTURED IN THE UNITED STATES OF AMERICA

The manuscript was put into production on January 13, 1969; this volume was published on August 29, 1969

ISBN 0-8053-2363-5
HIJKLMNO-HA-798

W. A. BENJAMIN, INC., Menlo Park, California

To Earl S. Dickerson,
who taught me that teaching
is an honorable profession.

Editor's Foreword

THOSE RESPONSIBLE for introducing physical chemistry to students must make difficult decisions in selecting the topics to be presented. Molecular physics and quantum mechanics have been added to the subject matter of physical chemistry in the past 35 years and are essential to the chemist's training. Yet many of the more classical areas of physical chemistry continue to be important, not only in chemistry, but also to an increasing extent in biology, geology, metallurgy, engineering, and medicine. Consequently there is pressure on teachers and textbook writers to cover more and more material, but the time available in the curriculum has not increased correspondingly, and there is a limit to the size of any textbook. Furthermore, it is difficult for any one author to write with authority about all of physical chemistry.

This text–monograph series is an attempt to make it easier to deal with this problem. The important basic topics of physical chemistry are covered at an introductory level in relatively brief, interrelated volumes. The volumes are written in such a way that if a topic of special interest to a student may not have been included in the course to which he has been exposed, he can learn about it through self-study. Consequently instructors can feel less reluctant to omit or condense material in their courses, and flexibility will be possible in the course plan, both from year to year and from institution to institution, in accordance with changing demands. The introductory presentation of physical chemistry in this form has the additional advantage that it permits a more detailed explanation of difficult points. It also permits the occasional inclusion of more advanced material to which the instructor can refer the more highly motivated students.

WALTER KAUZMANN

Princeton, New Jersey
March 1969

Authors' Foreword

PHYSICAL CHEMISTRY has been defined (by a practitioner) as "That part of chemistry which is fundamental, molecular, and interesting." Although this may be a useful guide in selecting a research problem, it is of very little help in planning an introductory course. Such a course should not be put together from a little bit of everything that has ever been called physical chemistry. Instead, it should concentrate on fundamentals so that wherever a student turns later, he can build on a secure foundation.

Physical chemistry has the general task of explaining the causes of chemical behavior. The essential, irreducible fundamentals of the subject are four in number:

1. Quantum mechanics: the mechanics of atoms and of their combinations in molecules;
2. Statistical mechanics: the framework by which molecular properties can be related to the macroscopic behavior of chemical substances;
3. Thermodynamics: the study of energy and order–disorder, and their connections with chemical changes and chemical equilibrium;
4. Kinetics: the study of the rates of chemical reactions and of the molecular processes by which reactions occur.

Many additional topics are found in introductory physical chemistry textbooks. These include methods of molecular structure determination, the several branches of spectroscopy, electrochemistry, surface chemistry, macromolecules, photochemistry, nuclear and radiation chemistry, and theories of condensed phases. These are essentially applications of the fundamental concepts, and in our books they are taught as such. The relative emphasis given to these topics will vary with the nature and level of the course, the needs of the students, and the inclination of the instructor. Yet a secure grounding in the four fundamentals will give a chemist not everything that he needs to know in physical chemistry, but the ability to recognize and learn what he needs to know as circumstances arise.

These books are an outgrowth of our experience in teaching the basic physical chemistry course over the past few years at the Universities of Illinois, Colorado, and Texas, and the California Institute of Technology. Each of us has written the part of the course that he knows best. Hopefully this approach will avoid the arid style of a book-by-committee, and yet allow each topic to be covered by an author who is vitally concerned with it. Although the present order—quantum mechanics, thermodynamics, kinetics—appears to us to be the most desirable pedagogically and the most obviously unified to the student, the material has been written so thermodynamics can precede quantum mechanics as in the more traditional course.

The authors feel strongly that basic physical chemistry should be presented as a unified whole rather than as a collection of disparate and difficult topics. We hope that our books will serve that end.

MELVIN W. HANNA
RICHARD E. DICKERSON
W. C. GARDINER, JR.

March 1969

Preface

THIS BOOK contains the thermodynamics part of the basic physical chemistry course as it is taught to juniors and seniors at the California Institute of Technology. It is preceded by an introduction to quantum mechanics, although the dependence of this book on quantum mechanics is small, and is followed by a unit on chemical kinetics. These three topics, quantum mechanics, thermodynamics, and kinetics, form the foundation on which all the rest of physical chemistry is built.

It has been a tradition to teach thermodynamics without specific reference to molecular theory. Thermodynamics, of course, does not depend for its validity on a particular molecular theory. It stands alone and is based on three great summaries of experience about the way in which the world operates. The feeling has been that it is poor practice to interweave thermodynamics and molecular theory, lest the student think that the former is in some sense derived from the latter. It is almost as though some great upheaval in molecular theory were anticipated, from the effects of which the independent study of thermodynamics should be shielded.

This I believe to be a mistake at the undergraduate level. The time for elegant and satisfying exercises in logic and reasoning is after the language has been mastered. I doubt very much that many people appreciate the beauty of the edifice of thermodynamics the first time through, simply because they are too busy memorizing the shapes of the bricks. A graduate course in pure classical thermodynamics will be all the more meaningful with a prior feeling for the meanings of quantities that are being manipulated in terms of the microscopic behavior of the real world.

It is true that you can reach most destinations on one leg. But if the object of the effort is to get somewhere and not primarily to be admired for your agility, then why hop when you can walk with both feet on the ground? The other leg in this analogy, of course, is molecular theory. The purpose of this book is to teach the student to use both feet equally without confusing the two.

I would like to thank Professors Walter Kauzmann, J. M. Anderson, Melvin Hanna, and William Gardiner for their detailed criticism of the manuscript. They have led to its being immeasurably strengthened in many places. I would also like to thank Mrs. Joyce Lundstedt, who helped me battle through several typed drafts, and my wife Lola, who kept the family at bay while the battle raged.

RICHARD E. DICKERSON

Pasadena, California
January 1969

Contents

Chapter 5 Thermodynamics of Phase Changes and Chemical Reactions **222**

Chapter 6 Solutions **300**

Chapter 7 Thermodynamics and Living Systems **387**

Chapter 1

INTRODUCTION: QUANTUM MECHANICS

1-1 FRAMEWORK OF QUANTUM MECHANICS

ALTHOUGH THIS book is designed to follow an elementary introduction to quantum mechanics, its dependence on that subject is brief and is summarized in the next few pages. Students who have had an introduction to the subject can consider this as a short review or can go immediately to Chapter 2. Those who have not had such an introduction will still have little trouble with the rest of the book if they can accept the results in this chapter at face value and look up the justification for these results at their own leisure.

Classical Newtonian mechanics is based on a differential equation that relates the behavior of a particle (or a system of particles) to its properties and to the forces exerted on it:

$$f = m\mathbf{a} = m\frac{d^2\mathbf{r}}{dt^2} \tag{1-1}$$

The more general quantum mechanics, which superseded classical mechanics and which contains it as a special case, is also based on a differential equation relating behavior to properties and to influences, now represented as a potential V rather than as a collection of forces f:

$$\frac{\partial^2\psi}{\partial x^2} + \frac{\partial^2\psi}{\partial y^2} + \frac{\partial^2\psi}{\partial z^2} + \frac{8\pi^2 m}{h^2}(E - V)\psi = 0 \tag{1-2}$$

This is Schrödinger's wave equation, and the wave functions ψ, which are the solutions of this equation, describe not the exact position of the particle, but the *probability* of the particle's being at any chosen point. Already, even at the level of a single particle, the old mechanistic sureness of nineteenth-century physics is slipping away.

The solutions of this equation under different conditions can be found in books such as Hanna's *Quantum Mechanics in Chemistry*; only the results are given here. Like any wave equation, the Schrödinger equation has an infinite number of possible solutions, only a certain set of which will be compatible with the physical boundary conditions of the experiment. For a particle in a state described by a particular wave function ψ, the probability of finding the particle within an element of volume $d\mathbf{r}$ is $|\psi|^2\,d\mathbf{r}$. This situation requires that ψ be a single-valued function of the position vector $\mathbf{r}$; there cannot be two simultaneous but different probabilities of finding a particle at a given point. It must be finite at all points as well; for if ψ went to infinity at any point, the probability of finding the particle anywhere else would be zero. Moreover, ψ must be a smooth continuous function of position; probabilities do not jump erratically. Finally, if two particles in a system of particles are absolutely undistinguishable, then a formal relabeling of these two particles must leave $|\psi|^2$ unchanged, and can at most only change the sign of ψ. Other specific boundary conditions arise from the conditions of the particular application.

Each wave function ψ is associated with a specific energy E; thus the fact that only certain wave functions are acceptable means that only certain energy states are available to the particle or system. This phenomenon of the quantization of energy states is the most striking difference between the old and the new mechanics. Quantization does occur in classical mechanics; the analogy of a vibrating string has been worked to death because it is a good analogy. The boundary condition that the ends of the string be motionless restricts its vibration to certain specific frequencies.

When using quantum mechanics, one must do the following.

(1) Formulate the proper boundary conditions for the mechanical system of interest.

(2) Solve the wave equation with these boundary conditions to find out what energies E and what wave functions ψ are permissible.

(3) Use these ψ and E values to calculate observable properties of the system.

In this book we are concerned only with the possible energy states that a particle or a system of particles can have.

In the solution of the wave equation often several different wave functions have the same energy E_j. These are called *degenerate wave functions*, and the number of such wave functions or states with the same energy is the degeneracy of the energy level g_j. When several states have a common quantum number, it is also possible to speak of the degeneracy of a particular quantum state. Thus, in the undisturbed hydrogen atom, the degeneracy of the nth principal quantum level, with energy E_n, is n^2, while within this level the degeneracy of any particular l state is $2l + 1$. The classical analog of the degeneracy is the statistical weight, and this concept is elaborated in Chapter 2.

Three solutions of the Schrödinger wave equation are particularly useful to us: the ones for a particle in a three-dimensional box, a rotating diatomic molecule, and the harmonic oscillator model of a vibrating molecule.

1-2 PARTICLE IN A THREE-DIMENSIONAL BOX

Suppose a particle to be confined to the interior of a rectangular box of dimensions a, b, and c cm so that the potential is zero inside the box but infinite outside it. Then the particle will be completely free to move within the box, but the walls will be infinitely hard. The solutions to this problem are a set of standing waves in ψ reminiscent of the vibrating string problem. Each such three-dimensional wave can be described by three integers or quantum numbers, which we shall call j, k, and l. These integers j, k, and l are each greater by one than the number of nodes encountered in the wave in the three principal directions a, b, and c of the box. The energy of any state $\psi_{j,k,l}$ is

$$\varepsilon_{j,k,l} = \frac{h^2}{8m}\left(\frac{j^2}{a^2} + \frac{k^2}{b^2} + \frac{l^2}{c^2}\right) \tag{1-3}$$

where h is Planck's constant and m is the mass of the particle. Throughout this book, ε will be used for molecular energies and E for molar or multimolar energies. There is only one state $\psi_{j,k,l}$ for quantum numbers j, k, and l; thus

$$g_{j,k,l} = 1 \tag{1-4}$$

1-3 ROTATING DIATOMIC MOLECULE

The wave functions for a rotating diatomic molecule (Figure 2-18) depend on two quantum numbers, J and m: $\psi_{J,m}$, where m can vary from $+J$ through zero to $-J$. States of the same J but different m differ only in orientation in space but not in energy, the latter being a function solely of the first quantum number.

$$\varepsilon_J = hBJ(J + 1) \tag{1-5}$$

Here the constant $B = h/8\pi^2 I$, where I is the moment of inertia of the molecule. If the perpendicular distance from a rotation axis of particle i of mass

m_i is r_i, then the moment of inertia of a system of particles about that axis is

$$I = \sum_i m_i r_i^2 \tag{1-6}$$

In Figure 2-18 the moment of inertia of the diatomic molecule about an axis perpendicular to the bond and through the center of gravity is $I = m_1 r_1^2 + m_2 r_2^2 = \mu r^2$. Here $r = r_1 + r_2$ is the total internuclear distance, and μ is a very convenient quantity known as the reduced mass, defined for diatomic molecules as

$$\mu = \frac{m_1 m_2}{m_1 + m_2} \tag{1-7}$$

Since all states of the same J but different m have the same energy, the degeneracy of quantum state J or of energy level ε_J is

$$g_J = 2J + 1$$

1-4 HARMONIC OSCILLATOR MODEL

As long as the energies involved are well below the dissociation energies, the vibration of a chemical bond can be approximated by a harmonic oscillator model in which the restoring force f is proportional to the displacement of the bond from equilibrium x

$$f = -kx \tag{1-8}$$

The quantity k is the force constant, and is a measure of the "stiffness" of the bond. In both classical and quantum mechanics, the natural vibration frequency of such a bond is ν_0, where for a diatomic molecule:

$$\nu_0 = \frac{1}{2\pi}\left(\frac{k}{\mu}\right)^{\frac{1}{2}} \tag{1-9}$$

Acceptable solutions of the Schrödinger equation lead to the restriction of the energy of a harmonic oscillator to

$$\varepsilon_v = (v + \tfrac{1}{2})h\nu_0 \tag{1-10}$$

where v is the only quantum number involved. There is no degeneracy; $g_v = 1$.

1-5 REAL MOLECULES

Some drastic approximations are hidden in the foregoing sections:

(1) It is proper to treat the vibratory behavior of a complex molecule as a collection of isolated ("normal") vibrations.

(2) The true potential curve of a diatomic molecule (or of a bond in a polyatomic molecule) can be approximated by a harmonic oscillator potential curve.

(3) Rotation and vibration do not interact and mutually perturb one another.

Still, this simple model is quite complex enough for the first time around, and if these approximations are kept in mind they can be remedied when necessary later.

REFERENCES AND FURTHER READING

M. W. Hanna, *Quantum Mechanics in Chemistry* (W. A. Benjamin, Inc., New York, 1969), 2nd ed. A good introduction at the same level as this book.

L. Pauling and E. B. Wilson, *Introduction to Quantum Mechanics* (McGraw-Hill, New York, 1935). Clear, well written. Has been under revision for over a decade, but in the interim the original version has much to offer.

E. D. Kaufman, *Advanced Concepts in Physical Chemistry* (McGraw-Hill, New York, 1966). The title is a complete misnomer. This is an elementary introduction to the subject on more or less the same level as Hanna's book.

J. M. Anderson, *Mathematics for Quantum Chemistry* (W. A. Benjamin, Inc., New York, 1966). A good treatment, at the elementary level, of the tools of mathematics needed in quantum mechanics.

G. M. Barrow, *Introduction to Molecular Spectroscopy* (McGraw-Hill, New York, 1962). A good treatment of rotation and vibration.

Chapter 2

STATISTICAL MECHANICS

2-1 INTRODUCTION

THE MATERIAL of this book is traditionally covered under thermodynamics and statistical mechanics, in that order. However, topics are treated from a somewhat different viewpoint here, and the customary order is reversed. The theme of this book could really be called "The Molecular and Statistical Picture of Macroscopic Chemical Phenomena." This molecular viewpoint is maintained even in those parts of thermodynamics where it is unnecessary, because in many ways thermodynamics is much clearer and more understandable if treated as a natural consequence of molecular theory.

Thermodynamics and statistical mechanics are the foundations on which all of chemistry is built. It is impossible to talk intelligently for ten minutes about chemistry without introducing terms that have their basis in thermodynamics or the statistics of molecules. Thermodynamics is historically the earlier of the two topics. It arose from the work of the great nineteenth-century engineers and scientists such as Rumford, Joule, Clausius, Carnot, and many others. These men concerned themselves with the nature of heat and work, interconversions between them, and the limits upon such interconversions, often with a very practical purpose in mind. But as L. J. Henderson once remarked, "Science owes much more to the steam engine than the steam engine owes to science." The outcome of this activity was more to the profit of the scientist than to the locomotive builder. The result of the work of these men is an elegant, logical, and powerful science of the nature of heat, energy, and work based upon two simple postulates.

The two great postulates of thermodynamics, the first and second laws, are statements about the relationships between work and heat in the real world in terms of two fundamental quantities, *energy* and *entropy*. The first and second laws are not in need of proof: they are not deductions from any prior postulates. They are simply summaries of observation. "This," they say, "is the way the world is." We can then use these laws as starting points for deriving powerful and elegant methods of treating chemical problems. *Why* these two laws are true is irrelevant.

And yet this frame of mind is not really satisfying. It is not enough merely to accept these principles as true. One would like to construct a picture of the world in which they would be reasonable as well. This is the task of statistical mechanics and the molecular theory of matter.

In our picture of statistical mechanics, the apparently continuous matter we see around us is assumed to be made of discrete individual particles, atoms and molecules. We need not know, and in fact quantum mechanics shows that we cannot know, the precise history of each and every molecule. But, if we do our averaging correctly, then we can calculate the average bulk properties that a large collection of these molecules might be expected to show. Then, almost miraculously, when these predictions are compared with real observed behavior, they are seen to be true.

Notice that we have only pushed the "Why?" question back one threshold. To the statement, "The observed experimental facts are predicted correctly because we have assumed the first and second laws of thermodynamics," we now add the further statement, "The first and second laws of thermodynamics are predicted correctly because we have assumed the kinetic molecular theory of matter and have done our statistics properly." But we could then ask why the kinetic molecular theory should hold; the answer would have to be that it holds as a consequence of our assumption of the validity of quantum mechanics. We could then ask why quantum mechanics should hold, and thus build up an infinite regression of "Why?" questions, each leading to a little better and more fundamental theory.

There is neither the space nor the necessity in this treatment for us to become overly philosophical. Realizing that the chain is endless, we can confine our attention to two particular links. First we want to use the kinetic molecular theory and statistical methods to explain the laws of thermodynamics; second we want to apply these laws to chemical problems. When we do not have the time to go too deeply into some applications, at least we will outline what is possible.

2-2 *ENERGY AND ENTROPY*

Thermodynamics is based on two statements, which can be simply if not rigorously paraphrased as follows:

(1) In any real process, the energy of the universe is constant.

(2) In any real process, the entropy of the universe always rises.

Many alternative statements of each law could be given. As an example, any of the following statements of the second law can be used as a starting point in place of statement (2) above.

(2a) Entropy S is a state function.

(2b) dS is an exact differential.

(2c) $1/T$ is an integrating factor for dq.

(2d) It is impossible to convert heat into work without wasting some at a lower temperature.

(2e) It is impossible to transfer heat from a cold body to a hot body without supplying work.

(2f) There are some states, infinitesimally close to any given state, which one cannot get to by an adiabatic process.

If any one of these statements is used as a starting point (of course, within the framework of the necessary definitions of the terms used), all the others can be shown to follow, and the choice is more or less a matter of personal inclination.

Everyone has an intuitive feel for the concept of energy. This concept, like force, existed before science took hold of it. And as with force, science took the existing concept, refined its definition, transformed it into a more precise term than it had been in everyday speech, and used it for a carefully defined idea.

There is no such intuitive grasp of entropy. This is a new word, coined for the purpose from a Greek root, with no equivalent in common language. The concept that entropy expresses is one of order, interconnection, arrangement, and nonrandomness. (Confusingly, entropy as defined in thermodynamics is a measure of the *lack* of these qualities.) These ideas are certainly familiar, but are seldom thought of in any rigorous or quantitative manner. When such a quantitative measure is attempted, the concept seems alien, and when such an unfamiliar label as "entropy" is placed on the concept it seems more alien still. In the molecular picture, however, energy and entropy are entirely comparable terms. Energy is a measure of molecular motion; entropy is a measure of molecular arrangement.

Why should motion be intuitive but arrangement not be? In part this is a consequence of the path that science took in the nineteenth century. During this period physics reigned supreme, and people were concerned by and large with inanimate objects—metals, minerals, inorganic compounds, mechanical devices, and simple heat engines. In such a device as a steam engine, energy is more obvious and more important than entropy. A steam engine must operate between two widely different temperatures, themselves rough measures of energy. The sole object of the engine is to convert one form of energy, heat, into another form, work. In contrast, the engine itself is made up of a relatively small number of pieces, of fixed size and shape, in a relatively elementary arrangement. Energy, to a locomotive engineer, is the very heart of the matter; entropy is a fuzzy concept of no real relevance.

Nearly the reverse is true in living systems. A living system is essentially an isothermal chemical machine. Although energy is constantly required, its main function is less concerned with conversion into mechanical work than with maintaining the order and arrangement of the structure of the organism. When we look at even the simplest organism, we are struck with the tremendous importance of maintaining the precise arrangement of all the components of the cell, an importance that overshadows the role of mechanical work. Thus if the biological sciences had developed in the nineteenth century more rapidly than the physical sciences, then perhaps today everyone would know intuitively what entropy was, but energy would be a confusing concept.

Entropy, as defined, is a measure of disorder. Yet disorder is felt to be a negative concept like cold. Cold is the absence of heat, and heat, being a measure of molecular motion, is the more fundamental concept. Similarly, order, being a measure of molecular arrangement, is the fundamental concept, and entropy is only the absence of order. Hence it is often convenient to use the term "negentropy," defined as the entropy with the sign reversed, as a measure of order. The higher the negentropy, the more ordered is the system, up to a maximum negentropy of a perfectly ordered system of zero. The laws of thermodynamics then state that in any real process in an isolated system, energy remains constant while negentropy falls. This terminology is particularly useful in discussing living systems.

The following chapters have two goals: to give a sufficiently rigorous introduction to thermodynamics to enable one to do something with it, and to provide an intuitive feel for the meaning of thermodynamic concepts so that they become not only manipulative devices, but a way of thought.

2-3 KINETIC MOLECULAR THEORY OF GASES

One of the first successful attempts to use a microscopic model to explain macroscopic behavior was the kinetic molecular theory of gases, associated with such names as Maxwell and Boltzmann. This model was based on as simple a molecular picture as could be imagined.

(1) Molecules are points (or at most, hard spheres).

(2) The only energy that molecules possess is kinetic energy of motion; moreover, any kinetic energy is possible.

(3) Molecules do not exert any influence upon one another except at the moment of impact; when they do collide, they do so elastically.

All three of these assumptions are wrong. The truth is more complicated.

(1) Molecules are grossly misshapen objects.

(2) They possess many and complex forms of internal energy; moreover, the available energy states are quantized.

(3) Molecules exert long-range forces on one another, for which "collision" is only a crude but useful description of those interactions that do not lead to reaction. Elastic collision is even more unrealistic.

Even with such a crude starting point it is surprising how much can be accounted for, including the following. (1) The phenomenon of pressure, and Dalton's law of partial pressures in gas mixtures. (2) The ideal gas law, and the nature of first order corrections such as are embodied in van der Waals' equation. (3) The effusion of gas from a small orifice. (4) Transport processes: viscosity, diffusion, and thermal conductivity. (5) The distribution of energy among molecules. Average speeds and other properties calculable from a knowledge of the energy distribution. (6) Collision frequencies, and hence some idea of reaction rates. We shall look at only one of these, the phenomenon of pressure.

Assume for simplicity a cubical box, of side a, containing n molecules moving at random. Assume their motion to be isotropic, with no preferred direction. Each molecule will then have a velocity vector, $\mathbf{c}$,[1] with x, y, and z components (u, v, w), so that $c^2 = u^2 + v^2 + w^2$. If there is no correlation between velocity components u, v, and w in different molecules, then the mean square velocity and those of the components are related by

$$\overline{c^2} = \overline{u^2} + \overline{v^2} + \overline{w^2} = \overline{3u^2} = \overline{3v^2} = \overline{3w^2}$$

What produces the phenomenon of pressure? It arises from the collision of a molecule with a wall, and the consequent transfer of momentum to the wall. Consider the collisions of just one molecule with the two box faces which are perpendicular to the x direction or to the u component of velocity. Since this velocity component is not influenced by the assumed elastic collisions of the molecule with the other two pairs of walls, we can consider the "x-wall" collisions separately from any other behavior of the molecule.

Every time that the molecule collides with one of the two "x walls," its momentum component changes from $+m|u|$ to $-m|u|$ or back again, and the net momentum change in magnitude is $\Delta p = 2m|u|$. The molecule will collide with one end wall or the other at the rate of $|u|/a$ collisions per second, and the *rate* of momentum transfer to the two end walls is

$$\frac{dp}{dt} = \frac{2mu^2}{a} = f = \text{force on end walls}$$

The *average* force per molecule is $\bar{f} = 2m\overline{u^2}/a$, and the total force for one mole of molecules is $F = 2mN\overline{u^2}/a$. The pressure on these two end walls is then

[1] Vectors will be represented by boldface symbols. The same symbol in italic will represent the *magnitude* of the corresponding vector, with or without magnitude brackets.

the force per unit area

$$P = \frac{F}{2a^2} = \frac{mN\overline{u^2}}{a^3} = \frac{mN\overline{c^2}}{3V}$$

The same arguments apply to the other two pairs of walls, and we can see that the independent, uncorrelated nature of the three velocity components leads us to the idea that pressure is an isotropic quantity, the same in all directions in a gas. The total kinetic energy of the mole of gas is $E = \frac{1}{2}Nm\overline{c^2}$, and hence the pressure is related to the energy[2] by

$$P = \frac{2E}{3V} \qquad \text{or} \qquad P\bar{V} = \frac{2}{3}\bar{E}$$

We see in Section 4-3 that the PV behavior, which is approached by all gases under certain limiting conditions and which can be assumed to hold for a hypothetical "ideal" gas, can be used to define a quantity known as temperature T by the relationship $P\bar{V} = RT$ for one mole of gas. With this definition of temperature, we now see that the energy of an ideal gas is a function simply of its temperature

$$\bar{E} = \tfrac{3}{2}RT$$

This is the goal of the derivation, and will be used later.

The simple kinetic theory, clearly inadequate though it may be, is at least the starting point for better treatments. We shall look at a considerably more sophisticated treatment, and focus our attention on the calculation of those properties which depend upon energy and how it is distributed among molecules. This is the domain of statistical mechanics.

2-4 *ELEMENTS OF STATISTICAL MECHANICS—BASIC DEFINITIONS*

Assume a collection of molecules—gas, liquid, or solid—each having available to it various molecular states describable by wave functions ϕ_k of various energies and spatial distributions. The measurable properties of the bulk system in which we shall be interested will depend upon how many molecules

[2] A bar will often be used over the symbol for an extensive variable to indicate the value *per mole*. Thus, to be perfectly consistent, $PV = nRT$ and $P\bar{V} = RT$. But this will be done as an aid to the student and not as a matter of absolute principle. No thermodynamic expression should ever be applied by rote. If, in a given expression, you cannot tell what variables would be affected by the amount of material present, then you are better off doing something else.

are occupying which of these states. If we knew the distribution of molecules among their possible states, we could calculate these properties. The second task, having this distribution, is to calculate properly the quantities we are seeking.

Many of these states will have the same energy (or more precisely, large numbers of these states will have so nearly the same energy that they can be grouped into one energy "level" and treated together). Let the number of wave functions having an energy ε_j be g_j. This g_j is known as the degeneracy or the statistical weight of energy level ε_j. An example of this with which you are familiar is the isolated unperturbed hydrogen atom. All wave functions with the same principal quantum number n have the same energy, so that the degeneracy g_n of the nth level is n^2. Part of this degeneracy is removed in alkali metal atoms, where *s*, *p*, *d*, and *f* orbitals have different energies and the degeneracy of a given level of second quantum number l is $2l + 1$. Even this degeneracy can be removed by placing the atom in an external magnetic field, causing the five *d* orbitals, for example, to be of different energies. Note that we cannot tell which of these five *d* orbitals an electron may be occupying without using some sort of asymmetric field or influence, whether it be an external magnetic field or a local electrostatic field produced by adjacent ligands in a crystal. But these influences discriminate between orbitals by giving them different energies. The act of distinguishing degenerate orbitals itself removes the degeneracy. Hence we must treat degenerate orbitals as in fact distinct but in practice indistinguishable.

Assume now that we have n_j molecules of energy ε_j, to be found in one or another of the g_j possible degenerate states designated by $\phi_{k(j)}$. The total number of molecules is $n = \sum_j n_j$, and the total energy of the collection of molecules is $E = \sum_j \varepsilon_j n_j$. A particular distribution of molecules among the various energy levels is described by giving the number of molecules in each energy level

$$(n_j) = (n_0, n_1, n_2, n_3, \ldots) \tag{2-1}$$

The properties in which we are interested will then be functions of the distribution (n_j).

Not all distributions are possible. There are physical limitations upon the ways in which molecules can be distributed among energy states (or ways in which energy can be partitioned among molecules). Moreover, not all *possible* distributions are equally *probable*. Our search therefore has two parts:

(1) Which of the many conceivable distributions (n_j) are physically possible?

(2) What are the relative probabilities of occurrence of those distributions that *can* exist?

To find out what kinds of distributions are possible, we must return to quantum mechanics. Consider a system of three particles, each with individual wave functions ϕ_k. Then to a first approximation the wave function of the three-particle system can be built up as a sum of triple products of individual wave functions

$$\psi_{123} = \sum_j \sum_k \sum_l a_{jkl} \phi_{j(1)} \phi_{k(2)} \phi_{l(3)} \qquad (2\text{-}2)$$

Since the individual particles do not have labels, any acceptable system wave function must be of such form that $|\psi|^2$ is unchanged by a formal exchange of numbering of any two particles. This means that the wave function itself must be either symmetrical or antisymmetrical with respect to such an exchange

$$\psi_{132} = \pm\psi_{123} \qquad \text{etc.}$$

The fundamental particles—electrons, neutrons, and protons—are all antisymmetric, the best demonstration of this being the existence of a periodic table. Composite particles—nuclei, atoms, molecules—are antisymmetric if they are built up from an odd number of fundamental particles and symmetric if from an even number. It is easy to see why: although each particle wave function changes sign upon exchange of labels, an odd number of particles means that the composite particle wave function changes sign and an even number means that it remains unchanged.[3]

In summary, then, electrons, protons, neutrons, and Li^7 nuclei are antisymmetric and are known as Fermi–Dirac particles or "fermions." Deuterium nuclei, He^4 nuclei, and (for what may seem here like no particularly good reason) photons are symmetric Bose–Einstein particles or "bosons." For most of the time, for particles heavier than electrons and at normal temperatures, the differences between behavior of the two classes of particles are totally negligible. Both are approximated quite well by the mathematically simpler (although fictional) particles known as Maxwell–Boltzmann particles or "boltzons." Boltzons will be reintroduced later, but for the moment the approximations involved with them will be avoided.

Two fermions obviously cannot occupy the same wave function. If they did, then an interchange of the labels designating these two particles would leave the system wave function completely unchanged, and its behavior would be symmetrical. Therefore, we can say that each wave function $\phi_{k(j)}$ of an energy level ε_j is either empty or filled with no more than one particle. The total number of particles in this energy level n_j must be equal to or less than the degeneracy of the level g_j. For bosons no such restriction exists, and any number of bosons may be piled into one single wave function.

[3] See Hanna, Chapter 6.

The physical limit upon possible distributions, then, is simple. Any distribution of fermions that places two fermions in the same wave function or in which the number of particles occupying any energy level exceeds the degeneracy of that level, must be thrown out at once. No corresponding restriction holds for bosons. How, then, does one establish the relative probabilities of the still enormous number of possible distributions?

Consider one particular energy level ε_j, with n_j molecules. There will be many different ways of allocating these n_j molecules to the g_j wave functions, all giving rise to one and the same macroscopically distinguishable situation. For recall that we cannot distinguish between degenerate wave functions without perturbing their energies; and once we have done this they are no longer degenerate and the problem is changed. *Each* different way of assigning n_j indistinguishable particles among g_j logically and physically distinct but indistinguishable wave functions is a "microscopically separate state" or *microstate*. All of these individual microstates produce the same macroscopically observable state or *macrostate*. These macrostates are just our distributions (n_j).

The fundamental principle of statistical mechanics is that, energy considerations aside, one microstate is just as probable as any other. Therefore, the probability of occurrence of any given macrostate (n_j) is proportional to the number of possible microstates leading to it. We can write, as the probability,

$P_{(n_j)} \propto W_{(n_j)} =$ number of microstates per macrostate = total number of ways of arranging indistinguishable particles to produce the distribution (n_j)

Our next task is to calculate $W_{(n_j)}$ for fermions and for bosons.

2-5 *FERMI–DIRAC STATISTICS*

The restriction on fermions is that no two such particles can be in the same quantum state. Let us begin with a three-particle system as an illustration. Assume three identical particles, and assume that the energy states available to each of them are a series of equally spaced levels of energy $\varepsilon_j = J$ units. Assume, moreover, that the degeneracy of the levels resembles that of an isolated hydrogen atom: $g_j = J^2$. Finally, assume that the three-particle system possesses a total of six units of energy. Under these conditions, what distributions of energy among the particles are possible? An energy level diagram is shown in Figure 2-1, with the degeneracies of the levels indicated. The lowest energy level is unoccupied; there are no wave functions with zero energy. If $n_1 = 1$, that is, if only one of the three particles has just one unit of energy, then five units are left, and the only possible allocation is two units to one particle and three to the last one. Then distribution (a) can be written

$$\text{(a)} \qquad (n_j) = (n_0, n_1, n_2, n_3, n_4, \ldots) = (0, 1, 1, 1, 0, 0, 0, \ldots)$$

If there are *no* molecules with just one unit of energy, then each of the three must have at least two, and the only possible distribution is

$$\text{(b)} \qquad (n_j) = (0, 0, 3, 0, 0, 0, \ldots)$$

If $n_1 = 2$, then four units of energy remain for the third molecule, and the only distribution is

$$\text{(c)} \qquad (n_j) = (0, 2, 0, 0, 1, 0, 0, \ldots)$$

But distribution (c) is illegal, because two particles have been assigned to an energy level where only one wave function exists. The number of particles in level 1 exceeds the degeneracy, an impossible situation for fermions.

Which of the remaining distributions, (a) or (b), is more probable? In distribution (a), the molecule in level 1 has only one wave function or state open to it, the molecule in level 2 can be in any one of four degenerate states, and that in level 3 has nine wave functions available. The total number of microstates leading to the macrostate (0, 1, 1, 1, 0, 0, ...) is then $W_a = 1 \times 4 \times 9 = 36$. In distribution (b), the number of permutations of three molecules among four boxes is four; each of the boxes in turn could be the empty one. Hence $W_b = 4$. By this analysis, assuming the equal *a priori* probability of all microstates, distribution (a) will be nine times as probable of occurrence as distribution (b). Note that this agrees with one's intuitive feeling that spread out or disordered arrays are more likely, other things being equal, than confined or structured arrangements. This is the beginning of the concept of entropy.

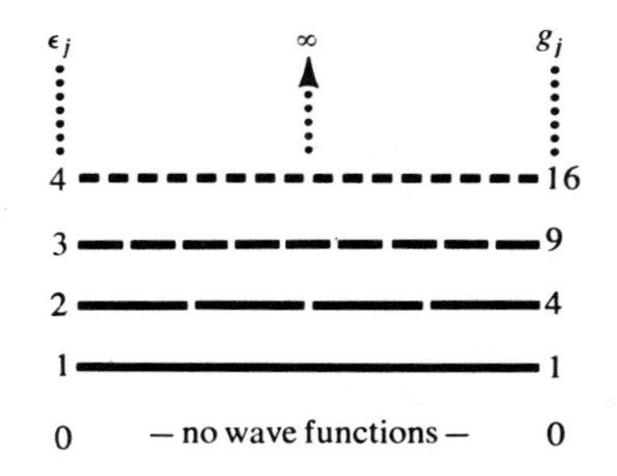

FIGURE 2-1. *Energy level diagram for a model system with levels of energy proportional to J and degeneracy equal to J^2. Each bar in a given level represents one degenerate state.*

Now let the total energy be nine units instead of six. The possible distributions are listed in Table 2-1. The factor of 120 in W_c is just the number of

TABLE 2–1 POSSIBLE DISTRIBUTIONS OF THREE FERMIONS AMONG THE ENERGY LEVELS OF FIGURE 2–1

$j =$	0	1	2	3	4	5	6	7	8	9	⋯	*Relative*
$g_j =$	0	1	4	9	16	25	36	49	64	81	⋯	*probabilities*
(a)	0	1	1	0	0	0	1	0	0	0	⋯	$W_a = 1 \times 4 \times 36 = 144$
(b)	0	1	0	1	0	1	0	0	0	0	⋯	$W_b = 1 \times 9 \times 25 = 225$
(c)	0	1	0	0	2	0	0	0	0	0	⋯	$W_c = 1 \times 120 \quad = 120$
(d)	0	0	2	0	0	1	0	0	0	0	⋯	$W_d = 6 \times 25 \quad = 150$
(e)	0	0	1	1	1	0	0	0	0	0	⋯	$W_e = 4 \times 9 \times 16 = 576$
(f)	0	0	0	3	0	0	0	0	0	0	⋯	$W_f = 84 \quad = 84$

ways of permuting two indistinguishable objects among 16 boxes, with never more than one object per box. The first object can go into any one of 16 boxes; the second has only 15 left open to it. The product, 16 × 15, must then be divided by two to compensate for the fact that when we spoke of "first" and "second" objects we were giving them a distinguishability they did not possess. The states, "electron 1 in the p_x orbital and electron 2 in p_z" and, "electron 2 in the p_x orbital and electron 1 in p_z" are not only phenomenologically identical on the macroscopic level, but are logically identical on the microscopic, where the only proper description is "one electron in p_x and one in p_z." There is a real conceptual difference between p_x and p_z orbitals even in situations where we cannot experimentally distinguish between them; there is no such conceptual difference between electrons.

In a similar fashion, the 6 in W_d is the number of permutations of two indistinguishable objects among four boxes, subject to the Fermi–Dirac restriction, and the 84 of W_f is the like number of permutations of three objects among nine boxes. The intuitive calculations of the preceding paragraph become clumsy with three or more objects, and a general expression is needed.

Now the randomness principle mentioned in the previous example is even more noticeable. The overwhelmingly probable distribution W_e is one in which all three particles are spread out over different energy levels. The second most probable distribution is also one with particles spread widely; it is less probable only because the levels which the particles occupy have less favorable degeneracies. The most orderly state of all, with all three molecules in the same energy level, W_f, is again the least probable of all.

Fermi–Dirac Permutations—General Case

How many ways can one distribute n_j identical objects among g_j boxes, so that each box is either empty or filled with one object? To begin with, there will be $g_j!/(g_j - n_j)!$ permutations as the objects are dropped into the boxes, one at a time. The first object has g_j possible destinations, the second object has only $g_j - 1$ since one box has been taken, the third has $g_j - 2$, and so on to the last, or n_jth object, which will have $g_j - n_j + 1$ openings. The total number of permutations will then be

$$\omega_j' = g_j(g_j - 1)(g_j - 2) \cdots (g_j - n_j + 1) = \frac{g_j!}{(g_j - n_j)!} \tag{2-3}$$

This is not the whole story, however. We have recommitted the error of labeling electrons. Since each object is in a separate box, we can correct ω_j' by dividing by the number of ways of permuting the labels upon objects, or by $n_j!$. The final expression for the number of possible permutations is

$$\omega_j = \frac{g_j!}{n_j!(g_j - n_j)!} \tag{2-4}$$

There is another formal way of deriving the above expression which will help bridge the step to bosons. We can consider that we have g_j boxes, each one to be wrapped around either one of n_j different objects (full box) or $g_j - n_j$ "nonobjects" (empty box). There are then $g_j!$ different ways of permuting our g_j boxes among our g_j objects and nonobjects. But we have erred in labeling our n_j objects, requiring a correction for overcounting in the denominator of $n_j!$. And since nonelectrons are even more indistinguishable than electrons, we must correct the overcounting again with a factor of $(g_j - n_j)!$.

The total number of ways of obtaining a distribution of many levels is the product of the ways of obtaining each of the levels. Thus the total number of ways of arriving at a given distribution $(n_j) = (n_0, n_1, n_2, \ldots)$ of fermions, or the number of microstates per macrostate, is

$$W_{(n_j)} = \prod_j \frac{g_j!}{n_j!(g_j - n_j)!} \tag{2-5}$$

2-6 *BOSE–EINSTEIN STATISTICS*

The mathematics of the derivation of the corresponding expression for bosons is identical, although the frame of mind which makes this so will seem very strange at first. In how many ways can we distribute n_j indistinguishable objects among g_j boxes with *no* restrictions as to the number in any given

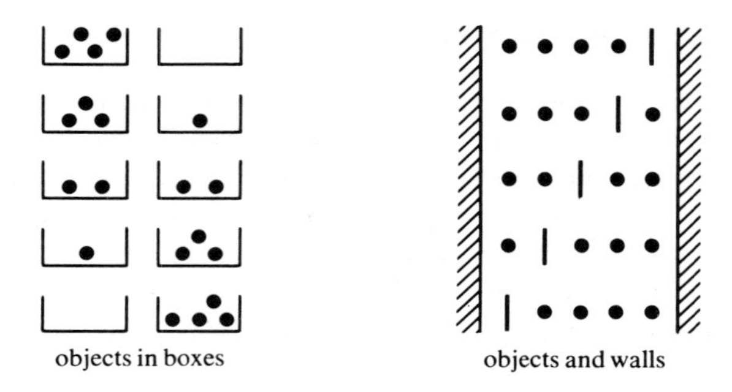

FIGURE 2-2. *Alternate ways of describing the permutations of four bosons among two boxes. Left: objects in boxes; right: objects and walls.*

box? One can think of the boxes as being lined up in a row and sharing adjacent walls. An array of n_j objects in g_j boxes, some empty and some with several objects in them, can then be thought of as a linear array of n_j objects and $g_j - 1$ walls, as in Figure 2-2. One then has a row of $g_j + n_j - 1$ "stations," each one of which is occupied either by an object or by a wall. Considering both objects and walls as distinguishable, there would be $(g_j + n_j - 1)!$ possible permutations. The corrections for the indistinguishability of the n_j objects and the $g_j - 1$ walls make the actual expression for the number of permutations in any one energy level

$$\omega_j = \frac{(g_j + n_j - 1)!}{n_j!(g_j - 1)!} \tag{2-6}$$

The total number of microstates per macrostate for a given distribution of bosons is then

$$W_{(n_j)} = \prod_j \frac{(g_j + n_j - 1)!}{n_j!(g_j - 1)!} \tag{2-7}$$

2-7 *DISTINGUISHABLE PARTICLES*

As is so often done in science, we are soon going to abandon the real but complicated in favor of the simple but fictional, and turn to Maxwell–Boltzmann statistics. Let us now assume that our n particles which we have distributed, n_j to a level of energy ε_j and degeneracy g_j, really do have labels. Initially there would be $n!$ possible permutations of objects. But each level will have to be corrected by its own factor $n_j!$ to account for the fact that the *order* of even distinguishable objects in the same hopper is irrelevant. The preliminary expression for the number of microstates is then

$$W_{(n_j)} = \frac{n!}{\prod_j n_j!} \tag{2-8}$$

One more complication remains. If the jth level has g_j possible states, then each object has g_j alternates open to it, and the total number of different arrangements is $g_j^{n_j}$. The final expression is then

$$W_{(n_j)} = n! \prod_j \left(\frac{g_j^{n_j}}{n_j!}\right) \tag{2-9}$$

2-8 DILUTE SYSTEMS—CORRECTED BOLTZONS

Systems in which the degeneracy of each level far exceeds the number of objects in that level, or for which $g_j \gg n_j$, are known as dilute systems. Under these conditions, although many bosons *could* occupy the same wave function, it is very unlikely that they would ever do so with all the other wave functions available. The distinction between bosons and fermions then disappears.

For bosons,

$$\begin{aligned}\omega_j &= \frac{(g_j + n_j - 1)!}{n_j!(g_j - 1)!} \\ &= \frac{(g_j + n_j - 1)(g_j + n_j - 2)(g_j + n_j - 3) \cdots (g_j + n_j - n_j)}{n_j!}\end{aligned} \tag{2-10}$$

Each term in the numerator is equal to or greater than g_j; thus for bosons,

$$\omega_j \geq \frac{g_j^{n_j}}{n_j!}$$

For fermions,

$$\omega_j = \frac{g_j!}{n_j!(g_j - n_j)!} = \frac{g_j(g_j - 1)(g_j - 2) \cdots (g_j - n_j + 1)}{n_j!} \tag{2-11}$$

Each term in the numerator is equal to or less than g_j; thus for fermions

$$\omega_j \leq \frac{g_j^{n_j}}{n_j!}$$

In summary, the numbers of microstates for fermions and bosons are related as

$$W_{\mathrm{FD}} \leq \prod_j \frac{g_j^{n_j}}{n_j!} \leq W_{\mathrm{BE}}$$

The equalities hold only in the limit of infinite dilution. The central expression is just that derived for distinguishable objects divided by $n!$. But division by $n!$ is the crudest sort of correction for indistinguishability. It applies precisely only when all objects are in different energy levels. Where several particles are of the same energy, their interpermutations have already been allowed for and the $n!$ factor overshoots the mark.

Nevertheless, because of the appearance of the central term in the expression above, it is useful to discuss a fictitious kind of particle, a corrected Maxwell–Boltzmann particle or "boltzon," for which the number of microstates per macrostate is

$$W_{\mathrm{MB}} = \prod_j \frac{g_j^{n_j}}{n_j!} \tag{2-12}$$

Then we can say that, in the limit of high dilution, fermions and bosons both approach boltzons in behavior; if it proves mathematically advantageous to deal with these new particles, we are free to do so.

2-9 MOST PROBABLE DISTRIBUTION—MAXIMIZATION OF W

Now that we know how to find all *possible* distributions for the different kinds of particles, we must look for the most *probable* distribution, and then see how much more probable it is than any other. In order to find the most probable distribution, we must maximize $W_{(n_j)}$ with respect to changes in populations of the energy levels, dn_j, subject to the two boundary conditions of a fixed number of particles and a fixed total amount of energy.

$$\sum_j n_j = n \quad \text{or} \quad \sum_j dn_j = 0$$

$$\sum_j \varepsilon_j n_j = E \quad \text{or} \quad \sum_j \varepsilon_j \, dn_j = 0$$

It proves to be simpler to maximize ln W instead of W because all products become sums in the logarithm, but the maximum point is the same.

For fermions,

$$\ln W = \sum_j \{\ln (g_j!) - \ln (n_j!) - \ln [(g_j - n_j)!]\} \tag{2-13}$$

In the limit of large numbers of particles per level and large degeneracies, one can use Stirling's approximation

$$\ln n! = n \ln n - n$$

Then,

$$\ln W = \sum_j [g_j \ln g_j - n_j \ln n_j - (g_j - n_j) \ln (g_j - n_j)]$$

$$d \ln W = \sum_j \left(-\ln n_j - \frac{n_j}{n_j} + \ln (g_j - n_j) + \frac{(g_j - n_j)}{(g_j - n_j)} \right) dn_j$$

$$d \ln W = \sum_j \left\{ \ln \left(\frac{g_j - n_j}{n_j} \right) \right\} dn_j \qquad (2\text{-}14)$$

For bosons,

$$\ln \quad = \sum_j [(n_j + g_j - 1) \ln (n_j + g_j - 1) - n_j \ln n_j - (g_j - 1) \ln (g_j - 1)]$$

$$d \ln W = \sum_j \left(\ln (n_j + g_j - 1) + \frac{n_j + g_j - 1}{n_j + g_n - 1} - \ln n_j - \frac{n_j}{n_j} \right) dn_j$$

$$d \ln W = \sum_j \left[\ln \left(\frac{g_j + n_j - 1}{n_j} \right) \right] dn_j \qquad (2\text{-}15)$$

For boltzons,

$$\ln W = \sum_j (n_j \ln g_j - n_j \ln n_j + n_j)$$

$$d \ln W = \sum_j \left(\ln g_j - \ln n_j - \frac{n_j}{n_j} + 1 \right) dn_j$$

$$d \ln W = \sum_j \left[\ln \left(\frac{g_j}{n_j} \right) \right] dn_j \qquad (2\text{-}16)$$

Each of the above expressions holds only when both n_j and g_j are large numbers. Under these conditions the 1 in the boson expression can be neglected, and all three expressions can be written together as

$$d \ln W = \sum_j \left[\ln \left(\frac{g_j + kn_j}{n_j} \right) \right] dn_j \qquad (2\text{-}17)$$

The constant k is -1 for fermions, 0 for boltzons, and $+1$ for bosons.

Method of Lagrangian Multipliers

The straightforward process for finding an extremum (maximum, minimum, or inflection point) is to find the point at which the derivatives with respect to all independent variables are zero.

$$\frac{\partial \ln W}{\partial n_j} = \ln\left(\frac{g_j + kn_j}{n_j}\right) = 0 \qquad \text{for all } n_j$$

If there are J energy levels, then these derivatives form a set of J simultaneous equations in J unknowns, $n_0, n_1, \ldots, n_{J-1}$. But the addition of the two boundary conditions reduces the number of independent variables to $J - 2$. The straightforward but impossibly tedious method of solution would be to use the two boundary conditions to eliminate two of the variables from the equations and then to solve for the other $J - 2$, but the easier path is to use the method of Lagrangian multipliers. An example or two will help to convince you that the method is general and not the product of the imagination of statistical mechanicians:

EXAMPLE 1. Suppose that one has a Gaussian-shaped mountain whose height at any point is given by $Z = \exp[-(x^2 + y^2)]$, and is asked to find the highest point on the mountain. The answer is trivial, but a formal treatment would be as follows:

$$-\ln Z = x^2 + y^2$$

Minimizing $-\ln Z$ is equivalent to maximizing Z.

$$d(-\ln Z) = (2x)dx + (2y)dy = 0$$

$$2x = 0 \qquad 2y = 0$$

Hence the highest point occurs at the origin, (0, 0).

EXAMPLE 2. Suppose now that travel on the mountain were confined to the straight line path described by $x + y = 4$. The direct method of solution would be as follows:

$$y = 4 - x$$

$$-\ln Z = x^2 + (4 - x)^2 = 2x^2 - 8x + 16$$

$$\frac{d(-\ln Z)}{dx} = 4x - 8 = 0 \qquad x = 2$$

From the original boundary conditions, y must be 2 as well, and the highest point traversed by this path is at (2, 2).

The Lagrangian method is somewhat different. The minimization equation, without constraints, is

$$d(-\ln Z) = 2x\,dx + 2y\,dy = 0$$

The equation of constraint, $x + y = 4$, can be written

$$dx + dy = 0$$

Now if the two expressions above are each zero, then their sum or difference is also zero, and, in the most general case, the first equation plus the product of the second with any constant whatever is likewise zero

$$\begin{array}{r} 2x\,dx + 2y\,dy = 0 \\ \alpha(dx + dy) = 0 \\ \hline (2x + \alpha)\,dx + (2y + \alpha)\,dy = 0 \end{array}$$

This new minimization equation, for the first time, includes explicit reference to the boundary constraints. Minimization now requires that the new differential coefficients each be zero, or

$$2x + \alpha = 0 \qquad 2y + \alpha = 0$$

$$x = -\frac{\alpha}{2} \qquad y = -\frac{\alpha}{2}$$

Comparison of these results with the original boundary conditions shows that the undetermined Lagrangian multiplier α is equal to -4, or that the maximum point is at (2, 2). We have actually solved a more general problem, and have shown that for *any* path described by $x + y = \text{const}$, the high point occurs at the point where $x = y$.

EXAMPLE 3. Lastly, let the mountain path now be the hyperbolic one described by $xy = 4$. What is the high point of a path subject to these constraints? The boundary equation can be written

$$y\,dx + x\,dy = 0$$

This expression, multiplied by the arbitrary undetermined multiplier α, can now be added to the original equation of minimization without constraints to obtain the constrained expression

$$(2x + \alpha y)\, dx + (2y + \alpha x)\, dy = 0$$
$$2x + \alpha y = 0 \qquad 2y + \alpha x = 0$$

Using the original boundary condition in the form $y = 4/x$ and $x = 4/y$ yields

$$x = (-2\alpha)^{\frac{1}{2}} \qquad y = (-2\alpha)^{\frac{1}{2}}$$

This is compatible with $xy = 4$ only if $\alpha = -2$, and the maximum point is at (2, 2). This method should be compared with the straightforward one of substituting $4/x$ for y in the $-\ln Z$ equation and minimizing with respect to one variable, x.

Note the difference in philosophy of the two methods. The familiar substitution method copes with the added boundary condition by eliminating one independent variable and one equation. The method of undetermined multipliers keeps the original number of variables and equations but introduces a new degree of freedom into the system by means of the multiplier α, whose value must be determined later from the specific problem at hand. For two variables the undetermined multiplier method has no advantages; for three or more variables the retention of the symmetry of the original equations makes it by far the preferred method.

Application to ln W

The minimization equation without constraints is

$$\sum_j \ln\left(\frac{g_j + kn_j}{n_j}\right) dn_j = 0 \tag{2-18}$$

The equation of constraint arising from the fixed number of particles is

$$\sum_j dn_j = 0 \tag{2-19}$$

The equation of constraint arising from the fixed amount of energy available is

$$\sum_j \varepsilon_j\, dn_j = 0 \tag{2-20}$$

If the undetermined multipliers for the two constraint equations are $-\alpha$ and $-\beta$, respectively, then the minimization equation *with* constraints is

$$\sum_j \left[\ln\left(\frac{g_j + kn_j}{n_j}\right) - \alpha - \beta\varepsilon_j\right] dn_j = 0 \tag{2-21}$$

At the extremum (maximum in this case), each coefficient is individually zero

$$\ln\left(\frac{g_j + kn_j}{n_j}\right) = \alpha + \beta\varepsilon_j \tag{2-22}$$

Solving for n_j, we find that the most probable distribution, or that for which $\ln W$ (and therefore W as well) is at a maximum, is that for which

$$n_j = \frac{g_j}{e^{\alpha}\, e^{\beta\varepsilon_j} - k} \tag{2-23}$$

For fermions,

$$(n_j)_{\text{FD}} = \frac{g_j}{e^{\alpha}\, e^{\beta\varepsilon_j} + 1} \tag{2-24}$$

For boltzons,

$$(n_j)_{\text{MB}} = g_j e^{-\alpha} e^{-\beta\varepsilon_j} \tag{2-25}$$

For bosons,

$$(n_j)_{\text{BE}} = \frac{g_j}{e^{\alpha} e^{\beta\varepsilon_j} - 1} \tag{2-26}$$

Note that $(n_j)_{\text{FD}} < (n_j)_{\text{MB}} < (n_j)_{\text{BE}}$. There are fewer fermions in a given state than there would have been bosons, because the exclusion principle prevents multiple occupancy of any one quantum state.

In order to discuss the applicability of Fermi–Dirac, Maxwell–Boltzmann, and Bose–Einstein statistics it is necessary to know the identity of α and β. These, as in the simple Gaussian mountain problem, are mathematically undetermined and can be given values only by looking at the physical conditions of the problem. For the moment, let us continue with boltzons only.

2-10 EVALUATION OF THE MULTIPLIER α

For Maxwell–Boltzmann particles, the number in energy level j in the most probable distribution is

$$n_j = g_j e^{-\alpha} e^{-\beta\varepsilon_j} \tag{2-27}$$

The total number of particles is

$$n = \sum_j n_j = e^{-\alpha} \sum_j g_j e^{-\beta \varepsilon_j} \tag{2-28}$$

Hence $e^{-\alpha} = n/q$, where q is defined by

$$q \equiv \sum_j g_j e^{-\beta \varepsilon_j} \tag{2-29}$$

The new quantity q is the weighted sum over states available to a molecule, or the molecular partition function. It will have extensive use later.

The fraction of molecules in the jth energy level is then

$$\frac{n_j}{n} = \frac{g_j e^{-\beta \varepsilon_j}}{q} \tag{2-30}$$

This is the most probable distribution. But how much *more* probable is it than any neighboring distribution? Does the probability curve peak sharply at the most probable value, or is it a slowly varying function? We prove in Section 2-13 that the most probable distribution is so vastly more probable than any other that all others may be neglected entirely.

2-11 EVALUATION OF THE MULTIPLIER β

The multiplier β is obtained from the physical boundary conditions, and not from the mathematics of solution of the Lagrangian equations. The physical property that we need is the average molecular energy $\langle \varepsilon \rangle$. Let F be *any* property dependent only upon the distribution of molecules among energy levels, and let F_j be the value of the property if a molecule is in energy level ε_j. The average value of this property is then

$$\langle F \rangle = \frac{\sum_j F_j n_j}{\sum_j n_j} = \frac{\sum_j F_j g_j e^{-\beta \varepsilon_j}}{q} \tag{2-31}$$

If F_j is the molecular energy ε_j itself, then the average energy per molecule is

$$\langle \varepsilon \rangle = \frac{1}{q} \sum_j \varepsilon_j g_j e^{-\beta \varepsilon_j} \tag{2-32}$$

But

$$\frac{\partial q}{\partial \beta} = -\sum_j \varepsilon_j g_j e^{-\beta \varepsilon_j}$$

Hence

$$\langle \varepsilon \rangle = -\frac{1}{q}\frac{\partial q}{\partial \beta} = -\frac{\partial \ln q}{\partial \beta} \tag{2-33}$$

An ideal gas can be treated as a collection of free particles in an infinite potential well. For a three-dimensional potential well of dimensions a, b, and c, the possible energy levels are given by[4]

$$\varepsilon_{j,k,l} = \frac{h^2}{8m}\left(\frac{j^2}{a^2} + \frac{k^2}{b^2} + \frac{l^2}{c^2}\right) \tag{2-34}$$

Here j, k, and l are the integral quantum numbers for translational motion. The molecular partition function is

$$q = \sum_j \sum_k \sum_l \exp\left[-\frac{\beta h^2}{8m}\left(\frac{j^2}{a^2} + \frac{k^2}{b^2} + \frac{l^2}{c^2}\right)\right]$$

$$q = \left[\sum_j \exp\left(-\frac{\beta h^2}{8m}\frac{j^2}{a^2}\right)\right]\left[\sum_k \exp\left(-\frac{\beta h^2}{8m}\frac{k^2}{b^2}\right)\right]\left[\sum_l \exp\left(-\frac{\beta h^2}{8m}\frac{l^2}{c^2}\right)\right] \tag{2-35}$$

Now assume that the energy levels are sufficiently close together that the summations can be replaced by integrations. As we shall see later, this assumption will lead to trouble from some forms of energy, but is certainly true for translational energy. Then,

$$q = \left[\int_{j=0}^{\infty} \exp\left(-\frac{\beta h^2}{8ma^2} j^2\right) dj\right]\left[\int_{k=0}^{\infty} \exp\left(-\frac{\beta h^2}{8mb^2} k^2\right) dk\right] \times \left[\int_{l=0}^{\infty} \exp\left(-\frac{\beta h^2}{8mc^2} l^2\right) dl\right]$$

$$q = \left(\frac{2\pi m}{\beta h^2}\right)^{\frac{3}{2}} abc = \left(\frac{2\pi m}{\beta h^2}\right)^{\frac{3}{2}} V \tag{2-36}$$

$$\left(\frac{\partial q}{\partial \beta}\right)_V = -\frac{3}{2}\left(\frac{2\pi m}{h^2}\right)^{\frac{3}{2}} \frac{V}{\beta^{\frac{5}{2}}} = -\frac{3}{2}\frac{q}{\beta} \tag{2-37}$$

And the mean molecular energy is

$$\langle \varepsilon \rangle = \frac{3}{2}\frac{1}{\beta} \tag{2-38}$$

[4] See Hanna, Chapter 3.

But we have seen earlier that, according to the simple kinetic theory of gases, the energy per mole is $\bar{E} = \frac{3}{2}RT$ and the mean energy per molecule is $\langle \varepsilon \rangle = \frac{3}{2}kT$, where k is the Boltzmann constant, $k = R/N$. Hence, from the physical reality of the problem rather than from any mathematical necessity, we have established that $\beta = 1/kT$ for boltzons. It is equally true for fermions and for bosons; the proof of this can be found in texts in statistical mechanics such as that by Davidson.

2-12 *CONDITIONS FOR APPLICABILITY OF MAXWELL–BOLTZMANN STATISTICS*

Dilute Systems

We know that, for dilute systems, $g_j > n_j$ or that the dilution ratio g_j/n_j is large. Further,

$$\frac{g_j}{n_j} = \frac{q}{n} e^{\varepsilon_j/kT} \geqq \frac{q}{n}$$

because the exponential term is always greater than one. Then we can say

$$\frac{q}{n} = \frac{g_0}{n_0} < \frac{g_1}{n_1} < \frac{g_2}{n_2} < \frac{g_3}{n_3} < \cdots$$

In other words, in the most probable distribution, the higher up the energy ladder one goes, the more states are available per particle or the greater the dilution. If the dilution ratio of the ground state is great enough that one can forget about fermions and bosons and use Maxwell–Boltzmann statistics, then the same will also be true for all the higher energy levels. Therefore, we must find the conditions for which the ratio q/n is large, for Maxwell–Boltzmann statistics to be applicable.

$$\frac{q}{n} = \frac{\text{sum over states}}{\text{sum over particles}}$$

Ideal Monatomic Gas

We have already demonstrated that for a monatomic gas

$$q = \frac{(2\pi mkT)^{\frac{3}{2}}}{h^3} V \tag{2-36}$$

Therefore,

$$\frac{q}{n} = \frac{(2\pi mkT)^{\frac{3}{2}}}{h^3}\frac{V}{n} \tag{2-39}$$

where m is in grams, kT is in ergs, and V is in cubic centimeters. A unit analysis of Eq. 2-39 shows that q/n is a dimensionless quantity

$$\frac{q}{n} : \frac{(\text{g erg})^{\frac{3}{2}}(\text{cm})^3}{(\text{erg sec})^3} : \frac{(\text{g}^3\ \text{cm}^3/\text{sec}^3)\text{cm}^3}{(\text{g}^3\ \text{cm}^6/\text{sec}^6)\text{sec}^3} : 1$$

The quantity q/n is then an intensive quantity like pressure, and independent of the amount of material present. Before proceeding with our dilution studies, it is convenient to modify Eq. 2-39 by making the following changes (where M is molecular weight, and N is Avogadro's number):

$$m \rightarrow \frac{M}{N}$$

$$V_{(\text{cm}^3)} \rightarrow 1000\ V'_{(\text{liters})}$$

$$n_{(\text{molecules})} \rightarrow Nn'_{(\text{moles})}$$

Thus,

$$\frac{V}{n} = \frac{1000}{N}\frac{V'}{n'} = \frac{1000}{N}\frac{R'T}{P}$$

with P in atm and R' in l atm/°K. Finally,

$$\frac{q}{n} = \frac{(2\pi MRT)^{\frac{3}{2}}}{N^3h^3}\frac{1000}{N}\frac{R'T}{P} \tag{2-40}$$

where R' is in l atm/°K and R is in erg/°K. Introducing the values of the constants gives

$$\frac{q}{n} = 0.0257\frac{M^{\frac{3}{2}}T^{\frac{5}{2}}}{P} \tag{2-41}$$

where M is the molecular weight, T is in °K, and P is in atm.

EXAMPLE 4. Helium and Argon: 1 atm, 300°K. Under these conditions, can we use Maxwell–Boltzmann statistics for both gases? The dilution

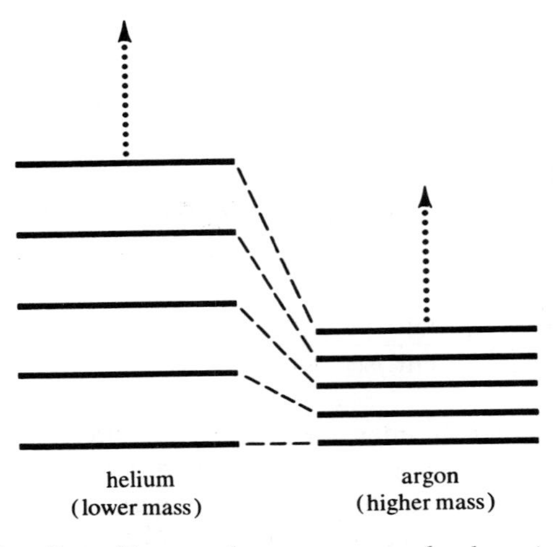

FIGURE 2-3. *The effect of increased mass on energy level spacings. For a particle in a box, the spacing between levels is inversely related to the mass of the particle.*

ratios are calculated from Eq. 2-41 to be

$$\text{He} : \frac{q}{n} \cong 320{,}000 \qquad \text{Ar} : \frac{q}{n} \cong 10^6$$

Clearly, in terms of states/molecule, both gases are extremely dilute systems at 1 atm and 300°K, and Maxwell–Boltzmann statistics will hold for both. Argon has more states than helium because its energy levels are more closely spaced than are those of helium. In general, other factors being comparable, heavier particles will have more closely spaced energy levels, an illustration of this being the dependence of energy levels of a particle in a box upon $1/m$. The energy levels of helium and argon are represented symbolically in Figure 2-3.

EXAMPLE 5. Helium and Argon: Low Temperatures. As the temperature is lowered, the total energy of the particles is lowered, thus forcing them to trickle down from upper energy levels and fill up the bottom ones, as illustrated in Figure 2-4. Therefore, by lowering the temperature, we should be able to reduce the q/n ratio as much as we please. How low a temperature is required to reduce the dilution ratio for helium and argon each to a value of ten? Using Eq. 2-41, we find that to obtain $q/n = 10$ for helium, the required temperature is 4.75°K. The boiling point of helium is about 4°K, so that it is possible to reduce the dilution ratio of helium to such a value that Maxwell–Boltzmann statistics fail in the region where the helium condenses. This is the basis of the superfluid behavior of helium around 4°K. To obtain

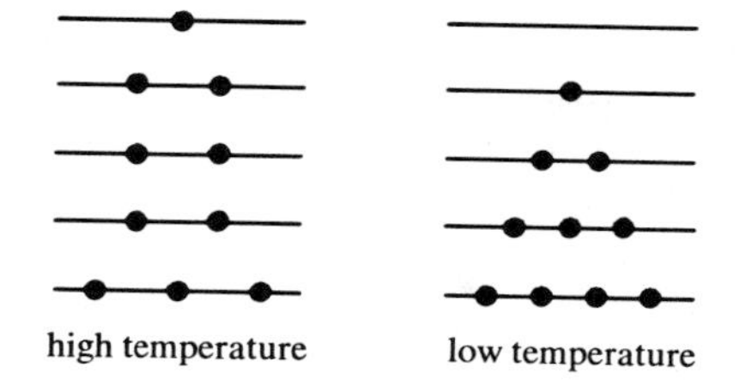

FIGURE 2-4. *The effect of lowering the temperature on the distribution of particles among energy levels. At very low temperatures, the lower energy states of very light atoms can become so populated that Maxwell–Boltzmann statistics are no longer applicable.*

$q/n = 10$ for argon requires a temperature of 1.2°K. But the boiling point of argon is 87°K; thus long before Maxwell–Boltzmann statistics begin to fail, argon liquefies and the nature of the problem changes.

EXAMPLE 6. Helium and Argon: High Pressures. Increasing the pressure while holding total energy (and hence temperature) constant results in a decrease in volume. A decrease in volume causes an increase in the spacing between energy levels, so that under conditions of constant total energy, the particles are forced to drop down into lower quantum states. This situation is illustrated in Figure 2-5, where it can be seen that as one goes to higher pressures, a molecule with a given energy which was in the jth energy level, now with the same energy can only occupy the $(j - k)$th energy level. As a

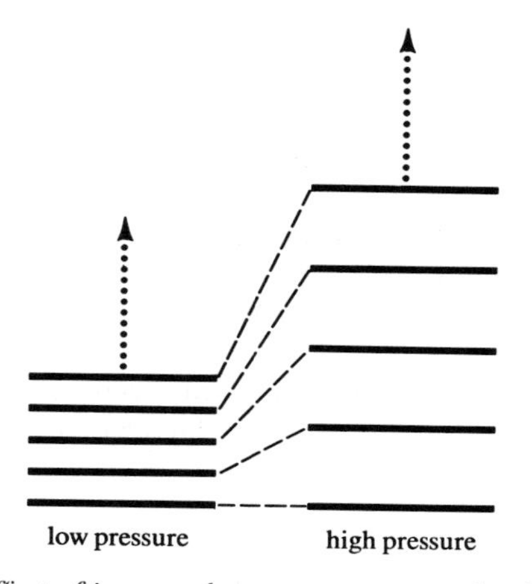

FIGURE 2-5. *The effect of increased pressure on energy level spacings. Increasing the pressure on a gas at constant temperature decreases the volume, and opens up the energy levels since for a particle in a box the energy level spacing varies inversely with the square of the linear dimensions.*

result of this change the molecules pile down into the lower quantum levels and decrease the dilution ratios of those levels. At sufficiently high pressures, it should be possible to separate the levels to the point that Maxwell–Boltzmann statistics fail. In order to make $q/n = 10$ for helium at 300°K one would have to obtain a pressure of 32 kb (1 kilobar = 1000 atm), which is conceivable. However, the same dilution ratio for argon requires a pressure of 100 kb, which is quite impractical. In any case, long before such a pressure was reached, the lack of ideality of the dense gas/liquids would complicate the experiment.

In summary, Maxwell–Boltzmann statistics will be expected to hold for all substances except a few very light atoms—hydrogen, helium, neon—and to hold even for these except at very low temperatures.

Electron "Gas" in Metals

One theory of metals states that a metal can be considered to be a stationary lattice of positive nuclei in a sea of electrons which move about in much the same way as do conventional gas molecules. Solid sodium metal, for example, has one free electron per sodium ion. X-ray diffraction studies show that sodium crystallizes in a cubic lattice containing two atoms per unit cell of volume 76.1×10^{-24} cm³. Therefore, the volume per electron in sodium is 38×10^{-24} cm³. At 300°K, the dilution ratio is

$$\frac{q}{n} = \frac{(2\pi mkT)^{\frac{3}{2}}}{h^3} 38 \times 10^{-24} = 0.00048$$

Clearly, any contention that sodium is a dilute system is absurd, and Fermi–Dirac statistics are mandatory. In fact, a dilution ratio of less than one is itself absurd, because one cannot have more than one particle per state; thus the calculation based on Maxwell–Boltzmann statistics was wrong to begin with. In order to bring the q/n ratio up to a value of ten where one might consider using Maxwell–Boltzmann statistics, one would have to raise the temperature of solid sodium to 2.2×10^{5}°K; again this condition is physically absurd.

What causes the big difference between electrons and helium atoms at 1 atm and 300°K, which would produce a factor of 10^9 difference in dilution ratios?

$$\text{He} : \frac{q}{n} \cong 3 \times 10^5 \qquad e^- : \frac{q}{n} \cong 5 \times 10^{-4}$$

The volume ratio is not the sole culprit; it furnishes only three of the nine orders of magnitude

$$\frac{(V/n)_{\text{He}}}{(V/n)_{\text{e}}} \cong \frac{4 \times 10^{-20}\ \text{cm}^3/\text{atm}}{4 \times 10^{-23}\ \text{cm}^3/\text{atm}} \cong 10^3$$

The real culprit is mass

$$\left(\frac{m_{He}}{m_e}\right)^{\frac{3}{2}} \cong \left(\frac{8000}{1}\right)^{\frac{3}{2}} \cong 10^6$$

If the electron had the mass of a proton, then the dilution ratio would be larger by a factor of $(1850)^{\frac{3}{2}}$, and q/n would be nearer 40. Thus the real reason why electrons are fermions at room temperature and helium atoms are boltzons is that electrons are so much lighter. A good demonstration of electrons being fermions is the existence of the periodic table. Try to think what the periodic table would be like if electrons were bosons. Chemistry would be a lot duller. (The point is deeper still. There would then be no differentiated material universe to produce intelligent beings capable of appreciating the dullness.)

An interesting situation occurs when one compares the behavior of boltzons, bosons, and fermions at absolute zero. At 0°K, boltzons and bosons will all pile up in the lowest energy state, ε_0. Fermions cannot do this because there can be no more than one fermion per state. Therefore, at absolute zero, not all fermions will have zero energy. All of the available microstates in the lowest energy levels will be filled with fermions, but then additional ones will have to go into the next higher energy level, and so forth. A plot of the concentration ratio n_j/g_j for fermions is given in Figure 2-6. Indicated in the figure is the so-called Fermi level, the energy level that is reached by the time all the molecules have been fed in. Increasing the temperature promotes some fermions to higher energy levels and results in a distribution indicated by the dotted line in Figure 2-6.

2-13 SHARPNESS OF THE MOST PROBABLE DISTRIBUTION

Is it enough to be able to calculate the most probable distribution, or must we also have other distributions that lie near it? If there are other distributions that are almost as probable, then having only the most probable distribution tells us little. If a plot of probability versus distribution were like that shown in Figure 2-7a, then whenever we wanted to calculate a property of our system we would have to consider the contributions to that property from distributions other than just the most probable, and average over all possible distributions. Fortunately, this is not necessary. The *most probable* distribution, for a macroscopic system with an appreciable

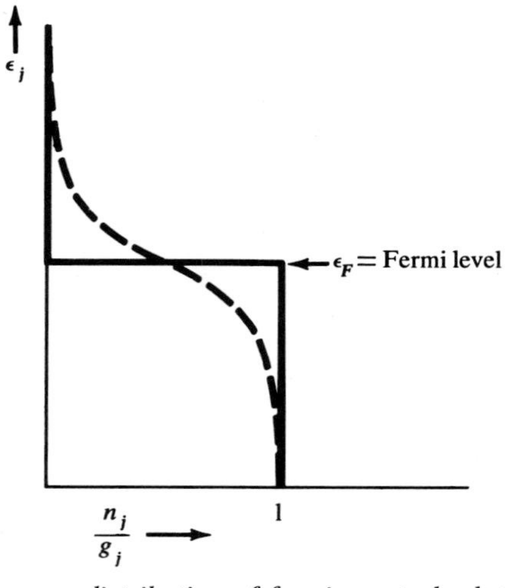

FIGURE 2-6. *The energy distribution of fermions at absolute zero (solid line) and at some higher temperature (dashed line). g_j is the number of states of a given energy ε_j, and n_j is the number of particles having that energy. For fermions, when $n_j/g_j = 1$, the energy level is completely filled, and further particles must occupy higher energy states. At absolute zero, when all particles have the lowest possible energies, all levels are completely filled up to energy ε_F, which is called the "Fermi level." At higher temperatures, some particles will have energies greater than ε_F, and the same number of states below the Fermi level will be unoccupied.*

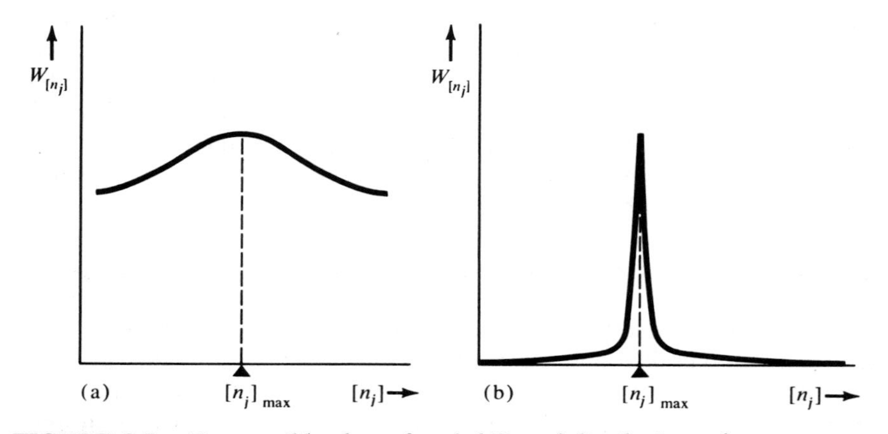

FIGURE 2-7. *Two possible plots of probability of distributions of energy among particles. The left-hand plot would require a complicated averaging process to obtain likely values for measurable quantities; the right-hand plot would indicate that only the* most probable *distribution of energy need be considered when predicting properties of a system. Section* 2-13 *demonstrates that the plot on the right is the true situation for macroscopic quantities of matter.*

number of particles, is so overwhelmingly favored that contributions from all less probable distributions can be neglected. A plot of probability versus distribution would appear more like Figure 2-7b rather than 2-7a, as a specific example will show.

Assume a system with equally spaced energy levels for simplicity, initially with its particles in a Maxwell–Boltzmann most probable distribution. Then let one molecule per million be excited from the first excited state to the second, and, to keep the total energy constant, let one molecule per million be dropped from that same first excited state to the ground state. The level populations before and after can then be represented as follows, with $\alpha = 10^{-6}$.

Level	Start	End
2	c	$c + \alpha b$
1	b	$b - 2\alpha b$
0	a	$a + \alpha b$

This new distribution, with only two molecules in 10^6 altered, could certainly be considered a neighboring distribution. How much less likely is it than the initial one?

The ratio of the probability at the start to that at the end, assuming Maxwell–Boltzmann statistics, is

$$R = \frac{W_{\text{start}}}{W_{\text{end}}} = \frac{(a + \alpha b)!\,(b - 2\alpha b)!\,(c + \alpha b)!}{a!\,b!\,c!}$$

Introducing the approximation that

$$\frac{(n + x)!}{n!} = (n + x)(n + x - 1) \cdots (n + 2)(n + 1) \cong \left(n + \frac{x}{2}\right)^x$$

yields

$$R \cong \left(\frac{(a + \alpha b/2)(c + \alpha b/2)}{(b - \alpha b)^2}\right)^{\alpha b}$$

Now for the simple case of equally spaced levels, the Boltzmann distributions will be

$$\frac{c}{b} = \frac{n_2}{n_1} = e^{-\Delta E/kT} = e^{-D} \qquad \text{where } D = \frac{\Delta E}{kT}$$

$$\frac{b}{a} = \frac{n_1}{n_0} = e^{-\Delta E/kT} = e^{-D}$$

or

$$c = be^{-D}$$
$$a = be^{D}$$

Introducing these relationships into our last expression for R and simplifying,

$$R \cong \left(\frac{(e^{D} + \alpha/2)(e^{-D} + \alpha/2)}{(1-\alpha)^2}\right)^{\alpha b} \cong \left(1 + \frac{\alpha}{2}e^{D} + \frac{\alpha}{2}e^{-D}\right)^{\alpha b}$$

At this point we must particularize our example, and apply it to the vibrational energy levels of HCl at 300°K.

$$D_{\text{HCl}} = \frac{h\nu_0}{kT} = 14 \text{ at } 300°K$$
$$e^{D} = 1.2 \times 10^{6}$$
$$e^{-D} = 0.83 \times 10^{-6}$$
$$R_{\text{HCl}} \cong [1 + \tfrac{1}{2} \times 10^{-6}(1.2 \times 10^{6})]^{\alpha b} = 1.6^{\alpha b}$$

Now b is the number of molecules per mole in the first excited energy level, approximately Ne^{-D} molecules per mole, or 5×10^{17} molecules. Then αb is approximately 5×10^{11}. This gives

$$R_{\text{HCl}} \cong 1.6^{5 \times 10^{11}} \cong 10^{10^{11}}$$

Therefore, the ratio of the probability of the most likely state to the probability of a state with two molecules in a million disarranged is of the order of $10^{10^{11}}$. This, for a concrete case of HCl at room temperature, is how sharp the distribution is, and nothing can compensate for this overwhelming exponential factor. We are perfectly safe in saying that the only distribution that need be considered is the most probable one.[5]

2-14 STATISTICAL THEORY OF GHOSTS

The probability argument of the previous section can be used to illustrate an interesting paradox. Consider Lake Tahoe, outwardly a quiet body of water, but in reality one in which the water molecules are moving about with considerable velocity. Only because the motion of the molecules is random

[5] The process of considering only the most probable distribution is the simpler of the two approaches to statistical mechanics. You will probably encounter the other approach of averaging over ensembles later.

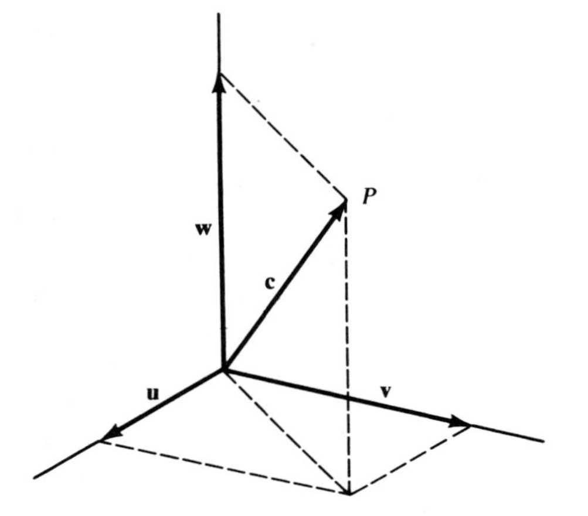

FIGURE 2-8. *Coordinate system for a velocity space plot. Each particle in a system is represented by a point in velocity space at coordinates* (u, v, w), *where these numbers are the components of the velocity of the particle in the x, y, and z directions. The speed c of the particle is given by the distance of the velocity space point from the origin;* $c^2 = u^2 + v^2 + w^2$.

do we see no net motion and a calm lake. A velocity space plot of the velocities of the molecules in the lake can be used to represent the motion of the molecules, using the coordinates of Figure 2–8. At any given instant each molecule can be represented by a point in this velocity space plot such that its three Cartesian coordinates (u, v, w) give its velocity components in the x, y, and z directions in space, and the distance of the point from the origin gives the speed c of the molecule. All points lying within a spherical shell of radius c and thickness dc will represent molecules having a speed between c and $c + dc$ but different directions. To say that the molecules are moving at random is then to say that the points representing all the molecules are distributed uniformly around the origin of velocity space with no bunching up in any one direction. Under normal conditions, that is, a calm lake, the points will be fairly uniformly distributed in velocity space. The individual points will move in velocity space with time as molecules collide and change velocity and direction. Just as with our energy levels, the overwhelmingly most probable distributions of molecule points in velocity space is one in which the points are spread out rather than bunched up.

What would be the macroscopic result if the points in velocity space were to congregate at one place? Lake Tahoe would then take off over the Sierra Nevadas at something like 1000 mph (based on the average speed of H_2O molecules[6] at room temperature as derived from simple kinetic theory).

[6] This assumption about the composition of Lake Tahoe is unfortunately becoming steadily poorer.

Unlikely, you say. Certainly. But it could happen. The improbable is not impossible.

Enough of science; now for the paradox. What if some unscientific observer happened to be present at the chance moment when all the points in velocity space did bunch together and Lake Tahoe suddenly did vanish into the blue? What would he think? He would probably invoke miracles or something else outside the ordinary frame of natural occurrences, although in fact he would have witnessed a natural, if exceedingly unlikely, phenomenon. In the early part of the century when probability theory was being developed, some people suggested in all seriousness that this might be the explanation of what we call psychic phenomena. Lacking any understanding of probability, they said, people erroneously attribute these rare occurrences to some sort of psychic effect. The only trouble is that the frequency of reports of such psychic phenomena (*reports*, not necessarily occurrences) greatly exceeds the probability of their natural occurrence by statistical fluctuation. The roots of such phenomena may be found in psychology, but not in statistics.

What if you were there and saw Lake Tahoe suddenly vanish in the direction of San Francisco? You understand probabilities. Would you say that an improbable but quite possible event finally came due? Probably not, because of the way in which we each decide upon truth. When we say that something is true, what we mean is that, based upon our prior experience, our estimate of the probability of its being true is high. If you were to see the lake suddenly vanish, and if you were able to think logically afterwards, your analysis of what happened might be as follows: "There is a probability P_e that the unlikely event in question really occurred, but there is also a certain probability P_h that I have been the victim of an optical hallucination and that it did not occur at all. Since P_h is much larger than P_e, I can simply reject the event as not having taken place."

Thus in spite of our supposedly objective frame of mind, we would be no more willing to accept the event as a natural physical occurrence than would the man who invoked miracles or poltergeists. And this is because we recognize the contingent nature of truth. We do not know anything for certain. All we do every day of our lives is to estimate probabilities, although usually from experience and not from statistical analysis.

2-15 VELOCITY DISTRIBUTION IN A GAS

In the story of Lake Tahoe we saw that particles moving in straight lines at constant speeds can be represented as points in velocity space. Since changes in the speeds and directions of the particles results in corresponding changes in the lengths and directions of their velocity vectors, we can map the distribution of velocities among molecules as a distribution of particle points in velocity space. Using this concept of velocity space and the

Boltzmann deduction that the probability of the existence of a given state is exponentially proportional to the energy of that state, $P_{(j)} \propto e^{-\varepsilon_j/kT}$, we can derive many useful thermodynamic expressions from studies of velocity distributions in gases.

One Dimension

The molecular energy of a one-dimensional gas is simply

$$\varepsilon_j = \tfrac{1}{2}mu_j^2 \tag{2-42}$$

A plot of velocity probability versus velocity is given in Figure 2-9. We know that the velocities for a particle in a box are quantized, so the axis in Figure 2-9 should be a collection of closely spaced points corresponding to the various energy levels. However, the energy levels for translational motion of macroscopic molecules are so close together that, for practical purposes, we can treat the series of energy states as a continuum and integrate over the energy levels

$$P_{(u)} \propto e^{-mu^2/2kT}$$

$$P_{(u)}\,du = Ae^{-mu^2/2kT}\,du \tag{2-43}$$

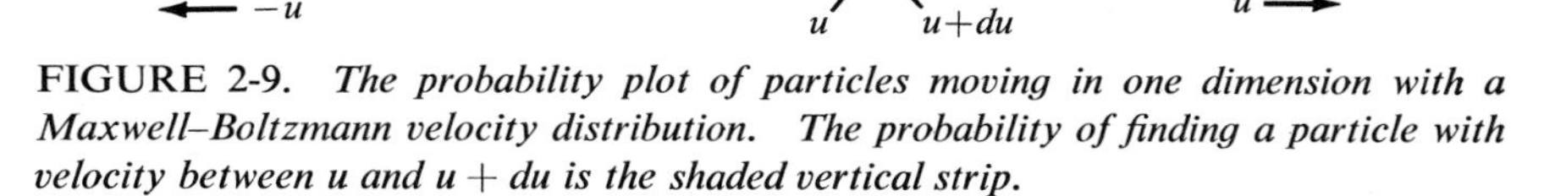

FIGURE 2-9. *The probability plot of particles moving in one dimension with a Maxwell–Boltzmann velocity distribution. The probability of finding a particle with velocity between u and u + du is the shaded vertical strip.*

$$P_{(u)}\,du = \left(\frac{m}{2\pi kT}\right)^{\frac{1}{2}} \exp\left(-\frac{m}{2kT}u^2\right)du$$

The most probable *velocity is zero, as is the* average *velocity, since as many particles are traveling to the right as to the left. But it would be quite wrong to calculate most of the properties of the collection of particles under the assumption that only the most probable velocity need be considered. Compare Figure* 2-7.

The constant A is found by normalizing

$$A \int_{-\infty}^{+\infty} e^{-mu^2/2kT}\, du = 1$$

$$A = \left(\frac{m}{2\pi kT}\right)^{\frac{1}{2}}$$

Finally, the normalized probability of finding a molecule with its velocity between u and $u + du$ is

$$P_{(u)}\, du = \left(\frac{m}{2\pi kT}\right)^{\frac{1}{2}} \exp\left(-\frac{m}{2kT}u^2\right) du \tag{2-44}$$

The average *velocity* is

$$\langle u \rangle = \int_{-\infty}^{\infty} P_{(u)} u\, du \bigg/ \int_{-\infty}^{\infty} P_{(u)}\, du = \frac{0}{1} = 0 \tag{2-45}$$

The trouble with Lake Tahoe in Section 2-14 was that the average velocity was not zero; but in the most probable state as many particles go in any one direction as go in the opposite direction. The entire system stays put.

The average *speed* per molecule is

$$\langle |u| \rangle = \int_{-\infty}^{\infty} P_{(u)} |u|\, du = 2 \int_{0}^{\infty} P_{(u)}\, u\, du$$

$$\langle |u| \rangle = 2\left(\frac{m}{2\pi kT}\right)^{\frac{1}{2}} \int_{0}^{\infty} \exp\left(-\frac{m}{2kT}u^2\right) u\, du = \left(\frac{2kT}{\pi m}\right)^{\frac{1}{2}} = \left(\frac{2RT}{\pi M}\right)^{\frac{1}{2}} \tag{2-46}$$

Equation 2-46 gives us a useful way of finding the mean speed of molecules when we know their molecular weight and the temperature.

The root mean square speed u_{rms} is a useful quantity. The mean square speed is

$$\langle u^2 \rangle = 2\left(\frac{m}{2\pi kT}\right)^{\frac{1}{2}} \int_{0}^{\infty} \exp\left(-\frac{m}{2kT}u^2\right) u^2\, du = \frac{kT}{m} = \frac{RT}{M} \tag{2-47}$$

Therefore, the root mean square speed is

$$u_{\text{rms}} = (\langle u^2 \rangle)^{\frac{1}{2}} = \left(\frac{kT}{m}\right)^{\frac{1}{2}} = \left(\frac{RT}{M}\right)^{\frac{1}{2}} \tag{2-48}$$

The most probable speed u_{max}, which corresponds to the highest point in the probability curve of Figure 2-9, is zero. That is, the *most likely* state is a molecule at rest. However, in this case, states other than the most probable state are of considerable importance; calculations based solely on properties of the most probable state would be grossly wrong. Nonetheless $u_{max} = 0$, as can be formally verified by setting $dP/du = 0$ and evaluating for u.

The mean molecular energy will be useful. Since $\varepsilon = \frac{1}{2}mu^2$, the mean molecular energy $\langle \varepsilon \rangle$ is given by

$$\langle \varepsilon \rangle = \tfrac{1}{2}m\langle u^2 \rangle = \tfrac{1}{2}kT \tag{2-49}$$

From Eq. 2-49 we immediately obtain the expression for the molar energy:

$$\bar{E} = N\langle \varepsilon \rangle = \tfrac{1}{2}RT \tag{2-50}$$

Two Dimensions

Adding a second dimension to our velocity introduces the subject of multiplicities and statistical weights. Our two-dimensional velocity vector $\mathbf{c}_{(u,v)}$ is related to its components by $c^2 = u^2 + v^2$, and the energy is

$$\varepsilon = \tfrac{1}{2}mc^2 = \tfrac{1}{2}m(u^2 + v^2) \tag{2-51}$$

The probability of finding the molecule with a certain velocity is related to the energy by

$$P_{(u,v)} \propto \exp\left(-\frac{m}{2kT}(u^2 + v^2)\right)$$

A plot of the probability of finding our gas molecule with a given u and v component of velocity is shown in Figure 2-10. By extension of the one-dimensional relationships, the probability of finding a molecule whose velocity space point lies in an element of area between u and $u + du$ and v and $v + dv$ is given by

$$P_{(u,v)}\, dudv = A' \exp\left(-\frac{m}{2kT}(u^2 + v^2)\right) dudv \tag{2-52}$$

Normalizing to find A',

$$A' \int_{u=-\infty}^{\infty} \int_{v=-\infty}^{\infty} \exp\left(-\frac{m}{2kT}(u^2 + v^2)\right) dudv = 1$$

$$A' = \frac{m}{2\pi kT}$$

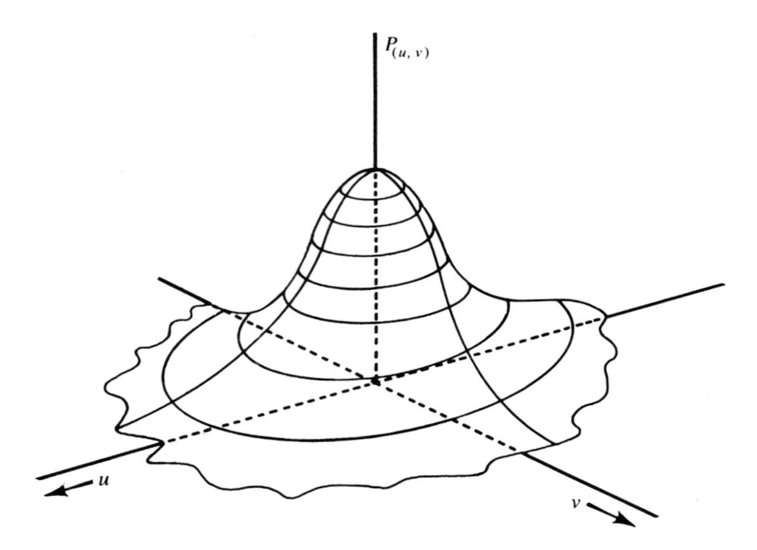

FIGURE 2-10. *The probability plot of particles moving in two dimensions with a Maxwell–Boltzmann velocity distribution. The probability of finding a particle with an x component of velocity between u and u + du,* and *a y component between v and v + dv is*

$$P_{(u,v)}\,dudv = \left(\frac{m}{2\pi kT}\right)\exp\left(-\frac{m}{2kT}(u^2+v^2)\right)dudv$$

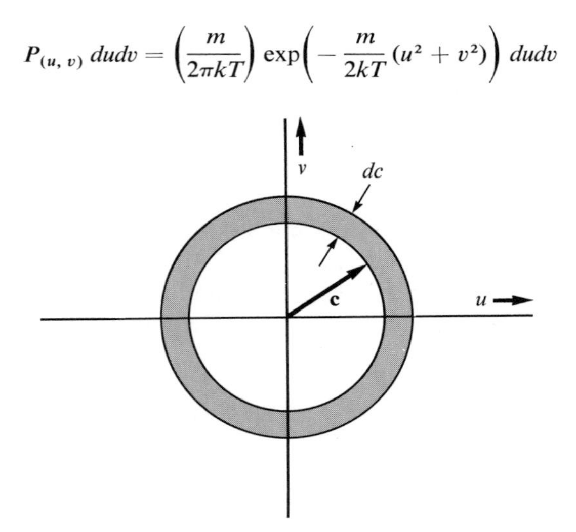

FIGURE 2-11. *A view of the basal plane of Figure* 2-10. *All particles having a speed between c and c + dc, regardless of direction, will have velocity space points lying in the shaded annulus, of area $2\pi cdc$. If* possible *states are distributed uniformly over the u - v plane, which is true for a particle in a two-dimensional potential well, then the statistical weight or degeneracy of the state with speed between c and c + dc is just $2\pi cdc$.*

This gives a normalized probability

$$P_{(u,v)}\,dudv = \left(\frac{m}{2\pi kT}\right)\exp\left(-\frac{m}{2kT}(u^2+v^2)\right)dudv \tag{2-53}$$

By introducing polar coordinates, we can easily calculate the probability of finding a particle with a speed within a certain range regardless of its direction. In essence we want to know the probability of finding a particle whose velocity point lies within the annulus shown in Figure 2-11. Transforming from Cartesian coordinates to polar coordinates,

$$u = c\cos\theta$$
$$v = c\sin\theta$$
$$dudv = cdcd\theta$$

Integrating over θ to remove the angular dependence,

$$P_{(c)}\,dc = \int_{\theta=0}^{2\pi}\left(\frac{m}{2\pi kT}\right)\exp\left(-\frac{m}{2kT}c^2\right)c\,dcd\theta$$

$$P_{(c)}\,dc = \left(\frac{m}{2\pi kT}\right)\exp\left(-\frac{m}{2kT}c^2\right)2\pi cdc \tag{2-54}$$

The area element $2\pi cdc$ in Eq. 2-11, performs the function of a statistical weight, as does the element of length du in the one-dimensional problem and the area element $dudv$ for the two-dimensional problem in Cartesian coordinates.

The following expressions, presented without proof, are useful in dealing with two-dimensional velocity problems.

Average speed in the x and y directions:

$$\langle|u|\rangle = \langle|v|\rangle = \left(\frac{2RT}{\pi M}\right)^{\frac{1}{2}} \tag{2-55}$$

Average speed, without regard to direction:

$$\langle|c|\rangle = \left(\frac{\pi kT}{2m}\right)^{\frac{1}{2}} = \left(\frac{\pi RT}{2M}\right)^{\frac{1}{2}} \tag{2-56}$$

Root mean square speed:

$$c_{\text{rms}} = \left(\frac{2kT}{m}\right)^{\frac{1}{2}} = \left(\frac{2RT}{M}\right)^{\frac{1}{2}} \tag{2-57}$$

Most probable speed:

$$c_{\max} = \left(\frac{kT}{m}\right)^{\frac{1}{2}} = \left(\frac{RT}{M}\right)^{\frac{1}{2}} \tag{2-58}$$

Mean molecular energy:

$$\langle \varepsilon \rangle = kT \tag{2-59}$$

Molar energy:

$$\bar{E} = RT \tag{2-60}$$

You should be able to derive all these relationships by the averaging process.

Three Dimensions

The three-dimensional velocity $\mathbf{c}_{(u, v, w)}$ is related to its components by

$$c^2 = u^2 + v^2 + w^2$$

The probability of finding a particle within a given volume element as that shown in Figure 2-12, or the probability of finding its x component of velocity between u and $u + du$, its y component between v and $v + dv$, and its z component between w and $w + dw$, is given by

$$P_{(u, v, w)}\, dudvdw = A'' \exp\left[-\frac{m}{2kT}(u^2 + v^2 + w^2)\right] dudvdw \tag{2-61}$$

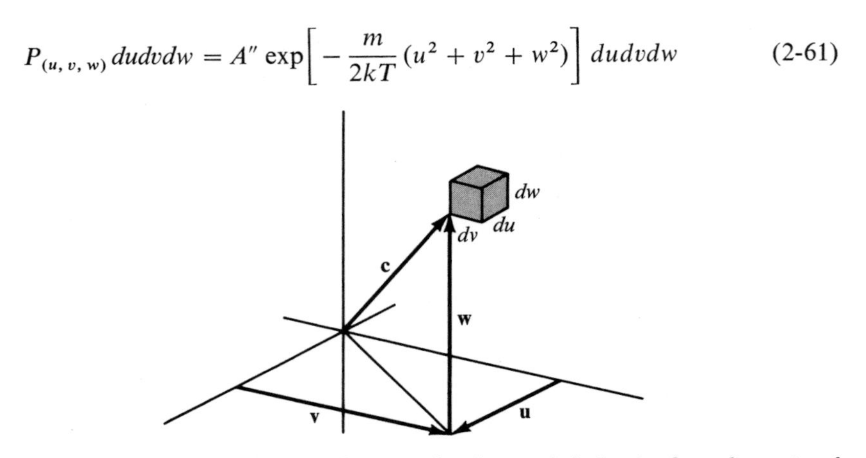

FIGURE 2-12. *The differential element of volume, dudvdw, in three-dimensional velocity space.*

Normalizing Eq. 2-61, we obtain

$$A'' = \left(\frac{m}{2\pi kT}\right)^{\frac{3}{2}}$$

which gives a normalized probability of

$$P_{(u,\,v,\,w)}\,dudvdw = \left(\frac{m}{2\pi kT}\right)^{\frac{3}{2}} \exp\left[-\frac{m}{2kT}(u^2+v^2+w^2)\right] dudvdw \quad (2\text{-}62)$$

With Eq. 2-62 in mind, what is the probability of finding a particle with a speed between c and $c + dc$ regardless of which direction it is going? In other words, what is the probability of finding a particle with a velocity point within the shell of thickness dc shown in Figure 2-13? Using the direct approach, we could convert from Cartesian coordinates to spherical polar coordinates and then integrate out the angular dependence, just as for the two-dimensional case. But now that we know the relationship between statistical weight and the volume element of the shell (Figure 2-13), we are able to remove the angular dependence merely by inspection. The area of the shell surface is $4\pi c^2$ and the volume of the shell is $4\pi c^2 dc$. The desired

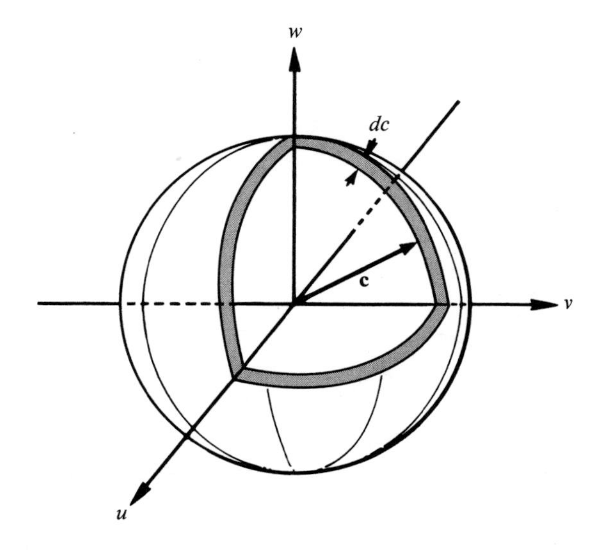

FIGURE 2-13. *All molecules having speed between c and c + dc, regardless of direction, will have velocity space points lying in the spherical shell about the origin of radius c and thickness dc. The statistical weight or degeneracy of this speed range will be equal to the volume of the shell,* $4\pi c^2 dc$.

probability expression is then

$$P_{(c)}\,dc = \left(\frac{m}{2\pi kT}\right)^{\frac{3}{2}} \exp\left(-\frac{m}{2kT}c^2\right) 4\pi c^2\,dc \tag{2-63}$$

The average speed is found by integrating $\int_0^\infty P_{(c)}\,c dc$

$$\langle c\rangle = \left(\frac{8kT}{\pi m}\right)^{\frac{1}{2}} = \left(\frac{8RT}{\pi M}\right)^{\frac{1}{2}} \tag{2-64}$$

The root mean square speed is

$$c_{\text{rms}} = (\langle c^2\rangle)^{\frac{1}{2}} = \left(\frac{3RT}{M}\right)^{\frac{1}{2}} \tag{2-65}$$

A plot of $P_{(c)}$ versus c, as in Figure 2-14, introduces the importance of the statistical weight. Figure 2-14 includes plots of the $4\pi c^2$ term and the exponential term of Eq. 2-63 along with a composite trace which gives the actual probability of a given velocity. The probability of having a particle in a given energy state decreases as we go to higher energies, but at the same time there are a great many more such states of the same speed $|c|$. The most probable speed c_{max} is found by setting $dP_{(c)}/dc = 0$ and evaluating for c

$$c_{\text{max}} = \left(\frac{2RT}{M}\right)^{\frac{1}{2}} \tag{2-66}$$

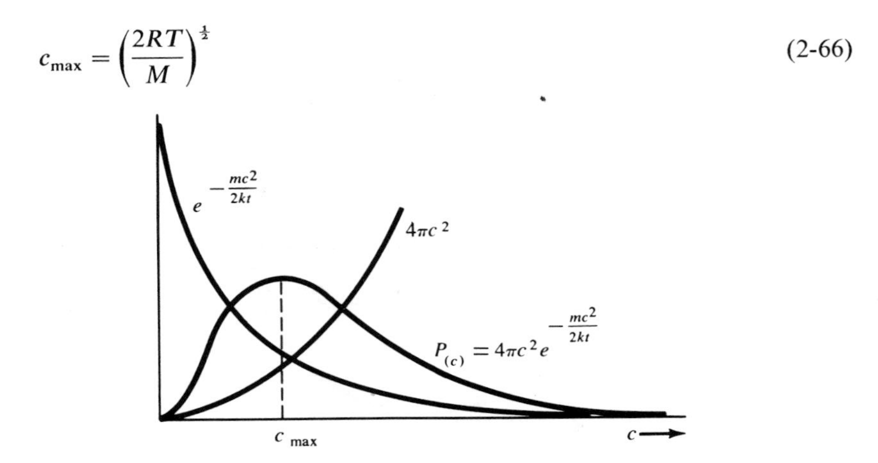

FIGURE 2-14. *The probability distribution of speeds c for gas molecules in three dimensions passes through a maximum at c_{max} because it is the product of two opposing factors. As the speed increases, the intrinsic Maxwell–Boltzmann probability of a state decreases exponentially, but the number of different states which show such a speed rises as $4\pi c^2$.*

Comparing $\langle c\rangle$ and c_{rms} with $c_{\max}$, we can see that both are greater than $c_{\max}$

$$\left(\frac{2RT}{M}\right)^{\frac{1}{2}} < \left(\frac{8RT}{\pi M}\right)^{\frac{1}{2}} < \left(\frac{3RT}{M}\right)^{\frac{1}{2}}$$

$$c_{\max} < \langle c\rangle < c_{\mathrm{rms}}$$

The mean molecular energy is

$$\langle\varepsilon\rangle = \tfrac{3}{2}kT \tag{2-67}$$

and the molar energy is

$$\bar{E} = \tfrac{3}{2}RT \tag{2-68}$$

When we compare the molar energies of the one-, two-, and three-dimensional problems, we see an illustration of the principle of equipartition of energy which will be taken up again later. At the moment, it appears to be possible for a mole of gas to have $\frac{1}{2}RT$ of energy per degree of freedom.

2-16 FIRST THERMODYNAMIC FUNCTION: ENERGY E

Energy

The first thermodynamic function to be expressed in terms of molecular and statistical models is energy. The molecular partition function q and the mean molecular energy $\langle\varepsilon\rangle$ have already been introduced in Section 2-10 and 2-11, but one expression needs development.

The molecular partition function

$$q = \sum_j g_j \exp(-\beta\varepsilon_j) \qquad \beta = \frac{1}{kT} \tag{2-29}$$

can also be written as

$$q = \sum_{j'} \exp(-\beta\varepsilon_{j'}) \tag{2-69}$$

where the degenerate states are now included in the formal summation, and an explicit degeneracy factor is unnecessary. This is only a change in nomenclature. It has been shown that

$$\langle\varepsilon\rangle = -\left(\frac{\partial \ln q}{\partial\beta}\right) = kT^2\left(\frac{\partial \ln q}{\partial T}\right) \tag{2-33}$$

Hence the *molar* energy is

$$\bar{E} = RT^2\left(\frac{\partial \ln q}{\partial T}\right)_V \tag{2-70}$$

In order for Eq. 2-70 to be valid, the energy levels must remain fixed, so that implicit in this derivation is the necessity of holding the volume constant. Furthermore, in going from the molecular to the molar level, we have assumed that the value of a function for N atoms is just N times the value of the function for each atom. This assumption is true for the energy, but will not be true for many other properties such as entropy.

To be perfectly correct, we should consider all the molecules at once when computing a *molar* sum over states, because any one state involves a description of the conditions of *all* the molecules in the mole of material. If this is done, then the ambiguities of the previous paragraph do not arise. Suppose that there are N particles, each of which has available to it the same set of individual molecular states:

particles: $1, 2, 3, 4, \ldots, N$

states: $a, b, c, d, e, \ldots$

With no restriction of the number of particles per state, any particular molar state can be described as a combination of individual molecular states:

$$\psi_i = a_{(1)}b_{(2)}g_{(3)}f_{(4)}c_{(5)}j_{(6)}b_{(7)}h_{(8)}\cdots$$

The energy of this state is then

$$\bar{E}_i = \varepsilon_a + \varepsilon_b + \varepsilon_g + \varepsilon_f + \varepsilon_c + \cdots$$

The N-molecule partition function is the sum over all possible states:

$$Q_N = \sum_i e^{-\bar{E}_i/kT} \tag{2-71}$$

Then the average energy of the system of N molecules is obtained by averaging over all possible energies of all possible molar states:

$$\bar{E} = \langle \bar{E}_i \rangle = kT^2\left(\frac{\partial \ln Q_N}{\partial T}\right)_V \tag{2-72}$$

Having introduced the molar partition function Q_N, how do we relate it to the molecular partition function q? The following example of two particles in two states will illustrate the relationship. For a two-particle, two-state (a, b) system there are only four possible system states with four corresponding energies:

state	aa	ab	ba	bb
energy	$2\varepsilon_a$	$\varepsilon_a + \varepsilon_b$	$\varepsilon_b + \varepsilon_a$	$2\varepsilon_b$

The molar partition function for the system is

$$Q_2 = e^{-\beta(2\varepsilon_a)} + e^{-\beta(\varepsilon_a+\varepsilon_b)} + e^{-\beta(\varepsilon_b+\varepsilon_a)} + e^{-\beta(2\varepsilon_b)}$$

or

$$Q_2 = (e^{-\beta\varepsilon_a} + e^{-\beta\varepsilon_b})^2 = q^2$$

and, in general, for distinguishable particles,

$$Q_N = q^N \tag{2-73}$$

When N is Avogadro's number, the molar partition function is written as Q without the subscript. But Eq. 2-73 contains a built-in fallacy. For gases, states ab and ba are the same; the particles are indistinguishable. The correction factor for superfluous permutations in very dilute systems of indistinguishable particles is $1/N!$. Therefore, for *ideal gases*, the correct relationship between Q_N and q is

$$Q_N = \frac{q^N}{N!} \tag{2-74}$$

In dealing with ideal crystals, in which the molecules are distinguishable, Eq. 2-73 is valid. Liquids are more difficult to characterize, and the correct Q_N for liquids lies somewhere between q^N and $q^N/N!$.

Given the correct partition function for a physical system, one can calculate any thermodynamic function desired. However, the toughest part of statistical mechanics is trying to find the right partition function for a given situation.

Heat Capacity

The heat capacity of a substance measures its ability to store energy for a given rise in temperature, and is simply the rate of change of its energy with temperature. The amount of energy which a substance can store before the temperature rises by a given amount depends upon the conditions under which the addition of energy is carried out. Two common conditions are those of constant volume or constant pressure. For ideal gases, the molar heat capacity in a constant-volume process is

$$C_V = \left(\frac{\partial E}{\partial T}\right)_V = \tfrac{3}{2}R \tag{2-75}$$

If the gas is allowed to expand at constant pressure during the addition of energy, some of this energy can be diverted into pressure-volume work on the surroundings, and the heat capacity is greater. We see in Chapter 3 that for an ideal gas at constant pressure

$$C_P = \left(\frac{\partial E}{\partial T}\right)_P = \tfrac{5}{2}R$$

2-17 *SECOND THERMODYNAMIC FUNCTION: ENTROPY S*

Mention was made earlier of the idea that motion and arrangement are two complementary properties of a system. The previous section on energy dealt with the motion of particles; now we are concerned with the arrangement of particles in a system.

Definition and Properties

We shortly define the entropy S of an arrangement of particles, to be a measure of the uniqueness of the arrangement. We know that the probability of finding a given distribution is proportional to the number of ways of obtaining that distribution

$$P_{[n_j]} \propto W_{[n_j]}$$

To form the probability of a distribution in the usual sense of the term we should divide the number of ways of producing the given distribution by the total number of ways of producing all possible distributions. It does a poker player very little good to know that there are 3744 different ways of obtaining a full house unless he also knows that there are 2,598,960 different possible five-card hands, or that the relative probability of being dealt a full house is 0.00144.

Yet for our purposes it is more informative to know the absolute likelihood of occurrence of a distribution, as measured simply by the number of different ways by which the distribution can be obtained, $W_{[n_j]}$. We shall define a quantity, known as the *entropy* S proportional to the logarithm of the number of ways of obtaining a given distribution

$$S_{[n_j]} = k \ln W_{[n_j]} \tag{2-76}$$

Such a treatment has the advantage of avoiding enormous exponential quantities. Also, since probabilities are multiplicative, the logarithms of these probabilities are additive. Being an additive property of the amount of matter in the system, the entropy is thus an extensive property like energy. The constant in Eq. 2-76 could be any quantity, but for convenience later we will let k be Boltzmann's constant, $k = R/N$. Such a choice makes the unit of entropy cal/deg mole. The standard unit of entropy is

$$1 \text{ cal/deg mole} = 1 \text{ entropy unit} = 1 \text{ e.u.}$$

Entropy and Disorder

There is only one way of constructing a perfectly ordered system, $W = 1$, and the entropy of such a system is zero. At the other extreme, the number of ways of obtaining a totally disordered system tends toward infinity for systems containing many particles. As the number of particles increases, the entropy of such a system also tends to infinity. Clearly, entropy can never be negative, because a negative entropy is meaningless in terms of probabilities.

The crystal structure of solid carbon monoxide is a good example of a system of partial disorder and, hence, of intermediate probability. The two ends of the CO molecule are almost identical, and CO is very close to being a homonuclear molecule. If we let Figure 2-15 represent the crystal lattice of CO, then the difference in energy between the next molecule coming into the lattice as C—O or O—C is very small.

In the limit of negligible energy difference between the two possible orientations, any given molecule is just as likely to come down lined up as reversed. But in such a case of negligible energy difference, there still exists a possibility of configurational disorder. Each molecule has two possible ways of entering the crystal lattice. Therefore, for a system of n molecules, there would

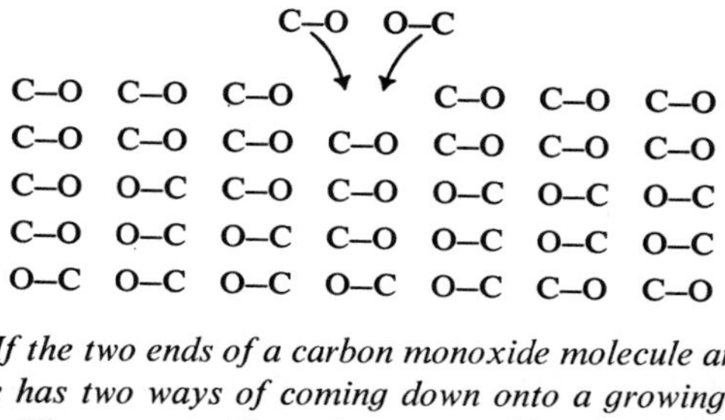

C–O O–C

C–O C–O C–O ↘ ↙ C–O C–O C–O
C–O C–O C–O C–O C–O C–O C–O
C–O O–C C–O C–O O–C O–C O–C
C–O O–C O–C C–O O–C O–C O–C
O–C O–C O–C O–C O–C C–O C–O

FIGURE 2-15. *If the two ends of a carbon monoxide molecule are very similar, then each new molecule has two ways of coming down onto a growing crystal, of roughly equal probability. The energy will not be appreciably affected, but the entropy will. The representation of the solid* C—O *crystal lattice is only schematic.*

be 2^n possible crystal arrangements. Hence the entropy is

$$S = nk \ln 2$$

and if n is Avogadro's number

$$\bar{S} = R \ln 2 = 1.38 \text{ cal/deg mole} = 1.38 \text{ e.u.} \tag{2-77}$$

For the general case, consider a mixture of two kinds of molecules a and b, such that

$$n_a + n_b = n$$

$$W = \frac{n!}{n_a!\, n_b!}$$

The entropy of this system, using Stirling's approximation for large numbers, is

$$\begin{aligned} S = k \ln W &= k(\ln n! - \ln n_a! - \ln n_b!) \\ &= k(n \ln n - n_a \ln n_a - n_b \ln n_b - n + n_a + n_b) \end{aligned} \tag{2-78}$$

Noting that n_a/n and n_b/n are just the mole fractions of a and b, we obtain

$$S = -k(n_a \ln X_a + n_b \ln X_b) \tag{2-79}$$

And if $n = N$, we have the entropy per mole

$$\bar{S} = -R(X_a \ln X_a + X_b \ln X_b) \tag{2-80}$$

In general, for a system of j components,

$$\bar{S}_{(\text{per mole})} = -R \sum_j X_j \ln X_j \tag{2-81}$$

The entropy just calculated is a form of configurational entropy, the entropy of mixing. Since the configurational entropy of the pure substances before mixing is zero, and since the mole fractions of components are always less than one, we can see from Eq. 2-81 that the entropy of mixing is always positive.

In any mixing operation, the entropy or the disorder of the system increases, so that mixing always leads to more probable states. This fact leads to the conclusion that mixing should be irreversible. As we shall see, these statements are two substatements of the second law.

Entropy and Partition Functions

Just as earlier we were able to use partition functions to obtain an expression for energy, it is now possible to obtain a corresponding expression for the entropy. We know that for a system of boltzons, with which we will be dealing from now on, the entropy of a given distribution is

$$S_{[n_j]} = k \ln W_{[n_j]} = k \ln \left(\prod_j \frac{g_j^{n_j}}{n_j!} \right) \tag{2-82}$$

For the most probable distribution,

$$n_{j\max} = \frac{n}{q} g_j e^{-\beta \varepsilon_j}$$

Therefore, the entropy of the most probable distribution, using Stirling's approximation, is

$$S_{\max} = S = k \sum_j (n_j \ln g_j - n_j \ln n_j + n_j) \tag{2-83}$$

Simplifying,

$$S = k\left(n + \sum_j n_j \ln \frac{g_j}{n_j}\right) \qquad \frac{g_j}{n_j} = \frac{q}{n} e^{\beta \varepsilon_j} \tag{2-84}$$

$$S = k\left(n + \sum_j n_j \ln \frac{q}{n} + \beta \sum_j n_j \varepsilon_j\right) \tag{2-85}$$

Rearranging Eq. 2-85,

$$S = k\left(n - n \ln n + n \ln q + \beta \sum_j n_j \varepsilon_j\right) \tag{2-86}$$

Using Stirling's formula in reverse, $n - n \ln n = -\ln n!$. Hence we can write Eq. 2-86 as

$$S = k \ln\left(\frac{q^n}{n!}\right) + \sum_j n_j \varepsilon_j \bigg/ T \tag{2-87}$$

If $n = N$, then the first term in Eq. 2-87 collapses into the molar partition function Q

$$\frac{q^N}{N!} = Q \qquad \text{(ideal gas)}$$

If we had begun with an ideal crystal, the results would have been the same. We would have had an $n!$ term,

$$W_{\text{ideal crystal}} = n! \prod_j \frac{g_j^{n_j}}{n_j!}$$

This would ultimately have cancelled the $n!$ in the denominator of Eq. 2-87, and the first term would still be $k \ln Q$.

The second term of Eq. 2-87 is the energy of the system $\bar{E}$, divided by T. Hence Eq. 2-87 can be written as

$$\bar{S} = k \ln Q + \frac{\bar{E}}{T} \tag{2-88}$$

Equation 2-72a allows the calculation of the energy if the partition function is known:

$$\bar{E} = kT^2 \left(\frac{\partial \ln Q}{\partial T} \right)_V \tag{2-72a}$$

Now from Eq. 2-88, the entropy can be calculated as well. These last two equations are two of the most fundamental expressions of statistical mechanics.

Ideal Monatomic Gas

For an ideal monatomic gas, $Q = q^N/N!$. Using Stirling's approximation for $\ln N!$, $N! \cong (N/e)^N$ or

$$Q \cong \left(\frac{qe}{N} \right)^N \tag{2-89}$$

Hence Eq. 2-88 becomes

$$\bar{S} = \frac{\bar{E}}{T} + R \ln \frac{qe}{N} \tag{2-90}$$

where

$$q = \frac{(2\pi mkT)^{\frac{3}{2}} V}{h^3}$$

$$\bar{E} = \tfrac{3}{2} RT$$

Then

$$\bar{S} = \tfrac{3}{2}R + R \ln\left(\frac{(2\pi mkT)^{\frac{3}{2}}eV}{Nh^3}\right) \tag{2-91}$$

and, since $\ln e = 1$,

$$\bar{S} = R\left[\frac{5}{2} + \ln\left(\frac{(2\pi mkT)^{\frac{3}{2}}V}{Nh^3}\right)\right] \quad \text{(Sackur–Tetrode equation)} \tag{2-92}$$

Thus we are now able to calculate the entropy of an ideal monatomic gas from first principles. The important unknowns in the Sackur–Tetrode equation are the molecular weight, temperature, and volume. Everything else is constant. We can emphasize the unknowns by rewriting Eq. 2-92 as

$$\bar{S} = \tfrac{3}{2}R \ln M + \tfrac{3}{2}R \ln T + R \ln V + K' \tag{2-93}$$

where the constants are collected into the K' term. Other things being constant, as volume increases, the entropy of the system rises. This is reasonable, because the more room there is available to the molecules, the more different arrangements are possible. From a quantum-mechanical viewpoint, as the volume increases, the energy levels settle downwards, the dilution ratio (Section 2-12) increases, and, with more states now available to the particles, W increases as well. The entropy also rises if the temperature is increased, which is certainly logical. A temperature increase results in an increase in molecular motion, a condition that enhances the chances of particle mixing. Finally, entropy increases as molecular weight increases. This last fact is harder to see intuitively, but a good analogy might be the fact that an elephant causes more disorder than a mouse. In fact, the mass increase causes the energy levels to be more closely spaced (think of the particle in a box), and both the dilution ratio and W increase.

By letting $V = 1000RT/P$, we can rewrite Eq. 2-93 as

$$\bar{S} = \tfrac{3}{2}R \ln M + \tfrac{5}{2}R \ln T - R \ln P + K'' \tag{2-94}$$

From this expression, we can see that when the pressure increases with the temperature held constant, the entropy decreases. This is obvious, since an increase in pressure with temperature held constant results in a decrease in volume and a more crowded system.

It may be that the molecule at rest is not the most convenient point from which to measure energy, and that the energy of this state of rest is more conveniently given the value ε_0. Then the molecular energy on this scale is $\varepsilon = \varepsilon_t + \varepsilon_0$, where ε_t is the translational energy relative to rest. Then the partition function on this new energy scale is

$$Q = Q_t e^{-\varepsilon_0/kT} \tag{2-95}$$

The numerical value of the molar energy, of course, is different:

$$\bar{E} = kT^2\left(\frac{\partial \ln Q_t}{\partial T}\right)_V + \bar{E}_0 = \bar{E}_t + \bar{E}_0 \tag{2-96}$$

where $\bar{E}_0 = N\varepsilon_0$.
But that of the entropy is not different; the contributions of $\bar{E}_0$ to the two terms cancel:

$$\bar{S} = k \ln Q_t - \frac{\bar{E}_0}{T} + \frac{\bar{E}_t + \bar{E}_0}{T} = k \ln Q_t + \frac{\bar{E}_t}{T} \tag{2-97}$$

The energy $\bar{E}_t = \bar{E} - \bar{E}_0$ is often called the thermal energy. Note that if entropy truly is a measure of the *arrangement* of a system, then it must not be sensitive to such externally imposed factors like the choice of a zero point from which to measure energy.

2-18 INTERNAL DEGREES OF FREEDOM

In the previous sections, emphasis was placed on the importance of being able to find the partition function, the sum over states. Now what energy states in a molecule are available? We know that molecules move, vibrate, rotate, and change their electronic distributions. Previously, we have considered only translational quantum states, but now we deal with the other modes of energy storage as well.

The n-Atom Molecule

No matter what coordinate system is chosen, $3n$ coordinates are needed to describe the positions of all n particles in a molecule. In Cartesian coordinates, three numbers, x, y, and z, are needed for each particle. Another choice of coordinates is that of Figure 2-16. Here the coordinates are

(1) the coordinates (X, Y, and Z) of the center of gravity of the molecule;
(2) three angles (ϕ, θ, ψ) giving the orientation of the collection of particles in space, taken as a unit;
(3) all of the structural parameters—the bond lengths and angles—necessary to build up a rigid unit.

All told, there are three positional, three rotational, and $3n - 6$ structural parameters required. Figure 2-17 shows the nine structural parameters needed to describe chloromethane. Note that only two H—C—H angles are needed, because the third one is determined by the other two and the other structural parameters. (If you do not believe this, try building the molecule from toothpicks and clay.) Adding the three center-of-mass coordinates and the three orientational ones, we obtain a total of $3n$ or 15 coordinates.

3 positional	X, Y, Z
3 orientational	θ, ϕ, ψ
9 structural	$r_1, r_2, r_3, r_4, \beta_1, \beta_2, \alpha_1, \alpha_2, \alpha_3$

Changes in positional parameters are commonly referred to as translations. Changes in orientational parameters are known as rotations, and changes in structural parameters can be described as vibrations. Changes in the basic pattern of electron distribution around atomic nuclei can also occur with electronic excitation, although we will seldom be concerned with this.

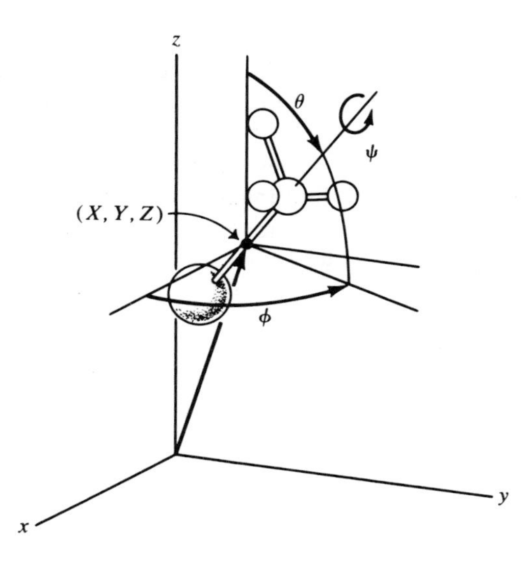

FIGURE 2-16. *Six parameters are required to specify the position and orientation of an asymmetric object in space. Here the ones chosen are the coordinates of the center of mass of the object, (X, Y, Z), the orientation of an arbitrary rotation axis in spherical coordinates (θ, φ) and the rotation about that axis (ψ).*

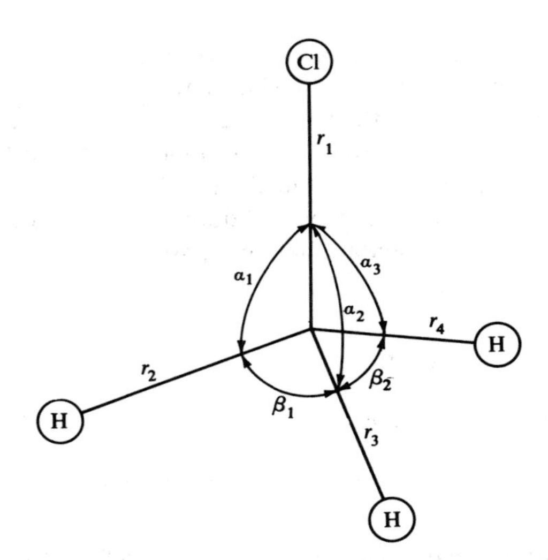

FIGURE 2-17. *The bond lengths* (r_j) *and bond angles* (α_j, β_j) *needed to specify the structure of a five-atom molecule such as chloromethane,* CH_3Cl. *Nine parameters are required. Note that the third* Cl—C—H *angle, between bonds* r_2 *and* r_4, *is redundant.*

Diatomic Molecule

A diagram of a typical diatomic molecule is given in Figure 2-18 in which the center of gravity is at (X, Y, Z) and the two rotational parameters are θ and ϕ. A linear molecule has only two rotational degrees of freedom in contrast to the three rotational modes for nonlinear molecules, but in compensation has one additional vibrational mode. Of course, our diatomic molecule has just one vibrational mode. Considering the allowed energy states for each mode of energy, we have

Translation:

$$\varepsilon_{jkl} = \frac{h^2}{8m}\left(\frac{j^2}{a^2} + \frac{k^2}{b^2} + \frac{l^2}{c^2}\right) \qquad g_{jkl} = 1$$

Rotation[7]:

$$\varepsilon_j = hBJ(J+1) \qquad g_J = 2J+1$$

$$B = \frac{h}{8\pi^2 I} \qquad I = \text{moment of inertia}$$

[7] See Hanna, Chapter 5.

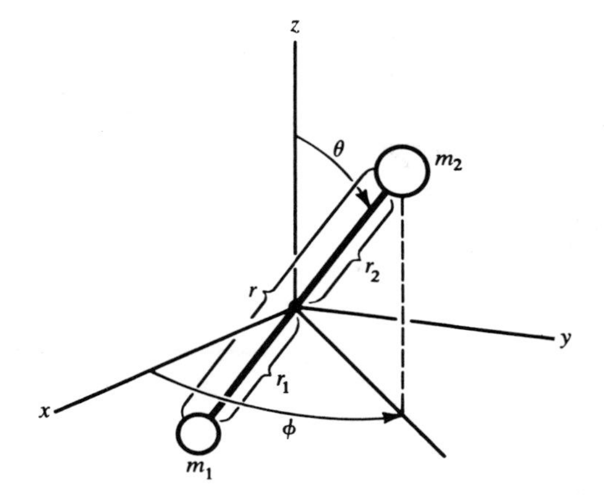

FIGURE 2-18. *No matter how they are chosen, six parameters are needed to fix a diatomic molecule completely in space. Here they are the coordinates of the center of mass (which serves as the origin for the local coordinate system), the orientation of the internuclear bond* (θ, ϕ)*, and the internuclear separation* r*.* r_1 *and* r_2 *are the distances of the masses* m_1 *and* m_2 *from the center of mass.*

Vibration[7] (in terms of the harmonic oscillator):

$$\varepsilon_v = (v + \tfrac{1}{2})h\nu_0 \qquad g_v = 1$$

$$\nu_0 = \frac{1}{2\pi}\left(\frac{k}{\mu}\right)^{\frac{1}{2}}$$

where k is the force constant and μ is the reduced mass.

This ε_v is relative to the energy of the *bottom* of the potential well. Relative to the ground vibrational energy level, $\varepsilon_v' = vh\nu_0$.

Electronic: Neglect for now.

An energy level diagram for all these modes of energy is shown in Figure 2-19. The energy needed to excite the molecule from P_1 to level P_2 is simply the sum of the electronic, vibrational, and rotational excitations.[8] The zero-point energy ε_0 just locates the lowest energy state relative to the chosen baseline for energy measurement. If we include translational energy, then the total energy of the molecule at P_2 is

$$\varepsilon = \varepsilon_0 + \varepsilon_e + \varepsilon_v + \varepsilon_r + \varepsilon_t$$

The corresponding energy per mole is

$$\bar{E} = \bar{E}_0 + \bar{E}_e + \bar{E}_v + \bar{E}_r + \bar{E}_t$$

[8] This principle of the separation of energies is itself only approximately true, but is assumed valid in this elementary treatment.

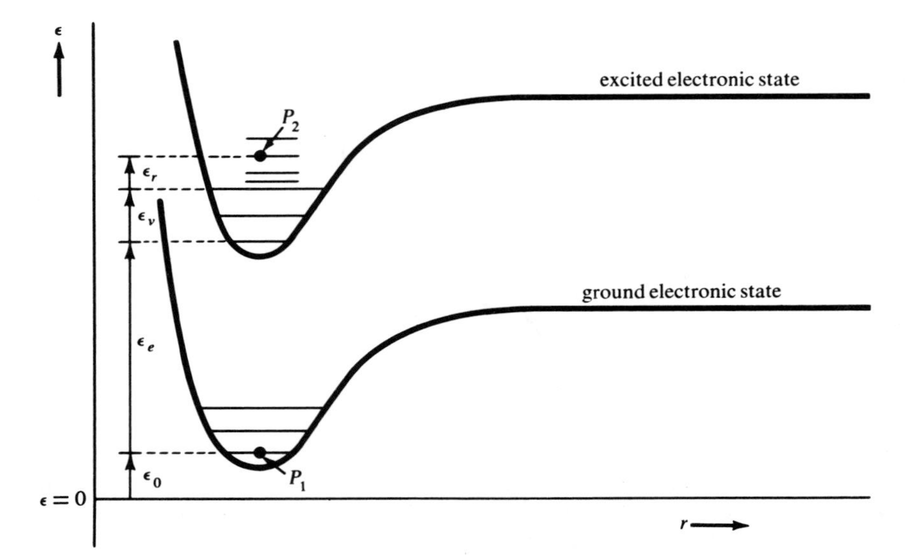

FIGURE 2-19. *Energy level diagram for a diatomic molecule. ε_0 is the energy of the motionless, nonvibrating, nonrotating molecule in its ground electronic state, measured relative to some convenient zero of energy. ε_e is the energy needed to take the molecule to the excited electronic state, still without vibrating or rotating. ε_v is the energy of vibration, and ε_r is the rotational energy. The translational energy ε_t is not shown. At absolute zero, the molecule would be in energy state P_1, with energy ε_0.*

Partition Function for Internal Degrees of Freedom

For internal degrees of freedom, the molecular partition function is proportional to the energy through an exponent, $q_j \propto e^{-\beta\varepsilon_j}$; because energies are additive, this proportionality makes the partition functions multiplicative. The total molecular partition function, including the formal zero-point partition function as well as those for translation, vibration, rotation, and electronic excitation, is

$$q = q_0 q_t q_v q_r q_e \qquad q_0 \equiv e^{-\varepsilon_0/kT} \tag{2-98}$$

The most convenient reference for the zero point energy of a molecule, at least in terms of chemical reactions, is the ground state energies of the *dissociated atoms* at 0°K. In this definition, as shown in Figure 2-20, ε_0 is a negative quantity because the energy of the molecule at 0°K is less than the energy of the dissociated atoms. The total *molar* partition function is

$$Q = \frac{q_t^N}{N!} q_v^N q_r^N q_e^N e^{-E_0/kT} \tag{2-99}$$

where N is Avogadro's number. In Eq. 2-99, it is necessary to make the factorial correction only once, and it is reasonable to associate the $N!$ with translation because it is the fact that the molecules move that makes them indistinguishable.

Translation

The translational relationships have been derived previously:

$$q_t = \frac{(2\pi mkT)^{\frac{3}{2}} V}{h^3} \tag{2-36}$$

$$\bar{E}_t = \tfrac{3}{2}RT \tag{2-68}$$

$$\bar{S}_t = R\left[\frac{5}{2} + \ln\left(\frac{(2\pi mkT)^{\frac{3}{2}} V}{Nh^3}\right)\right] \tag{2-92}$$

Rotation

The molecular partition function for closely spaced rotational states is

$$q_r = \sum_{J=0}^{\infty} (2J + 1) \exp[-(hB/kT)(J^2 + J)]$$

$$\cong \int_{J=0}^{\infty} (2J + 1) \exp[-(hB/kT)(J^2 + J)]\, dJ \tag{2-100}$$

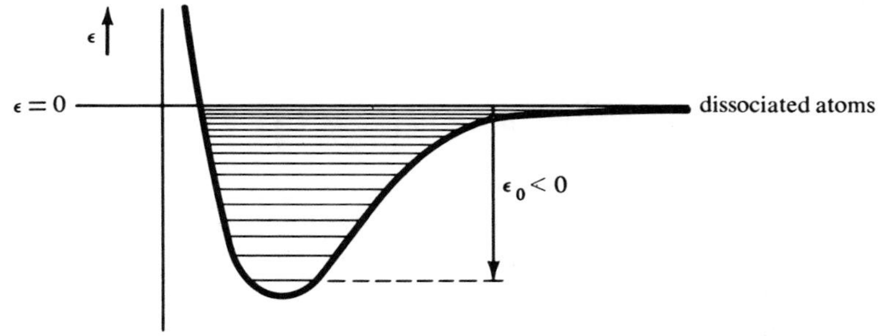

FIGURE 2-20. *The most common reference zero for energy measurement in a diatomic molecule is the state with the atoms dissociated and infinitely far apart.* ε_0 *is then negative, and the dissociation energy of the diatomic molecule at* 0°K *is* $-\varepsilon_0$. *Near the bottom of the potential well for a diatomic molecule, the potential function and the energy levels approximate those of a simple harmonic oscillator. But at higher energies, the restoring force of the bond weakens as the internuclear distance increases. The potential function flattens out to the right as dissociation of the molecule occurs, and the energy levels are more and more closely spaced until a continuum of possible kinetic energies is reached at the point of dissociation.*

Letting $z = J^2 + J$, and integrating,

$$q_r = \int_{z=0}^{\infty} e^{-(hB/kT)z}\,dz = \frac{kT}{hB} = \frac{8\pi^2 IkT}{h^2} \tag{2-101}$$

This last expression is not quite correct, in that we must make a symmetry correction to account for the fact that rotating a homonuclear diatomic molecule through a 180° angle does not bring about a physically distinguishable change.[9] The correct partition function, including the symmetry correction factor σ, is

$$q_r = \frac{8\pi^2 IkT}{\sigma h^2} \tag{2-102}$$

$\sigma = 1$ for heteronuclear diatomic molecule

$\quad = 2$ for homonuclear diatomic molecule

Hence the rotational energy is

$$\bar{E}_r = RT\left(\frac{\partial \ln q}{\partial T}\right)_V = RT^2\left(\frac{1}{T}\right) = RT \tag{2-103}$$

We see here further evidence for the principle of equipartition of energy. The diatomic molecule has two rotational degrees of freedom, θ and ϕ, with $\frac{1}{2}RT$ per mode.

The entropy is

$$\bar{S}_r = R + R\ln\left(\frac{8\pi^2 IkT}{\sigma h^2}\right) = R\left[1 + \ln\left(\frac{8\pi^2 IkT}{\sigma h^2}\right)\right] \tag{2-104}$$

This equation is rather cumbersome, but we can modify it by collecting the various constants into a single term. Then we obtain

$$\bar{S}_r = R\ln I + R\ln T - R\ln\sigma + K \tag{2-105}$$

This indicates that, other things being constant, the entropy increases with increasing moment of inertia. This situation is analogous to the fact that in translations, the heavier the molecule the greater the entropy. As with translation, the entropy also rises with temperature. The σ-containing term,

[9] A more precise explanation is that, for homonuclear diatomic molecules, only all the even or all the odd quantum states are ordinarily available to a given molecule. Thus the sum over states q must be reduced by half.

which is zero for a heteronuclear diatomic molecule such as HCl, but is $-R \ln 2$ for a homonuclear molecule like O_2, is a reflection of the fact that a molecule with both ends alike is intrinsically a more ordered object than one which lacks this symmetry.

Vibration

The partition function for vibrational states is given by

$$q_v = \sum_{v=0}^{\infty} e^{-vx} \qquad x = \frac{h\nu_0}{kT} \tag{2-106}$$

It is often convenient to define a "characteristic temperature" θ of a vibration by

$$\theta \equiv \frac{h\nu_0}{k} \tag{2-107}$$

so that $x = \theta/T$, and

$$q_v = \sum_{v=0}^{\infty} e^{-v(\theta/T)}$$

Assuming that Eq. 2-106 can be integrated

$$q_v = \int_0^{\infty} e^{-vx}\, dv = \frac{1}{x} = \frac{kT}{h\nu_0} = \frac{T}{\theta} \tag{2-108}$$

Having the partition function, the energy follows immediately

$$\bar{E} = RT^2\left(\frac{\partial \ln q}{\partial T}\right)_V = RT^2\left(\frac{1}{T}\right) = RT \tag{2-109}$$

Here, for the first time, we have come upon a potential term in our means of storing energy. In neither translation nor rotation was there any such thing as potential energy. In vibratory motion, in contrast, energy can be stored either as kinetic or as potential energy, and the equipartition principle associates with *each* form an energy of $\frac{1}{2}RT$. Once we have the energy, the entropy follows.

$$\bar{S}_v = R - R \ln \frac{\theta}{T} \tag{2-110}$$

Expressing Eq. 2-110 in a more understandable form,

$$\bar{S}_v = R \ln T - R \ln \nu_0 + K' \tag{2-111}$$

Again it is obvious that increasing temperature creates more disorder in the system, but a closer look at the vibration frequency proves interesting. The higher the natural vibration frequency of the molecule, the lower the entropy. In other words, as the reduced mass decreases, so does the entropy; this is the same old principle of a heavy system being less ordered than a light system. But, as the vibrational force constant decreases, the entropy increases. A weaker spring permits a wider range of motion and more disorder than does a tight spring.

Electronic Excitation

The partition function for electronic energy storage is

$$q_e = g_0 + g_1 e^{-\beta\varepsilon_1} + g_2 e^{-\beta\varepsilon_2} + \cdots \tag{2-112}$$

For most cases at ordinary temperatures, the energy of the first excited electronic state is very large, and as a result $e^{-\beta\varepsilon_1}$ is very small. Thus to a first approximation, the q_e can be simplified to

$$q_e \cong g_0 \tag{2-113}$$

If this is the case, then the electronic contribution to the total energy is zero. The molecule is not electronically excited at all:

$$\bar{E}_e = 0 \tag{2-114}$$

The entropy is also a simple expression:

$$\bar{S}_e = R \ln g_0 \tag{2-115}$$

And if there is only one state corresponding to the ground state electronic energy level, the electronic contribution to the entropy will also be zero.

Summary—Diatomic Molecule

The total molecular partition function is just the product of $q_t q_r q_v q_e q_0$:

$$q = \frac{(2\pi mkT)^{\frac{3}{2}} V}{h^3} \frac{8\pi^2 IkT}{\sigma h^2} \frac{kT}{h\nu_0} q_e e^{-\varepsilon_0/kT} \tag{2-116}$$

The total energy, corresponding to each of the terms in Eq. 2-116, is

$$\bar{E} = \tfrac{3}{2}RT + RT + RT + 0 + \bar{E}_0 = \bar{E}_0 + \tfrac{7}{2}RT \tag{2-117}$$

or, for the thermal energy,

$$\bar{E} - \bar{E}_0 = \tfrac{7}{2}RT \tag{2-118}$$

Recall that the thermal energy for a monatomic gas was $\frac{3}{2}RT$. You can feed more heat into a mole of diatomic molecules before it rises one degree in temperature than you can the same number of monatomic molecules, simply because there are more internal modes in which the diatomic molecule can soak up heat before its translational motion increases to the point that the temperature rises one degree. The molar heat capacity at constant volume is therefore $\frac{7}{2}R$, compared to $\frac{3}{2}R$ for the monatomic molecule:

$$C_V = \left(\frac{\partial E}{\partial T}\right)_V = \tfrac{7}{2}R \tag{2-119}$$

The total entropy is

$$\bar{S} = R\left[\frac{5}{2} + \ln\left(\frac{(2\pi mkT)^{\frac{3}{2}}V}{Nh^3}\right)\right] + R\left[1 + \ln\left(\frac{8\pi^2 IkT}{\sigma h^2}\right)\right] + R\left[1 - \ln\left(\frac{h\nu_0}{kT}\right)\right] + R \ln g_0 \tag{2-120}$$

Experimental Measurements of Heat Capacities

The sole justification for any theory is its success in predicting the results of experiment, and it is time to test the conclusions of the preceding theories. A relatively easy experimental property of gases to measure is their heat capacity. Ignoring deviations from ideal gas behavior (or better, correcting for them), the heat capacity at constant volume of a monatomic gas is predicted to be $\frac{3}{2}R$. Figure 2-21 shows that helium, neon, argon, and even mercury vapor fit the theory very well. However, for a diatomic gas the heat capacity is expected to be $\frac{7}{2}R$, and the data for oxygen gas deviate widely from this figure. It is to be expected that the fault lies with one of the two new factors, rotation or vibration.

At sufficiently high temperatures, C_v for oxygen approaches $\frac{7}{2}R$. At room temperature it appears to be approaching $\frac{5}{2}R$, and at even lower temperatures it would approach the monatomic gas curve. Apparently one of the two new energy storage mechanisms, rotation or vibration, fails

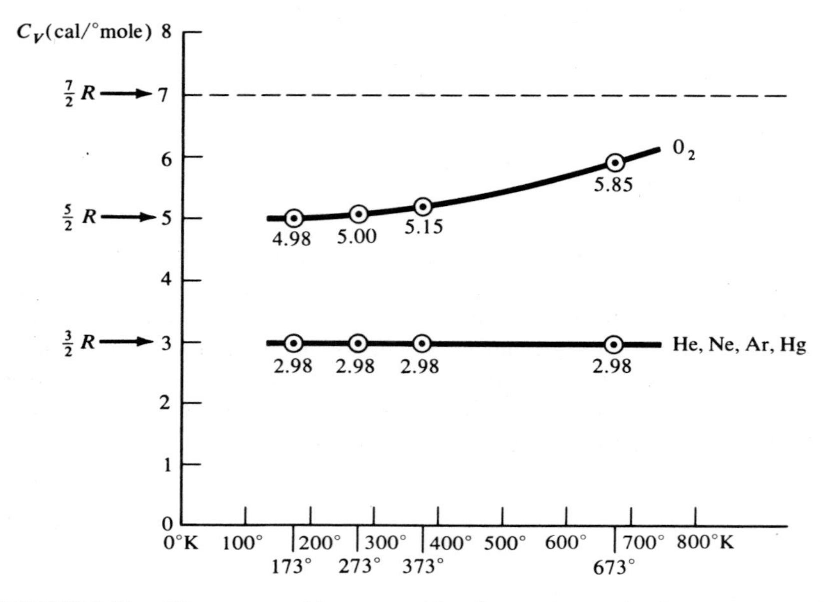

FIGURE 2-21. *The measured heat capacities for most monatomic gases agree well with their expected values of $\frac{3}{2}R$. But those even for diatomic molecules such as O_2 deviate widely from the predicted value of $\frac{7}{2}R$, and the deviation is worse at lower temperatures. For explanation, see text.*

above room temperature, and the other fails somewhere below room temperature. Which is which?

In calculating the partition functions which lead to a theoretical heat capacity, we assumed that the energy levels were close together compared with the average thermal energy per molecule, of the order of kT; sufficiently close that we could integrate rather than summing. This procedure fails, however, as soon as the typical energy level spacing becomes of the same order of magnitude as kT. For which form of energy storage will this approximation fail first as the temperature is lowered, rotation or vibration? As a concrete example, consider the energy difference between ground state and first excited states for carbon monoxide gas at 300°K in a box 10 cm on a side[10]:

(1) Average thermal energy, $kT = 0.59$ kcal/mole

(2) Translational energy, $\Delta E_t \cong \dfrac{h^2}{8ma^2} = \dfrac{47.8 \times 10^{-20}}{M}$ kcal/mole

$$\Delta E_t \cong 1.7 \times 10^{-20} \text{ kcal/mole}$$

[10] For spectral data, see Appendix 2 and Herzberg.

(3) Rotational energy, $\Delta E_r \cong 2hB = hc\ (2B/c) = 0.01104$ kcal/mole
[$hc = 0.00286$ kcal cm/mole, $B/c = 1.931\ \text{cm}^{-1}$ (Appendix 2)]
(4) Vibrational energy, $\Delta E_v = h\nu_0 = hc\omega_e = 6.21$ kcal/mole
($\omega_e = 2170\ \text{cm}^{-1}$)
(5) Electronic energy, $\Delta E_e = 186.0$ kcal/mole
(lowest electronic level above ground at 65,075 cm^{-1})

The spacing between vibrational levels is almost 11 kT. We had assumed that we could integrate instead of summing over the vibrational levels, when in fact the gap over which we were integrating was an order of magnitude greater than the average energy per molecule. Thus at room temperature the translational and rotational modes are fully excited, but the vibrational mode is only slightly excited and contributes practically nothing in the way of energy storage. Looking at the gap between electronic levels, we see that it is ridiculous to integrate over these levels, and of course we did not. In fact we omitted the upper electronic levels altogether.

An interesting comparison comes to light when one considers what part of the electromagnetic spectrum, and hence what kinds of instrument, would have to be used in order to do a complete spectroscopic study of CO at room temperature. Given the energy level spacings for CO as presented above, the following wavelengths would be necessary to measure changes in the various modes of energy.

Mode	Wavelength	Portion of spectrum
electronic	$\lambda_e = 154\ \text{m}\mu$	uv
vibrational	$\lambda_v = 4610\ \text{m}\mu$	near ir
rotational	$\lambda_r = 0.26$ cm	far ir or microwave
translational	$\lambda_t = 0.2$ light years	?

It is obvious from this calculation why no one ever does translational spectroscopy. A 500-line translational diffraction grating would reach from earth to Arcturus. Even the most liberal research grant agency would draw the line at a proposal for a grant to construct a translational spectrometer.

Proper Treatment of Vibrational Energy

Going back to Eq. 2-106, and this time summing over the energy levels instead of integrating,[11] we obtain the correct partition function

$$q_v = \sum_{v=0}^{\infty} e^{-xv} = \frac{1}{1 - e^{-x}} \qquad x = \frac{h\nu_0}{kT} = \frac{\theta}{T} \tag{2-121}$$

[11] The easiest way to demonstrate that the summation over v does equal $1/(1 - e^{-x})$ is to divide 1 by $1 - e^{-x}$ by long division.

Using this equation to calculate the energy,

$$\bar{E}_v = RT^2\left(\frac{\partial \ln q}{\partial T}\right)_V = RT^2 \frac{1}{q}\frac{\partial q}{\partial T} = RT^2 \frac{1}{q}\frac{\partial q}{\partial x}\frac{\partial x}{\partial T}$$

$$\bar{E}_v = RT^2 \frac{1}{q}\left[\frac{-e^{-x}}{(1 - e^{-x})^2}\right]\left(-\frac{x}{T}\right) = RT\left(\frac{x}{e^x - 1}\right) \tag{2-122}$$

The correction factor $x/(e^x - 1)$ approaches unity as $T \to \infty$, and $\bar{E}_v$ approaches the fully excited value of RT. At finite temperatures, $0 \leq x/(e^x - 1) < 1$ and $0 \leq \bar{E}_v < RT$.

As an example, the characteristic temperature θ for CO is $\theta = hc\omega_e/k = 3120°$, since $hc/k = 1.4388$ cm deg and $\omega_e = 2170\ \text{cm}^{-1}$. At 300°K, $x = 10.4$ and $x/(e^x - 1) = 0.00033$. The vibrational mode of CO at 300°K is only 0.033% excited, and $\bar{E}_v = 0.00033\ RT$. Even at the characteristic temperature when $x = 1$, $\bar{E}_v = 0.58\ RT$.

The heat capacity now is

$$C_V = \left(\frac{\partial E}{\partial T}\right)_V = \left(\frac{\partial E}{\partial x}\right)_V\left(\frac{\partial x}{\partial T}\right)_V = R\frac{x^2 e^x}{(e^x - 1)^2} \tag{2-123}$$

For carbon monoxide at 300°K, $C_V = 0.0034\ R$, and when $x = 1$ (corresponding to a temperature of 3120°K), C_V is still only $0.92\ R$.

The proper expression for entropy is

$$\bar{S}_v = \frac{\bar{E} - \bar{E}_0}{T} + R\ln q = \frac{Rx}{e^x - 1} - R\ln(1 - e^{-x}) \tag{2-124}$$

In Appendix 3 are tables of $x/(e^x - 1)$, $x^2e^x/(e^x - 1)^2$, and $\ln(1 - e^{-x})$, which make the calculation of the vibrational contributions to $\bar{E}$, C_V, and $\bar{S}$ rapid.

2-19 *SUMMARY OF INTRODUCTION TO STATISTICAL MECHANICS*

After initially studying the properties of two types of real particles, fermions and bosons, we found that under usual conditions they both act like a third, hypothetical kind of particle, boltzons. From that point on, we used Maxwell–Boltzmann statistics to derive an expression for the number of ways of getting a given arrangement of boltzons. We found that the number of ways of getting a state is proportional to the probability of its occurrence. We found that the most probable distribution was so overwhelmingly probable that no other distribution need be considered; all the properties of macroscopic systems can be calculated by assuming the sole existence of only the Boltzmann distribution.

We saw that energy is a measure of motion in a system and entropy is a measure of arrangement. For mathematical convenience we introduced the molar partition function Q, and from it we derived two important expressions for the energy and entropy:

$$\bar{E} = kT^2\left(\frac{\partial \ln Q}{\partial T}\right)_V \tag{2-72}$$

$$\bar{S} = k \ln Q + \frac{\bar{E}}{T} \tag{2-88}$$

Using these equations, we discussed velocity distributions in an ideal gas and internal degrees of freedom in ideal monatomic and diatomic molecules. We saw that when a molecule has all states fully excited, it contains $\frac{1}{2}RT$ of energy per mole of energy storage, as expressed by the principle of equipartition of energy.

To this point we have made quantitative measurements of energy and entropy, but we still cannot use them because we are missing many useful relationships between motion and arrangement. To find these relationships we must turn to thermodynamics, which is the older of the two fields but which logically follows statistical mechanics.

REFERENCES AND FURTHER READING

N. Davidson, *Statistical Mechanics* (McGraw-Hill, New York, 1962). A thorough treatment with much the same viewpoint as this chapter. The logical next step in the subject.

F. T. Wall, *Chemical Thermodynamics* (W. H. Freeman, San Francisco, 1958). A better treatment of elementary statistical thermodynamics than is to be found in most thermodynamics texts. A first line of defense if you cannot fathom this chapter.

F. C. Andrews, *Equilibrium Statistical Mechanics* (Wiley, New York, 1963).

D. Ter Harr, *Elements of Thermostatics* (Holt, Rinehart and Winston, New York, 1966), 2nd ed.

J. G. Aston and J. J. Fritz, *Thermodynamics and Statistical Thermodynamics* (Wiley, New York, 1959). The above three books present slightly different viewpoints of statistical mechanics at a level not too far above that of this chapter.

EXERCISES

2-1. (a) Assume a system of five particles, all bosons. Open to these particles are energy levels $e_J = hBJ(J + 1)$, with degeneracy $g_J = 2J + 1$, and with J from zero to infinity. For simplicity, use hB as the unit of energy. If the five-particle system has a total energy of exactly 50 such units, then one possible distribution is

$$(n_j) = (2, 1, 1, 0, 0, 0, 1, 0, 0, \ldots)$$

List all the possible distributions, and calculate the number of microstates of each. Which distribution is the most probable?

(b) Repeat the problem with the particles being fermions instead.

2-2. Consider a collection of hypothetical particles, "hypothons," individually indistinguishable except for sex, and distributed among the various degenerate compartments of the available energy levels subject only to the following restrictions:

(1) In energy level j, of energy e_j, there are a_j male and b_j female particles distributed among g_j compartments.

(2) No two males can tolerate being confined within the same compartment.

(3) Females, being more sociable, can occupy a compartment in unlimited numbers.

(4) House rules forbid the occupancy of any one compartment by both sexes.

(a) Write the expression for the number of ways of producing a given distribution (a_j, b_j).

(b) Using Lagrangian multipliers, and subject to the constraints of a fixed number of each kind of particle and a fixed total energy, derive the expressions for the most probable distributions of male and female particles, a_j and b_j.

(c) What types of statistics result when only one sex or the other is present? Demonstrate, with the results of part (b).

Note: For simplicity, assume as rapidly as possible that g_j, a_j, and b_j are all much greater than 1. Also, make the following substitutions quickly to keep the mathematics tractable:

$$e^{\alpha}e^{\gamma\varepsilon_j} = A \qquad e^{\beta}e^{\gamma\varepsilon_j} = B$$

2-3. Calculate the dilution ratio q/n for mercury vapor at 1 atm pressure and 2000°K, assuming ideal gas behavior. *If* mercury remained a perfect gas at as low a temperature as was required, calculate the temperature to which the gas would have to fall before the dilution ratio fell to 10.

2-4. (a) An electron moves in a cubical box 10^{-8} cm on a side. How many quantum states are there with energy less than kT at 300°K?

(b) How many quantum states would there be with energy less than kT if the electron moved in a cubical box 1 cm on a side?

(c) The density of sodium metal is 0.97 gm/cm³ and its atomic weight is 23. Each atom has a single valence electron. How many valence electrons are there in 1 cm³ of sodium metal?

(d) The valence electrons of sodium metal are believed to move about more or less as if they were a gas. Because of the Pauli principle and because of the two spin states of an electron, two electrons can be placed in each translational energy state. How does the number of electrons per cm^3 in sodium metal compare with the number of quantum states available to electrons with energy less than kT at 300°K?

(e) At what temperature would the number of quantum states available to electrons with energy less than kT be equal to the number of valence electrons in 1 cm^3 of sodium metal?

2-5. (a) Show that the mean velocity of a nitrogen molecule at 300°K is 476 m/sec.

(b) What is the ratio of the probability of finding a nitrogen molecule whose velocity is 476 m/sec at 300°K to the probability of finding a nitrogen molecule with the same velocity at 310°K?

(c) What is the corresponding ratio for a nitrogen molecule with a speed in the vicinity of $3 \times 476 = 1428$ m/sec?

2-6. Derive expressions for the following quantities for a two-dimensional Maxwell–Boltzmann gas:

(a) mean speed
(b) root mean square speed
(c) most probable speed
(d) mean molecular energy
(e) energy per mole

2-7. Prove that in a three-dimensional gas with a Maxwell–Boltzmann distribution of molecular velocities the fraction of molecules whose kinetic energy $\varepsilon = \frac{1}{2}mu^2$ lies between ε and $\varepsilon + d\varepsilon$ is proportional to

$$\varepsilon^{1/2}\exp\left(-\frac{\varepsilon}{kT}\right) d\varepsilon$$

2-8. In order for any object to escape from the earth's gravitational field it must have a velocity greater than 11 km/sec. What fraction of the nitrogen molecules of the air at 300°K have a velocity greater than the escape velocity? If the earth's surface temperature were suddenly raised to 1300°K, what fraction of the nitrogen molecules in the air would have velocities greater than the escape velocity?

2-9. A spherical pot of honey hangs from a tree. The radius of the pot is a. After some time, insects are attracted to the pot and it is found that the average number of insects per unit volume at a distance r from the center of the pot is given by K/r^n, where K and n are empirical constants and $r > a$. The total number of insects flying around the pot is N.

(a) Show that K, N, a, and n are related by $K = (n - 3)Na^{n-3}/4\pi$.

(b) What is the minimum possible value of n if the above relation is to hold at all values of r from $r = a$ to $r = \infty$?

(c) Show that the probability of finding a particular insect in the region between r and $r + dr$ is $(n - 3)(a/r)^{n-2}(dr/a)$.

(d) Find $\bar{r}$, the mean distance of an insect from the pot.

(e) Find r_{max}, the most probable distance between an insect and the pot.

2-10. (a) Assuming that 75% of the atoms of Cl have mass number 35 and that 25% have mass number 37, what fraction of the molecules of Cl_2 will have masses 70, 72, and 74 if pairs of atoms are taken at random?

(b) For a total of 1 mole of molecular chlorine, what is the entropy of mixing of the three kinds of molecules?

(c) How does the answer to part (b) differ from the entropy of mixing of the same total number of atoms (2 moles) if we assume that no diatomic molecules exist?

2-11. Calculate the value of the constant K in the expression for the entropy of the ideal monatomic gas

$$S = \tfrac{5}{2}R \ln T - R \ln P + \tfrac{3}{2}R \ln M + K$$

Use this to calculate the entropy of neon gas at 298°K. Compare your answer with the measured value (listed as S^0) in the tables of Appendix 4.

2-12. (a) Integration over energy levels begins to fail when the energy spacings approach kT. At what temperature will kT equal the spacing between the lowest two levels for Cl_2 for

(1) vibration
(2) rotation
(3) translation

Assume ideal gas translation behavior at all temperatures, and assume that the gas is in a cubical container 1 cm on a side. Other necessary data are available in tables in this chapter or the Appendices.

(b) Repeat the problem for H_2. What trends do you see?

2-13. The vibrational frequency of H_2 is 4395 cm^{-1}, and that of I_2 is 214 cm^{-1}.

(a) Calculate the separation of the vibrational levels in calories for each molecule.

(b) Calculate the ratio of the number of molecules in the first excited vibrational state to the number in the ground state for each molecule at 100°C.

(c) Would you expect the vibrational contribution to the heat capacity to be large or small at 100°C for each of these molecules?

2-14. (a) Calculate the vibrational contribution to the heat capacity of Cl_2 at 100, 500, 1000, and 5000°K. Plot the heat capacity C_V against temperature. What heat capacity would be expected from a fully excited vibrational mode?

(b) We shall see later that, for an ideal gas, the heat capacity at constant pressure C_P is related to that at constant volume with which we have been dealing, C_V, by $C_P = C_V + R$. Calculate the rotational and translational contributions to the heat capacity of Cl_2 at 500°K, add them to the vibrational contribution, and compare the total C_P to that obtained by the power series approximation in Table 3-5. What is the discrepancy, and how might it be accounted for?

2-15. In Appendix 2 there are listed the rotational and vibrational properties of several diatomic molecules.

(a) Why should the rotational constant B/c decrease in the series H_2, HD, D_2?

(b) Can you correlate the fundamental vibration frequencies of the diatomic molecules in the first row of the periodic table, Li_2–F_2, with what you know about the bonds involved?

(c) Does this explanation account for the internuclear separations?

(d) Why, in this same row of elements, is there also a correlation between trends in vibration frequency and rotation constant?

(e) Why should the vibration frequencies decrease in the diatomic halogen series F_2–I_2?

2-16. (a) Calculate the translational entropy of N_2 at 298°K and 1 atm.

(b) Compare your answer with the third law thermal entropy S^0 in Appendix 4. Assume that the discrepancy is entirely due to rotation. Calculate the moment of inertia of the N_2 molecule from this. How does this compare with the true moment of inertia, obtainable from the data in Appendix 2?

(c) Calculate the rotational entropy of nitrogen in the proper way from its true moment of inertia.

(d) Calculate the vibrational entropy of nitrogen, from data in Appendices 2 and 3.

(e) Add the translational, rotational, and vibrational entropies for N_2 and compare them with the third law entropy. How much error is left for electronic contribution?

Chapter 3

FIRST LAW AND THERMOCHEMISTRY

THE LEITMOTIF of this treatment of thermodynamics will be a search for a potential function which, when minimized, allows us to locate the position of chemical equilibrium in a system, in a manner analogous to minimizing the gravitational potential energy to find the position of mechanical equilibrium.

We already have two clues to aid our search. First, we know that in any choice between states of equal statistical weight there is a spontaneous tendency to go to the state of lowest energy. Also, in choosing among states of the same energy, there is a natural tendency for the system to assume the state of highest entropy. These two facts suggest that the proper chemical potential function would involve energy and entropy terms and be of the general form

$$\chi = K'E - K''S$$

This function would be minimized by simultaneously minimizing energy and maximizing entropy. The next two chapters on thermodynamics will be concerned with finding suitable potential functions and defining the conditions under which they are applicable.

3-1 EQUILIBRIUM

When a macroscopic system attains equilibrium, all activity apparently ceases. On a microscopic level, however, things are still happening, for what appears to be immobility at the macroscopic level is really just the balance of opposing processes at the lower level.

In a mechanical system, an equilibrium state is one in which there are no net forces on the system or one in which the potential function is at a minimum. In general, the relationship between force and potential is

$$F_j = -\frac{\partial V}{\partial x_j}$$

From this equation, we can say that the equilibrium state in a system is the state for which the potential is minimized or for which the slope of the potential, the force, is zero for every parameter that could affect the equilibrium. Such a system is diagrammed in Figure 3-1. Some examples of systems that attain equilibrium by minimizing a potential function are the following:

(1) A ball on an irregular surface rolls until it finds a low spot, in which its gravitational potential energy has a local minimum.

(2) A compass needle placed in a magnetic field minimizes its potential function by aligning itself with the field.

(3) An electrostatic charge in an electric field, if not subject to other constraints, will find its equilibrium position at the point of minimum electrical potential.

(4) If a hot beaker of water is placed upon a cold bench top, heat flows until the temperature difference $|\Delta T|$ is minimized at zero.

The situation of particular interest to chemists is that of a chemical reaction. Starting with pure reactants, products will form until the equilibrium position is reached, where the forward and reverse reactions proceed at the same rate. For such systems there are several potential functions whose minimization corresponds to the point of chemical equilibrium, and the function which must be used depends upon experimental conditions. It is the goal of thermodynamics to find these chemical potential functions and to define the conditions under which particular ones are valid.

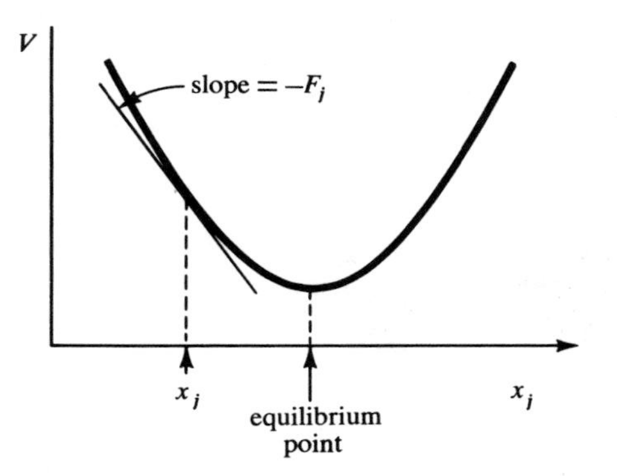

FIGURE 3-1. *The force on a particle, F_j, is the negative of the gradient of the potential V. At equilibrium the potential is at a minimum and the force is zero.*

3-2 *FUNCTIONS AND VARIABLES OF STATE*

In discussing chemical systems, it is advantageous to describe experiments only in terms of variables that will have the same values when someone else repeats the experiment. The weather, the time of day, and the frame of mind of the experimenter are not in any sense variables of state. A variable of state, or a state function, is a variable whose change during any process depends only upon the initial and final states of the system and *not* upon its intermediate history. A ball rolling down an inclined plane provides a good example. As it rolls, its total change in altitude does not depend upon the particular path it follows. Altitude is a state function, path length is not.

Another example which is closer to the thermodynamic application is that of a man making a trip to the top of Mt. Wilson. (See Figure 3-2.) There are many ways he could make the trip and many variables by which it could be described. He could hike up the toll road with a lunch, or without a lunch. He could go in hot weather or in cold weather. He could carry a pack or go without one. Or, he could take the easy way out and drive up the Angeles Crest highway. Now the amount of oxygen that he would consume during this trip would certainly depend upon the conditions of travel. His weight loss during the hike would depend on how much weight he was carrying, whether he ate on the way, and whether he threw his pack away half way up. Total fatigue and loss of shoe leather would depend very much on the path. Obviously none of these variables is a state function. However, his altitude would be a variable of state, because change in altitude depends only on initial and final position.

If, from the results of one man's trip, we were to make the statement "When he climbed Mt. Wilson he wore off 0.5 mm of shoe leather," we would not be making a general statement, for it would be incorrect to con-

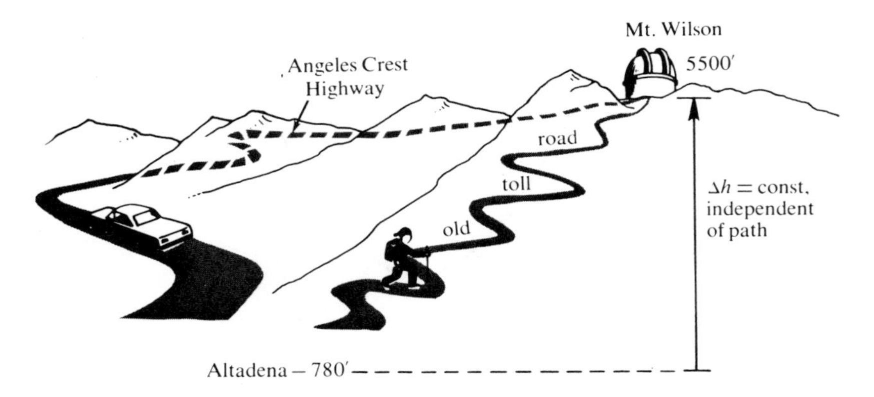

FIGURE 3-2. *The altitude increase Δh during a trip from Altadena to Mt. Wilson, is a state function of the journey, and is not a function of the path (or road) employed.*

clude that all people climbing Mt. Wilson lose 0.5 mm of shoe leather. On the other hand, the remark "When he climbed Mt. Wilson, he gained 5000 ft" could be made into a general statement applicable to all climbers because in this case we are describing the changes in a function whose value does not depend on the history of any one particular trip.

You can see that state functions and state variables are particularly useful in describing a physical experiment. We return to the mathematical properties of state functions after the first law.

3-3 *TEMPERATURE*

The concept of temperature can be developed from nothing more than the idea of equilibrium. Without assuming the prior existence of any temperature scales we can acknowledge the intuitive recognition of "hot" and "cold" objects, and the validity of the statement that if two objects are each in thermal equilibrium with a third object, then they themselves are in mutual thermal equilibrium.

Is there a potential function for the type of thermal equilibrium just described? Yes, there is; it can be found without any preconceived notions of heat. For example, by placing together a cold block and a hot block, as shown in Figure 3-3a, one might observe that when the two were in contact the colder block expanded and the warmer block contracted, and after a while there was no further change in the dimensions of the two. There are physical ways to determine when a system has reached a state of thermal equilibrium.

The physical properties of the block—length, breadth, height, color, density, electrical conductivity, and so on—all can be used to describe what happens to the blocks. Let all the measurable properties of the blocks be listed.

block 1 $\quad a_1, b_1, c_1, d_1, \ldots$

block 2 $\quad a_2, b_2, c_2, d_2, \ldots$

When we say that we intend to treat equilibrium as the minimization of a potential function, we mean that there exists a function of the general form

$$F_{(1,2)} = F_{(a_1,b_1,c_1,d_1,\ldots,\,a_2,b_2,c_2,d_2,\ldots)}$$

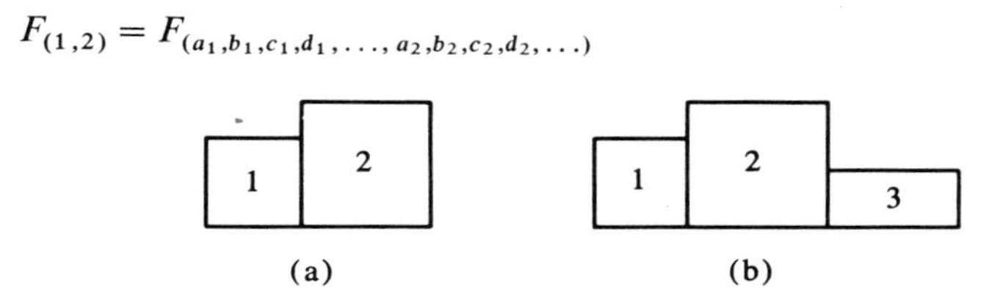

FIGURE 3-3. *If two objects are each in thermal equilibrium with a third object, then they must be in thermal equilibrium with one another.*

whose minimum corresponds to the state of thermal equilibrium between the two blocks. In other words, at thermal equilibrium all $F'_{(1,2)} = 0$.

If we introduce a third block as shown in Figure 3-3b, the condition of thermal equilibrium now requires that the analogous function $F_{(2,3)}$ must be minimized along with $F_{(1,2)}$. From our understanding of equilibrium, we know that having blocks 1 and 2 in thermal equilibrium and blocks 2 and 3 in thermal equilibrium implies that blocks 1 and 3 must also be in thermal equilibrium with each other. Therefore, without knowing what the function F is, we do know that if $F'_{(1,2)} = 0$ and $F'_{(2,3)} = 0$, then $F'_{(1,3)}$ must also be zero.

A function that satisfies the above condition must have a first derivative that is expressible as the difference between two functions, each of which depends only on the properties of one block

$$F'_{(i,j)} = f_{(i)} - f_{(j)}$$

We do not know the potential function itself, and as it turns out we will not need to determine it. We can say, however, that each block has a property, which, although not directly measurable, is expressible in terms of the other measurable properties of the block, such that when this first property is the same for both blocks, thermal equilibrium exists. This statement then forms a general definition of temperature: temperature is that property which when equal in both blocks results in a state of thermal equilibrium in the system.

It was stated that the temperature of a given block is a function of the properties of that block. The choice of properties by which one defines a temperature scale is arbitrary. Some common examples include the following.

(1) length of a rod—thermostat
(2) volume of a liquid—mercury thermometer
(3) electrical conductivity—resistance thermometer
(4) bimetallic junction potential—thermocouple
(5) color—ceramic temperature gauge
(6) phase transitions—freezing and boiling points of water

Water is a good example with which to elaborate on the method of developing a temperature scale. We could define a scale such that the temperature was 0° at the freezing point of water and 100° at the boiling point, with 100 equal divisions of scale in between. At first glance, our scale looks something like a centigrade scale, but it is not, because it is not completely defined. We said that we divided the scale into 100 equal increments, but equal with respect to what? Do we mean equal volume increments? This implies that we think that the thermal expansion coefficient of water is constant within the range $t = 0°–100°$. Do we mean equal increments of heat? This is cheating a bit because we have not yet defined heat; such a scale would imply, however, that the specific heat of water was constant within the range of the scale.

Actually it is easily shown that either scale would lead to trouble. If three arbitrary substances are defined to be at 0° when at equilibrium with an ice/water bath and at 100° when in equilibrium with boiling water, it is not possible in general to divide each substance's temperature scale into 100 equal intervals based on length, heat, or any other property such that the three substances are at thermal equilibrium with each other at every corresponding point on their *individual* temperature scales. The upper and lower limits would represent points of thermal equilibrium because they were defined that way, but thermal equilibrium would not exist at intermediate gradations because the physical properties of the three substances differ between the freezing and boiling points of water.

To make a useful temperature scale, we need some property common to all substances (or at least common to a large class of substances). One such property is the volume behavior approached by all gases under conditions of low pressure and high temperature, the so-called ideal gas behavior. Ideal gas behavior is a norm for all gases and is the logical choice on which to base a temperature scale. When an unknown quantity of heat is added to two ideal gases which are in thermal equilibrium with each other, both gases experience the same *fractional* change in volume.

An ideal gas temperature scale can be defined as follows:

$t = 0$ at freezing point of water

$t = 100$ at boiling point of water

$t = 1, \ldots, 99$ *equal increments* of volume

By this definition, the volume at any temperature t is given by

$$V_t = V_0 + kt \tag{3-1}$$

where the constant k is

$$k = \frac{V_{100} - V_0}{100}$$

This temperature scale is better than any based on the properties of metal blocks, because the similarities among metals are in no way as close as the similarities among gases. No one gas will ever give this scale, but all gases approach such behavior in the limit of ideality.

Equation 3-1 can be written as

$$V_t = V_0(1 + \alpha t) \tag{3-2}$$

where α is the coefficient of thermal expansion. By measurement, it is found that for an ideal gas, $\alpha = 1/273$. Then Eq. 3-2 can be written as

$$V_t = V_0\left(1 + \frac{t}{273}\right) \tag{3-3}$$

This equation can be simplified by changing the zero point on the scale. By making $T = t + 273$, one obtains

$$V_T = \frac{V_0}{273}T = k'T \tag{3-4}$$

Equation 3-4, in which volume is directly proportional to temperature, is a concise definition of the absolute temperature scale based on ideal gas behavior. From experiments by Boyle and others, we know that $k' = R/P$ and we have, for one mole of gas,

$$T = \left(\frac{P}{R}\right)\bar{V} \tag{3-5}$$

Having thus defined a temperature scale with wide applicability, we can go back and recalibrate all other practical temperature scales in terms of the ideal gas scale. The intervals between the freezing and boiling points of water can now be marked off with the gradations calculated from an ideal gas thermometer.

3-4 HEAT

What do we mean by heat? What is it that flows between cold block 1 and hot block 2 in the process of attaining thermal equilibrium? Is the temperature rise of the cold block just matched by the temperature fall of the hot block?

$$T_2 - T_f \stackrel{?}{=} T_f - T_1 \tag{3-6}$$

There are several reasons why this equation is not correct, and why the final temperature is not usually the average of the two initial temperatures. Block 2 may not be of the same mass as block 1. But even correction for differences in mass will not usually give a correct statement about the temperature changes leading to thermal equilibrium:

$$m_2(T_2 - T_f) \neq m_1(T_f - T_1) \tag{3-7}$$

where m is the mass. This expression is only correct if both blocks are made of the same material.

The easiest way to make the equation valid is to introduce two constants which serve to balance Eq. 3-7:

$$m_2 c_2(T_2 - T_f) = m_1 c_1(T_f - T_1) \tag{3-8}$$

This is certainly no profound maneuver; fitting the equation to the experimental data by way of unspecified constants is an old trick. However, if we do this experiment over and over, we soon make the discovery that every time we use material 2, constant c_2 is the same. Every time we use material 1, even if we use it with something other than material 2, c_1 is the same. It turns out that the constants, c_j's, are *intensive* properties of the substances used and are independent of experimental conditions. These constants are defined as *specific heats* or *heat capacities* and are in units of heat/gm degree.

With the aid of Eq. 3-8 we can now give a formal definition of heat. The heat change by a substance is a product of its mass times its specific heat times the temperature change. For our example of the two blocks,

$$\text{heat gained by block 1} = q_1 = m_1 c_1(T_f - T_1)$$
$$\text{heat gained by block 2} = q_2 = m_2 c_2(T_f - T_2)$$

(Since $T_2 > T_f$ and heat is *lost* by block 2, q_2 is a negative number.) Equation 3-8 shows that the heat gained by the cold block matches the heat lost by the hot block

$$q_1 + q_2 = 0 \tag{3-9}$$

Equation 3-9 is a simple conservation equation. It tells us that heat is conserved in the process of attaining thermal equilibrium in an isolated system.

We have not defined any absolute heat content as of yet. But we know that the heat *change* in a process can be defined as

$$q = mc\Delta T \tag{3-10}$$

where heat is conventionally measured in ergs, calories, liter atmospheres, joules, and so on. For heat in erg, c is in units of erg/deg g. One could also write

$$q = nc'\Delta T \tag{3-11}$$

where n represents the number of moles of substance. Then with q in ergs, c' would be in units of ergs/deg mole. From both Eqs. 3-10 and 3-11 it is apparent that

$$c = \frac{dq}{dT} \tag{3-12}$$

This equation is not quite the same as the earlier definition of heat capacity,

$$C_V = \left(\frac{\partial E}{\partial T}\right)_V \tag{2-75}$$

but we see later that these two are in fact equivalent.

3-5 WORK

Work is a mechanical concept, not a thermal one. In general, work is a result of action against an opposing force and is commonly written

$$dW = Fds$$

where ds is the path length over which action is taken and F is the force against which one has to work. If there is no opposing force, then the motion itself produces no work. On the other hand, in the absence of motion, even the strongest opposing force generates no work. There are many kinds of work: mechanical, electrical, work against a magnetic field, work against a gravitational field, and others. Initially, we will consider only mechanical work.

The world of thermodynamics is full of frictionless pistons, volumeless molecules, ideal gases, and other such esoteric chemical hardware. We shall make extensive use of gas expansions against frictionless pistons, not because of any lingering allegiance to engineering, but simply because they convert expansion to a one-dimensional process. Consider, then, an ideal gas at pressure P_{int}, expanding against an outside pressure P_{ext}, as in Figure 3-4. The direction of movement depends on our choice of appropriate internal

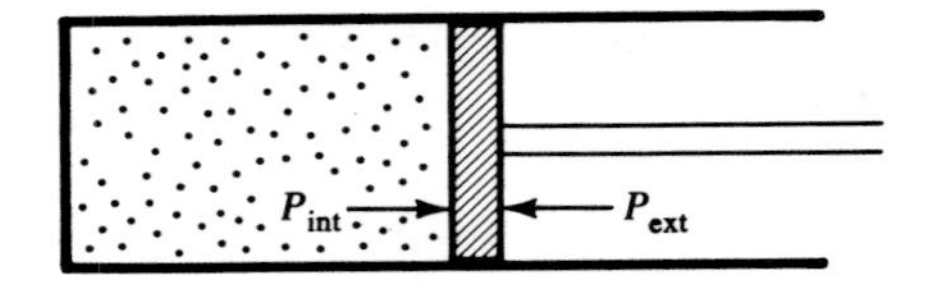

FIGURE 3-4. *The motion of the (frictionless) piston will be dictated by the relative magnitudes of the pressure of the gas inside the cylinder,* P_{int}, *and the external pressure* P_{ext}.

and external pressures:

$P_{ext} < P_{int}$: expansion—work against P_{ext}
$P_{ext} > P_{int}$: compression—work against P_{int}
$P_{ext} = P_{int}$: equilibrium

For an expansion, the work done *against the environment* is

$$dw' = F\,ds = P_{ext} A\,ds = P_{ext}\,dV$$

where A is the area of the piston and $A\,ds$ is the volume change during the expansion. In other words, dw' is the work done by the gas on its surroundings. We are usually interested in the work performed on the gas by its surroundings, or in the case of an expansion process

$$dw = -dw' = -P_{ext}\,dV$$

We will always define *the work done on the gas by its surroundings* as dw.

$$\begin{aligned} &\text{for an expansion:} \quad dw = -P_{ext}\,dV \quad (dV > 0,\, dw < 0) \\ &\text{for a compression:} \quad dw = -P_{int}\,dV \quad (dV < 0,\, dw > 0) \end{aligned} \tag{3-13}$$

From Eq. 3-13, the work done in a finite process is given by

$$w = -\int_{V_1}^{V_2} P\,dV = -\int_{V_1}^{V_2} P(V)\,dV \tag{3-14}$$

This integral cannot be evaluated unless we know the dependence of pressure on volume. Since this dependence determines the path taken in going from V_1 to V_2, the numerical value of the work done is directly related to the path. Work is not a state function.

The following three examples illustrate the influence of path on the work done.

(1) Expansion into a vacuum. In this case the gas is expanding against a zero external pressure and hence it can do no work. $P = 0$, $w = 0$.

(2) Expansion against a constant external pressure. A constant pressure has no volume dependence, and hence the work done on the gas is simply $w = -P_{ext}(V_2 - V_1) = -P_{ext}\Delta V$.

(3) Expansion such that $P_{int} = P_{ext}$. Such a process, corresponding to an equilibrium situation at every instant of the expansion, would require an infinite time to accomplish. Of course, if it were true that $P_{int} = P_{ext}$, there would be no expansion at all. Any reversible process is the un-

attainable limit of a series of increasingly efficient (and slower) real irreversible processes. But reversible reactions, like ideal gases, are useful working concepts. For such a reversible expansion,

$$P_{ext} = P_{int} = \frac{RT}{V} \text{ (if } P_{ext} > P_{int}, \text{ compression occurs)}$$

or

$$w = -\int_{V_1}^{V_2} \frac{RT}{V} dV = -RT \ln \frac{V_2}{V_1} \qquad (3\text{-}15)$$

We see from these three examples that the work the system performs depends upon the conditions of expansion. Furthermore, the reversible process in example (3) performs more work than either of the other two irreversible processes. In fact, because it is the work done against the greatest opposing force, the work obtained from a reversible process is the greatest possible work obtainable from a given expansion. We will frequently make use of this fact. A reversible process is always more efficient than any corresponding irreversible process.

Work can be represented particularly clearly in terms of areas on a *PV* diagram.

(a) *Typical expansion.* The *PV* plot of the expansion of a gas from volume V_a to volume V_b with pressure varying in a general manner is shown in Figure 3-5a. In this expansion the system does work on the surroundings. This work on the surroundings is a positive quantity and is equal to the area under the curve. The work that the surroundings do *on* the system is equal

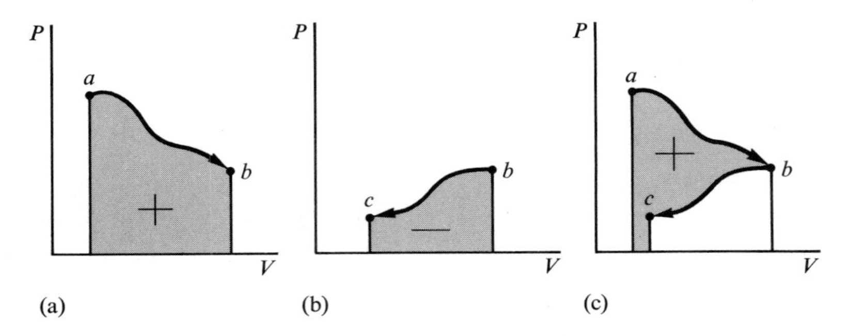

FIGURE 3-5. *The integral $\int_a^b P\,dV$, representing the area under the curve from a to b, is positive if V increases and negative if it decreases. The value of the integral from a to b to c (right) is the algebraic sum of the areas swept out in the two separate steps, and is equal to the area shaded.*

therefore to the negative of this area

$$w = -(\text{area under curve}) = -\int_a^b P\,dV$$

(b) *Positive and negative areas.* We define the *area* under a PV curve as being positive when the integration is performed in the direction of increasing volume, as shown in Figure 3-5a. The area under a PV curve will be negative for an integration in which one goes from a larger to a smaller volume, as in Figure 3-5b. Having established these definitions, it is an easy matter to evaluate more complex PV areas such as that shown in Figure 3-5c. In this plot the PV integral from point a to point b is positive and that from point b to point c is negative. Therefore, the total area under the integral curve in going from a to b to c (shaded in Figure 3-5c) is simply the sum of the two integrals: $\text{area}_{(a\to c)} = \text{area}_{(a\to b)} + \text{area}_{(b\to c)}$, where $\text{area}_{(b\to c)}$ is negative.

(c) *Compression.* The work done on the system by the surroundings during a compression is illustrated in Figure 3-5b. By earlier conventions, the work is the negative of the area, but in the case of a compression, the area itself is negative, so that the work is positive.

$$w = -\int_b^c P\,dV = -\text{area}$$

area < 0; therefore, $w > 0$.

(d) *Choice of path.* Further proof that work is not a state function. Figure 3-6 shows four examples that illustrate the importance of the choice of path in determining the work done in a typical compression process.

(1) Figure 3-6a shows a typical isothermal compression. By our definition the area under the curve is negative. Hence, the work done on the system is positive.

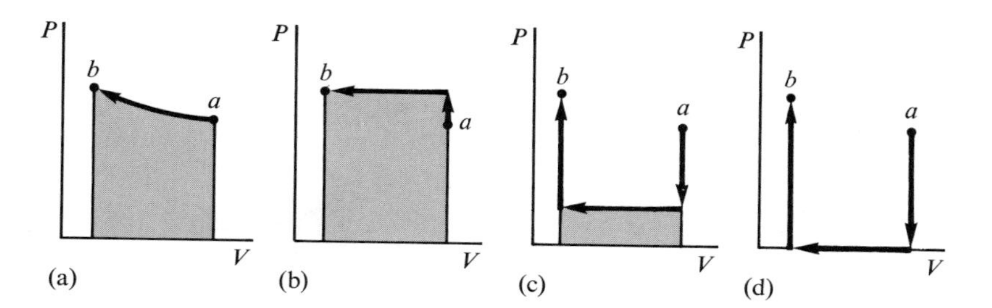

FIGURE 3-6. *Work is not a state function, for* $-w = \int_a^b P\,dV$*, and the area under the path depends very much upon the nature of the path from a to b, or upon the function* $P(V)$.

(2) Figure 3-6b shows a different path between V_a and V_b. The first step of this process is taken by increasing the pressure from P_a to P_b while holding volume constant. From the equation of state $P\overline{V} = RT$ we can see that this process is accompanied by a rise in temperature. Then by carefully cooling the system while holding pressure constant, we can effect a compression from V_a to V_b. The area under the curve in process (2) is greater than that in the isothermal process, and the work done on the gas is likewise greater.

(3) A third path from point a to b is shown in Figure 3-6c. The gas is first cooled at constant volume until the pressure falls to some value P'. It is then cooled at constant P' until the volume falls to V_b, and finally heated at constant V_b until the pressure P_b is reached. In this case, the work done is much less than in either of the first two cases.

(4) Figure 3-6d shows the limiting case in which the gas is cooled at volume V_a down to absolute zero, at which point the pressure would be zero. Then we can simply push the piston in to volume V_b and heat the system until pressure P_b is obtained. In this case no work has been done, and it looks as if we can carry out a simple compression of a gas to any desired state without doing any work on it. This, of course, is nonsense, and we must eventually get a statement of this fact into our thermodynamic system.

(e) *Compound paths.* Two examples of compound paths are depicted in Figure 3-7. Figure 3-7a shows a compression from a to b followed by an expansion from b to c. In this case we have done work on the gas in going from a to b, and the gas has done work in going from b to c. The shaded area represents the net work done by the gas on the surroundings in this two-step process and is negative in sign, indicating that net work has been done *on* the gas.

In Figure 3-7b the paths of integration cross. Again, the net work done on the surroundings by the gas is the sum of that done in compression from a to b (a negative quantity) plus that done in expansion from b to c (a positive quantity).

(f) *Cyclic paths.* The net work done on the surroundings by the gas in the course of a cyclic process is describable by the area enclosed by the cyclic

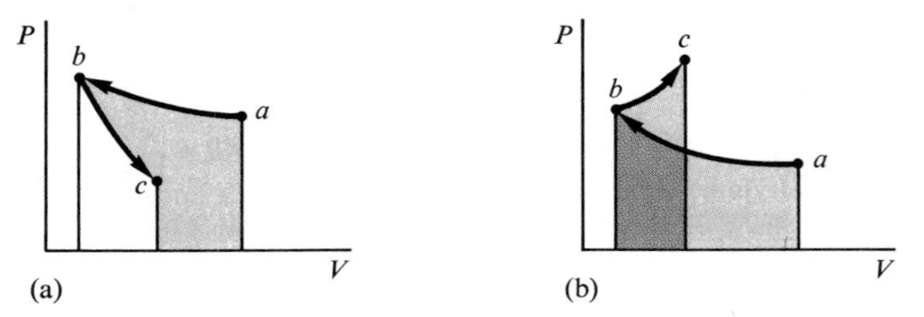

FIGURE 3-7. *When paths of integration double back or cross, careful account must be kept of the* signs *of the areas swept out under the segments of the total path.*

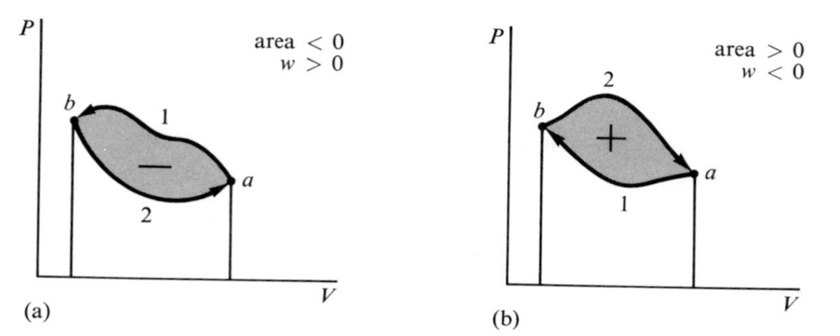

FIGURE 3-8. *In a cycle running counterclockwise on a PV plot (a), net work is done on the gas. In a clockwise cycle (b), net work is done by the gas on its surroundings.*

curve, as in the two examples illustrated in Figure 3-8. In Figure 3-8a the net work done on the gas is greater than zero, while the example shown in Figure 3-8b shows a process in which the gas does work on its surroundings. This second diagram is an example of what happens in a heat engine: the system does work on its surroundings and the working substance is returned to its original state. Of course, nothing is free, and to get this work out we must drive the engine by feeding in heat.

All of the PV plots in the previous section implied that in the process of going from one volume to another the pressure at each instant was uniform through the system. In fact, for every process conducted at any finite rate, a pressure gradient will exist in the system. In principle one must wait an infinite time for this gradient to fall to zero, but in practice these gradients fall to negligible values in a relatively short time.

The theoretical process in which every intermediate state is an equilibrium state is called a reversible process. Such a process can be represented by a line on a PV diagram, because at every instant the internal pressure is uniform throughout the gas, and $P_{int} = P_{ext}$. Since work is a function of the pressure against which the expansion takes place, the work in an irreversible process is always equal to or less than the work in any reversible process between the same two points.

An irreversible process cannot be represented by a solid line on a PV plot because there is no one uniform pressure throughout the system. We will use a dotted line such as that shown in Figure 3-9 to represent an irreversible process between two specified states.

3-6 CYCLES AND THE FIRST LAW

In a cyclic process such as that of Figure 3-8b, an expansion followed by a compression to the initial state, work is done by the gas on its surroundings. We can repeat this cycle as often as we like, and do as much work on the environment as we choose. But this work must be paid for, and the payment is in the form of heat supplied to make the cycle operate. The steam engine and the internal combustion engine are simple examples of such a cyclic machine; muscle tissue or a reproducing living species are more complex (steam engines do not evolve) but obey the same thermodynamic constraints.

In the simplest type of cyclic process, we can add or remove heat, and do work *on* or *by* the system. These are not independent; within limits, heat and work are interconvertible. Nowadays it is almost impossible to see anything remarkable in such a statement. Yet a century and a half ago, before Rumford's cannon-boring experiments, this was not at all self-evident; today's paradoxes become tomorrow's platitudes.

If all of the net heat which is fed into a cyclic process is converted into work, then the heat gain is exactly balanced by the work loss. Conversely, if one does work on a system and can convert all of the work into heat, then again the work gain is balanced by the heat loss. In both cases, $q + w = 0$. But there are other processes in which the work done is greater or less than the heat fed in, or in which the heat drained off is greater or less than the work done on the system. How do we handle this imbalance of q and w?

The law of conservation of energy will always hold, because if we ever find a breakdown in the law we will simply invent a new type of energy to make the equations balance. This is what we are going to do now. In order to make these equations balance, even in those cases where the heat supplied does not balance the work obtained, we can define a purely bookkeeping number called the internal energy E. Nobody has ever seen internal energy. Heat can be measured, work can be measured, but internal energy is more elusive. What we can say is that the difference between the heat fed in and the work done, or the heat taken out and the work done on the system, is the *change* in this mysterious quantity.

The first law says that this bookkeeping function, internal energy, is more than merely a device for tallying the difference between heat and work in a given experiment. The internal energy is a function, the change in which

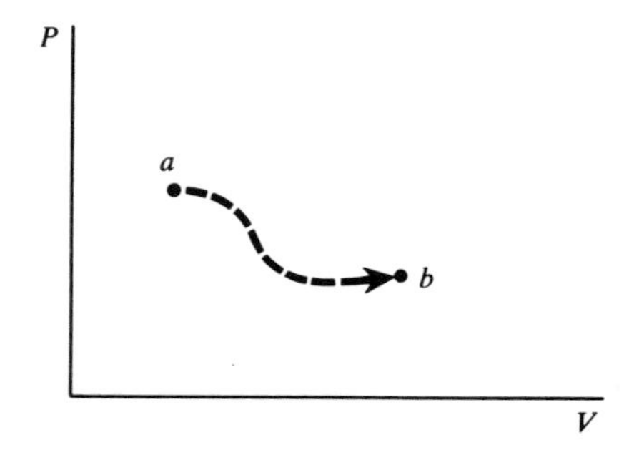

FIGURE 3-9. *Conventional representation of a PV plot of an irreversible process between states a and b.*

in going from state A to state B depends only upon the *conditions* of state A and of state B and not upon the path taken between the two points. If the system goes from state A to state B by a dozen different paths, and if E is only a bookkeeping function, then there is no particular reason to expect that the change in internal energy in all these processes would be the same, any more than the heat involved or the work done would be. But the first law of thermodynamics is an assertion that it will be the same.

The change in internal energy is simply the difference between the initial and final values of the energy. There is no equivalent "heat content" or "work content," the difference in which would give the heat absorbed or work done in the process.

The equation $q + w = \Delta E$ only defines internal energy. The first law of thermodynamics is the additional statement that this internal energy is a state function.

3-7 STATE FUNCTIONS AND EXACT DIFFERENTIALS

A *state function* E is a function, the change in which during any process depends only upon the nature of the initial and final states and not upon the details of the course of the process. The differential of such a function, dE, is an *exact differential*:

$$\int_a^b dE = E_b - E_a$$

Exact differentials are usually integrable by inspection.[1] In contrast, an inexact differential such as dw

$$\int_a^b dw = -\int_a^b P_{(V)}\, dV$$

is not the differential of a state function and cannot be integrated without knowing the dependence of the pressure on volume.

Examples of Exact Differentials

(1) $dz = ydx + xdy$. Here dz is the exact differential of (xy) and can be integrated at once:

$$\int_a^b dz = \int_a^b d(xy) = x_b\, y_b - x_a\, y_a$$

[1] The amount of inspection, of course, will vary with the complexity of the differential.

(2) $dz = 2xdx + 2xydy + y^2\,dx$. Though it is not obvious, dz here is again an exact differential, the differential of $x^2 + xy^2$. Again it is integrable by inspection:

$$\int_a^b dz = \int_a^b d(x^2 + xy^2) = x_b^{\,2} + x_b\,y_b^{\,2} - x_a^{\,2} - x_a\,y_a^{\,2}$$

(3) *Graphical representation of* $d(xy)$. From example (1), we saw that $dz = ydx + xdy = d(xy)$ was an exact differential. However, each of its components $dz' = ydx$ and $dz'' = xdy$ is inexact. Figure 3-10 shows how these two inexact differentials add to give an exact differential. Letting the Roman numerals represent the absolute areas of the four sections of the x-y plane, the integrals of ydx and xdy are

$$\int_a^b ydx = \text{I} + \text{II}$$

$$\int_a^b xdy = -\text{I} - \text{III}$$

Both integrals contain area I and are inexact because in each case the path from a to b forms one boundary of area I. But we can see that if we add the two inexact integrals, the dependence on area I vanishes.

$$\int_a^b ydx + \int_a^b xdy = \text{I} + \text{II} - \text{I} - \text{III} = \text{II} - \text{III}$$

Another approach would be to integrate dz directly

$$\int_a^b (xdy + ydx) = x_b\,y_b - x_a\,y_a = \text{II} + \text{IV} - \text{III} - \text{IV} = \text{II} - \text{III}$$

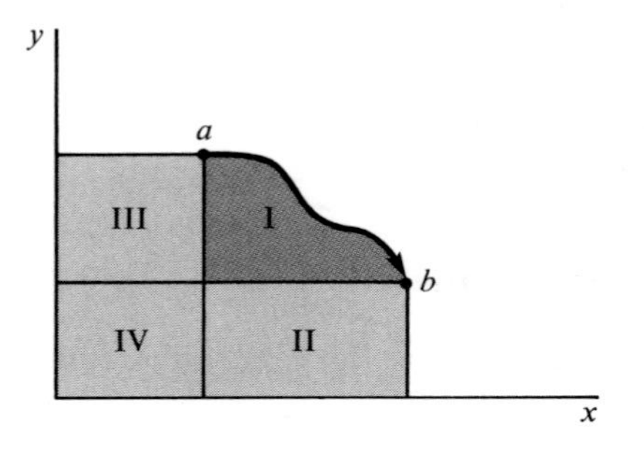

FIGURE 3-10. *Each of the integrals:* $\int_a^b ydx$ *and* $\int_a^b xdy$ *is undefined unless the nature of the path from a to b is specified, and both ydx and xdy are inexact differentials. But the sum* $\int_a^b ydx + \int_a^b xdy$ *= area II − area III and does not depend on the path from a to b. Accordingly,* $ydx + xdy$ *is an exact differential.*

and the same result is obtained.

By viewing Figure 3-10 as a PV plot instead of an xy plot, one can see why PdV and VdP are inexact differentials, although their sum is the exact differential of the quantity PV.

(4) $dz = 2ydx + xdy$. This differential is inexact; it is not the differential of any state function. However, it can be made exact by multiplying by x.

$$xdz = 2xydx + x^2\,dy = d(x^2y)$$

Here an inexact differential has been multiplied by a quantity known as an integrating factor and made into an exact differential. We see later that when the inexact differential dq is multiplied by $1/T$, the result is the exact differential dq/T. This last statement is actually one form of the second law.

Differential Expressions

If a state function is a function of independent variables, as $E = E(x, y, z, \ldots)$, where $x, y, z, \ldots$ are independent, then we can write

$$dE = M_x\,dx + M_y\,dy + M_z\,dz + \cdots \tag{3-16}$$

where

$$M_x = \left(\frac{\partial E}{\partial x}\right)_{y,z,\ldots} \qquad \text{etc.}$$

We can test for the exactness of the differential dE by applying to it the Euler reciprocity theorem. If dE is an exact differential, then

$$\frac{\partial M_x}{\partial y} = \frac{\partial M_y}{\partial x}$$

(valid for all possible combinations of variables). This equivalence is true because the order of differentiation is irrelevant, and both partials are equal to $\partial^2 E/\partial x dy$.

As a test, we can apply Euler's theorem to the example given earlier:

$$dz = (2x + y^2)\,dx + (2xy)\,dy$$

$$\frac{d(2x + y^2)}{dy} = \frac{d(2xy)}{dx} = 2y$$

Therefore dz is exact. For another earlier example,

$$dz = (2y)\,dx + (x)\,dy$$

$$2 \neq 1$$

Therefore, dz is inexact. But the integrating factor x makes the differential exact:

$$x\,dz = (2xy)\,dx + (x^2)\,dy$$
$$2x = 2x$$

Therefore $x\,dz$ is exact.

A typical thermodynamic function such as internal energy will be a function of pressure, volume, and temperature. But since pressure, volume, and temperature are related through an equation of state (such as $P\overline{V} = RT$ for ideal gases), only two of the three are truly independent. The choice of which two to use is a matter of convenience.

$$\text{(a)}\qquad E = E(P, T):\ dE = \left(\frac{\partial E}{\partial P}\right)_T dP + \left(\frac{\partial E}{\partial T}\right)_P dT \qquad (3\text{-}17)$$

$$\text{(b)}\qquad E = E(V, T):\ dE = \left(\frac{\partial E}{\partial V}\right)_T dV + \left(\frac{\partial E}{\partial T}\right)_V dT \qquad (3\text{-}18)$$

$$\text{(c)}\qquad E = E(P, V):\ dE = \left(\frac{\partial E}{\partial P}\right)_V dP + \left(\frac{\partial E}{\partial V}\right)_P dV \qquad (3\text{-}19)$$

It should be noted that $(\partial E/\partial V)_T$ and $(\partial E/\partial V)_P$, representing the change of energy with volume under conditions of constant temperature and pressure, respectively, are two quite different quantities.

Cyclic Processes

The energy change in a cyclic process, such as that shown in Figure 3-11, is zero.

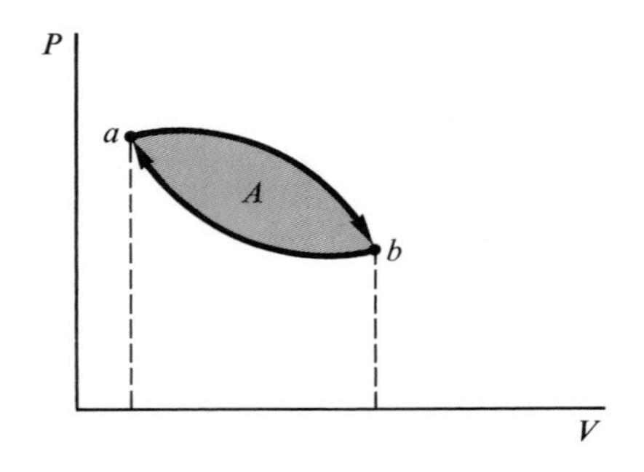

FIGURE 3-11. *In a cyclic process, ΔE is zero, and the work done by the gas on its surroundings must be compensated for by an absorption of heat; $\Delta E = q + w$.*

$$\Delta E = \oint dE = \int_a^b dE + \int_b^a dE = 0$$

If in any cyclic process there is no net change in E, then any work performed by the system must be paid for by supplying an equivalent amount of heat. In other words,

$$\Delta E = q + w = 0 \tag{3-20}$$

3-8 CONSTANT PRESSURE PROCESSES AND ENTHALPY

Since most chemical reactions are conducted under conditions of constant pressure and not constant volume, it is convenient to introduce a new energy function, called the enthalpy. In the course of a reaction in a bomb calorimeter, with fixed volume, no PV work can be done. In such a process, the heat absorbed is equal to the change in the internal energy.

$$\Delta E_V = q_V \tag{3-21}$$

By the first law, the sum of heat and work is a state function. Here, for the special case in which work is not permitted, heat itself is a state function.

The question naturally arises whether there is a general state function which at *constant pressure* is a measure of heat change in the system. For a constant pressure process, the first law gives

$$\Delta E = E_b - E_a = q_P - P(V_b - V_a) \tag{3-22}$$

Rearranging,

$$q_P = (E_b + PV_b) - (E_a + PV_a) \tag{3-23}$$

This suggests a new function[2] which we shall call the enthalpy H.

$$H \equiv E + PV \tag{3-24}$$

We know that both E and PV are state functions (V is a state function and P is an intensive property). Hence H is a state function as well, and at constant pressure

$$\Delta H_P = q_P \tag{3-25}$$

[2] Enthalpy is sometimes called "heat content" in older texts. This confusing notation will be avoided.

Once again, although in general it is the combination of q with another non-state-function $\int PdV$ which is a function of state and not q alone, in the special case where pressure is kept constant, heat becomes a state function. Another way of appreciating this is to realize that to specify constant pressure is to specify the path of integration in the integral of $dw = -PdV$ and to make work done also depend only on the initial and final states of the system.

3-9 USEFUL EXPRESSIONS FOR E, H, AND q

Both internal energy and enthalpy can be expanded in terms of any two of the three variables: pressure, volume, and temperature. Since we have seen that E is associated with constant volume processes, it is logical to express internal energy as $E(V, T)$:

$$dE = \left(\frac{\partial E}{\partial T}\right)_V dT + \left(\frac{\partial E}{\partial V}\right)_T dV \tag{3-26}$$

But

$$dq = dE - dw = dE + PdV \tag{3-27}$$

Using Eqs. 3-26 and 3-27,

$$dq = \left(\frac{\partial E}{\partial T}\right)_V dT + \left[P + \left(\frac{\partial E}{\partial V}\right)_T\right] dV \tag{3-28}$$

A similar derivation holds with enthalpy, $H = H(P, T)$:

$$dH = \left(\frac{\partial H}{\partial T}\right)_P dT + \left(\frac{\partial H}{\partial P}\right)_T dP \tag{3-29}$$

But

$$dH = dE + PdV + VdP = dq - PdV + PdV + VdP \tag{3-30}$$

or

$$dq = dH - VdP \tag{3-31}$$

Inserting Eq. 3-29 into Eq. 3-31,

$$dq = \left(\frac{\partial H}{\partial T}\right)_P dT + \left[\left(\frac{\partial H}{\partial P}\right)_T - V\right] dP \tag{3-32}$$

Note that Eqs. 3-28 and 3-32 are both expressions for dq; hence the right-hand sides of both equations are equal to one another. We could use this identity to derive some useful relationships, but, as we will see, there are easier ways to develop these formulas.

3-10 HEAT CAPACITY

With the expressions for q that were presented in the previous section, we can now treat heat capacities in a systematic manner. We know that $C \equiv dq/dT$, so that from Eqs. 3-28 and 3-32, respectively, the following expressions for the heat capacity result:

$$C = \left(\frac{\partial E}{\partial T}\right)_V + \left[\left(\frac{\partial E}{\partial V}\right)_T + P\right]\frac{dV}{dT} \tag{3-33}$$

$$C = \left(\frac{\partial H}{\partial T}\right)_P + \left[\left(\frac{\partial H}{\partial P}\right)_T - V\right]\frac{dP}{dT} \tag{3-34}$$

As was mentioned in Section 2-16, the numerical value of the heat capacity depends upon the conditions under which the experiment is carried out. At constant volume or at constant pressure, the expressions for C that have just been presented reduce to

$$C_V = \left(\frac{\partial E}{\partial T}\right)_V \tag{3-35}$$

$$C_P = \left(\frac{\partial H}{\partial T}\right)_P \tag{3-36}$$

The difference $C_P - C_V$ is easily obtained by making Eq. 3-33 apply under constant pressure conditions and then subtracting Eq. 3-35 from it.

$$C_P - C_V = \left[\left(\frac{\partial E}{\partial V}\right)_T + P\right]\left(\frac{\partial V}{\partial T}\right)_P \tag{3-37}$$

The same treatment can be used to obtain a similar expression from Eq. 3-34.

$$C_P - C_V = \left[V - \left(\frac{\partial H}{\partial P}\right)_T\right]\left(\frac{\partial P}{\partial T}\right)_V \tag{3-38}$$

This last equation is seldom used, but we will make frequent use of Eq. 3-37.

The last two equations raise an interesting question. Why should the heat storage ability of a substance under conditions of constant pressure be different from that under conditions of constant volume? The answer is not difficult. Under constant volume conditions, all heat fed into the system goes either into internal energy modes such as rotation or vibration or into increasing the molecular kinetic energy, and thereby raises the temperature. At constant pressure, these storage modes are available, but additional heat is used when the system does work of expansion. There are two types of work that can be done by an expanding gas. The first is the obvious one: work against the surroundings. This external work gives us part of Eq 3-37.

$$(C_P - C_V)_{\text{ext}} = \left(\frac{\partial w_{\text{ext}}}{\partial T}\right)_P = P\left(\frac{\partial V}{\partial T}\right)_P \tag{3-39}$$

But an expanding gas must also do work against its own internal cohesive forces, the attractive forces between molecules. (This statement holds for real gases, but ideal gases have no such attractive forces.) The work necessary to overcome this "internal pressure" during an expansion manifests itself as a change in the internal energy of the system, $(\partial E/\partial V)_T\, dV$. Now for real gases, this term adds to Eq. 3-39 to give

$$C_P - C_V = \left[\left(\frac{\partial E}{\partial V}\right)_T + P\right]\left(\frac{\partial V}{\partial T}\right)_P \tag{3-40}$$

From kinetic theory, we found that $C_V = \frac{3}{2}R$ and $(\partial E/\partial V)_T = 0$. In other words, an ideal gas has no intermolecular forces, a fact first demonstrated by Joule in 1843. Joule tested this by a free expansion of a gas into an evacuated chamber in a water bath calorimeter. He found that in the limit of low pressure, when gas behavior approached ideality, there was no warming or cooling effect when the gas expanded. Since the gas was expanding into a vacuum, there was no external force on which it could do work. The only possible way for a temperature change to occur would have been for the gas to do work on its own cohesive forces, thus lowering its own internal energy and cooling. The experimental result that $(\partial E/\partial V)_T = 0$ for ideal gases is evidence for the absence of such cohesive forces.

For one mole of an ideal monatomic gas,

$$C_P = C_V + P\left(\frac{\partial V}{\partial T}\right)_P = \frac{3}{2}R + P\frac{R}{P} \tag{3-41}$$

or

$$C_P = \tfrac{5}{2}R$$
$$C_P - C_V = R \tag{3-42}$$

3-11 *JOULE–THOMSON EXPERIMENT—DEVIATIONS FROM IDEALITY*

Joule's original free expansion experiment was followed by a better experiment carried out in collaboration with Thomson. Their experiment involved not a free expansion into a vacuum, but a throttled expansion, an expansion under conditions of constant enthalpy.

The throttling process is explained diagramatically in Figure 3-12. This process allows one to set up a pressure difference between two adjacent chambers while still allowing the interchange of molecules between them. Figure 3-12a shows an unstable condition in which the pistons will move until there is a uniform pressure throughout the chamber. A plot of chamber pressure versus cylinder length for the instant at which the pistons are released is given in Figure 3-12b. If a thin barrier is placed in the middle of the chamber (Joule and Thomson first used a pocket handkerchief), it will disrupt the pressure gradient by hindering the free movement of molecules from one side of the chamber to the other. A plot for this situation is given in Figure 3-12c, which shows that the pressure change across each of the two chambers has been reduced by introducing a certain pressure gradient within the barrier. Meerschaum, the porous clay used in pipes, and sintered glass are both more efficient barriers than cloth. They are sufficiently thick that the molecules diffusing through from one side can be retained within the barrier until they come to equilibrium with the other side of the chamber, as shown in Figure 3-12d. In this case, the entire pressure differential exists within the

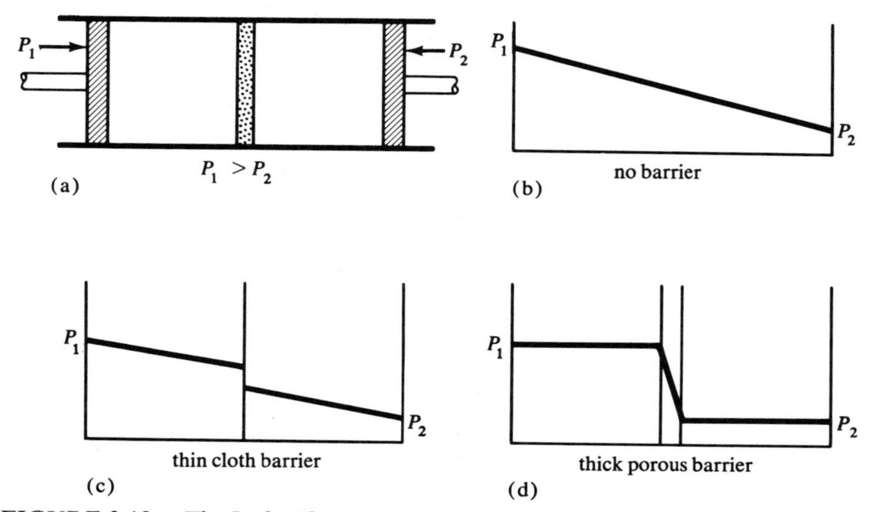

FIGURE 3-12. *The Joule–Thomson expansion apparatus (a), and pressure gradients down the chamber for (b) no barrier, (c) a thin cloth barrier, and (d) a thicker porous barrier such as sintered glass or meerschaum clay.*

barrier, with essentially uniform pressures existing on both sides and the gas is said to be throttled.

The actual Joule–Thomson experiment, carried out under adiabatic conditions, is depicted in Figure 3-13. The initial situation, shown in Figure 3-13a, has the left-hand chamber at a given pressure, temperature, and volume and the right-hand chamber with a given piston pressure and temperature, but a zero volume. In the final position, Figure 3-13c, the situation is reversed and the volume of the left-hand chamber is zero. Since $P_2 < P_1$, the volume at the completion of the process will be greater than the initial volume.

We can easily justify this rather unusual experiment by proving that these conditions produce an isoenthalpic expansion, or that ΔH for the process is zero. It was initially stated that the experiment was conducted under adiabatic conditions, so that $q = 0$. But when $q = 0$, $\Delta E = w$, which means that w is a state function for this process.

$$w = -\int_{V_1}^{0} P_1\, dV - \int_{0}^{V_2} P_2\, dV = P_1 V_1 - P_2 V_2 \tag{3-43}$$

Then,

$$\Delta E = E_2 - E_1 = P_1 V_1 - P_2 V_2 \tag{3-44}$$

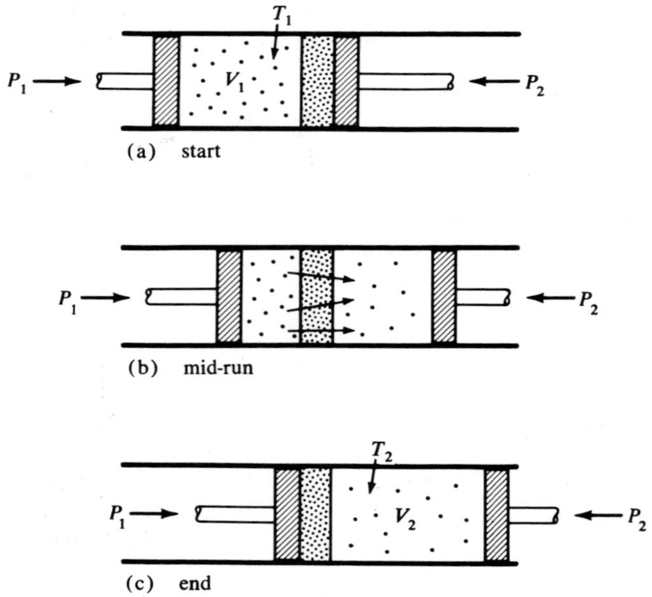

(a) start

(b) mid-run

(c) end

FIGURE 3-13. *The Joule–Thomson expansion. (a) Start, with all of the gas in chamber 1; (b) mid-run; and (c) end, with all of the gas at lower pressure in chamber 2.*

Rearranging,

$$E_1 + P_1 V_1 = E_2 + P_2 V_2$$

Therefore,

$$H_1 = H_2$$
$$\Delta H = 0$$

Q.E.D.

Joule-Thomson Coefficient

The purpose of the throttling process is to provide a situation where one can observe the change in temperature with pressure under conditions of constant enthalpy. This purely experimental quantity is known as the Joule–Thomson coefficient

$$\mu_{JT} = \left(\frac{\partial T}{\partial P}\right)_H \tag{3-45}$$

Most common gases cool as they expand at constant enthalpy at room temperature, and have positive Joule–Thomson coefficients (Figure 3-14). But H_2 and He show the opposite effect; they are warmed by expansion. At suitably low temperatures ($-80°C$ for H_2), these gases, too, will be cooled upon expansion.

The general behavior for real gases is to show warming upon isoenthalpic expansion, and a negative Joule–Thomson coefficient, above a characteristic temperature, and cooling and a positive Joule–Thomson coefficient below it. This characteristic temperature is the Joule–Thomson inversion temperature T_{JT}. At sufficiently low temperatures and high pressures the Joule–Thomson coefficient may again become negative, as shown for N_2 in Figure 3-14.

The positive Joule–Thomson coefficient at low temperatures arises from the necessity of doing work against intermolecular attractions in an adiabatic expansion. As the mean kinetic energy of the molecules rises with temperature, this effect should become less and less important; in fact, in Section 4–6 we show that for a van der Waals gas, this contribution to the Joule–Thomson coefficient varies inversely with temperature.

The negative coefficient at high temperatures represents the overpowering of this attractive work effect by a more subtle effect of opposite sign. Suppose that the molecules are hard spheres rather than points as in the ideal gas assumptions, so that the volume available to the centers of the gas molecules is less than the total volume by a molar factor b. The equation of state of this hard sphere gas would then be

$$P(\overline{V} - b) = RT$$
$$P\overline{V} = RT + bP \tag{3-46}$$

By Eq. 3-44, the energy change in the Joule–Thomson expansion can be written

$$\Delta E = C_V(T_2 - T_1) = P_1 V_1 - P_2 V_2$$
$$= R(T_1 - T_2) + b(P_1 - P_2)$$

(*Note*: As an exercise, use Eq. 4-59 to prove that $dE = C_V\, dT$ for a hard

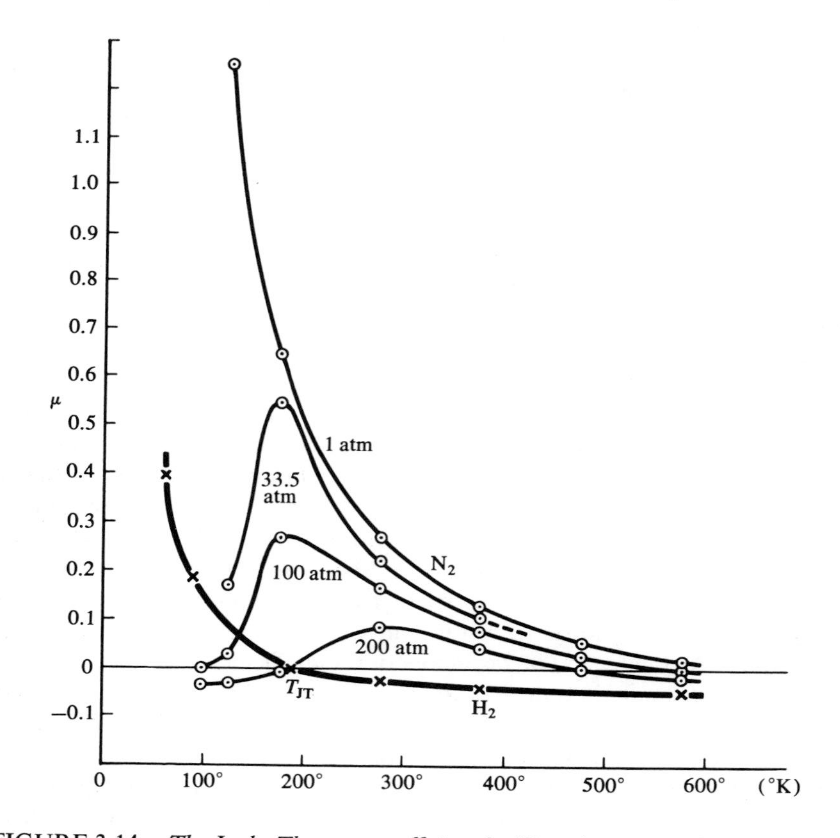

FIGURE 3-14. *The Joule–Thomson coefficient for* H_2 *at* 1 atm, *and for* N_2 *at* 1 atm *and elevated pressures. The point at which each curve crosses the* $\mu_{JT} = 0$ *axis is called the Joule–Thomson inversion temperature* T_{JT}.

sphere gas as well as for an ideal gas.) The temperature change is then given by

$$T_2 - T_1 = \frac{b}{C_V + R}(P_1 - P_2) \tag{3-47}$$

The temperature rises as the pressure falls, and the Joule–Thomson coefficient for a hard sphere gas is always negative and equal to $-b/C_P$. For a gas in which an appreciable fraction of the true volume is inaccessible to any given molecule (because of the finite volumes of the other molecules), the work fed in on the high pressure side of the barrier is greater than that taken out during the expansion at lower pressure, and since the process is adiabatic, this excess work can only serve to increase the internal energy and the temperature.

Joule–Thomson Coefficient and Enthalpy: μ_{JT} *and* $(\partial H/\partial P)_T$

The Joule–Thomson coefficient provides a measure of the deviation of a gas from ideal behavior because of finite molecular volumes and intermolecular attractions. For ideal gases, both H and E are functions only of temperature, and it should not be surprising to find a close relationship for real gases between μ_{JT} and $(\partial H/\partial P)_T$. The derivation of the relationship between μ_{JT} and $(\partial H/\partial P)_T$ will provide an example of the manipulation of exact differentials to produce a useful result. Since the desired expression involves P and T as variables, it is logical to begin by expanding enthalpy in terms of them:

$$H = H(P, T)$$

$$dH = \left(\frac{\partial H}{\partial P}\right)_T dP + \left(\frac{\partial H}{\partial T}\right)_P dT \tag{3-48}$$

The Joule–Thomson coefficient $(\partial T/\partial P)_H$ can be brought into the expression by taking the partial of the above equation with respect to pressure under conditions of constant enthalpy:

$$\left(\frac{\partial H}{\partial P}\right)_H = \left(\frac{\partial H}{\partial P}\right)_T + \left(\frac{\partial H}{\partial T}\right)_P \left(\frac{\partial T}{\partial P}\right)_H \tag{3-49}$$

Substituting $(\partial H/\partial P)_H = 0$, $C_P = (\partial H/\partial T)_P$, and $\mu_{JT} = (\partial T/\partial P)_H$,

$$0 = \left(\frac{\partial H}{\partial P}\right)_T + C_P \mu_{JT} \tag{3-50}$$

Rearranging terms,

$$\left(\frac{\partial H}{\partial P}\right)_T = -\mu_{\text{JT}} C_P \tag{3-51}$$

For an ideal gas, we know that $(\partial E/\partial V)_T = 0$; and, although we did not say as much, we implied that $(\partial E/\partial P)_T = 0$. Therefore, if $\bar{H} = \bar{E} + P\bar{V}$, or in the case of an ideal gas, $\bar{H} = \bar{E} + RT$, enthalpy is also only a function of temperature, and $(\partial H/\partial P)_T$ must be zero:

$$dH = dE + d(PV) = dE + d(RT)$$

Therefore,

$$\left(\frac{\partial H}{\partial P}\right)_T = \left(\frac{\partial E}{\partial P}\right)_T + \left(\frac{\partial RT}{\partial P}\right)_T = 0 + 0 = 0 \tag{3-52}$$

Since $(\partial H/\partial P)_T = 0$ for an ideal gas, we see from Eq. 3-51 that the Joule–Thomson coefficient for an ideal gas is always zero. This means that one can use the Joule–Thomson coefficient as a convenient experimental measure of the deviations from ideality of real gases.

Refrigeration

Gases with positive Joule–Thomson coefficients find practical use in refrigeration units. A schematic drawing of a typical gas refrigeration unit is shown in Figure 3-15. Expansion coils are located within the area to be cooled,

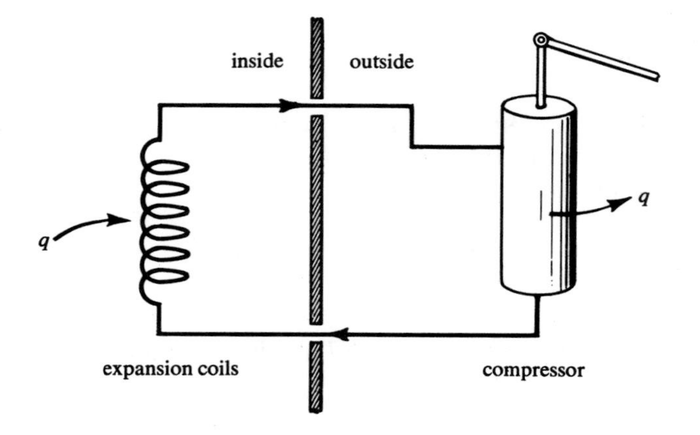

FIGURE 3-15. *Schematic of a refrigeration system using Joule–Thomson cooling.*

and a compressor on the outside. When a gas with a positive μ is compressed, its temperature rises, and heat flows from the gas to the surroundings. After cooling, the compressed gas passes into the refrigerator and is allowed to expand, with a drop in temperature. Accordingly, heat flows from the refrigerator to the gas in the coils. The now warm and uncompressed gas then reenters the compressor and the cycle is repeated. Heat is pumped from a cold region to a warmer one at the price of doing work on the gas in compressing it.

The Joule–Thomson effect is also made use of in liquefying gases such as nitrogen and oxygen. Compressed gases are cooled by passing them over the expansion coils of gases at a more advanced stage of the process. Eventually, the point is reached in this cyclical compression-expansion process where oxygen begins to liquefy, and at a later stage liquid nitrogen begins to form. Hydrogen cannot be liquefied by this process until it has been cooled below its Joule–Thomson inversion point $T = -80°C$. It must be cooled below $-80°C$ by some other means, and only when it has a positive Joule–Thomson coefficient can it be cycled down and liquefied.

3-12 FIRST LAW AND IDEAL GASES

The application of the first law to ideal gases is particularly straightforward, and the results, although wrong in detail for real gases, will certainly indicate trends in real gas behavior. For one mole of an ideal gas,

$$PV = RT$$

Energy and enthalpy are expressed by

$$dE = \left(\frac{\partial E}{\partial T}\right)_V dT + \left(\frac{\partial E}{\partial V}\right)_T dV = C_V\,dT \tag{3-53}$$

$$dH = \left(\frac{\partial H}{\partial T}\right)_V dT + \left(\frac{\partial H}{\partial P}\right)_T dP = C_P\,dT \tag{3-54}$$

with both of the above partials at constant T going to zero. The first equation is true for all gases at constant volume, the second for all gases at constant pressure, and both are true for ideal gases under all conditions.

Since we know C_P and C_V, we can obtain the useful ratio γ:

$$C_V = \tfrac{3}{2}R \qquad C_P = \tfrac{5}{2}R \qquad \gamma = \frac{C_P}{C_V} = \frac{5}{3}$$

Isothermal Processes with Ideal Gases

For an isothermal process, there is no temperature change, and therefore (by Eq. 3-53) no energy change. Any work that is done must be exactly counterbalanced by the absorption of an equivalent amount of heat from the surroundings.

$$\Delta E = 0 \qquad dq = -dw = PdV$$

(a) Reversible expansion. For a reversible expansion, $P_{ext} = P_{int}$ at all times, so that $P_{ext} = RT/V$. Since $dq = -dw$, integration yields

$$q = -w = \int_{V_1}^{V_2} PdV = RT \int_{V_1}^{V_2} \frac{dV}{V} = RT \ln \frac{V_2}{V_1} \tag{3-55}$$

Or, since the ratio of pressures varies inversely as the ratio of volumes:

$$q = -w = RT \ln \frac{P_1}{P_2} \tag{3-56}$$

(b) Irreversible expansion. In this case $P_{ext} < RT/V$ and

$$q = -w < RT \ln \frac{V_2}{V_1} = RT \ln \frac{P_1}{P_2} \tag{3-57}$$

For this isothermal process, $\Delta E = 0$ and the work done is still balanced by the heat absorbed. But since the irreversible process does *less* work than a comparable reversible process, less heat is absorbed. The extreme case is reached in the limit of a free expansion into a vacuum, in which case $w = 0$ and $q = 0$.

Adiabatic Processes

An adiabatic process is one in which there is no heat flow in or out of the system. In this case $q = 0$ and $\Delta E = w$, which means that any work done by the gas must be paid for by a drop in the internal energy of the system. Conversely, if work is done *on* the system, the exclusion of heat loss to the surroundings means that the internal energy rises by an equivalent amount. In general,

$$dE = C_V\, dT = dw = -PdV$$

or for an ideal gas,

$$PdV + C_V\, dT = 0 \tag{3-58}$$

(a) Reversible expansion. Here $P_{ext} = P_{int} = RT/V$. Substituting this identity into Eq. 3-58 and dividing by T, we obtain

$$R \frac{dV}{V} + C_V \frac{dT}{T} = 0 \tag{3-59}$$

Integrating,

$$R \ln \frac{V_2}{V_1} + C_V \ln \frac{T_2}{T_1} = 0 \tag{3-60}$$

Converting to the exponential form,

$$\left(\frac{V_2}{V_1}\right)^R = \left(\frac{T_1}{T_2}\right)^{C_V} \tag{3-61}$$

Equation 3-61 leads to a useful set of relationships for reversible adiabatic processes with ideal gases. First we see that

$$VT^{C_V/R} = \text{const} \tag{3-62}$$

By substituting $V_2/V_1 = (T_2/T_1)P_1/P_2$ into Eq. 3-61, we obtain an equivalent expression in terms of P and T:

$$\left(\frac{P_1}{P_2}\right)^R = \left(\frac{T_1}{T_2}\right)^{C_P} \tag{3-63}$$

or

$$\frac{T^{C_P/R}}{P} = \text{const} \tag{3-64}$$

A third expression, in terms of V and P, is derived from Eq. 3-61 by substituting $T_1/T_2 = (P_1/P_2)V_1/V_2$.

$$\left(\frac{P_2}{P_1}\right)^{C_V} = \left(\frac{V_1}{V_2}\right)^{C_P} \tag{3-65}$$

or

$$PV^\gamma = \text{const} \qquad \gamma = \frac{C_P}{C_V} \tag{3-66}$$

It is this last expression that finds most frequent application in problems dealing with adiabatic processes.

Equation 3-66 points out the essential difference between isothermal and adiabatic processes. For an isothermal process, $PV = \text{const}$, while for an adiabatic process, $PV^{\gamma} = \text{const}$. The physical meaning of this difference is illustrated in Figure 3-16, where a comparison is made between an isothermal and adiabatic expansion for an ideal gas ($\gamma = 5/3$). In both processes, PV work is done on the surroundings as the gas expands. In the isothermal process, the temperature and the internal energy are held fixed, so that an equivalent amount of heat is absorbed by the system to offset the work done. However, in the adiabatic process there is no heat available and the system must offset the work done by lowering its internal energy and hence its temperature. Because of this drop in temperature, PV cannot be constant for the adiabatic process, but $PV^{5/3}$ is constant instead. The 5/3 power compensates for the fact that, with the fall in T, V need not increase as much as in an isothermal expansion.

(b) Irreversible expansion. A typical irreversible adiabatic expansion is shown in Figure 3-17. With the piston initially held in place and a pressure differential such that $P_{\text{int}} > P_{\text{ext}}$, the stop is removed and the gas is allowed to expand adiabatically until $P_{\text{int}} = P_{\text{ext}}$. From Eq. 3-58,

$$C_V\, dT = -P_{\text{ext}} dV \tag{3-67}$$

or

$$C_V(T_b - T_a) = P_{\text{ext}}(V_a - V_b) \tag{3-68}$$

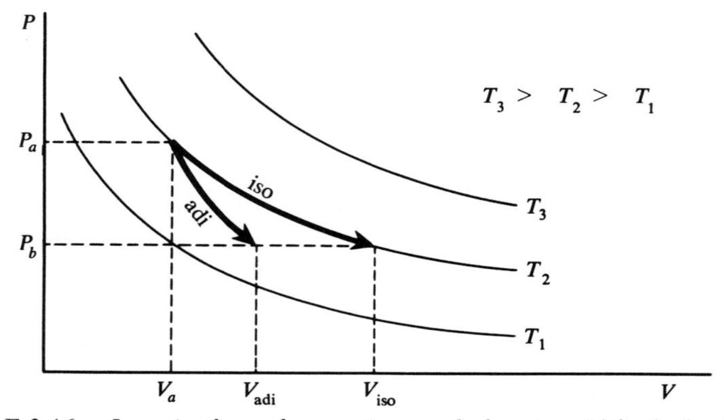

FIGURE 3-16. *In an isothermal expansion, work done is paid for by heat absorbed. In an adiabatic expansion, this is not possible, and the work done is compensated for by a drop in internal energy of the gas, and hence of its temperature. The volume then does not increase as much for a given pressure drop.*

(The pressure *against which* the gas does work remains P_{ext}.) From the equation of state for an ideal gas, we can obtain the expression for the initial and final volumes:

$$V_a = \frac{RT_a}{P_{int}} \qquad V_b = \frac{RT_b}{P_{ext}}$$

Substituting these volume relationships into Eq. 3-68 and simplifying, we obtain an expression for the temperature change during this expansion:

$$C_V(T_b - T_a) = RT_a \frac{P_{ext}}{P_{int}} - RT_b \tag{3-69}$$

$$T_b = T_a\left(\frac{C_V + (P_{ext}/P_{int})R}{C_P}\right) \tag{3-70}$$

Isothermal–Adiabatic Comparison

Now that we have the equations which describe the behavior of a gas under various isothermal and adiabatic processes, a practical comparison is in order. For our model, we will take 1 mole of gas at 300°K and 10 atm and expand it to 1 atm both reversibly and irreversibly under isothermal and adiabatic conditions.

$$\text{standard conditions}\begin{cases}\text{initial: } P_i = 10 \text{ atm} \qquad T_i = 300°\text{K} \qquad V_i = V_i \\ \text{final: } P_f = 1 \text{ atm}\end{cases}$$

(a) Reversible expansion.

Isothermal: $T_f = 300°\text{K}$, $\Delta E = 0$.

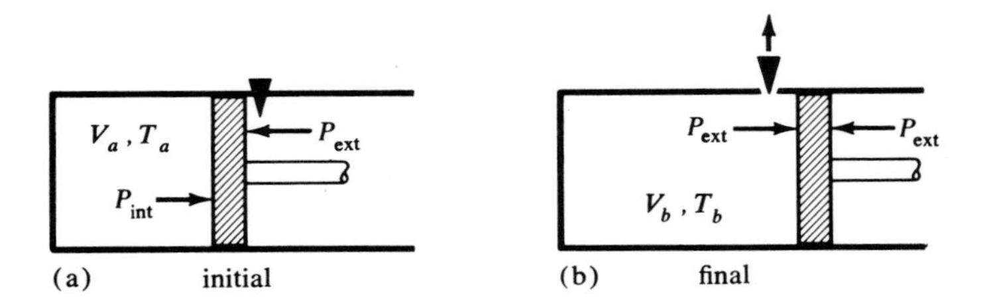

FIGURE 3-17. *Apparatus for carrying out an irreversible expansion against constant pressure. (a) Start, with piston held in place by a pin, and $P_{int} > P_{ext}$. (b) End, when P_{int} has fallen to match P_{ext}.*

From Eq. 3-56,

$$w = -RT \ln \frac{P_i}{P_f} = -RT \ln \frac{10}{1} = -1.99 \times 300 \times 2.303 = -1378 \text{ cal}$$

$$q = -w = +1378 \text{ cal}$$

$$V_f = \frac{P_i}{P_f} V_i = 10V_i$$

Adiabatic: $q = 0, \qquad \Delta E = w.$

From Eq. 3-63,

$$T_f = T_i\left(\frac{P_f}{P_i}\right)^{R/C_P} = 300\left(\frac{1}{10}\right)^{\frac{2}{5}} = 120°K$$

From Eq. 3-58,

$$\Delta E = w = C_V \int_{Ti}^{Tf} dT = \tfrac{3}{2} R(T_f - T_i)$$

$$= \tfrac{3}{2} \times 1.99 \times (120 - 300) = -540 \text{ cal}$$

$$\frac{V_f}{V_i} = \frac{T_f}{T_i}\frac{P_i}{P_f} = \frac{120}{300}\frac{10}{1} = 3.99$$

$$V_f \cong 4V_i$$

A PV plot and a tabular summary of the differences between these two reversible processes are given in Figure 3-18 and Table 3-1.

TABLE 3–1

	P_f	T_f	V_f	w	q	ΔE
Isothermal reversible	1 atm	300°K	$10V_i$	−1378 cal	+1378 cal	0
Adiabatic reversible	1 atm	120°K	$4V_i$	−540 cal	0	−540 cal

(b) Irreversible expansion. For our example, assume that the piston expands against a constant external pressure of 1 atm.

Isothermal: $T_f = 300°\text{K}$, $\Delta E = 0$.

$$V_f = \frac{RT}{P_{ext}} \qquad V_i = \frac{1}{10}\frac{RT}{P_{ext}}$$

therefore,

$$V_f = 10V_i$$

$$w = -P_{ext}(V_f - V_i) = -RT\left(1 - \frac{1}{10}\right) = -0.9RT = -537 \text{ cal}$$

$$q = +537 \text{ cal}$$

Adiabatic: $q = 0$, $\Delta E = w$.

From Eq. 3-70,

$$T_f = T_i\left(\frac{C_V + (P_f/P_i)R}{C_P}\right) = 300\left[\frac{3}{5} + \frac{1}{10}\frac{2}{5}\right] = 192°K$$

$$\Delta E = w = C_V(T_f - T_i) = \tfrac{3}{2} \times 1.99(192 - 300) = -322 \text{ cal}$$

$$V_f = V_i\frac{T_f}{T_i}\frac{P_i}{P_f} = V_i\frac{192}{300}\frac{10}{1} = 6.4V_i$$

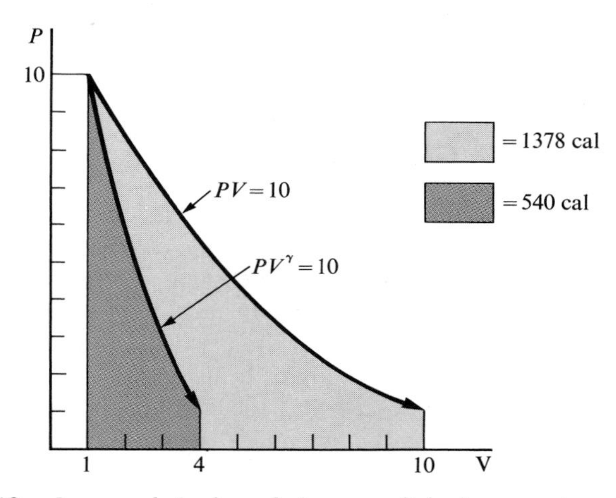

FIGURE 3-18. *Less work is done during an adiabatic expansion than during an isothermal expansion to the same final pressure.*

TABLE 3–2

	P_f	T_f	V_f	w	q	ΔE
Isothermal reversible	$\frac{1}{10}P_i$	300°K	$10V_i$	−1378 cal	+1378 cal	0
Isothermal irreversible	$\frac{1}{10}P_i$	300°K	$10V_i$	−537 cal	+537 cal	0
Adiabatic reversible	$\frac{1}{10}P_i$	120°K	$4V_i$	−540 cal	0	−540 cal
Adiabatic irreversible	$\frac{1}{10}P_i$	192°K	$6.4V_i$	−322 cal	0	−322 cal

A final comparison of all four processes is given in Table 3-2. In reversible expansions, more work was done when the process was carried out isothermally than adiabatically, because the volume increase was greater. The isothermal process was paid for by pulling in heat; the adiabatic process was paid for by dropping the internal energy and hence the temperature.

The irreversible processes did less work than the comparable reversible processes, because the system did not have to expand against as great an external pressure. The result in one case is that less heat was absorbed; in the other, that the temperature drop was less.

All of the above comparisons assumed ideal gas behavior. Real gases will deviate from these numerical values, but the comparisons between isothermal and adiabatic, reversible and irreversible behavior are qualitatively applicable.

3-13 HEATS OF REACTION

The chemist is seldom concerned with the work done during a chemical reaction, but is very often interested in measuring, calculating, and predicting the heat involved in a chemical process. The prime source of heat of reaction, of course, is the making and breaking of chemical bonds. Reactions which liberate heat are called exothermic, and those which absorb heat, endothermic. It was once thought that a study of heats of reaction would lead to a measure of the tendency to react—that the guiding principle would be a minimization of energy or enthalpy, and that those reactions would be spontaneous which were exothermic. As you undoubtedly know and as we show later, this is not true. Nevertheless, a knowledge of heats of reaction tells us much about what is going on in the reacting system.

Not all heats of reaction are equally easy to measure. An ideal reaction should be fast, effectively complete, and clean, with no significant side reactions. Combustion reactions satisfy the above requirements for a great many substances:

$$\text{reactant} + xO_2 \rightarrow \text{products } (H_2O \text{ and } CO_2 \text{ for hydrocarbons})$$

Reaction is carried out at constant volume in a bomb calorimeter. The heats of combustion are typically hundreds of kilocalories, and accuracies of better than 0.01% can be achieved. One problem is that if differences between heats are desired for related reactions, the heats of combustion themselves are so large and their differences so small that accuracy is lost in the subtraction process. Hydrogenation reactions, involving the addition of hydrogen across multiple bonds in unsaturated systems, involve less heat.

The above reactions are commonly carried out at constant volume. Inorganic reactions in aqueous solution, in contrast, take place at constant pressure. These include heats of neutralization, solution, and complex formation.

From the first law, we know that when a reaction is carried out at constant volume, the heat absorbed is equal to the rise in internal energy of the reaction system. Under constant pressure conditions the heat absorbed is equal to the increase in enthalpy:

$$\text{at constant } V\text{: } q_V = \Delta E$$

$$\text{at constant } P\text{: } q_P = \Delta H$$

An endothermic process at constant pressure is sometimes referred to as "endoenthalpic."

Enthalpies of Reaction

As an example, consider the following combustion reaction:

$$C_3H_{8(g)} + 5O_{2(g)} \rightarrow 3CO_{2(g)} + 4H_2O_{(l)} \tag{3-71}$$

The heat of this reaction at 298°K is $\Delta H = -530.61$ kcal, and the reaction is strongly exothermic. Comparable heats of combustion for the first few hydrocarbons are given in Table 3-3. Note the very slight dependence upon structure in normal and isobutane, and the regular increase in heat of about 155 kcal per added —CH_2—.

TABLE 3-3 REPRESENTATIVE HEATS OF COMBUSTION

Reactant	H_2	CH_4	C_2H_6	C_3H_8	$n-C_4H_{10}$	$i-C_4H_{10}$
ΔH^0_{298} (kcal/mole)	−68.3	−212.8	−372.8	−530.6	−688.0	−686.3
Increment in ΔH		−144.5	−160.0	−157.8	−157.4	

Factors Affecting ΔH and ΔE

The exact value of the heat of reaction depends upon several factors, the most important of which are the following:

(1) Stoichiometry. The ΔH value associated with a reaction represents the heat change per unit of reaction *as written*. The heat of combustion of benzene, $\Delta H = -781.0$ kcal, is per mole of benzene.

$$C_6H_{6(l)} + 7\tfrac{1}{2}O_{2(g)} \rightarrow 3H_2O_{(l)} + 6CO_{2(g)} \tag{3-72}$$

We could just as well have written

$$2C_6H_{6(l)} + 15O_{2(g)} \rightarrow 6H_2O_{(l)} + 12CO_{2(g)} \tag{3-73}$$

to eliminate the fractional coefficient of oxygen, but the heat of combustion for this reaction, as written, would be $\Delta H = -1562$ kcal. It is a common mistake to adjust the stoichiometry of a reaction and then forget to adjust the heat of reaction by the same factor.

(2) Physical state of reactants and products. If the reaction of Eq. 3-72 were carried out but with the production of water vapor instead of liquid water, the heat of reaction would be only -749.4 kcal. Less heat would be given off, because the heat formerly available from the condensation of the water produced would not be liberated. The state of each reactant and product must be specified carefully, whether solid (s), liquid (l) or gas (g). The symbol (aq), for "aqueous" is often used to represent a component in aqueous solution. When a substance is dissolved in water or when an existing solution is diluted further, heat may be given off or absorbed. However, for very dilute solutions, the addition of more water causes a negligible heat change. The symbol (aq) applied to a solute indicates that the substance is dissolved in so much water that the addition or subtraction of a little water causes no heat change. This will be referred to as "infinite dilution."

(3) Temperature and pressure. The heat given off or absorbed in the course of a reaction depends upon the temperature at which the reaction is

carried out and, to a considerably smaller extent, upon the pressure. The temperature is designated by a subscript, ΔH_{298°.

(4) Path of reaction? The values of ΔH and ΔE, H and E being state functions, do *not* depend upon the path of the reaction. ΔH and ΔE are the same whether the reaction is carried out at one fixed temperature or allowed to go to any intermediate reaction temperature and then brought back to the initial temperature at the end. They will also be the same, whether the reaction is carried out at fixed pressure, or merely brought back to the initial pressure at the end. The heat and the work involved will differ with choice of intermediate conditions, but if the final pressure is returned to its initial value, then

$$\Delta H = q_P$$

and the enthalpy of reaction will be equal to the heat change which would have been involved had a constant pressure been maintained.

From here on, when state functions are being discussed, the terms "at constant P," "at constant V" or "at constant T" should be interpreted as meaning only "with the same final as initial P, or V, or T."

Standard States

Because of the dependencies of the preceding section, standard states have been chosen for reporting and comparing heats of reaction. For solids and liquids, the standard state is usually the *most stable* form at 298°K and an external pressure of one atmosphere. The standard state of carbon is then graphite and not diamond, and that of sulfur is the rhombic crystal form and not the monoclinic. For gases, the standard state is 298°K and a partial pressure (*not* total pressure) of 1 atm. A standard enthalpy at 298°K is designated by

$$\Delta H^0_{298}$$

Relationship between ΔH and ΔE

At constant pressure,

$$\Delta H = \Delta E + P\,\Delta V \tag{3-74}$$

If the volumes of solid and liquid components can be neglected, and if the same molar volume V is assumed for all gases, then this becomes

$$\Delta H = \Delta E + \Delta nPV = \Delta E + \Delta nRT \tag{3-75}$$

assuming ideal gas behavior. The quantity Δn is the net change in moles of gas when one unit of reaction goes. For Eq. 3-71, $\Delta n = -3$. If this same reaction were run in such a manner as to produce water *vapor*, then Δn would be $+1$ instead. For Eq. 3-72, $\Delta n = -\frac{3}{2}$. If the enthalpy of this reaction at 298°K is -781.0 kcal/mole of benzene, then the energy of reaction under the same conditions is

$$\Delta E = -781.0 + \tfrac{3}{2}(1.987 \times 10^{-3})(298) \text{ kcal/mole}$$

$$\Delta E = -781.0 + 0.9 = -780.1 \text{ kcal/mole}$$

Of the total drop in enthalpy, 780 kcal came from internal energy drop and 1 kcal from the work done on the system by its surroundings when the volume fell by $\frac{3}{2}$ mole's worth per unit of reaction.

Additivity of Heats of Reaction

Because enthalpies are state functions, heats of reaction are additive in the same way that the reactions themselves are. Take the following example of heats of combustion of methane, hydrogen, and graphite.

$$\text{(a)}\ CH_{4(g)} + 2O_{2(g)} \rightarrow CO_{2(g)} + 2H_2O_{(l)} \qquad \Delta H^0_{298} = -212.8 \text{ kcal} \quad (3\text{-}76)$$

$$\text{(b)}\ H_{2(g)} + \tfrac{1}{2}O_{2(g)} \rightarrow H_2O_{(l)} \qquad \Delta H^0_{298} = -68.3 \text{ kcal} \quad (3\text{-}77)$$

$$\text{(c)}\ C_{\text{graphite}} + O_{2(g)} \rightarrow CO_{2(g)} \qquad \Delta H^0_{298} = -94.1 \text{ kcal} \quad (3\text{-}78)$$

From these three reactions, we can calculate the heat of formation of methane from its elements.

$$2(b) + (c) - (a)\text{: } 2H_{2(g)} + \cancel{O}_{2(g)} + C_{(gr)} + \cancel{O}_{2(g)} + \cancel{CO}_{2(g)} + 2\cancel{H_2O}_{(l)} \rightarrow \cancel{CO}_{2(g)} + 2\cancel{H_2O}_{(l)} + CH_{4(g)} + \cancel{2O}_{2(g)}$$

Combining the ΔH's in the same way,

$$\Delta H^0_{298} = 2(-68.3) + (-94.1) - (-212.8)$$

Simplifying,

$$2H_{2(g)} + C_{(gr)} \rightarrow CH_{4(g)} \qquad \Delta H^0_{298} = -17.9 \text{ kcal} \qquad (3\text{-}79)$$

Clearly, it is not necessary to measure and tabulate heat of reaction data for every chemical reaction. By obtaining data on a few key reactions, we can combine them appropriately to obtain results for many others.

The treatment of enthalpies just used to obtain Eq. 3-79 is often known as Hess's law of heat summation: when reaction equations are combined to give a desired over-all reaction, the heats of reaction can be combined in the same way to produce the over-all heat of reaction. But this is only an unnecessary naming of an obvious consequence of the first law. A cyclic representation of the processes used to obtain the first over-all reaction is given in Figure 3-19a. The first law tells us that it makes no difference which path around the cycle is used to make methane from graphite and hydrogen gas; the heat effect remains the same. Another method of diagramming this reaction is shown in Figure 3-19b, where the enthalpies of reactants, products, and

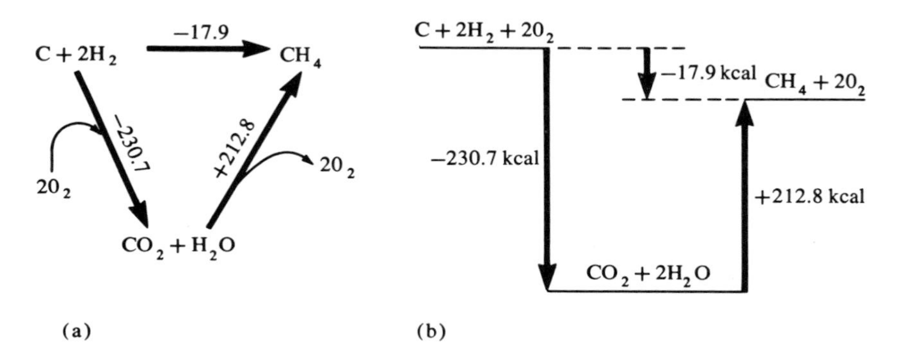

FIGURE 3-19. *Two ways of representing the application of the first law (or Hess's law of heat summation) to chemical reactions. (a) a cycle with equivalent paths. (b) an energy level diagram.*

intermediates are plotted on an energy scale. The fact that a system state can be represented by a level on an energy diagram at all is a reflection of the dependence of the energy of the system only upon its current state and not its history.

3-14 HEATS OF FORMATION ΔH_f^0

From Hess's law, we saw that ΔH data from selected key reactions can be combined in order to obtain ΔH data for any other chemical reaction. The key reactions which are tabulated for this purpose are those which give the heat of formation of a compound from its elements in their standard states. Equations 3-77–3-79 are examples of such formation reactions. Such heats of formation are often designated by a subscript f, ΔH_f^0.

The most easily measurable heats are often heats of combustion. But these can then be combined with the standard heats of formation of the products to yield the standard heats of formation of reactants as defined above.

"Latent Heat"—Enthalpy of Phase Change

For historical reasons, the heat change that accompanies the phase transition of a substance is called the latent heat. As an example, the ΔH for converting diamond into graphite or the latent heat of the phase transition can be calculated in the following manner from the respective heats of combustion.

(a)	$C_{(gr)} + O_{2(g)} \rightarrow CO_{2(g)}$	$\Delta H = -94.052$ kcal
(b)	$C_{(dia)} + O_{2(g)} \rightarrow CO_{2(g)}$	$\Delta H = -94.505$ kcal
(b)–(a)	$C_{(dia)} \rightarrow C_{(gr)}$	$\Delta H = -0.453$ kcal

Since graphite is the standard state for carbon, $\Delta H^0_{298} = +0.453$ kcal for diamond and $\Delta H^0_{298} = 0$ for graphite.

The latent heat of vaporization of H_2O at standard conditions can be obtained in a similar way from standard heats of formation of liquid and gaseous H_2O.

$$\left.\begin{array}{l} H_2O_{(g)}: \Delta H_f^0 = -57.798 \text{ kcal} \\ H_2O_{(l)}: \Delta H_f^0 = -68.317 \text{ kcal} \end{array}\right\} H_2O_{(l)} \rightarrow H_2O_{(g)} \qquad \Delta H^0_{298} = +10.521 \text{ kcal}$$

Of course, this ΔH was calculated for 298°K. At the boiling point of H_2O, the heat of vaporization is smaller; $\Delta H_{373} = +9.7$ kcal. This difference merely illustrates the fact that the heat of vaporization, like all heats of reaction, is a function of temperature.

There are other possible latent heats, such as the heat of fusion for the change between solid and liquid states, and heat of sublimation for the heat change on going from solid to gas phase. For example,

Heat of fusion

$$H_2O_{(s)} \rightarrow H_2O_{(l)} \qquad \Delta H_{273} = 1.44 \text{ kcal/mole}$$

Heat of sublimation

$$\Delta H_{\text{sub}} = \Delta H_{\text{fus}} + \Delta H_{\text{vap}} \qquad \text{all measured at } \textit{the same temperature}$$

Two of the most extensive compilations of heats of formation and other thermodynamic functions are the National Bureau of Standards tables edited by Frederick D. Rossini, entitled "Selected Values of Chemical Thermodynamic Properties" (1952) and "Selected Values of the Properties of Hydrocarbons." More convenient tables, drawn largely from these sources, are to be found in Appendix 4, in the Chemical Rubber Company *Handbook of Chemistry and Physics* and Lang's *Handbook of Chemistry.*

Generalized Notation for a Chemical Reaction

A general chemical reaction can be written in the form

$$n_a A + n_b B \rightarrow n_c C + n_d D \tag{3-80}$$

The heat of this reaction expressed in terms of standard heats of formation is

$$\Delta H^0 = n_c \Delta H^0_{f(C)} + n_d \Delta H^0_{f(D)} - n_a \Delta H^0_{f(A)} - n_b \Delta H^0_{f(B)} \tag{3-81}$$

This expression can be condensed somewhat by combining products and reactants into two terms:

$$\Delta H_0 = \sum_{\text{products}} n_P \Delta H^0_{f(P)} - \sum_{\text{reactants}} n_R \Delta H^0_{f(R)} \tag{3-82}$$

A still more compact notation for a chemical reaction is the following:

$$0 = n_c C + n_d D - n_a A - n_b B \tag{3-83}$$

In this notation the products have positive coefficients and the reactants, which are being used up, have negative coefficients. Carrying the process one step further, we can obtain a generalized expression for a chemical reaction in which chemical species A_j reacts in relative molar amount ν_j.

$$0 = \sum_j \nu_j A_j \qquad \begin{aligned} \nu_j &= + \text{ for products} \\ &= - \text{ for reactants} \end{aligned} \tag{3-84}$$

The coefficient ν_j is known as the *stoichiometry coefficient* or *stoichiometry number* of component A_j. The stoichiometry numbers of the components of Eq. 3-72 are

$$\nu_{CO_2} = 6 \qquad \nu_{H_2O} = 3 \qquad \nu_{O_2} = -\tfrac{15}{2} \qquad \nu_{C_6H_6} = -1$$

The stoichiometry numbers for Eq. 3-73 are twice these values.

This notation also gives a more general expression for the heat of a reaction

$$\Delta H^0 = \sum \nu_j \Delta H^0_{f(j)} \tag{3-85}$$

If this equation is understood, it is virtually impossible to go astray when calculating heats of reaction.

3-15 HEATS OF SOLUTION

The heat of solution, analogous to the latent heat of phase change, is very important in chemistry because most reactions occur in solutions.

Integral Heat of Solution

The integral heat of solution is that heat change encountered when a specified amount of solute is added to a specified amount of solvent under conditions of constant temperature and pressure. A typical example of heat of solution is the mixing of HCl and H_2O.

$$HCl_{(g)} + n \text{ aq} \rightarrow HCl \cdot n \text{ aq}$$

where aq represents the solvent, in this case H_2O, and n represents a specified number of moles of solvent. The amount of heat given off by this reaction depends upon the magnitude of n, and a plot of heat versus n for this reaction is given in Figure 3-20. In each case, the heat given off per finite amount of solvent added is an integral heat of solution for those conditions. It is the total heat given off in carrying out that particular process. In many cases, the integral heat of solution for a very dilute solution is particularly useful. This ΔH is called the heat of solution at infinite dilution and represents the upper limit of the heat released when a substance is dissolved in a large excess of solvent.

Integral Heat of Dilution

From the data contained in Figure 3-20 it is possible not only to calculate the integral heat of solution, but also to calculate the heat released on going from one solution concentration to another. The heat change that accompanies a change in solute concentration is called the integral heat of dilution. For example, a typical dilution might be the addition of more water to an aqueous HCl solution:

(a) $HCl_{(g)} + 200 \text{ aq} \rightarrow HCl \cdot 200 \text{ aq} \qquad \Delta H_{298} = -17.735 \text{ kcal}$

(b) $HCl_{(g)} + 10 \text{ aq} \rightarrow HCl \cdot 10 \text{ aq} \qquad \Delta H_{298} = -16.608 \text{ kcal}$

(a)–(b) $HCl \cdot 10 \text{ aq} + 190 \text{ aq} \rightarrow HCl \cdot 200 \text{ aq} \qquad \Delta H_{298} = -1.127 \text{ kcal}$

The integral heat of dilution for this process is thus -1.127 kcal.

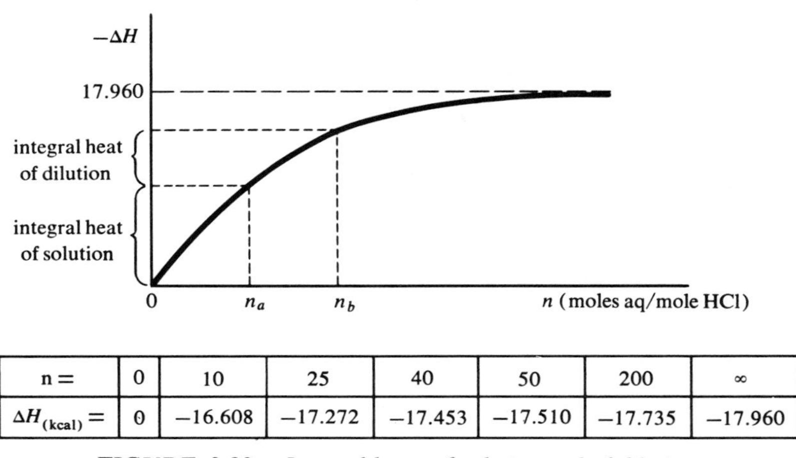

n =	0	10	25	40	50	200	∞
$\Delta H_{(\text{kcal})}$ =	0	−16.608	−17.272	−17.453	−17.510	−17.735	−17.960

FIGURE 3-20. *Integral heats of solution and of dilution.*

Differential Heat of Solution

A plot of enthalpy versus amount of HCl dissolved in a fixed amount of solvent is shown in Figure 3-21, in which H^0 is the enthalpy of pure solvent. Note that the addition of more and more HCl produces successively smaller heat effects. The integral heat of solution at any point a is $(\Delta H)_a/n_a$, which is simply the slope of the line from the origin to point a.

In contrast, the differential heat of solution at point a is defined as the slope of the experimental curve at that point in a plot of this type. It is the rate of change of the heat of solution with respect to moles of solute, moles of

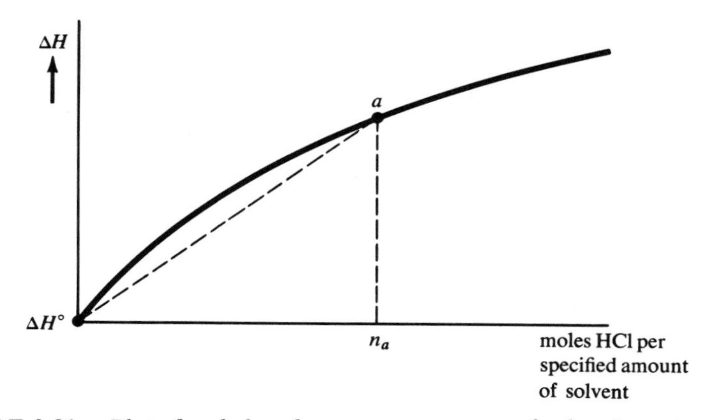

FIGURE 3-21. *Plot of enthalpy change versus amount of solute in a given amount of solvent, to illustrate integral and differential heats of solution.*

solvent being constant. Put another way, the differential heat of solution of HCl is the heat effect produced when 1 mole of HCl is added to such a large amount of solution that the concentration of HCl is essentially unchanged. In mathematical terms, the differential heat of solution is

$$\left(\frac{\partial \Delta H}{\partial n_{\text{solute}}}\right)_{n_{\text{solvent}}}$$

We see later that it is also known as the partial molar heat of solution, and in this context is written as

$$\Delta H_1 = \left(\frac{\partial \Delta H}{\partial n_1}\right)_{n_2} \qquad \begin{array}{l} n_1 = \text{solute} \\ n_2 = \text{solvent} \end{array}$$

3-16 TEMPERATURE DEPENDENCE OF ΔH

Standard heat of formation data limit one to working at a temperature of 298°K; since not all reactions are run at this temperature by any means, it is necessary to know how to go from ΔH^0_{298} to ΔH^0 for any other temperature.

The heat of reaction varies with temperature, but not because of any temperature effect on the energy involved in making and breaking of bonds. Rather, the variation arises from the difference in heat capacity behavior of the reactants and products with temperature.

As an example, let us look at the heat capacity behavior of the reaction for the formation of water vapor from hydrogen and oxygen gas:

$$2H_{2(g)} + O_{2(g)} \rightarrow 2H_2O_{(g)}$$

Assuming ideal gas behavior and the full excitation of vibrational energy states, molecular theory leads to the heat capacity values of Table 3-4. Note that the error, in the last line of the table, can be ascribed almost entirely to the relatively small excitation of vibrational modes. As a first approximation to the truth, then, let us take the heat capacities as constant within our temperature range, and $\frac{7}{2}R$ for hydrogen and oxygen and $\frac{8}{2}R$ for water vapor.

The standard heat of reaction ΔH^0_{298} is -115.60 kcal per unit of reaction. What is the heat of reaction at some other temperature T? The over-all reaction

$$[2H_{2(g)} + O_{2(g)}]_T \rightarrow [2H_2O_{(g)}]_T$$

can just as well be written as a three-step process involving cooling of reactants, reaction at 298°K, and reheating of products.

TABLE 3-4 STATISTICAL MECHANICAL CALCULATION OF HEAT CAPACITIES

Component	$H_{2(g)}$	$O_{2(g)}$	$H_2O_{(g)}$
Degrees of freedom	6	6	9
Translational	3	3	3
Rotational	2	2	3
Vibrational	1	1	3
Energy contributions			
Translational	$\frac{3}{2}RT$	$\frac{3}{2}RT$	$\frac{3}{2}RT$
Rotational	RT	RT	$\frac{3}{2}RT$
Vibrational	RT	RT	$3RT$
Total energy	$\frac{7}{2}RT$	$\frac{7}{2}RT$	$\frac{12}{2}RT$
Total enthalpy	$\frac{9}{2}RT$	$\frac{9}{2}RT$	$\frac{14}{2}RT$
Heat capacity C_P	$\frac{9}{2}R$	$\frac{9}{2}R$	$\frac{14}{2}R$
C_P in cal/mole	8.94	8.94	13.90
Experimental C_P at 398°K	6.86	7.00	8.70
Error in theory	2.08	1.94	5.20

$$[2H_{2(g)} + O_{2(g)}]_T \rightarrow [2H_{2(g)} + O_{2(g)}]_{298^\circ}$$
$$[2H_{2(g)} + O_{2(g)}]_{298^\circ} \rightarrow [2H_2O_{(g)}]_{298^\circ}$$
$$[2H_2O_{(g)}]_{298^\circ} \rightarrow [2H_2O_{(g)}]_T$$

By the first law or by Hess' law, as you prefer, the enthalpy change in the two processes is the same. These processes are diagrammed in Figure 3-22.

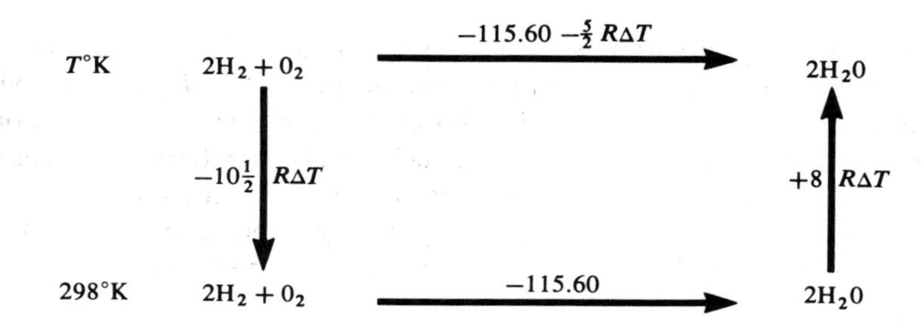

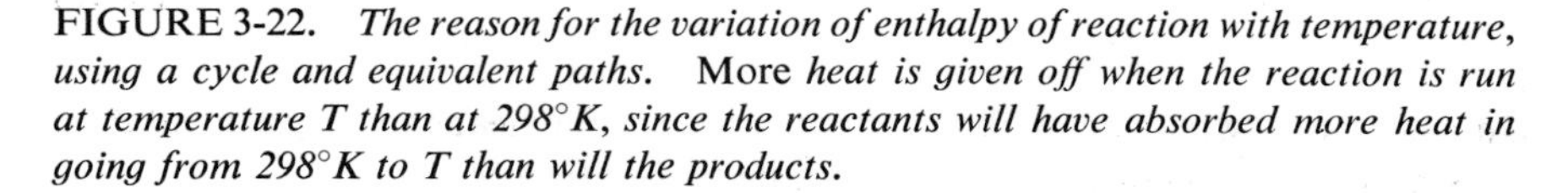

FIGURE 3-22. *The reason for the variation of enthalpy of reaction with temperature, using a cycle and equivalent paths.* More *heat is given off when the reaction is run at temperature T than at 298°K, since the reactants will have absorbed more heat in going from 298°K to T than will the products.*

The enthalpy change in cooling the reactants is

$$\Delta H = 2 \int_T^{298} C_{P(H_2)}\, dT + \int_T^{298} C_{P(O_2)}\, dT$$

$$\Delta H_R = -3 \times \tfrac{7}{2} R\, \Delta T$$

where $\Delta T = T - 298°$. Less heat is required to heat the products from 298° to T

$$\Delta H_P = 2 \int_{298}^{T} C_{P(H_2O)}\, dT = 8R\, \Delta T$$

because two moles of H_2O have fewer modes of energy storage than two moles of H_2 and one of O_2, if only translation and rotation are considered. (Note how the situation is reversed if vibration is fully activated.)

The heat of reaction at temperature T is then

$$\Delta H_T = \Delta H_{298} - \tfrac{5}{2} R(T - 298°)$$

General Case

For a constant pressure process, enthalpy is related to the heat capacity by

$$dH = C_P\, dT$$

Upon integration, this yields

$$\Delta H = H_{T_2} - H_{T_1} = \int_{T_1}^{T_2} C_{P(T)}\, dT \tag{3-86}$$

where the heat capacity itself will generally be a function of temperature. In this equation, we have introduced the absolute enthalpies H_{T_2} and H_{T_1} when in fact thermodynamics only deals with enthalpy changes. At this point, however, it is permissible to talk about such absolute enthalpies, because in the end they all drop out of the equations and only ΔH terms remain.

Continuing with Eq. 3-86, the enthalpy at any given temperature can be expressed as the enthalpy at 298°K plus the change with temperature.

$$H_T = H_{298} + \int_{298}^{T} C_P\, dT \tag{3-87}$$

In most cases, where an analytical expression for the heat capacity is available, it is more convenient to use an indefinite integration:

$$H_T = H_0 + \int C_P \, dT \tag{3-88}$$

where H_0 is only an integration constant. In either case, once we know $C_{P(T)}$ and the enthalpy at any one temperature within the range of validity of the $C_{P(T)}$ data, we can find the enthalpy at any *other* temperature within this range.

In Section 3-14, we saw that a general chemical reaction could be written as

$$0 = \sum_j \nu_j A_j \tag{3-84}$$

The heat of reaction for this process at temperature T is then

$$\Delta H_T = \sum \nu_j H_{j(T)} \tag{3-89}$$

(Why is Eq. 3-89 equivalent to Eq. 3-85 in practice? What happens to the absolute enthalpies of the elements in their standard states?) Bringing in the heat capacity expression, Eq. 3-89 can also be written

$$\Delta H_T = \Delta H_{298} + \int_{298}^{T} \Delta C_P \, dT, \tag{3-90}$$

where

$$\Delta C_P = \sum_j \nu_j C_P$$

The most common tabulation of heat capacities as functions of temperature is that which approximates C_P within a given temperature range by means of a three-term power series[3]

$$C_P = a + bT + cT^2 \tag{3-91}$$

Coefficients (a, b, and c) for this expression for several common gases are given in Table 3-5, and more extensive tables are to be found in the references listed. Note the general correlation of a with molecular weight and with the number of atoms in the molecule. (It is actually C_P itself which rises with molecular weight as the translational contribution increases.) Note as well the increase of b with the number of vibrational modes in the series: H_2O, NH_3, and CH_4, and hence the more rapid rise in C_P with temperature as

[3] Some tables use $C_P = a' + b'T + c'/T^2$ as the data-fitting series.

TABLE 3-5 HEAT CAPACITY COEFFICIENTS

$C_P = a + bT^1 + cT^2$ (cal/mole)
valid between 300 and 1500°K[a]

1. *Monatomic gases with no appreciable electronic excitation*: $C_P = 4.97$
2. *Other gases*:

Gas	a	b ($\times 10^{-3}$)	c ($\times 10^{-7}$)
H_2	6.947	−0.200	4.81
N_2	6.524	1.250	−0.01
O_2	6.148	3.102	−9.23
Cl_2	7.576	2.424	−9.65
Br_2	8.423	0.974	−3.56
CH_4	3.381	18.044	−43.00
NH_3	6.189	7.887	−7.28
H_2O	7.256	2.298	2.83
H_2S	6.385	5.704	−12.10
CO	6.420	1.665	−1.96
CO_2	6.214	10.396	−35.45
SO_2	6.147	13.844	−91.03
HCl	6.732	0.435	3.70
HBr	6.578	0.955	1.58

[a] H. M. Spencer, *J. Am. Chem. Soc.* **67**, 1859 (1945); *Ind. Eng. Chem.* **40**, 2152 (1948); K. K. Kelley, *U. S. Bur. Mines Bull.* **476** (1949); **584** (1960).

vibrational modes become activated. The heat capacity change of the system per unit of reaction is then

$$\Delta C_P = \Delta a + \Delta bT + \Delta cT^2 \tag{3-92}$$

where

$$\Delta a = \sum_j \nu_j a_j \qquad \Delta b = \sum_j \nu_j b_j \qquad \Delta c = \sum_j \nu_j c_j$$

EXAMPLE 1. As an example, consider the following reaction:

$$CO_{(g)} + \tfrac{1}{2}O_{2(g)} \rightarrow CO_{2(g)} \qquad \Delta H_T = ?$$

From heat of formation tables, $\Delta H^0_{298} = -26.416$ kcal/mole for $CO_{(g)}$ and -94.052 kcal for $CO_{2(g)}$. Then

$$\begin{aligned}\Delta H_{298} = \sum \nu_j H_{j(298)} &= (-94{,}052) - (-26{,}416) - (0)\\ &= -67{,}636 \text{ cal/mole}\end{aligned}$$

Evaluating Δa, Δb, and Δc in the same manner,

$$\Delta a = \sum v_j a_j = 6.214 - 6.420 - \tfrac{1}{2}(6.148)$$
$$= -3.280 \text{ cal/deg mole}$$
$$\Delta b = \sum_j v_j b_j = [10.396 - 1.665 - \tfrac{1}{2}(3.102)] \times 10^{-3}$$
$$= +7.180 \times 10^{-3} \text{ cal/deg}^2 \text{ mole}$$
$$\Delta c = \sum_j v_j c_j = [(-35.45) - (-1.96) - \tfrac{1}{2}(-9.23)] \times 10^{-7}$$
$$= -28.87 \times 10^{-7} \text{ cal/deg}^3 \text{ mole}$$

Substituting this data in Eq. 3-90 gives

$$\Delta H_T = -67{,}636 + \int_{298}^{T} (-3.280 + 7.18 \times 10^{-3}T - 28.87 \times 10^{-7}T^2)\, dT$$

Integrating

$$\Delta H_T = -67{,}636 + [-3.280T + 3.59 \times 10^{-3}T^2 - 9.62 \times 10^{-7}T^3]_{298}^{T}$$

Since the lower limit of integration is a known quantity, we can evaluate this part of the expression and then combine it with the first constant:

$$\Delta H_T = -67{,}636 + 684.1 + (-3.280T + 3.59 \times 10^{-3}T^2 - 9.62 \times 10^{-7}T^3)$$
$$\Delta H_T = -66{,}952 - 3.280T + 3.59 \times 10^{-3}T^2 - 9.62 \times 10^{-7}T^3$$

The $-66{,}952$ kcal in the last equation is ΔH_0, an integration constant as in Eq. 3-88. If the C_P power series approximation held all the way to 0°K (which it does not), ΔH_0 would be the enthalpy of reaction at absolute zero. In practice it is simply a constant to be evaluated from a knowledge of the enthalpy of reaction at some one temperature such as 298°K.

With this last equation one can calculate ΔH for the reaction at any temperature within the range of validity of the C_P data. For example, at $T = 308$°K, $\Delta H_{308} = -67{,}650$ cal. Here 14 cal of additional heat are given off when the temperature is raised 10°. If the object of this reaction were the production of heat, it would be advantageous to carry it out at a high temperature.

In general, we can write

$$\Delta H_T = \Delta H_0 + \int (\Delta a + \Delta b T + \Delta c T^2)\, dT \tag{3-93}$$

This can be integrated directly to

$$\Delta H_T = \Delta H_0 + \Delta a T + \frac{\Delta b}{2} T^2 + \frac{\Delta c}{3} T^3 \tag{3-94}$$

and then ΔH_0 evaluated from a knowledge of ΔH_{298}.

3-17 MOLECULAR PICTURE OF E AND H

We know that a polyatomic molecule of N atoms has $3N$ degrees of freedom, of which three are translational, three are rotational (two if the molecule is linear), and $3N - 6$ are vibrational ($3N - 5$ for a linear molecule). The translational energy is $\frac{3}{2}RT$, the rotational energy is $\frac{3}{2}RT$ (or RT for a linear molecule), and the vibrational energy is the sum of the individual vibrational contributions of all different modes of vibration:

$$E_{\text{vib}} = \sum_{j=1}^{3N-6} RT\left(\frac{x_j}{e^{x_j}-1}\right) \qquad x_j = \frac{h\nu_j}{kT}$$

As an example, the nonlinear molecule NO_2 at 298°K yields the following energy values:

$$E_{\text{trans}} = \tfrac{3}{2}RT = 888 \text{ cal/mole}$$

$$E_{\text{rot}} = \tfrac{3}{2}RT = 888 \text{ cal/mole}$$

E_{vib} = sum of contributions from three fundamental vibration frequencies

$\omega_e = \nu_0/c$	750 cm^{-1}	1323 cm^{-1}	1616 cm^{-1}
x	3.63	6.40	7.80
$RT\left(\frac{x}{e^x - 1}\right)$	60 cal	7 cal	2 cal

$$E_{\text{vib}} = 60 + 7 + 2 = 69 \text{ cal}$$

The theoretical value for fully excited vibrations would have been $3RT$ or 1776 cal.

Note that for E_{vib} the wider the spacing between energy levels, the less completely the vibration is activated and the smaller its contribution to the total vibrational energy.

Summing over the separate contributions to the energy,

$$E - E_0 = E_{trans} + E_{rot} + E_{vib} = 888 + 888 + 69 = 1845 \text{ cal/mole}$$

Or

$$E = E_0 + 1845 \text{ cal/mole}$$

where E_0 is the energy at 0°K, and $E - E_0$ is the "thermal energy."

For the enthalpy of an ideal gas, we already have the following relationships:

$$H = E + PV \cong E + RT$$

For the NO_2 molecule at 298°K

$$H = E_0 + 2437 \text{ cal/mole}$$

and

$$H - E_0 = 2437 \text{ cal/mole} = \text{"thermal enthalpy"}$$

Inherent in the definitions of thermal enthalpy and thermal energy is the fact that for an ideal gas the difference between enthalpy and energy vanishes as one approaches absolute zero.

3-18 BOND ENERGIES

Enthalpies of reaction reflect two things: the energy needed to make and break bonds and the ability of a substance to soak up heat as it goes from absolute zero to the temperature at which the reaction is conducted. In theory, if we conducted all reactions at 0°K, we would eliminate the thermal energy arising from differences in heat capacities among reactants and products. Under these conditions, the energy change in a reaction would be solely that of making and breaking bonds.

Therefore, at least in principle, when constructing a table of bond energies we should correct all of our heat of reaction data to 0°K. In practice, this extrapolation is not necessary. Bond energy tables are only approximate to begin with. The assumption of the existence of bonds between pairs of atoms, uninfluenced by the rest of the molecule, is not all that good. Moreover, the typical ΔH of a reaction is large in comparison with the correction necessary because of the difference in heat capacities of reactants and products between 0°K and 298°K, and it is reasonably satisfactory to construct tables of bond energies at 298°K and forget about these differences.

Bonds between Pairs of Atoms

What prompts us to say that there are bonds between pairs of atoms? In molecular orbital theory one can avoid such an assumption and can describe a molecule in terms of a series of full-molecule orbitals which extend over the entire collection of atoms in the molecule. An example of full-molecule orbitals for water is given in Figure 3-23. In this case, all of the molecular orbitals extend over the entire molecule, and one cannot say that any two atoms form a bond. Though this approach is perhaps the most elegant way of representing the state of things, it is unnecessarily complicated for most practical chemical applications. The molecular orbital method itself shows that localized orbitals provide a representation that is almost as good as that given by a full orbital treatment. It is reasonable to talk about individual bonds between atoms even though the full molecular wave-function picture may be more accurate.

The validity of the localized orbital treatment was historically demonstrated when it was found that the energy needed to break a particular kind of bond was more or less constant regardless of the compound in which the bond was found. It seemed as though molecules did behave as though they had individual bonds between atomic pairs and the energy needed to break the separate bonds was, to a first approximation, independent of the structure of the molecule. It is necessary to postscript this statement with the comment

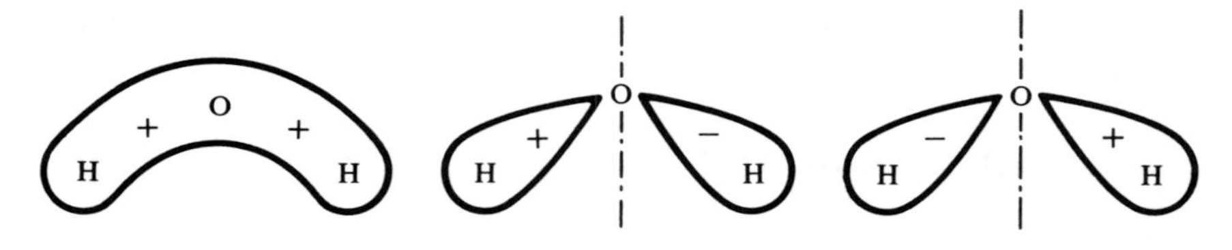

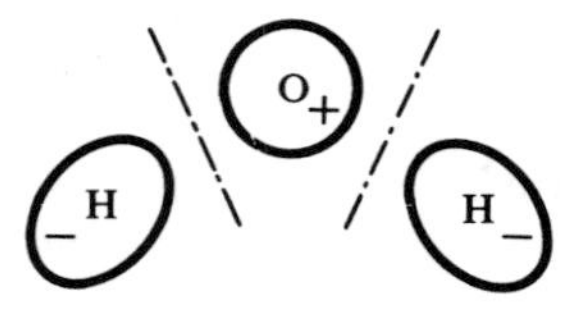

FIGURE 3-23. *Bonding in the water molecule can be described perfectly well in molecular orbital theory by wave functions which extend over the entire molecule. The energy of these full-molecular orbitals rises with the number of nodes in the wave-function. The alternate description of bonding in terms of two individual* O—H *bonds is then a matter only of convenience.*

that while this localized treatment is widely applicable, there are several interesting cases when a strict localized approach gives grossly incorrect values for the bond energies.

The bond energies of the four C—H bonds in methane are not to be assumed to be the energies of successive removal of four protons.

$$CH_4 \rightarrow \cdot CH_3 + H\cdot \qquad \Delta H = +102 \text{ kcal/mole}$$
$$\cdot CH_3 \rightarrow \cdot\dot{C}H_2 + H\cdot \qquad \Delta H = +105 \text{ kcal/mole}$$
$$\cdot\dot{C}H_2 \rightarrow \cdot\underset{\cdot}{\dot{C}}H + H\cdot \qquad \Delta H = +108 \text{ kcal/mole}$$
$$\cdot\underset{\cdot}{\dot{C}}H \rightarrow \cdot\underset{\cdot}{\dot{C}}\cdot + H\cdot \qquad \Delta H = +\ 83 \text{ kcal/mole}$$

Here we see evidence that the energy needed to break one C—H bond after another is not constant; the stability of each intermediate must be taken into consideration. Nevertheless, if we add the energies of this stepwise process and divide by 4, in a formal sense the *average* energy of a C—H bond is 99.5 kcal/mole.

The ΔH of the degradation of methane to atoms as described above applies to the over-all reaction

$$CH_{4(g)} \rightarrow C_{(g)} + 4H_{(g)} \qquad \Delta H = 398 \text{ kcal/mole}$$

Since carbon gas and hydrogen atoms are by no means convenient reference materials, we need to know how much energy is needed to convert the elements in their standard states into gaseous atoms. Once these data are obtained, the thermal data relating to heats of formation of compounds can be used to determine individual bond energies.

For methane, we need to know

$$C_{(gr)} \rightarrow C_{(g)} \qquad \Delta H_{\text{sublimation}} = ?$$
$$\tfrac{1}{2}H_{2(g)} \rightarrow H_{(g)} \qquad \Delta H_{\text{dissociation}} = ?$$

The heats of sublimation of metals and the heats of dissociation of simple diatomic gases are easily obtained by spectroscopic and thermal methods. The heat of sublimation of graphite is considerably harder to measure. In the 1930's the accepted ΔH_{subl} for graphite was 123 kcal/mole; today's value is 171 kcal/mole. The presently accepted heats of atomization (per mole of atoms) for hydrogen, carbon, and oxygen measured at 298°K are

H	52.09 kcal
C	171.70 kcal
O	59.16 kcal

These values and others used in this chapter come from L. Pauling.

C—H *Bond Energy*

We can use the ΔH_f^0 for CH_4 along with the ΔH of atomization for graphite and H_2 to obtain the average bond energy (B.E.) of a C—H bond.

$$
\begin{array}{ll}
CH_{4(g)} \rightarrow C_{(gr)} + 2H_{2(g)} & \Delta H = -\Delta H_f^0 = +\ 17.89 \\
C_{(gr)} \rightarrow C_{(g)} & \Delta H = +171.70 \\
2H_{2(g)} \rightarrow 4H_{(g)} & \Delta H = +208.36 \\
\hline
CH_{4(g)} \rightarrow C_{(g)} + 4H_{(g)} (4\ \text{C—H bonds}) & \Delta H = 397.95\ \text{kcal}
\end{array}
$$

Assuming that the C—H bond energy is one-fourth of the heat of dissociation of methane into its atomic components, we arrive at a value of 99.49 kcal/per C—H bond. The bond energies listed in standard tables have been adjusted to give values that apply to the largest possible number of compounds, and the adjusted value for the C—H bond energy is 98.8 kcal. Already we have a clue as to how approximate these bond energies are going to be, the methane model gives a C—H bond energy that differs by 700 cal from the "accepted" value.

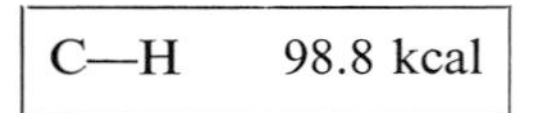

C—C *Bond Energy*

The energy of a C—C bond can be found by applying the method of the previous section to ethane.

$$
\begin{array}{ll}
H_3C\text{—}CH_3 \rightarrow 2C_{(gr)} + 3H_{2(g)} & \Delta H = -\Delta H_f = \ +20.24 \\
2C_{(gr)} \rightarrow 2C_{(g)} & \Delta H = +343.40 \\
3H_{2(g)} \rightarrow 6H_{(g)} & \Delta H = +312.54 \\
\hline
H_3C\text{—}CH_3 \rightarrow 2C_{(g)} + 6H_{(g)} & \Delta H = +\ 676.18\ \text{kcal} \\
\text{less 6 C—H bonds } (98.8 \times 6): & \Delta H = -592.8 \\
\hline
\text{1 C—C bond:} & \Delta H = \ +83.38\ \text{kcal}
\end{array}
$$

The adjusted value for a C—C bond is slightly lower than that calculated from ethane.

C—C	83.1 kcal

Propane

Having calculated C—H and C—C bond energies, let us see how our values hold up when compared with experimental data for propane. The heat of atomization of propane is easily obtained from data already at hand.

```
    H  H  H
    |  |  |
H–C–C–C–H  → 2 C–C + 8 C–H
    |  |  |
    H  H  H
```

$$\Delta H_{\text{atom}} = 2(\text{B.E.}_{\text{C-C}}) + 8(\text{B.E.}_{\text{C-H}})$$
$$\Delta H_{\text{atom}} = (2 \times 83.1) + (8 \times 98.8) = 956.6 \text{ kcal/mole}$$

The reverse reaction, for formation of 1 mole of propane from its atoms, would proceed with the emission of 956.6 kcal of heat.

$$3C_{(g)} + 8H_{(g)} \rightarrow C_3H_{8(g)} \qquad \Delta H = -956.6 \text{ kcal/mole} \qquad (3\text{-}95)$$

The energy needed to convert 3 moles of carbon (graphite) to vapor and 4 moles of H_2 gas to atomic hydrogen is easily calculated.

$$\begin{array}{ll} 3C_{(gr)} \rightarrow 3C_{(g)} & \Delta H_{\text{atom}} = 3 \times 171.70 = 515.1 \text{ kcal} \\ 4H_{2(g)} \rightarrow 8H_{(g)} & \Delta H_{\text{atom}} = 8 \times 52.09 = 416.8 \\ \hline 3C_{(gr)} + 4H_{2(g)} \rightarrow 3C_{(g)} + 8H_{(g)} & \Delta H = 931.8 \end{array} \qquad (3\text{-}96)$$

The relationship between Eqs. 3-95 and 3-96 is shown in Figure 3-24. Adding the two equations eliminates the atomic intermediates and gives the standard

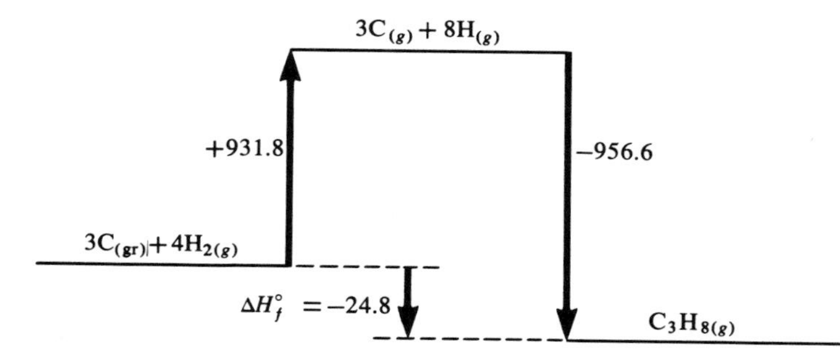

FIGURE 3-24. *The formation of propane from graphite and hydrogen gas, using the hypothetical intermediate step of atomic carbon gas and hydrogen gas for purposes of calculation.*

heat of formation of propane,

$$\Delta H_f^0 = -24.8 \text{ kcal/mole}$$

As is so often the case, the thermodynamically measurable quantity is a small difference between two very large numbers. The amazing thing is that when one compares this calculated value with the observed value, their agreement is so good. The observed ΔH_f^0 of propane is -24.82 kcal. This close agreement between the observed ΔH_f^0 for propane and that calculated from bond energies is somewhat fortuitous, but it is not unique. Heat of reaction data calculated from bond energies often agree within 1 kcal with the observed data.

Butane: Limits of Method

Some of the limitations of calculating heat of reaction data from bond energies become apparent when one compares the calculated and observed results from heats of formation of *n*-butane and *i*-butane. Both compounds have three C—C bonds and ten C—H bonds, and from simple bond theory the heats of formation should be the same for both compounds. A graphical presentation showing the calculated ΔH_f^0 is given in Figure 3-25

```
  H H H H
  | | | |
H-C-C-C-C-H
  | | | |
  H H H H
```

```
  H H H
  | | |
H-C-C-C-H
  | | |
  H | H
    |
  H-C-H
    |
    H
```

Bond energies		Heats of atomization	
3C—C	249.3 kcal	4C	686.80 kcal
10C—H	988.0 kcal	10H	520.90 kcal
	1237.3 kcal		1207.70 kcal

$$\Delta H_f^0 = (\text{Heats of atomization}) - (\text{bond energies}) = -29.6 \text{ kcal/mole}$$

The observed ΔH_f^0 for *n*-butane is -29.81 kcal and is in good agreement with the calculated value. However, the observed ΔH_f^0 for *i*-butane is -31.45 kcal, a difference of almost 2 kcal from the calculated result. Evidently this simple theory works for straight chains but begins to fail where branched chains are concerned.

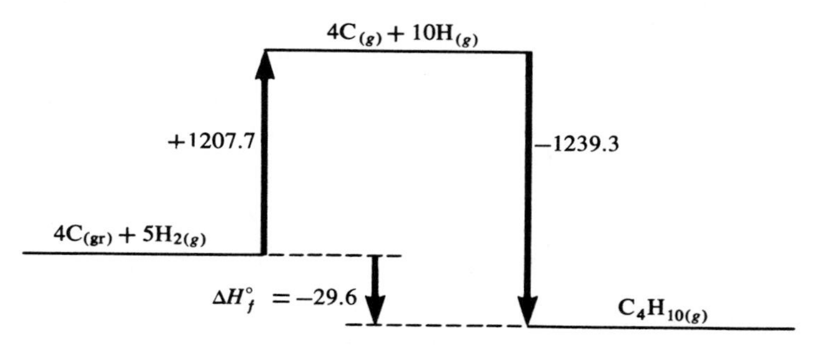

FIGURE 3-25. *A heat of formation diagram for butane. The method of using heats of atomization and bond energies makes no distinction between stereo-isomers such as normal and isobutane.*

C═C *Bond Energy*

The energy of a C═C bond can be found if we add the ΔH_f^0 of ethylene to our previous knowledge of the energy of a C—H bond.

$H_2C{=}CH_2 \rightarrow 2C_{(gr)} + 2H_{2(g)}$	$\Delta H = -\Delta H_f = \ -12.50$
$2C_{(gr)} \rightarrow 2C_{(g)}$	$\Delta H = +343.40$
$2H_{2(g)} \rightarrow 4H_{(g)}$	$\Delta H = +208.36$
$H_2C{=}CH_2 \rightarrow 2C_{(g)} + 4H_{(g)}$	$\Delta H = +539.26$ kcal
less 4C—H bonds (98.8×4)	$\Delta H = -395.2$
1 C═C bond	$\Delta H = +144.06$ kcal

Again, the adjusted value for a C═C is slightly different from that calculated from this particular compound.

C═C	147.0 kcal

Benzene and Resonance

We can calculate the ΔH_f^0 of benzene in the same manner as for the simpler hydrocarbons. If we base our calculations on the assumed structure of benzene shown below, our calculations are as follows:

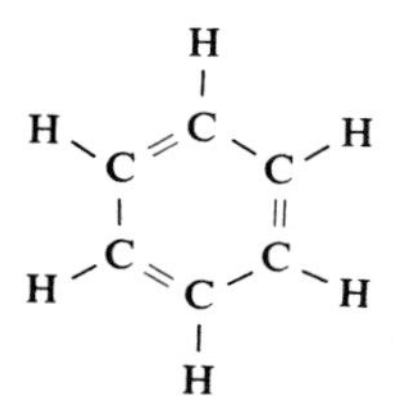

Bond energies		Heats of atomization	
6C—H	592.8	6C	1030.2
3C—C	249.3	6H	312.5
3C=C	441.0		
	1283.1 kcal		1342.7 kcal

$$\Delta H_f^0 = (\text{heats of atomization}) - (\text{bond energies}) = +59.6 \text{ kcal/mole}$$

This is plotted in Figure 3-26. But the observed ΔH_f^0 is only $+19.8$ kcal/mole. That is, the benzene molecule is 40 kcal/mole *more stable* than predicted by our simple bond model.

The reason for this unexpected stability rests in the fact that our crude benzene structure is not a true representation of molecular structure. The molecule is known to have sixfold symmetry and not an alternation of single and double bonds. The real state can be described in valence bond language as having the character of both the two Kekulé structures of Figure 3-27 without being describable precisely by either. The two extreme forms are called "resonance" structures, a term which unfortunately suggests a flipping back and forth which has no reality; other resonance structures than the Kekulé structures can be used to build up an even more accurate model of the real molecule. The molecular orbital interpretation would invoke full-molecule orbitals and delocalized electrons, and abandon the idea of bonds between pairs of atoms.

Yet another way of looking at the benzene molecule is to consider the carbon to carbon bonds as being neither pure single nor double bonds. Using this approach, one can quickly calculate the bond energy of these C to C bonds.

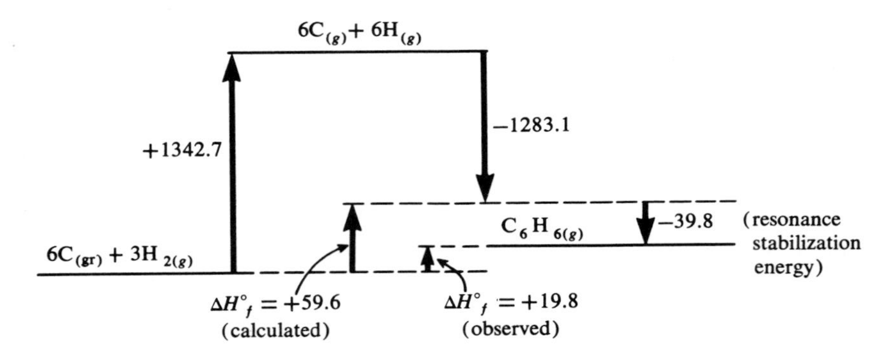

FIGURE 3-26. *The benzene molecule has an experimental heat of formation which is* less *than that predicted from heats of atomization and bond energies by* 40 kcal. *The extra stability of the molecule arises from the delocalization of the π electrons in the ring.*

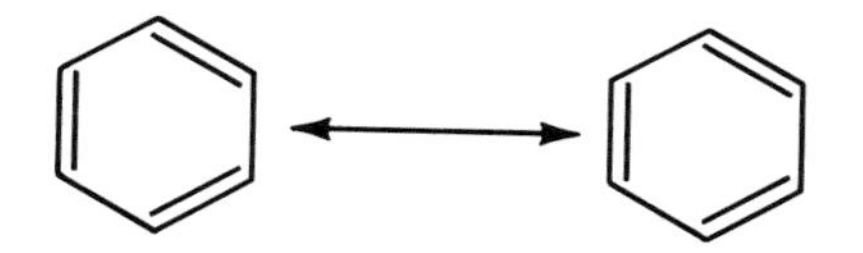

FIGURE 3-27. *The localized bond concept can be retained in benzene only by assuming that the true bonding is an intermediate structure which can be described as a combination of various localized-bond models. The two simplest "Kekulé" models for benzene are shown here.*

$C_6H_6 \rightarrow 6C_{(gr)} + 3H_{2(g)}$	$\Delta H = -\Delta H_f = -19.8$
$6C_{(gr)} \rightarrow 6C_{(g)}$	$\Delta H = +1030.2$
$3H_{2(g)} \rightarrow 6H_{(g)}$	$\Delta H = +312.5$
$C_6H_6 \rightarrow 6C_{(g)} + 6H_{(g)}$	$\Delta H = +1322.9$ kcal
less 6 C—H bonds (98.8 × 6)	$\Delta H = -592.8$
6C to C bonds	$\Delta H = -730.1$ kcal

A value of 730.1 kcal for the six C to C bonds in the ring allows us to assign a value of 121.7 kcal to each bond. Knowing that a pure C—C bond has an energy of 83.1 kcal and a pure C=C bond has an energy of 147.0 kcal lends weight to the idea that the bonds in benzene have a bond order intermediate between 1 and 2. The second bonds between carbons are not localized in three places, but are spread among all six C—C distances. The electrons involved are "delocalized." In general, if a molecule can be described as having delocalized electrons (or resonance forms in valence bond language), then it will usually be more stable than would be predicted by simple bond theory.

C=O, C—O, *and* O—H *Bond Energies*

Since we already know the energies of C—H and C—C bonds, we can obtain the energy of a C=O bond by adding the ΔH_f^0 of acetone to our data. Using $\Delta H_f^0 = -51.79$ kcal/mole for acetone and proceeding as in previous bond energy calculations, we obtain a value of 179.6 kcal/mole for the C=O bond in acetone. The adjusted value for the carbonyl group is 174.0 kcal.

The O—H bond energy is obtained by assuming that the O—H bonds in H_2O are representative examples of this type of bond. Such a treatment allows us to assign a bond energy of 110.6 kcal to the hydroxyl group.

Finally, by taking methanol, in which the only unknown bond energy is that of the C—O bond, we find that the C—O bond energy is of the order of 84.0 kcal.

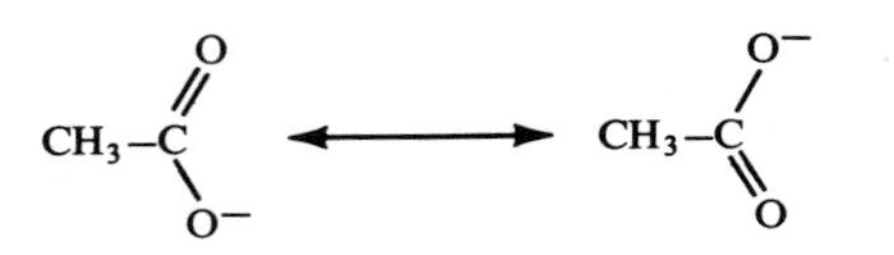

FIGURE 3-28. *As with benzene, the bond structure in the carboxyl group cannot be described accurately with any one localized bond model. The true bonding, with symmetrical C—O bonds, can be described as a combination of these two extremes.*

Carboxyl Group Resonance

A comparison of the calculated and observed heat of formation of acetic acid adds further evidence that the simplest way of writing a chemical structure can often be misleading. For acetic acid,

H
H − C − H ⸗O
O–H
H

Bond energies	
3C—H	296.4
1C—C	83.1
1C=O	174.0
1C—O	84.0
1O—H	110.6
	748.1 kcal

Heats of atomization	
2C	343.40
4H	208.36
2O	118.32
	670.08 kcal

$$\Delta H_f^0 = (\text{heats of atomization}) - (\text{bond energies}) = -78.0 \text{ kcal/mole}$$

However, the observed ΔH_f^0 for acetic acid is -104.72 kcal/mole, a difference of 27 kcal from the calculated value. Here again is a molecule that is more stable than would be predicted from a single valence bond structure. The observed ΔH_f^0 can be obtained if one considers the acetate anion to be describable by the resonance structures shown in Figure 3-28, in which the bond order of the C—O bonds is intermediate between 1 and 2. In fact, x-ray analysis of crystals of many carboxylic acids shows that the bond lengths of the two C—O bonds in the carboxyl group are indeed the same.

REFERENCES AND FURTHER READING

I. M. Klotz, *Chemical Thermodynamics* (W. A. Benjamin, Inc., New York, 1964). A very clear and readable book, but without the molecular or statistical interpretation. Especially good on standard states, fugacities, and activities.

L. Pauling, *The Nature of the Chemical Bond* (Cornell, Ithaca, 1960), 3rd ed. A good discussion of bond energies and resonance.

G. N. Lewis, M. Randall, K. S. Pitzer, and L. Brewer, *Thermodynamics* (McGraw-Hill, New York, 1961). A thoroughgoing classic which has been shaping the teaching of thermodynamics to chemists since 1923. Reworked and to some extent rewritten by Pitzer and Brewer in 1961.

F. D. Rossini, D. D. Wagman, W. H. Evans, S. Levine, and I. Jaffe, "Selected Values of Chemical Thermodynamic Properties," Nat. Bur. Std. (U.S.) Circ. 500, U.S. Government Printing Office, Washington, D.C., 1952. The standard source. But don't order this "circular" by air mail without checking the postage rates.

F. D. Rossini, "Selected Values of Physical and Thermodynamic Properties of Hydrocarbons," American Petroleum Institute Project 44, Carnegie Institute of Technology, 1953.

EXERCISES

3-1. (a) An aluminum rod is 0.23% longer at 100°C than at 0°C, whereas a rod of copper expands by 0.17% over the same temperature range. If rods of these metals were used for constructing thermometers, and if linear extrapolations were carried back to define "absolute zero" as the point of zero length, what would absolute zero be for each of these temperature scales?

(b) Liquid mercury has a volume 1.8% greater at 100°C than at 0°C. What would absolute zero be on a mercury scale? Note that the three figures are widely different and hence suggest nothing of general significance. Different gases, on the other hand, extrapolate back to the same point.

3-2. (a) At 1 atm pressure the volume of one mole of N_2 gas is 22.401 liters at 0°C and 30.627 liters at 100°C. Using only these data, what value would one calculate for absolute zero?

(b) At 0.1 atm pressure the molar volume of nitrogen is 224.13 liters at 0°C and 306.20 liters at 100°C. From these data what would one calculate for absolute zero?

(c) If the results of parts (a) and (b) are extrapolated to zero pressure, what is the best value for absolute zero?

3-3. (a) The restoring force f which a strip of rubber exerts when stretched is a function of its length L and its temperature T. Assuming that the energy of the rubber strip is similarly a function of T and L, show that the general expression for the heat capacity of the strip is

$$C = \left(\frac{\partial E}{\partial T}\right)_L + \left[\left(\frac{\partial E}{\partial L}\right)_T - f\right]\frac{dL}{dT}$$

(b) Defining a new quantity $h = E - fL$ (the counterpart of enthalpy in one dimension), show also that

$$C = \left(\frac{\partial h}{\partial T}\right)_f + \left[\left(\frac{\partial h}{\partial f}\right)_T + L\right]\frac{df}{dT}$$

(c) What are the heat capacities of constant length C_L and at constant force C_f?
(d) Write the first law of thermodynamics in a form applicable to the stretching.
(e) Why are the signs of the second terms in brackets apparently reversed as compared with the three-dimensional analogs of these equations?

3-4. An "ideal" strip of rubber has an equation of state of the form

$$f = T\phi_{(L)}$$

in which $\phi_{(L)}$ is a function only of the length. The energy E of such an ideal strip is a function only of the temperature T. Under these conditions, prove that

$$C_f - C_L = \frac{f^2}{T(\partial f/\partial L)_T}$$

3-5. Show that the differential equation for a reversible adiabatic extension or retraction of an ideal rubber strip can be written with variables separated inthe form

$$\left(\frac{C_L}{T}\right)dT = \phi_{(L)}\, dL$$

Recognizing that $\phi_{(L)}$ increases monotonically with L, show that a reversible adiabatic extension of a rubber strip is accompanied by heating.

3-6. Derive the following expression from the first law and related definitions:

$$C_V = -\left(\frac{\partial E}{\partial V}\right)_T\left(\frac{\partial V}{\partial T}\right)_E$$

3-7. The following experiment is a variant of the Joule and Joule–Thomson experiment. A tube of uniform cross section is equipped with a porous plug and provided with a frictionless piston on one side and a fixed end on the other. The space between the plug and the fixed end is evacuated, and the space between the plug and the piston contains a large quantity of an ideal gas at temperature T_1 and pressure P. The gas is allowed to seep through the plug while the pressure on the gas is kept constant at P, until the gas pressure on the originally evacuated side reaches P and the piston stops moving. Assuming that the system is insulated so that the process takes place adiabatically, and neglecting any heat conduction through the plug, show that the final temperature of the gas on the left-hand side is

$$T_2 = \gamma T_1$$

where γ is the ratio C_P/C_V.

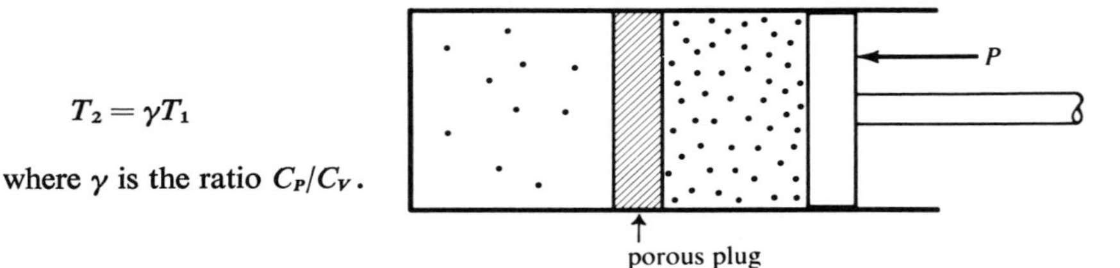

3-8. The heat capacity ratio γ for a gas can be determined by a simple experiment. A carboy is filled with the gas to a pressure of 1.10 atm (the laboratory pressure is 1.00 atm). The stopper of the carboy is then removed suddenly so that the gas expands virtually adiabatically. After a few seconds, the stopper is replaced and the gas is allowed to warm up to room temperature again, at which point the pressure inside the carboy is found to be 1.03 atm. Assuming the gas to be ideal, calculate C_P and C_V.

3-9. In an industrial accident, the neck of a tank of hydrogen gas under 2000 lb./in.^{2} pressure was snapped off, the gas rushed out and exploded. The obvious first explanation for the ignition is a static spark, but it has also been suggested that ignition could have been caused by Joule–Thomson heating upon expansion. Test this hypothesis by a series of approximate calculations.

(a) Calculate the heat which would be evolved (in cal/mole) if the expansion were carried out at a constant temperature of 300°K.

(b) Calculate the rise in temperature which such a quantity of heat would produce in a mole of hydrogen gas which was initially at 300°K.

Is the Joule–Thomson explanation correct? Are any of the approximations in the above calculations sufficiently severe to change your conclusions?

3-10. Substance X, an ideal gas containing N atoms per molecule. The following are observed:

(a) C_P for X and for nitrogen gas are the same at 0°C, where the vibrational contribution to the heat capacity is negligible.

(b) At a temperature high enough for the vibrational contribution to be fully excited, the difference between the C_P of X and of nitrogen is roughly 6 cal/deg mole.

Show what can be concluded from the above information about the number N and any other aspects concerning the structure of the molecule X.

3-11. One mole of an ideal, but not necessarily monatomic, gas is subjected to the following sequence of steps:

(a) It is heated at constant volume from 25 to 100°C.

(b) It is expanded freely into a vacuum in a Joule-type experiment to double its volume.

(c) It is cooled reversibly at constant pressure to 25°C.

Calculate ΔE, ΔH, q and w in calories for the *over-all* process. Note that it is not necessary to know the heat capacity of the gas. Explain why.

3-12. A five mole sample of an ideal gas with $C_V = 5.0$ cal per degree per mole, initially at 298°K and 1 atm pressure, is subjected to the following reversible steps:

(a) It is first heated at constant volume to twice its initial temperature.

(b) It is then expanded adiabatically until it is back to its initial temperature.

(c) It is then compressed isothermally until the pressure is 1 atm again.

Calculate q, w, ΔE, and ΔH for steps (a) and (b), and for the entire process.

3-13. One mole of an ideal monatomic gas is subjected to the following sequence of steps.

(a) The gas is heated reversibly at constant pressure (1 atm) from 25 to 100°C.
(b) Next, the gas is expanded reversibly and isothermally to double its volume.
(c) Finally, the gas is cooled reversibly and adiabatically to 35°C. Calculate ΔE, ΔH, q, and w in calories for the *over-all* process.

3-14. Ten grams of oxygen gas, initially at 25°C and 1 atm pressure, are compressed adiabatically until the temperature reaches 60°C. If it is approximately true that $C_V = \frac{5}{2}R$, and ideal gas behavior is assumed, calculate the amount of work done on the gas. What can be said about the final pressure and volume of the gas?

3-15. 0.35 mole of an ideal monatomic gas is expanded adiabatically from a volume of 1 liter at 400°K to a volume of 5 liters against a constant external pressure of 0.5 atm. What is the final temperature of the gas, and what is its enthalpy change in the process?

3-16. One mole of an ideal gas at 25°C is held in a cylinder by a piston at a pressure of 100 atm. The piston pressure is released suddenly in three stages, with time for equilibration between stages: first to 50 atm, then to 20 atm, and finally to 10 atm. Calculate the work done by the gas during these irreversible isothermal expansions and compare it with the work done in an isothermal reversible expansion from 100 to 10 atm at 25°C.

3-17. The heat of combustion of cyanamide

$$CH_2N_{2(g)} + \tfrac{3}{2}O_{2(g)} \rightarrow CO_{2(g)} + H_2O_{(l)} + N_{2(g)}$$

is $\Delta H_{298} = -177.20$ kcal. Calculate the standard enthalpy of formation of cyanamide.

3-18. The heat of formation of liquid ethanol is -66 kcal/mole, while the heat of combustion to CO_2 and liquid water of the isomeric CH_3—O—$CH_{3(g)}$ is -348 kcal/mole. The heat of formation of liquid water is -68 kcal/mole, and the heat of combustion of carbon to $CO_{2(g)}$ is -94 kcal/mole (all data for 25°C).
(a) Calculate ΔH_{298} for the isomerization reaction

$$C_2H_5OH_{(l)} = CH_3\text{—O—}CH_{3(g)}$$

(b) Calculate ΔE_{298} for the reaction.

3-19. The heats of combustion of cyclopropane, $(CH_2)_3$, graphite, and H_2 are -500.0, -94.0, and -68.0 kcal/mole, respectively, when burned to CO_2 and liquid water. The heat of formation of propylene, CH_3—CH═CH_2, is 4.9 kcal/mole.
(a) Calculate the heat of formation of cyclopropane.
(b) Calculate the heat of isomerization of cyclopropane to propylene.

3-20. The *adiabatic flame temperature* is the highest temperature which can be attained in a combustion process if the heat produced is used to heat the product gases. Calculate this temperature for the combustion of ethane with twice the quantity of air (1/5 oxygen, 4/5 nitrogen) which is required for total combustion to

CO_2 and H_2O. Use the power series expansion for heat capacities, but neglect the cT^2 terms. (*Note:* A standard thermodynamic trick when dealing with state functions is to think of an alternative way of doing a one-step process in two steps and to perform the calculations on these.)

3-21. From the heats of *formation* (from elements in their standard states) of the following solutions

Solution	ΔH_f^0 (kcal)
$H_2SO_4 \cdot 600$ aq	-212.35
$KOH \cdot 200$ aq	-114.82
$KHSO_4 \cdot 800$ aq	-274.3
$K_2SO_4 \cdot 1001$ aq	-336.75

calculate ΔH for the reactions

(a) $H_2SO_4 \cdot 600\text{ aq} + KOH \cdot 200\text{ aq} \rightarrow KHSO_4 \cdot 800\text{ aq} + H_2O_{(l)}$

(b) $KHSO_4 \cdot 800\text{ aq} + KOH \cdot 200\text{ aq} \rightarrow K_2SO_4 \cdot 1001\text{ aq}$

Note the potential trap in the way in which (b) is written, and the implied assumption that one water more or less makes no appreciable difference in the heat of formation of a dilute solution.

3-22. For the process

$$(NH_4)_2C_2O_{4(s)} + H_2O_{(l)} \rightarrow \text{equimolar solution}$$

the enthalpy change is 0.8 kcal as the equation is written. The heat capacities of the components are

ammonium oxalate$_{(s)}$	$C_P = 20$ cal/deg mole
water$_{(l)}$	$C_P = 1$ cal/deg gram
solution of one mole of each	$C_P = 40$ cal/deg

If 1 mole of ammonium oxalate and 1 mole of water, both at 25°C, are mixed in an insulated vessel, what will be the final temperature of the solution?

3-23. From the heats of solution

Reaction	ΔH (kcal)
$HCl_{(g)} + 100\text{ aq} \rightarrow HCl \cdot 100\text{ aq}$	-17.650
$NaOH_{(s)} + 100\text{ aq} \rightarrow NaOH \cdot 100\text{ aq}$	-10.12
$NaCl_{(s)} + 200\text{ aq} \rightarrow NaCl \cdot 200\text{ aq}$	1.016

and the standard heats of formation of $HCl_{(g)}$, $NaOH_{(s)}$, $NaCl_{(s)}$ and $H_2O_{(l)}$, calculate ΔH for the reaction

$$HCl \cdot 100\text{ aq} + NaOH \cdot 100\text{ aq} \rightarrow NaCl \cdot 200\text{ aq} + H_2O_{(l)}$$

What would have been different if you had forgotten about the 1 mole of water produced by the neutralization? Compare this error with the one that we have committed in failing to distinguish between $NaCl \cdot 200\,aq + H_2O_{(l)}$ and $NaCl \cdot 201\,aq$.

3-24. The following information is available.
(1) The heats of formation at 298°K of $CO_{2(g)}$ and $H_2O_{(g)}$ are -94 and -58 kcal/mole, respectively.
(2) The heats of dissociation of $H_{2(g)}$ and $O_{2(g)}$ at 298°K are 103 and 34 kcal/mole, respectively.
(3) Heat capacities C_P for all gases can be taken to be 7.00 cal/deg mole, and 2.5 cal/deg mole for graphite.
(4) The enthalpy of the water gas reaction at 298°K is 32 kcal per unit of reaction

$$H_2O_{(g)} + C \rightarrow H_{2(gr)} + CO_{(g)}$$

(a) Calculate the heat of combustion of graphite at 298°K.
(b) Calculate the heat of combustion, at 298°K, of the H_2 and the CO formed by the complete reaction of water with 1 mole of graphite.
(c) Calculate ΔE for the water gas reaction at 298°K.
(d) Calculate ΔH at 600°K for the water gas reaction.
(e) Calculate the H—O bond energy at 298°K.

3-25. (a) Steam reacts with coke at 1000°C according to the equation

$$C_{gr} + H_2O_{(g)} \rightarrow CO_{(g)} + H_{2(g)}$$

(1) Determine the heat of the reaction at 1000°C.
(2) What is the heat of combustion of the mixture of CO and H_2 at 25°C?

$$C_P \text{ for coke (graphite)} = 2.673 + 2.617 \times 10^{-3}T + 1.169 \times 10^{-7}T^2$$

(b) The reaction $C_{(gr)} + \frac{1}{2}O_{2(g)} \rightarrow CO_{(g)}$ is exothermic and $C_{(gr)} + H_2O_{(g)} \rightarrow CO_{(g)} + H_{2(g)}$ is endothermic. It is therefore theoretically possible to pass a mixture of air and steam over coke so that the temperature remains constant. Assuming complete reaction, determine the ratio of steam to air which will keep the coke at 1000°C. The gases are assumed to enter the reaction preheated to 1000°C and to leave at 1000°C.

3-26. The heats of formation at 298°K of gaseous CO_2, water vapor, and liquid CH_3COOH are -94.0, -57.8, and -116.4 kcal/mole, respectively. The heat combustion of methane gas to CO_2 and water vapor is -192.7 kcal/mole. The heat of vaporization of water at 100°C is 9.4 kcal/mole. Heat capacity values C_P in cal/deg mole are

$CH_{4(g)}$	9.0
$CH_3CHO_{(g)}$	12.5
$CO_{(g)}$	7.5
$H_2O_{(g)}$	7.3
$H_2O_{(l)}$	18.0

(a) Calculate the heat of formation of liquid water at 298°K.
(b) Calculate the enthalpy change at 298°K for the reaction

$$CH_3COOH_{(l)} = CH_{4(g)} + CO_{2(g)}$$

(c) Calculate the temperature at which ΔH for the reaction

$$CH_3CHO_{(g)} = CH_{4(g)} + CO_{(g)}$$

will be zero. ΔH_{298} is -4.0 kcal.

3-27. When normal hexane is passed over a catalyst bed at 500°C, benzene is formed by the reaction

$$C_6H_{14(g)} \rightarrow C_6H_{6(g)} + 4H_{2(g)}$$

(a) What would be the enthalpy of this reaction if it were carried out at 25°C?
(b) What is the enthalpy of the reaction at 500°C?

3-28. Using tabulated data, calculate the heat of the reaction

$$C_2H_{4(g)} + H_{2(g)} \rightarrow C_2H_{6(g)}$$

at 1500°K. Obtain the general expression for the heat of this reaction as a function of temperature.

3-29. From the following data at 25°C

Reaction	ΔH° (kcal)
$\frac{1}{2}H_{2(g)} + \frac{1}{2}O_{2(g)} \rightarrow OH_{(g)}$	$+10.06$
$H_{2(g)} + \frac{1}{2}O_{2(g)} \rightarrow H_2O_{(g)}$	$-\ 57.80$
$H_{2(g)} \rightarrow 2H_{(g)}$	$+104.178$
$O_{2(g)} \rightarrow 2O_{(g)}$	$+118.318$

calculate ΔH° for the following reactions
(a) $OH_{(g)} \rightarrow H_{(g)} + O_{(g)}$
(b) $H_2O_{(g)} \rightarrow 2H_{(g)} + O_{(g)}$
(c) $H_2O_{(g)} \rightarrow H_{(g)} + OH_{(g)}$

Assuming the gases to be ideal, compute the values of ΔE for these three reactions. [The enthalpy change for (a) is the bond energy of the OH radical; half the enthalpy change in (b) is the average O—H bond energy in water vapor; and the enthalpy change for (c) is the bond dissociation energy of the first O—H bond in water vapor. Note that these are three different quantities.]

3-30. (a) From the table of bond energies in Appendix 5, calculate the standard heat of formation of C_2H_6, CH_3SH, and HCOOH (all gases).

(b) Compare these with the measured thermodynamic values in Appendix 4.

3-31. Using bond energies, estimate the heat of formation of ethanol vapor at 25°C. How does this compare with the measured value?

3-32. Estimate the heat of the formation of 1 mole of water vapor from one mole of hydrogen gas and $\frac{1}{2}$ mole of oxygen gas, using bond energies (enthalpies) of Appendix 5. How does this compare with the observed value?

3-33. (a) Assume that the bond structure of carbon monoxide is C=O, and that that of carbon dioxide is O=C=O. Calculate the standard enthalpy of the reaction

$$CO_{(g)} + \tfrac{1}{2}O_{2(g)} \rightarrow CO_{2(g)}$$

Compare this with the observed value. How big is the error, and how good do your assumptions about the bond structure of these two molecules appear to be?

(b) Select two other reactions, one involving CO but not CO_2 and the other the opposite. Calculate the enthalpies of these two reactions, compare them with measured values, and decide which bond assumption is worse, that of C=O or O=C=O.

3-34. The heat of combustion of gaseous isoprene, C_5H_8, is -745.8 kcal/mole. Calculate the heat of formation and by comparison with a bond energy calculation estimate the resonance energy of isoprene. Draw at least two possible resonance structures for isoprene. (If you are uncertain about this last part, then see Pauling's book.)

3-35. Verify the value given for the N—H single bond in the bond energy table in Appendix 5 by assuming that the other bond energies are correct and choosing some simple nitrogen-containing compound for which data are available either in Appendix 4 or some other source. If you use other compilations of enthalpy of formation, give a reference to them.

3-36. Verify the bond energy of a C=C double bond in Appendix 5, assuming that all of the other values are correct. If, instead, you *assumed* that a C=C double bond had exactly twice the bond energy of a C—C single bond, then what value would you calculate for the C—H single bond in ethylene?

3-37. Explain why the C—C single bond energy in diamond (in kcal/mole) is *half* the value of the molar heat of sublimation of diamond. The structure of diamond, if you do not know it, can be found in Pauling's book and in many other places.

Chapter 4

SECOND LAW AND FREE ENERGY

4-1 RELEVANCE OF THE SECOND LAW

WHY DO we need the second law? This law, as you probably know, involves Carnot cycles, steam engines, pistons, and hot and cold reservoirs. What has this to do with the chemist? We must become involved with steam engines and cycles because our goal is still the finding of the proper potential function for chemical equilibrium, and we need the second law to lead us to it. This function is not energy, and not enthalpy. Although a law can never be *proven* by example, only one contradiction is needed to *disprove* it. We can quickly find one example, or with a little care many examples, of spontaneous processes which are accompanied by an uptake of energy or of enthalpy. If energy were the potential function to be minimized at chemical equilibrium, then a plot of E versus reaction coordinate (some measure of the concentration of reactants or products) like Figure 3-1 could be drawn. The energy curve would be at a minimum at equilibrium, and every spontaneous process going toward equilibrium in an irreversible manner would lead to a drop in energy. This is not always true. Most spontaneous reactions are exothermic, but there are a minority which are endothermic, and these are enough to disprove the whole argument.

EXAMPLE 1. At 25°C the vapor pressure of water, P_V, is 23.7 mm. This is the pressure of vapor in equilibrium with liquid water at 25°C. Therefore, if a dish of water is opened to a large environment where the vapor pressure is 10 mm there will be a spontaneous evaporation of the water as long as the partial vapor pressure is less than 24 mm. If the system is completely open, evaporation will continue until the water is gone. If the system is enclosed, then eventually the pressure may rise to 24 mm before all the water is evaporated, and again equilibrium will be established. This can be written as a chemical equation

$$H_2O_{(l,\,25°C)} \rightarrow H_2O_{(g,\,25°C,\,P=10\text{ mm})} \qquad \Delta E = +9.9 \text{ kcal}$$

The change in internal energy of this reaction is +9.9 kcal. The fact that the system absorbs energy as it evaporates does not prevent it from evaporating spontaneously under the right conditions.

EXAMPLE 2. Another example, involving enthalpy, is the solution of ammonium chloride in water. The process

$$NH_4Cl_{(s)} + \infty H_2O \rightarrow NH_4Cl_{(\text{aq})} \qquad \Delta H = +3.62 \text{ kcal}$$

involves an enthalpy change at 298°K of +3.62 kcal. But crystals of ammonium chloride will not sit undissolved in a beaker of water, and an ammonium chloride solution will not spontaneously separate out into crystals and pure water, even though in doing so it would go to a state of lower enthalpy. Thus there exist spontaneous, irreversible processes during which the internal energy rises, and others which are accompanied by a rise in enthalpy. A brief look through thermodynamic tables will show other examples, while only one is needed to destroy our theorem.

The question remains, then, what *is* the right potential function? The missing factor has already been suggested from statistical mechanics. Recall that if all states were equally probable then there would be a spontaneous tendency to go to the state of lowest energy. This is the basis for the assumption that E or H was the chemical potential function. But the missing factor is that of probability of states: if all states are of the same energy, then the most likely state is the state of greatest entropy. With time, since there are many more ways of getting a more likely state, there will be a spontaneous tendency of a system to go from an unlikely state to a likely state. In both of our examples above, there was a great increase in entropy as the reaction went to the right. Liquid molecules are more ordered than gaseous molecules, and a solid ammonium chloride crystal is much more ordered than a solution of ammonium and chloride ions in solution. What we need then is more information about entropy, and to get it we have to understand the nature of the steam engine.

4-2 CYCLIC PROCESSES

Let us begin by looking at an isothermal system consisting of a piston filled with an ideal gas at a certain pressure, and of volume V_1. Evacuate the surroundings so that there is no pressure on the back side of the piston, then suddenly allow the volume to expand to V_2. This process, known as a free isothermal expansion, is irreversible; we would never expect the piston to compress the gas spontaneously in the absence of a greater pressure on the outside. No work is done in this free expansion since there is no resisting force. Since the process is isothermal, the internal energy is constant. Therefore, by the first law, no heat flows in or out of the system.

Now let us return the system to the original set of conditions, doing as little work as possible. This means that we must push on the piston with a pressure infinitesimally greater than the pressure inside. The pressure against which the work is done is $P_{\text{int}} = RT/V$, and the work is $w = RT \ln(V_2/V_1)$. Again, the internal energy does not change, and by the first law, $q = -w$, or the heat loss just equals the work done on the system.

The cycle is now complete. The net effect is the conversion of a certain amount of work into heat; work has been done on the system, and heat has been given off. In general, any cyclic process of which one or more steps are irreversible will convert some work into heat. This is known as the principle of the degradation of energy. An irreversible process degrades work, an exchangeable form of energy, into heat, which is a poorer, less useful form.

Conversion of Heat to Work

Now let us look at the opposite process, the conversion of heat to work in a cyclic process. By the first law, there is no change in the energy in a cyclic process; hence any heat that is added is paid for by work done on the surroundings. Any device for converting heat into work can be called a thermodynamic "machine."

By experience, we know that any machine for converting heat to work must operate in the presence of two different temperatures. In a steam engine, for example, the upper temperature is the temperature of the steam, and the lower, that of the condensing coils. If the steam is no hotter than the condenser cooling water, the engine will not work; a hot condenser is similarly ineffective. But in the process of operating between these two temperatures, some heat is always lost to the condenser or out the exhaust. If an amount of heat q_u is absorbed at the upper temperature, if work $-w$ is done on the surroundings, and if a quantity of heat $|q_l|$ is lost at the lower temperature, then the first law says that for the entire cyclic process, $q_u + q_l + w = \Delta E = 0$. Note that q_l and w are both negative numbers (Figure 4-1). The efficiency of a heat engine is commonly defined as the ratio of the work obtained from the system, $-w$, to the heat taken from the hot reservoir:

$$e \equiv \frac{-w}{q_u} = \frac{q_u + q_l}{q_u} = 1 + \frac{q_l}{q_u} \tag{4-1}$$

Carnot Cycle

A particularly simple heat engine cycle to handle mathematically is the Carnot cycle. On a PV plot, this cycle is simply the path around the perimeter of a region bounded by two isotherms and two adiabats as shown in Figure 4-2. If the gas is ideal, then these curves will be given by $PV = \text{const}$ and $PV^{\gamma} = \text{const}$, respectively. Two temperatures are involved, t_u and t_l. These temperatures are not yet based on any sort of temperature scale; all that is known is that t_u is hotter than t_l.

The first step in a Carnot cycle is a reversible isothermal expansion at t_u, from state (P_1, V_1, t_u) to (P_2, V_2, t_u), or from point 1 to point 2 in Figure 4-2. This expansion could be achieved by expanding the gas in contact with a suitably large heat reservoir at t_u. A certain amount of work will be done on the surroundings, and if w_1 is the work done on the gas, then that done on the surroundings will be $-w_1$, with w_1 itself being a negative number. There will be an absorption of heat q_u, and probably a minor change in internal energy for a real gas. If the gas were ideal, there would be no such change, and it would be true that $q_u = -w_1$, or that the work done was exactly counterbalanced by the heat absorbed.

The second step is an adiabatic expansion from the state at point 2 to that at point 3, (P_3, V_3, t_l). Under these conditions, $q = 0$, and $\Delta E = w_2$; the internal energy change is the same as the work done on the gas. Since work

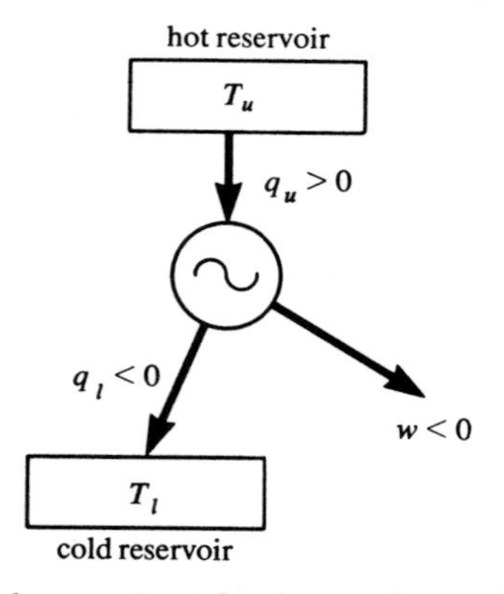

FIGURE 4-1. *A simple heat engine takes heat q_u from a high temperature reservoir, converts some of it to work against its surroundings, $-w$, and loses some of the heat $-q_l$ at a lower temperature. The quantities q_u, q_l, and w are positive when they represent heat fed* into *the engine or work* on *the engine.*

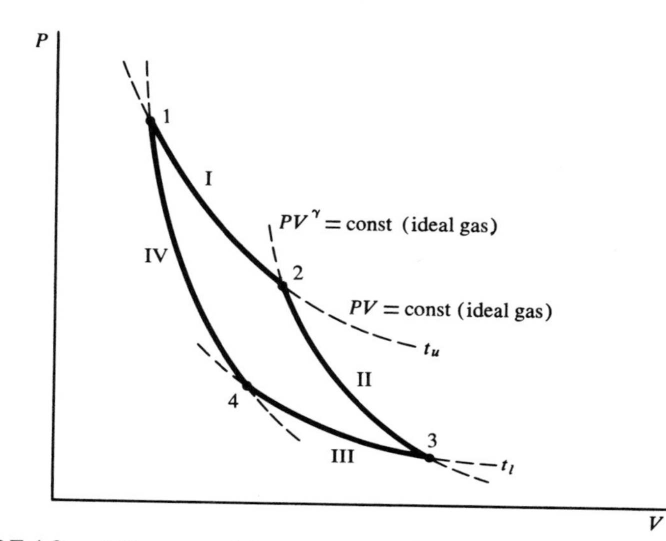

FIGURE 4-2. *A Carnot cycle is a process involving four steps: isothermal expansion (I), adiabatic expansion (II), isothermal compression (III), and adiabatic compression to the original state (IV). If the working gas is ideal, then the isothermals are given by the expression $PV = const$ and the adiabatics are given by the expression: $PV^{\gamma} = const$.*

is done on the surroundings, w_2 is negative and the internal energy must fall.

The third step, the isothermal compression, is continued just to the point (P_4, V_4, t_l) where a final adiabatic compression will bring the gas back to its starting conditions at point 1 on the PV plot. Work w_3 is done on the gas, and an amount of heat q_l is lost from the gas which compensates for this work exactly in an ideal gas and approximately in a real gas.

The final adiabatic compression to the starting point occurs with work w_4 done on the gas and an exactly matching rise in internal energy. (See Figure 4-3.)

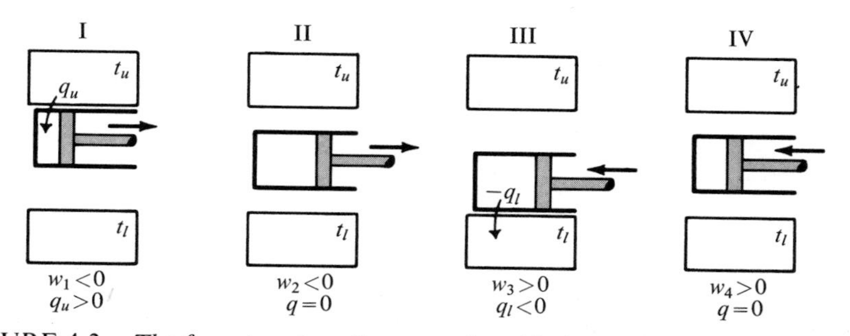

FIGURE 4-3. *The four steps in a Carnot cycle, with the working gas enclosed in a piston. Temperature t_u is greater than t_l. The work done on the gas and the heat absorbed by the gas in each step are shown at the bottom.*

For the entire cycle, $w = w_1 + w_2 + w_3 + w_4$, and $q = q_u + q_l$. As mentioned earlier, w_1, w_2, and q_l are negative, w_3, w_4, and q_u are positive. The total sum of heat and work is zero since the initial and final states are identical.

$$q_u + q_l + w_1 + w_2 + w_3 + w_4 = q + w = \Delta E = 0$$

The efficiency of the entire cycle in converting heat to work is

$$e = \frac{-w}{q_u} = \frac{q}{q_u} = \frac{q_u + q_l}{q_u} = 1 + \frac{q_l}{q_u} \tag{4-2}$$

Since q_l and q_u have opposite signs, the efficiency is less than 1. Since the process was carried out reversibly, against the maximum possible opposing force, this is the greatest possible efficiency.

Is it possible to have two *reversible* Carnot engines operating between the same two temperatures but with different efficiencies? There is no reason by the first law why this should not be so. Let us then assume that we can, and see what happens. Assume that we have two Carnot engines A and B, that A is 60% efficient and B is 50% (Figure 4-4). Now run engine A forward for one cycle. Some heat will be lost and some work will be obtained, which can then be used to run engine B backwards since we have assumed the engines to be reversible. If 100 cal of heat are absorbed at the upper temperature, 60% efficiency means that 40 cal will be wasted as heat

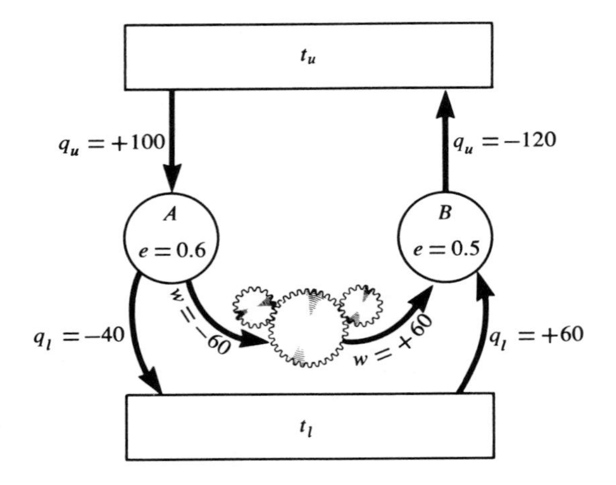

FIGURE 4-4. *If two Carnot cycle heat engines operating* reversibly *between the same two temperatures can have different efficiencies, then they can be linked together in such a way as to transfer heat from a cold object to a hot object* without *any work being done from the outside to make the flow occur. The statement that, in fact, this cannot be done is Clausius' version of the second law.*

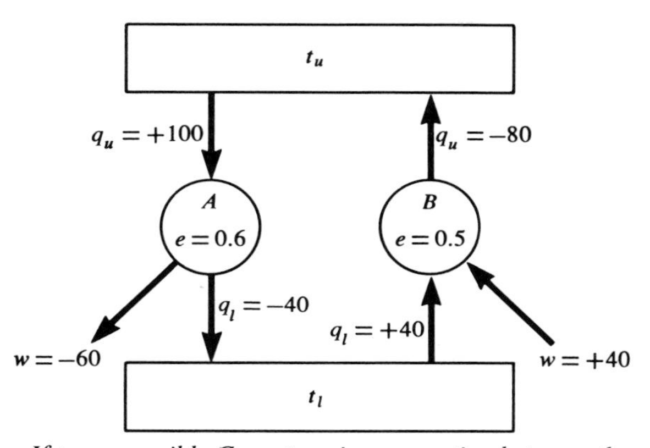

FIGURE 4-5. *If two reversible Carnot engines operating between the same two temperatures could have different efficiencies, then it would be possible to couple them so as to convert heat at a high temperature completely to useful work, with no losses at a lower temperature. The impossibility in reality of converting heat to work with 100% efficiency is Thompson's version of the second law. The conclusion is that all reversible Carnot engines operating between the same two temperatures must convert heat to work with the* same *efficiency.*

lost to the cold reservoir and 60 cal of useful work will result. Now if these 60 cal are used to operate the less efficient *B* in reverse as a heat pump, 60 cal of heat will be removed from the low temperature reservoir, and 120 cal will be transferred to the high temperature sink. No net work is done in this over-all process, but 20 cal of heat will have been taken from a lower temperature to a higher one. Heat will have flowed uphill.

Now try again. This time, run *A* as before, but run engine *B* only until all the heat given off to the cold reservoir is used up (Figure 4-5). 100 cal of heat fed into *A* will yield 60 cal of work and 40 cal of heat to the cold reservoir as before. Only 40 cal of work will be required to take those 40 cal back to the hot reservoir with engine *B*. The net result is that 20 cal of heat from the hot reservoir have been converted to 20 cal of work with 100% efficiency. The right combination of machines of 60% and 50% efficiency has produced a compound machine of 100% efficiency.

What is wrong with these two processes? As far as the first law is concerned, nothing is wrong. Yet we know from experience that heat will not flow uphill against a temperature gradient, and that no machine is 100% efficient. The incorporation of these new summaries of experience leads to the second law.

A theory is only as good as the facts that are fed in. Theory is somewhat like an aged machine; left to itself long enough it will run down and stop, and it requires a kick to get it going again. The kicks in science are experimental facts. The discussion of thermodynamics began with talk about equilibrium.

From equilibrium came the idea of temperature and heat, then work and energy, after which the machine died. We had defined energy, but to what purpose? We then introduced some physical information; we said that energy is a *state function.* Nothing that had been said up to that point required this to be so; rather, we looked at the physical world and found that energy indeed behaved as a function of state. This was the first law, and it gave us enough impetus to go on to talk about enthalpy, the additivity of heats of reaction, thermochemistry, and bond energies. The machine has now died once again, and we must give it a second experimental kick. This is the statement, not derivable from anything said so far, that heat does not flow uphill, or, equivalently, that one cannot convert heat to work without wasting some of the heat. These purely observational statements furnish the impetus which gets thermodynamics going again for a little while.

4-3 SECOND LAW

There are two equivalent forms of the second law, one proposed by Clausius and one by Thompson. Clausius' form says that it is impossible by a cyclic process to transfer heat from a lower temperature to a higher temperature without doing work on the system. Heat must be forced to flow from a cold object to a hot object. Alternatively, Thompson stated that it is impossible, by any cyclic process, to take heat at some high temperature and convert it into work without losing some heat at a lower temperature. These two statements are entirely equivalent, and if one is true the other must be as well. Their equivalence can be proven by using the simple heat engine scheme of Section 4-2, assuming one statement to be true and the other false, and showing that this leads to a contradiction.

Since both results of assuming variable efficiencies in heat engines operating between the same two temperatures have been shown to violate common experience as summarized in the second law, then it must be true that all reversible cycles have efficiencies which depend *only* upon the two temperatures between which they operate. This leads to the concept of a thermodynamic temperature scale.

Thermodynamic Temperature Scale

How does the efficiency depend on temperature? To find out, run two Carnot engines in succession (Figure 4-6). Let engine A run from t_u to a middle temperature t_m. It extracts a quantity of heat q_u from the upper reservoir and converts it to work w_1, with a heat loss $q_m < 0$ to the middle reservoir. Now take exactly this heat $q_m' = -q_m$ out of the middle reservoir and use it to run engine B, giving work w_2 and heat loss $q_l < 0$ at the lower temperature t_l.

From Eq. 4-2, we can calculate the efficiency of this process.

$$e_A = 1 + \frac{q_m}{q_u} \qquad e_B = 1 + \frac{q_l}{q_m'} = 1 - \frac{q_l}{q_m} \tag{4-3}$$

Now operate a third engine C directly between t_u and t_l, taking out q_u from the upper reservoir and getting out work w_3 and heat q_l', where these quantities are not for the moment necessarily identical to the net w and q_l resulting from the two-step process. The efficiency is

$$e_C = 1 + \frac{q_l'}{q_u}$$

However, the total efficiency in the two-step process is $e_{A+B} = -(w_1 + w_2)/q_u = 1 + q_l/q_u$. But from the second law, we know that the two efficiencies e_C and e_{A+B} must be identical. This immediately requires that $q_l' = q_l$ and $w_3 = w_1 + w_2 = w$.

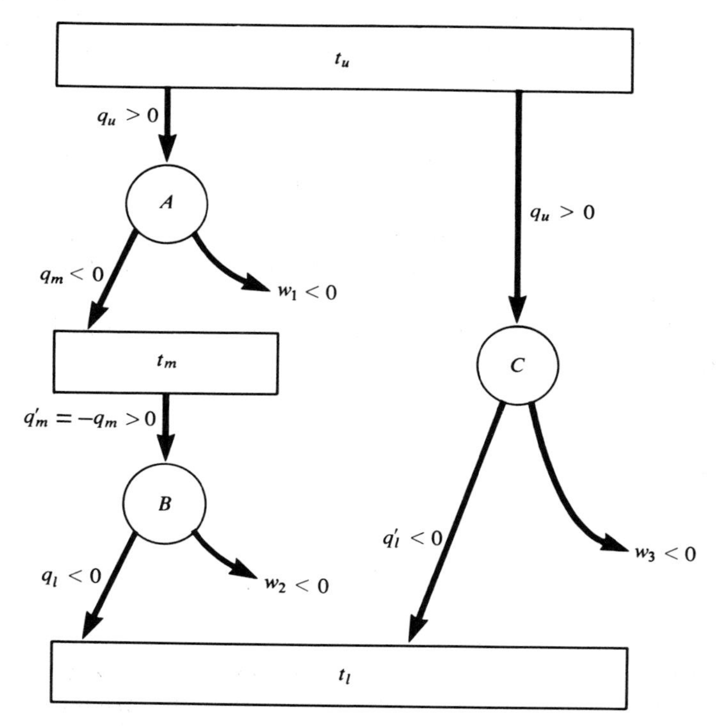

FIGURE 4-6. *The fact that a one-step heat engine and a two-step compound heat engine operating between the same two temperatures must have the same efficiency leads to the concept of a thermodynamic temperature scale.*

If efficiency is a function of temperatures, the heat ratios must also be so. From the two-step and one-step processes,

$$\frac{q_m}{q_u} = e_A - 1 = f(t_u, t_m)$$

$$\frac{-q_l}{q_m} = e_B - 1 = f(t_m, t_l) \tag{4-4}$$

$$\frac{q_l}{q_u} = e_C - 1 = f(t_u, t_l)$$

from which we get immediately

$$f(t_u, t_m)f(t_m, t_l) = -\frac{q_l}{q_u} = -f(t_u, t_l)$$

What kind of function f satisfies this sort of relationship, in which $f_{(a,b)} = -f_{(a,c)}f_{(c,b)}$? The function must be separable into a quotient of functions of the individual temperatures, that is,

$$f(t_u, t_m) = -\frac{\phi(t_m)}{\phi(t_u)} \qquad f(t_m, t_l) = -\frac{\phi(t_l)}{\phi(t_m)} \qquad f(t_u, f_l) = -\frac{\phi(t_l)}{\phi(t_u)} \tag{4-5}$$

The exact nature of the function ϕ is up to us; as a matter of convenience, we can define a *thermodynamic* temperature scale to be such that this function is just the temperature itself.

The efficiency can now be expressed in terms of the thermodynamic temperature scale. For

$$\frac{q_l}{q_u} = -\frac{t_l}{t_u} \tag{4-6}$$

and

$$e = 1 - \frac{t_l}{t_u} \tag{4-7}$$

For a heat engine to be 100% efficient, the hot reservoir would have to be infinitely hot (an impossibility), or the cold sink at absolute zero on the thermodynamic scale. The second law does not of itself say that a process cannot be 100% efficient in converting heat to work; it merely demands a cold sink at absolute zero for this to happen.

Rewriting Eq. 4-6 slightly, we see that, for the entire cycle,

$$\frac{q_u}{t_u} + \frac{q_l}{t_l} = 0 \tag{4-8}$$

or, for the *reversible Carnot* cycle as a whole,

$$\oint \frac{dq}{t} = 0 \tag{4-9}$$

This means that during this particular process dq/t is an exact differential, and that a new state function is lurking somewhere.

Carnot Cycle and the Ideal Gas Scale

Consider the Carnot cycle again, this time using an ideal gas. In the first step, the isothermal expansion, there is no energy change in an ideal gas. Then

$$-w_1 = q_u = RT_u \ln \frac{V_2}{V_1} \tag{4-10}$$

Note that T_u is the temperature based on the ideal gas scale, from the ideal gas relation $T = PV/R$.

The second step is an adiabatic expansion. Here $q = 0$, and the change in internal energy just matches the work done.

$$\Delta E_2 = w_2 = \int_{T_u}^{T_l} C_V \, dT \tag{4-11}$$

In the isothermal compression, the internal energy change is again zero, and

$$-w_3 = q_l = RT_l \ln \frac{V_4}{V_3} \tag{4-12}$$

The final adiabatic compression, if we have chosen point 4 (Figure 4-2) well, will bring us back to point 1. Again, $q = 0$, and

$$\Delta E_4 = w_4 = \int_{T_l}^{T_u} C_V \, dT = -w_2 \tag{4-13}$$

This integral is the inverse of the one in Eq. 4-11. For an ideal gas, the net work done in the two adiabatic steps is zero.

V_1, V_2, V_3, and V_4 are related, since they are the termini of two adiabats between the same two temperatures. Recalling a relationship derived earlier,

$$\frac{V_2}{V_3} = \left(\frac{T_l}{T_u}\right)^{C_V/R} = \frac{V_1}{V_4} \tag{4-14}$$

From this, we obtain

$$\frac{V_4}{V_3} = \frac{V_1}{V_2} \tag{4-15}$$

The net work of the four steps is $w = w_1 + w_3$. Substituting the results of Eqs. 4-10, 4-12, and 4-15 produces

$$w = -R(T_u - T_l) \ln \frac{V_1}{V_2} \tag{4-16}$$

The heat absorbed at the upper temperature is q_u (Eq. 4-10), so that the efficiency (from Eq. 4-7) is

$$e = \frac{-w}{q_u} = \frac{T_u - T_l}{T_u} = \frac{t_u - t_l}{t_u} \tag{4-17}$$

We see immediately that the ideal gas scale differs from the new thermodynamic scale by at most a constant multiplicative factor, or that the two are proportional. Since we may choose the constant of proportionality at our discretion, we will define it as 1, and from now on will use T for both ideal gas and thermodynamic temperature.

Generalization to Any Reversible Cycle

So far we have demonstrated the important result that, for the particular case of reversible Carnot cycles, although dq is not an exact differential, dq/T *is*, where T is the temperature measured on a new thermodynamic scale based upon heats absorbed or lost. Furthermore, this new scale has been shown to be strictly proportional to the earlier ideal gas scale. Therefore, for a *reversible Carnot* cycle,

$$\oint \frac{dq}{T} = 0 \quad \text{and} \quad \frac{dq}{T} \text{ is exact}$$

But is this generally true?

Any reversible cycle on a PV plane can be represented as the limit of an infinite number of Carnot cycles. To demonstrate this, consider such a reversible cycle as in Figure 4-7. On it draw a set of isotherms, and, superimposed on this, draw a set of adiabats. Now choose a set of tiny Carnot cycles which best covers the area of the given reversible cycle, and integrate any area-type quantity such as $w = -\oint P\,dV$ around each and add the results together. As we integrate around each one of these cycles, the integrals on the interior segments cancel because they are performed twice, once in each direction. The result is that the sum of the integrals over all these little Carnot cycles will just be equal to the integral around the perimeter of the tiny cycles. The finer the cycles are drawn, the closer the jagged Carnot cycle perimeter follows that of the given cycle. This particular integral is a measure of the *area* enclosed in the cycle, and in the limit of infinitesimally tiny Carnot cycles, the area enclosed in the zig-zag of Carnot cycles and the area of the original cycle match, although their perimeter lengths do not. In this sense any reversible cycle can be considered as the limit of many small Carnot cycles. The result is that dq/T is exact for each increment of path around *any reversible cycle*, and if a new function is defined by

$$d\mathscr{S} \equiv \frac{dq_{\text{rev}}}{T}$$

then $\mathscr{S}$ is a state function. To prove this, one must show that the change in $\mathscr{S}$ in going from an initial state A to a final state B is independent of path.

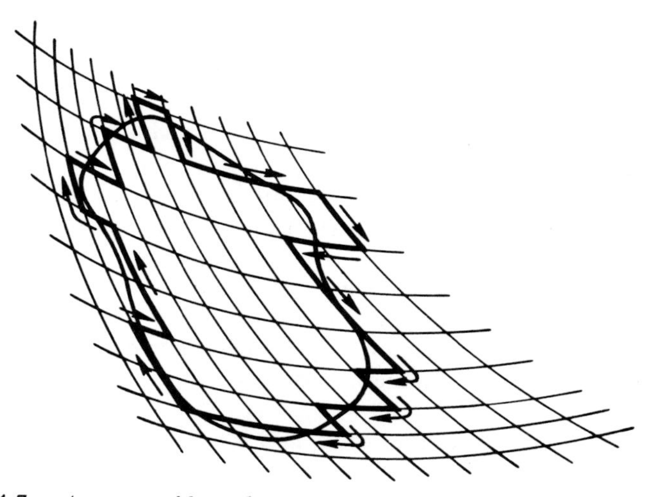

FIGURE 4-7. *Any reversible cycle may be represented on a PV plane as the limit of an infinite number of infinitesimal Carnot cycles.*

Assuming that the system can be taken from A to B reversibly, choose *any* two such reversible paths at random and form a cycle from them, traversing one from A to B and the other from B to A. We have shown that $\mathscr{S}$ during the entire reversible cycle is zero. $\mathscr{S}$ must therefore be the same for the traverse from A to B along either leg of the cycle, and since the reversible paths were chosen at random, $\mathscr{S}$ is a true state function.

This new state function will be called the *thermodynamic entropy*, and can be defined formally by

$$d\mathscr{S} \equiv \frac{dq_{\text{rev}}}{T} \qquad \Delta\mathscr{S} = \mathscr{S}_B - \mathscr{S}_A = \int_A^B \frac{dq_{\text{rev}}}{T} \tag{4-18}$$

The behavior of $\mathscr{S}$ during irreversible processes has not yet been specified. Moreover, at the moment, we do not know that this thermodynamic entropy has any connection with the entropy that we encountered in statistical mechanics. Hence this new function will be represented by a script $\mathscr{S}$ until we prove the equivalence of $\mathscr{S}$ and S.

There is an interesting way of looking at integral Eq. 4-18. One can regard the heat increments fed into a system reversibly, dq, as unnormalized, or somehow on different scales. Adding such misscaled bits of heat gives varying results depending on the conditions of the experiment. But if each bit of heat dq is first " normalized " by dividing by the temperature T at which it was added, then the sum over such properly scaled heat bits, or the integral of Eq. 4-18, is *independent* of the exact mode of going from initial to final states. Another way of describing Eq. 4-18 is to say that $1/T$ is an *integrating factor* for dq.

Summary

It is well to retrace the chain of logic which led to such an important concept as the thermodynamic entropy.

(1) Experience tells us that heat cannot be converted to work without losses, or that heat will not flow against a temperature gradient spontaneously.

(2) Therefore, *all* reversible Carnot cycles that are run between the same two temperatures have the same efficiency.

(3) This efficiency is a function only of these temperatures; $e = e(t_u, t_l)$, expressed in terms of a thermodynamic temperature scale based upon heats.

(4) For any reversible Carnot cycle, dq/t is an exact differential.

(5) $t \propto T$, where T is measured on the ideal gas scale.

(6) Hence $dq/T = d\mathscr{S}$ is an exact differential during any reversible Carnot cycle.

(7) Any reversible cycle can be represented as a sum of Carnot cycles.
(8) Therefore, $d\mathscr{S} = dq/T$ is an exact differential during the course of any reversible cycle, and $\Delta\mathscr{S} = \mathscr{S}_B - \mathscr{S}_A = \int_A^B dq/T$ for any reversible process.

4-4 *THERMODYNAMIC ENTROPY* $\mathscr{S}$

For any *reversible* process,

$$d\mathscr{S} = \frac{dq}{T} \tag{4-19}$$

from which

$$dq = Td\mathscr{S} \tag{4-20}$$

The first law then becomes

$$dE = Td\mathscr{S} - PdV \tag{4-21}$$

Notice that dV is exact because the volume change depends only on the initial and final volumes and not on how the expansion or contraction is carried out. But PdV is inexact, because to integrate over PdV one needs $P(V)$, pressure as a function of volume. Symmetrically, $d\mathscr{S}$ is exact, but is ruined by multiplying by temperature; $Td\mathscr{S}$ is inexact.

Consider now a Carnot cycle on a temperature–entropy plot instead of a pressure–volume plot (Figure 4-8). In the isothermal steps, T is constant.

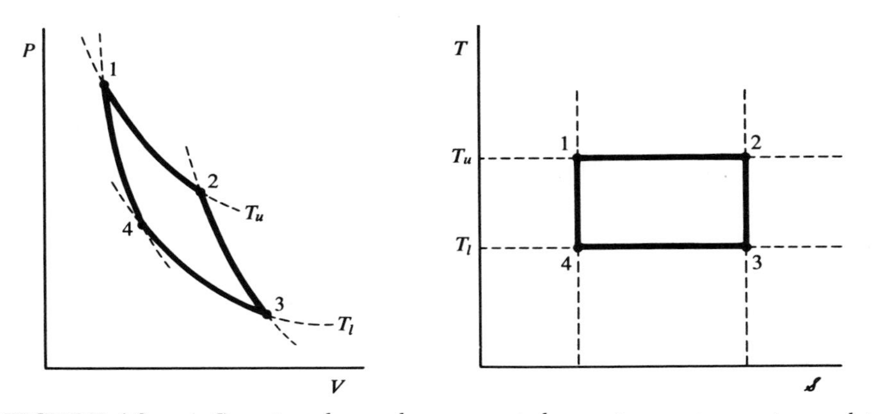

FIGURE 4-8. *A Carnot cycle can be represented on a temperature–entropy plot as well as on a pressure–volume plot. The area enclosed by the cycle in the PV plot is the work done on the surroundings by the gas, and the area in the T$\mathscr{S}$ plot is the heat absorbed by the gas. By the first law, these areas are equal.*

Therefore,

$$\Delta \mathscr{S}_{1 \to 2} = \frac{q_u}{T_u} \quad \Delta \mathscr{S}_{3 \to 4} = \frac{q_l}{T_l} \tag{4-22}$$

In the adiabatic steps, $dq = 0$, so that $\Delta \mathscr{S} = 0$, but the temperature is changing, and the cycle appears as in Figure 4-8b.

The area in a PV plot is the work done on the surroundings, or $-w$. But the area enclosed in a $T\mathscr{S}$ plot is the net heat absorbed. The area under the line from point 1 to point 2 is the heat absorbed in the isothermal expansion, $\int_1^2 T_u \, d\mathscr{S} = q_u$, while the area under line 3-4 is the heat released during the isothermal contraction, $\int_3^4 T_l \, d\mathscr{S} = q_l$. The difference, the net heat absorbed, is the area within the rectangle 1-2-3-4. Since from the first law there is no change in the internal energy in a cyclic process, this heat absorbed equals the work done, and the areas on the two planes are identical.

For ideal gases, one can obtain a general expression for the entropy by turning Eq. 4-21 around to give

$$T d\mathscr{S} = dE + PdV = C_V \, dT + PdV \tag{4-23}$$

$$d\mathscr{S} = \frac{C_V \, dT}{T} + \frac{R dV}{V} \tag{4-24}$$

We have already seen that the right side of this equation was zero in reversible adiabatic processes, giving the adiabatic relationships between P, V, and T. This equation is then a more general expression, of which the special case has already been encountered.

$$\Delta \mathscr{S} = \mathscr{S}_2 - \mathscr{S}_1 = C_V \ln \frac{T_2}{T_1} + R \ln \frac{V_2}{V_1} \tag{4-25}$$

One of the most interesting adiabatic systems is the entire material universe. The second law is sometimes simply stated as "The entropy of the universe is unchanged in all reversible processes," the first law being "The energy of the universe is constant." However, not all processes are reversible; in fact, no process is ever reversible in the strict sense of the word. What happens to entropy in an irreversible process?

Irreversible Processes

In the previous section we saw that entropy was defined for reversible processes in such a way that it was a state function. In other words, the change in entropy was the same if one took the system from a given initial state to a given final state by any *reversible* process. What happened in irreversible processes was not defined. But the great advantage of entropy is its state function property. It makes sense, therefore, to keep this property in our extended definition. We will define entropy to be a true state function in *any* process, reversible or not, so that the entropy change in any irreversible process is identical to the entropy change calculated in going from the same initial to the same final state in a reversible manner. With this definition, however, it is not true that $d\mathscr{S} = dq/T$ in an irreversible process.

Consider as a simple example the free expansion of an ideal gas into a vacuum. First expand the gas isothermally to twice its original volume in a *reversible* manner. Then from the previous sections,

$$\Delta\mathscr{S} = R \ln \frac{2}{1} = 1.38 \text{ e.u./mole}$$

Under these reversible conditions, the gas expands against the maximum external pressure, and $d\mathscr{S} = dq/T$.

Now suppose the expansion is to be carried out isothermally and irreversibly into a vacuum. Because the expansion is isothermal and the gas is ideal, there is no change in E. Because there is no opposing pressure, there is no work done. Hence by the first law there was no heat change as well, or $q = \Delta E - w = 0$. Then the integral of dq/T from state 1 to state 2 is zero. But from the previous definition of entropy, $\Delta\mathscr{S}$ must be the same as in the reversible process, 1.38 e.u./mole, so it cannot be true that $d\mathscr{S} = dq_{\text{irr}}/T$. In this case, and in general, $d\mathscr{S} > dq/T$, or equivalently, $dq < T\,d\mathscr{S}$. $T\,d\mathscr{S}$ forms an upper limit to the amount of heat that can be absorbed. If the process is carried out reversibly much heat is absorbed because much work is done; if irreversibly, less work is involved and hence less heat.

Another way of stating this is to say that the efficiency of an irreversible process is less than that of an equivalent reversible one.

$$e_{\text{irr}} = \left(\frac{q_u + q_l}{q_u}\right)_{\text{irr}} < \frac{T_u - T_l}{T_u} = e_{\text{rev}} \tag{4-26}$$

Rearranging the inequality, for irreversible processes,

$$\frac{q_u}{T_u} + \frac{q_l}{T_l} < 0 \tag{4-27}$$

or, for any cyclic process,

$$\oint \frac{dq}{T} \leqq 0 \tag{4-28}$$

This is known as Clausius' inequality; the equal sign holds only if the entire cycle is reversible. We can justify this intuitively from first law considerations, since we know that $-dw_{irr} < -dw_{rev}$ since the opposing back pressure is less during an irreversible expansion. Hence, since $dE_{irr} = dE_{rev}$, it must be true that $dq_{irr} < dq_{rev}$.

Let us use this definition of entropy for an irreversible process in a simple example. Suppose two blocks of metal, initially at two different temperatures, are brought together momentarily, long enough for heat q to flow between them from the hot to the cold block. Furthermore, suppose that the blocks are large enough that their temperatures remain essentially constant in this process. What is the entropy change in such a spontaneous, irreversible process? The problem is clearly one of finding an equivalent reversible process. We can do this by introducing a working substance such as an ideal gas which could draw the heat off from one block reversibly, take it to the other block, and feed it in again reversibly as in Figure 4-9.

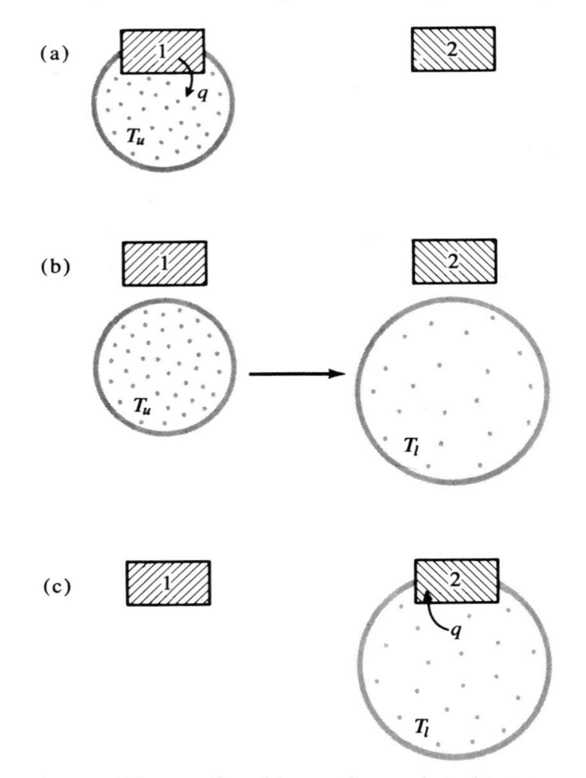

FIGURE 4-9. *A reversible transfer of heat q from a hot object to a cold one, using an ideal gas as a working substance.*

First, surround the hot block with the gas and allow the gas to expand isothermally and reversibly until it has absorbed heat q. The entropy change in the gas is $\Delta\mathscr{S} = \int dq/T_u = q/T_u$. Now isolate the gas from both blocks and expand it adiabatically to lower its temperature to T_l. With $dq = 0$, the entropy change is zero for the gas. Finally, place the gas in contact with the cold block and compress it isothermally until the heat q is lost to this block. The entropy change *of the gas* in this step is $\Delta\mathscr{S} = -q/T_l$. In this three-step process, the total entropy change of the gas is $q/T_u - q/T_l$. This quantity is less than zero; the gas loses entropy. But the system of gas + blocks underwent a reversible change in isolation, and its total entropy therefore remained unchanged. Thus from the additivity property of entropy, the entropy change *of the blocks* must be the negative of that of the gas, $\Delta\mathscr{S}_{\text{bl}} = q/T_l - q/T_u > 0$. By our definition of entropy it makes no difference whether we went from the initial to the final state reversibly using an ideal gas, or irreversibly by slapping the two blocks together momentarily. In both cases, the entropy of the metal blocks rises by $\Delta\mathscr{S}_{\text{bl}} = q/T_l - q/T_u$. This gain of entropy is a characteristic of irreversible processes in isolated systems.

Using Clausius' inequality, we can prove this behavior in the general case. Take an isolated system and let something happen irreversibly. If the system is isolated, this means that the irreversible step was adiabatic, $dq = 0$. Then return the system to the starting point reversibly, feeding in whatever heat and doing whatever work is necessary to get it there. Now consider the entropy change in this process. From Clausius' inequality, the total integral of dq/T around the cycle is negative, since there was an irreversible step. However, the dq/T integral for the irreversible step is zero, since this process was adiabatic. If the starting point was A and the irreversible process took the system to B, then

$$\oint \frac{dq}{T} = \int_A^B \frac{dq_{\text{irr}}}{T} + \int_B^A \frac{dq_{\text{rev}}}{T} < 0 \qquad (4\text{-}29)$$

but

$$dq_{\text{irr}} = 0$$

$$\int_A^B \frac{dq_{\text{irr}}}{T} = 0$$

$$\int_B^A \frac{dq_{\text{rev}}}{T} < 0$$

However, for reversible processes, we know that $dq_{rev} = T\,d\mathscr{S}$. Therefore,

$$\begin{aligned} \int_B^A d\mathscr{S} &< 0 \\ \mathscr{S}_A - \mathscr{S}_B &< 0 \\ \mathscr{S}_B &> \mathscr{S}_A \end{aligned} \tag{4-30}$$

The entropy of the terminus B of the irreversible step is thus seen to be greater than that of the starting point A.

Properties of $\mathscr{S}$

We have already seen that in the free expansion of an ideal gas against no opposing force, that is, into a vacuum, the work done, heat flow, and energy change are all zero. Note that the gas must be ideal, for only in this case is $(\partial E/\partial V)_T = 0$. However, the entropy change is not zero; $\Delta\mathscr{S} = R\ln(V_2/V_1)$, a positive quantity as expected.

We can also calculate the entropy change as a function of temperature, both for constant volume and constant pressure processes.

$$\begin{aligned} d\mathscr{S}_V &= \frac{dq_V}{T} = \frac{dE}{T} = \frac{C_V\,dT}{T} \\ d\mathscr{S}_P &= \frac{dq_P}{T} = \frac{dH}{T} = \frac{C_P\,dT}{T} \end{aligned} \tag{4-31}$$

Similarly, we can easily calculate the entropy of a phase change. At constant pressure, where $q_P = \Delta H$, the expression for the change in entropy is

$$\Delta\mathscr{S}_{\text{phase change}} = \frac{\Delta H_{\text{phase change}}}{T} \tag{4-32}$$

Note that ΔH, the heat of the phase change, may itself be a function of the temperature. The temperature T in the denominator is, of course, the temperature *at which* the phase change is occurring, which is not necessarily the same as the "normal melting point" or "normal boiling point" of the substance.

Calculation of Absolute Entropy

With a knowledge of the behavior of entropy with temperature for a given phase, and what happens at phase changes, it is possible to calculate the entropy of a substance relative to its entropy at absolute zero, using experimental thermal data alone.

As a first example, calculate the absolute entropy of methane gas. The data needed will be the heats of fusion and vaporization, and the heat capacities of solid, liquid, and gaseous methane at constant pressure as a function of temperature. All of these quantities can be measured, and if C_P/T is plotted against T, or C_P versus $\ln T$, $\int C_P/T\,dT$ can be evaluated graphically (see Figure 4-10). To this must be added the entropy changes involved in phase transitions.

The first entropy change is that involved in taking solid methane from 0°K to its melting point T_m:

$$\Delta\mathscr{S}_1 = \int_0^{T_m} \frac{C_P^s(T)}{T}\,dT$$

Then the solid is melted at this temperature:

$$\Delta\mathscr{S}_2 = \frac{\Delta H_f}{T_m}$$

where ΔH_f is the latent heat of fusion. Now the liquid is taken from its freezing point to its boiling point:

$$\Delta\mathscr{S}_3 = \int_{T_m}^{T_b} \frac{C_P^l(T)}{T}\,dT$$

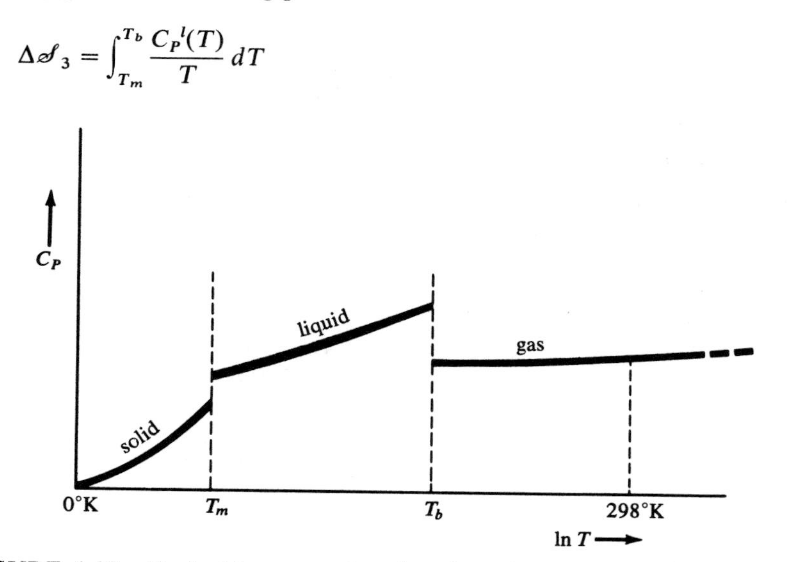

FIGURE 4-10. *Typical heat capacity plots for solid, liquid, and gas. The heat capacity of a substance ordinarily rises upon melting, for the new dimensions of motion in the liquid permit new ways of doing work against intermolecular attractive forces, and hence of storing heat. The heat capacity falls again upon vaporization, for the molecules in the gas phase are so far apart that intermolecular attractive forces are much weaker.*

The liquid is vaporized by feeding in the latent heat of vaporization ΔH_v:

$$\Delta \mathscr{S}_4 = \frac{\Delta H_v}{T_b}$$

In the final step the gas is taken from its boiling point to 298°K:

$$\Delta \mathscr{S}_5 = \int_{T_b}^{298^\circ\text{K}} \frac{C_P{}^g(T)}{T}\, dT$$

Since entropies are additive, the total entropy at 298°K is

$$\mathscr{S}_{298^\circ\text{K}} = \mathscr{S}_{0^\circ\text{K}} + \int_0^{T_m} \frac{C_P{}^s}{T}\, dT + \frac{\Delta H_f}{T_m} + \int_{T_m}^{T_b} \frac{C_P{}^l}{T}\, dT + \frac{\Delta H_v}{T_b} + \int_{T_b}^{298^\circ\text{K}} \frac{C_P{}^g}{T}\, dT \tag{4-33}$$

The only unknown term is $\mathscr{S}_{0^\circ\text{K}}$, the entropy at absolute zero. This cannot be evaluated without the third law, but our knowledge of the eventual equivalence of statistical and thermodynamic entropies should prepare us for the fact that $\mathscr{S}_{0^\circ\text{K}}$ will turn out to be zero.

Now look at a real example, oxygen gas. O_2 has three crystalline phases as well as liquid and gas phases. The same thermal data are needed as for methane—heats of all phase changes and all variations of C_P with T. For convenience, C_P is often plotted against $\log_{10} T$. The area under the curve is then the entropy change as given by $d\mathscr{S} = 2.303\, C_P\, d(\log_{10} T)$. It also turns out that it is technically difficult to measure heat capacities accurately below 10°K, but that heat capacities can be extrapolated fairly accurately using the Debye theory of heat capacity, according to which as absolute zero is approached, $C_V = kT^3$. The constant k can be evaluated by fitting the expression to the data at the lowest measurable temperatures. The calculations are shown in Table 4-1.

Note the relatively low entropy of melting, but the higher entropy of heating the liquid to its boiling point. Considering entropy as disorder, we can understand this. When the solid is first melted, not all of the crystalline order is destroyed; semiordered domains remain as with the water–ice system. Only at higher temperatures is this short range order totally destroyed. Compare as well the greater entropy increase in vaporizing the liquid than in melting the solid.

TABLE 4-1 CALCULATIONS INVOLVED IN OBTAINING THE ABSOLUTE THERMODYNAMIC ENTROPY OF OXYGEN GAS AT 298°K

Phases	*Method of obtaining* $\Delta\mathscr{S}$	$\Delta\mathscr{S}$ (e.u.)
I : 0° → 14°	Extrapolation	0.54
I : 14° → 23.66°	$\Delta\mathscr{S} = 2.303 \int_{14°}^{23.66°} C_P(T)\, d\log_{10} T =$ graph area I	1.50
I → II	$\Delta\mathscr{S} = \frac{\Delta H}{T} = \frac{22.42 \text{ cal}}{23.66\ °\text{K}}$	0.95
II : 23.66° → 43.76°	$\Delta\mathscr{S} =$ graph area II	4.66
II → III	$\Delta\mathscr{S} = \frac{177.6 \text{ cal}}{43.76°\text{K}}$	4.06
III : 43.76° → 54.39°	$\Delta\mathscr{S} =$ graph area III	2.40
III → l	$\Delta\mathscr{S} = \frac{106.3 \text{ cal}}{54.39°\text{K}}$	1.95
l : 54.39° → 90.13°	$\Delta\mathscr{S} =$ graph area l	6.46
$l \rightarrow g$	$\Delta\mathscr{S} = \frac{1628.8 \text{ cal}}{90.13°\text{K}}$	18.07
g : 90.13° → 298°	$\Delta\mathscr{S} =$ graph area g	3.91
		44.50

$$\mathscr{S}_{298°\text{K}} = \mathscr{S}_0 + 44.50 \text{ e.u.}$$

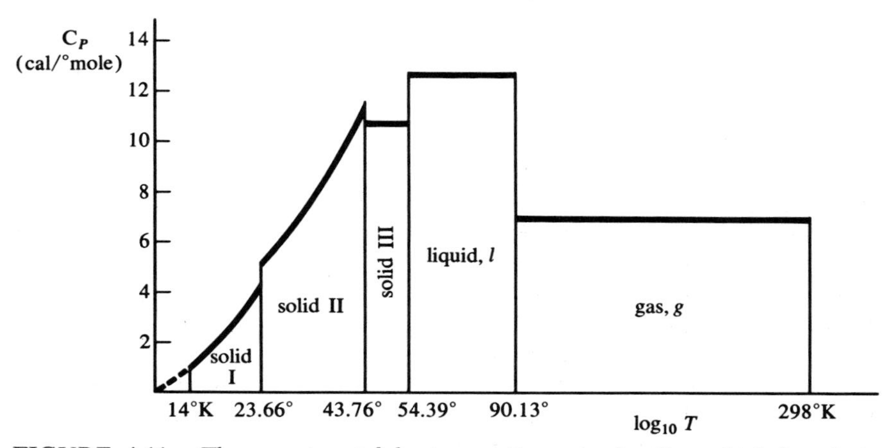

FIGURE 4-11. *The experimental heat capacity curve for* O_2*, which has three different crystal forms.*

4-5 *MOLECULAR INTERPRETATION OF THERMODYNAMIC ENTROPY*

We have already obtained an expression for the thermal energy, or the energy which a system possesses above that at absolute zero, in Chapter 2.

$$\bar{E} = kT^2\left(\frac{\partial \ln Q}{\partial T}\right)_V = RT^2\left(\frac{\partial \ln q}{\partial T}\right)_V \tag{2-72}$$

Can we find a similar expression for the *thermodynamic* entropy in terms of molecular partition functions? To do so will require a three-step derivation: first, work will be found in terms of partition functions, then heat will be calculated from $dq = dE - dw$, and finally entropy will be obtained from $d\mathscr{S} = dq/T$. The first step involves some new concepts, but the second and third steps are purely manipulative.

Molecular Form of the First Law

Suppose we have many particles with molecular energy levels ε_j and total energy $E = \sum_j n_j \varepsilon_j$. Then the overwhelmingly likely population distribution will be

$$\frac{n_j}{n} = \frac{g_j e^{-\varepsilon_j/kT}}{q} \tag{2-30}$$

In what ways can the energy be altered? The total energy can be altered either by changing the distribution of particles among the levels, or by shifting the levels up and down. Differentiating E,

$$dE = \sum_j \varepsilon_j \, dn_j + \sum_j n_j \, d\varepsilon_j \tag{4-34}$$

The first term represents a change in the population of the levels. How can one change the population of a system, if the levels themselves remain fixed? From Eq. 2-30, if the system of fixed levels remains at all times in its most probable state, this can be done only by changing the temperature, or by adding or extracting heat. Thus $\sum_j \varepsilon_j \, dn_j = dq$. The second term is the change in energy produced by leaving the particles in the same quantum levels, but changing the spacing between levels. We show in the next section that this corresponds to doing mechanical work on the system: $dw = \sum_j n_j \, d\varepsilon_j$.

A model system with a small number of particles may make these two terms clearer. Take a system of three energy levels, energies 0, 1, and 2 (Figure 4-12). Put five particles in the lowest level, four in the first level, and two in the highest level, giving $E = 8$ energy units. Now compress the gas adiabatically until the spacings have doubled, with energies 0, 2, and 4. The energy is now $E = 16$ units. One of the best experimental ways of causing an adiabatic compression is with a shock wave, where compression is so fast that heat does not have time to flow. In the absence of heat flow, the distribution of particles among levels is unchanged, and $dE = dw$. Now cool this system and extract eight units of energy as heat. The temperature will have risen in the adiabatic compression, and the restoration of the energy to eight units will bring the system down to its original temperature. This will be accomplished by having particles drop to lower energy levels.

There is a mild amount of oversimplification in this model. In an adiabatic process $dq = \sum_j \varepsilon_j \, dn_j = 0$, but this does not necessarily require that each individual $dn_j = 0$. There could be individual population changes of a sort which leave the sum unchanged. But it is proven in Exercise 26 of this chapter that this is not so; for an *isotropic* reversible adiabatic expansion of an ideal gas, the spacings between levels change but the populations of the levels remain the same. For less symmetrical expansions, such as an increase in the length of a rectangular chamber while the width and height remain unchanged, there are minor and compensating fluctuations in population (Exercise 27). Nevertheless, the picture of an adiabatic process as one involving alterations of the positions of energy levels without significant changes in populations of the levels remains valid.

The first law in its molecular form is

$$dE = dq + dw = \sum_j \varepsilon_j \, dn_j + \sum_j n_j \, d\varepsilon_j \tag{4-35}$$

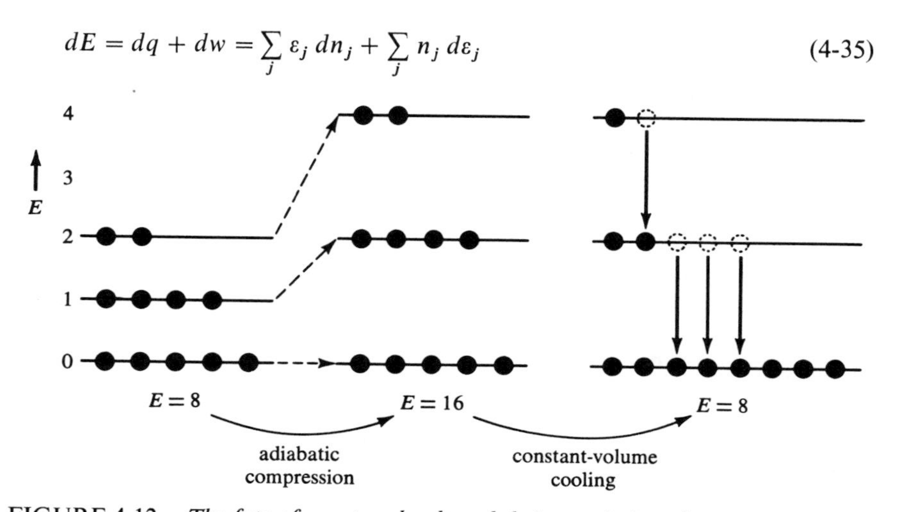

FIGURE 4-12. *The fate of quantum levels and their populations during an adiabatic compression and then a temperature drop at constant volume. Although there is rearrangement of molecules among the degenerate states of a given quantum level in an adiabatic compression, the populations of these quantum levels remain unchanged. See Exercises 4-26 and 4-27.*

Heat flow is reflected in an altered population of energy levels, and PV work done on or by the system is represented by a change in the energies of the levels.

Molecular Interpretation of Work

Let $x, y, z, u, v, w, \ldots,$ be all of the *independent* measurable parameters whose changes affect the volume of our system, and hence the energy levels. In the simplest case, they might be just the dimensions of the system. The volume is then a function of these parameters.

$$V = V(x, y, z, u, v, w, \ldots) = V_{(x_i)}$$

When any parameter x_i is changed, the energy levels shift because the volume changes. The force required to change parameter x_i for a molecule in the jth energy level is just the partial of ε_j with respect to x_i.

$$f_{j,i} = \frac{\partial \varepsilon_j}{\partial x_i} \tag{4-36}$$

The work done on the molecule is this force times dx_i.

$$dw_{j,i} = f_{j,i}\, dx_i = \frac{\partial \varepsilon_j}{\partial x_i}\, dx_\iota = d\varepsilon_{j,i} \tag{4-37}$$

This $d\varepsilon_{j,i}$ is the change in energy of a molecule in the jth level when parameter x_i is changed by dx_i.

As an example, consider a particle in a three-dimensional box. We can calculate $d\varepsilon_{j,i}$ from the expression for the energy levels.

$$\varepsilon_{j,k,l} = \frac{h^2}{8m}\left(\frac{j^2}{a^2} + \frac{k^2}{b^2} + \frac{l^2}{c^2}\right)$$

$$d\varepsilon_{j,k,l,a} = -\frac{h^2}{8m}\frac{2j^2}{a^3}\, da = dw$$

When a decreases the energy is increased, since the particle in the (j, k, l)th state now has a higher energy than it did before the compression.

Now look at the total energy change in the system due to changes in all the level spacings, or the total work done on the system by altering the parameter x_i.

$$dw_i = \sum_j n_j\, d\varepsilon_{j,i} = \frac{n}{q}\sum_j g_j e^{-\varepsilon_j/kT}\left(\frac{\partial \varepsilon_j}{\partial x_i}\right) dx_i \tag{4-38}$$

But the differential of the molecular partition function q with respect to parameter x_i is

$$\left(\frac{\partial q}{\partial x_i}\right) = -\frac{1}{kT}\sum_j g_j e^{-\varepsilon_j/kT}\left(\frac{\partial \varepsilon_j}{\partial x_i}\right) \tag{4-39}$$

Combining these last two expressions, we have

$$dw_i = \frac{-nkT}{q}\left(\frac{\partial q}{\partial x_i}\right)_T dx_i \tag{4-40}$$

or, if $n = N$ for one mole, then

$$dw_i = -RT\left(\frac{\partial \ln q}{dx_i}\right)_T dx_i \tag{4-41}$$

The constant temperature subscript has been added at this point to recognize explicitly the absence of heat effects and of redistribution of particles among levels. The total work done is the sum over all parameters x_i, or

$$dw = -RT\sum_i\left(\frac{\partial \ln q}{\partial x_i}\right)_T dx_i \tag{4-42}$$

There is a certain formal symmetry between this expression and the one for energy, which involved the rate of change of ln q with temperature at constant volume.

Now apply this last expression to the three-dimensional particle in a box, and see if the results so far are reasonable.

$$q = \frac{(2\pi mkT)^{\frac{3}{2}}}{h^3}abc$$

$$dw_a = -RT\left(\frac{\partial \ln q}{\partial a}\right)_T da$$

$$\frac{\partial q}{\partial a} = \frac{q}{a}$$

$$\frac{\partial \ln q}{\partial a} = \frac{1}{q}\frac{\partial q}{\partial a} = \frac{1}{a}$$

therefore

$$dw_a = -RT\frac{da}{a}$$

$$dw = -RT\left(\frac{da}{a} + \frac{db}{b} + \frac{dc}{c}\right)$$

But $V = abc$, so that $dV = bc\,da + ac\,db + ab\,dc$, $dV/V = da/a + db/b + dc/c$, or

$$dw = -\frac{RT}{V}\,dV = -P\,dV$$

which we already know is true. Thus our derivation for dw so far is consistent with our starting assumptions.

Calculation of Heat dq

With only 26 Roman letters, plus 24 Greek and a few more Cyrillic, confusion in notation in thermodynamics is not only possible but customary. For clarity in this section only, heat will be represented by q and the molecular partition function by q'.

We have already demonstrated one molecular form for dq, namely, $dq = \sum_j \varepsilon_j\,dn_j$, but this form is not particularly useful. The expression needed comes from the first law, $dq = dE - dw$. With expressions for dE and dw, obtaining dq is only a matter of mathematical manipulation. For simplicity we shall use $\beta \equiv 1/kT$ as in the original derivation of the Boltzmann function.

$$\bar{E} = RT^2\left(\frac{\partial \ln q'}{\partial T}\right)_V = -N\left(\frac{\partial \ln q'}{\partial \beta}\right)_V \tag{4-43}$$

$$d\bar{E} = -Nd\left(\frac{\partial \ln q'}{\partial \beta}\right)_V = -RT\beta d\left(\frac{\partial \ln q'}{\partial \beta}\right)_V \tag{4-44}$$

$$dw = -RT\sum_i\left(\frac{\partial \ln q'}{\partial x_i}\right)_T dx_i \tag{4-42}$$

Notice that this last expression for dw involves the volume parameters x_i. This is inconvenient, and we would like to replace the x_i's by volume itself.

The volume is a function of the volume parameters, $V = V(x_i)$, so that the partition function can be expressed alternatively as $q'(T, V)$ or $q'(\beta, x_i)$.

Then,

$$d \ln q' = \left(\frac{\partial \ln q'}{\partial \beta}\right)_V d\beta + \sum_i \left(\frac{\partial \ln q'}{\partial x_i}\right)_T dx_i \tag{4-45}$$

Substituting, we find

$$dw = RT\left(\frac{\partial \ln q'}{\partial \beta}\right)_V d\beta - RT\, d \ln q' \tag{4-46}$$

Then

$$dq = dE - dw = RT\, d \ln q' - RT\left(\frac{\partial \ln q'}{\partial \beta}\right)_V d\beta - RT\beta\, d\left(\frac{\partial \ln q'}{\partial \beta}\right)_V$$

$$dq = RT\, d \ln q' - RT\, d\left[\beta\left(\frac{\partial \ln q'}{\partial \beta}\right)_V\right] \tag{4-47}$$

But, from Eq. 4-43, we see that

$$\frac{\bar{E}}{T} = -\frac{N}{T}\left(\frac{\partial \ln q'}{\partial \beta}\right)_V = -R\beta\left(\frac{\partial \ln q'}{\partial \beta}\right)_V$$

$$dq = RT\, d \ln q' + Td\left(\frac{\bar{E}}{T}\right) = Td\left(R \ln q' + \frac{\bar{E}}{T}\right) \tag{4-48}$$

Calculation of Entropy

The last step in the derivation is the calculation of the thermal entropy, $d\mathscr{S} = dq/T$. From Eq. 4-48, we see immediately that

$$d\bar{\mathscr{S}} = d\left(R \ln q' + \frac{\bar{E}}{T}\right) \tag{4-49}$$

or, upon integrating,

$$\bar{\mathscr{S}} - \bar{\mathscr{S}}_0 = R \ln q' + \frac{\bar{E}}{T} \tag{4-50}$$

As we see in Section 4-10, this integration constant $\mathscr{S}_0$ will turn out to be the thermodynamic entropy per mole of a substance at 0°K. What this value is, or even whether it will differ from one substance to another, we cannot yet say.

Now recall the derivation of entropy in statistical terms in Chapter 2. Statistical entropy was defined as $S = k \ln W$, and when the expression for the most probable or Boltzmann distribution was substituted, the following equation resulted:

$$\overline{\mathscr{S}} = k \ln Q + \frac{\bar{E}}{T} = R \ln q' + \frac{\bar{E}}{T} \tag{2-88}$$

If we now compare statistical entropy S, which we know to be a measure of disorder, with the thermal entropy $\mathscr{S}$, which came from the second law and which has no *a priori* connection with disorder, we see that

$$\mathscr{S} = S + \mathscr{S}_0 \tag{4-51}$$

This $\mathscr{S}_0$ is at present an unknown quantity. It could be positive, negative, or zero, and could vary from one substance to another. For the moment, we can say that the statistical and thermodynamic entropies of a substance differ at most by an additive constant, and that this constant is the value of the thermodynamic entropy of the substance at absolute zero.

4-6 PROPERTIES OF ENTROPY FUNCTION $\mathscr{S}$

Remembering now that $\mathscr{S}$ is thermodynamic entropy, and that $S = \mathscr{S} - \mathscr{S}_0$ and therefore $dS = d\mathscr{S}$, we can derive some useful relationships between $\mathscr{S}$, P, V, and T, and can clean up some older relationships between E, H, P, V, and T.

Energy as a Function of T and V

Let us look first at the relationships between energy, volume, temperature, and entropy, considering energy as a function of temperature and volume. We can immediately say that

$$dE = \left(\frac{\partial E}{\partial T}\right)_V dT + \left(\frac{\partial E}{\partial V}\right)_T dV \tag{4-52}$$

and from the first law

$$dq = \left(\frac{\partial E}{\partial T}\right)_V dT + \left[P + \left(\frac{\partial E}{\partial V}\right)_T\right] dV \tag{4-53}$$

It is easy to prove by using Euler's test that dq is not an exact differential.

Applying it to Eq. 4-53, we find

$$\frac{\partial^2 E}{\partial T\, \partial V} \stackrel{?}{=} \left(\frac{\partial P}{\partial T}\right)_V + \frac{\partial^2 E}{\partial V\, \partial T} \tag{4-54}$$

This asserts that pressure is independent of temperature if volume is held constant, which in general is false. Hence the cross differentials are not equal and dq is not exact. From the relationship $d\mathscr{S} = dq/T$,

$$d\mathscr{S} = dS = \frac{1}{T}\left(\frac{\partial E}{\partial T}\right)_V dT + \left[\frac{P}{T} + \frac{1}{T}\left(\frac{\partial E}{\partial V}\right)_T\right] dV \tag{4-55}$$

But since $\mathscr{S}$ is a state function,

$$\mathscr{S} = \mathscr{S}(T, V)$$

$$d\mathscr{S} = \left(\frac{\partial \mathscr{S}}{\partial T}\right)_V dT + \left(\frac{\partial \mathscr{S}}{\partial V}\right)_T dV \tag{4-56}$$

and equating coefficients in this and in Eq. 4-55, we obtain two relationships

$$\left(\frac{\partial \mathscr{S}}{\partial T}\right)_V = \frac{1}{T}\left(\frac{\partial E}{\partial T}\right)_V = \frac{C_V}{T} \tag{4-57}$$

$$\left(\frac{\partial \mathscr{S}}{\partial V}\right)_T = \frac{1}{T}\left[P + \left(\frac{\partial E}{\partial V}\right)_T\right] \tag{4-58}$$

The first relation we have already seen, but the second is new.

Applying Euler's test to $d\mathscr{S}$, which is known to be an exact differential, the following must be true:

$$\left(\frac{\partial E}{\partial V}\right)_T = T\left(\frac{\partial P}{\partial T}\right)_V - P \tag{4-59}$$

This relationship between the volume dependence of energy in isothermal processes and the temperature dependence of pressure at constant volume is true for *all* substances under *all* conditions, and follows directly from the acceptance of the validity of the first and second laws. Such an expression is known as a *thermodynamic equation of state*. To illustrate how useful it is, let us apply it first to an ideal gas and then to a van der Waals gas.

We have argued intuitively that there is no connection between energy and volume for an ideal gas, and have even defined an ideal gas as one obeying the two relationships $P\overline{V} = RT$ and $(\partial E/\partial T)_V = 0$. But it turns out that this latter requirement is superfluous; the thermodynamic equation of state (Eq. 4-59) tells us that the former statement, $P\overline{V} = RT$, implies the latter:

$$\left(\frac{\partial P}{\partial T}\right)_V = \frac{R}{\overline{V}} = \frac{P}{T} \qquad \left(\frac{\partial E}{\partial V}\right)_T = T\left(\frac{\partial P}{\partial T}\right)_V - P = P - P = 0$$

Hence it was unnecessary to make the second statement about the ideal gas; it followed automatically. We could not say this, however, until we accepted the second law, and the fact that entropy is a state function.

Now consider a van der Waals gas, the first-order correction to an ideal gas. Although the true pressure and volume do not obey the expression $P\overline{V} = RT$, one can regard this as arising from the imperfections of the gas, and can imagine an *effective* pressure P' and an *effective* molar volume $\overline{V}'$ for which $P'\overline{V}' = RT$ still held. The effective volume would be less than the actual volume $\overline{V}$ because some of the space is occupied by the gas molecules themselves, giving $\overline{V}' = \overline{V} - b$. The effective internal pressure, however, would be larger than the externally measurable pressure because of attraction between the molecules, $P' = P + A$. Molecules in the center of the gas feel an isotropic attraction, but molecules near the walls feel a greater net attraction from molecules in the bulk of the gas, which tends to retard their flight toward the walls and hence to lower the externally measured pressure. The magnitude of this effect will be proportional to the number of molecules hitting the wall, and the degree to which each molecule near the wall is retarded. Both of these effects are proportional to the density of the gas, and hence inversely proportional to the volume. Thus $P' = P + A = P + a/\overline{V}^2$. The van der Waals approximation to real gas behavior is then

$$\left(P + \frac{a}{\overline{V}^2}\right)(\overline{V} - b) = RT \tag{4-60}$$

Which of these two effects is more important in causing the energy to be dependent on volume, intermolecular attraction or the bulk of the molecules? We can find this out from the thermodynamic equation of state:

$$\left(\frac{\partial P}{\partial T}\right)_V = \frac{R}{V - b} \qquad \left(\frac{\partial E}{\partial V}\right)_T = \frac{RT}{V - b} - P = \left(P + \frac{a}{V^2}\right) - P = \frac{a}{V^2}$$

The important factor is therefore intermolecular attraction and not the fact that molecules have finite volumes.

We can use the thermodynamic equation of state in still another way. When the expression for $(\partial \mathscr{S}/\partial V)_T$ was derived, it was left in a rather cumbersome form which can now be cleaned up. The use of Eq. 4-58 with Eq. 4-59 yields the very compact expression

$$\left(\frac{\partial \mathscr{S}}{\partial V}\right)_T = \left(\frac{\partial P}{\partial T}\right)_V \tag{4-61}$$

Look again at this last result. It looks suspiciously like the Euler reciprocity check result on some exact differential. This would be a function χ such that $d\chi = \mathscr{S} dT + P dV$, and if by Eq. 4-61, $d\chi$ is an exact differential, then χ is a state function. This is one of the two new state functions which will be found later in this chapter.

Enthalpy as a Function of T and P

An exactly similar analysis applicable to conditions of constant pressure can be made using enthalpy rather than energy. The mathematics are identical to the previous section—start with $H = H(T, P)$, use the exact differential properties of dH, use $dq = dH - V\,dP$, and differentiate. Including the results of the Euler cross differentials, the following expressions result:

$$\left(\frac{\partial \mathscr{S}}{\partial T}\right)_P = \frac{C_P}{T} \tag{4-62}$$

$$\left(\frac{\partial \mathscr{S}}{\partial P}\right)_T = -\left(\frac{\partial V}{\partial T}\right)_P \tag{4-63}$$

$$\left(\frac{\partial H}{\partial P}\right)_T = V - T\left(\frac{\partial V}{\partial T}\right)_P \tag{4-64}$$

This last expression is another thermodynamic equation of state, and can be used exactly like the first one. Equation 4-63 again suggests an exact differential $d\phi = -\mathscr{S}\,dT + V\,dP$, and a new state function ϕ.

We are now in the position to derive a useful expression for the Joule–Thomson coefficient μ_{JT}. It was defined as the pressure dependence of temperature under isoenthalpic conditions and was shown to be

$$\mu_{JT} = \left(\frac{\partial T}{\partial P}\right)_H = -\frac{1}{C_P}\left(\frac{\partial H}{\partial P}\right)_T \tag{3-51}$$

With the second thermodynamic equation of state, this becomes

$$\mu_{JT} = \frac{1}{C_P}\left[T\left(\frac{\partial V}{\partial T}\right)_P - V\right] \tag{4-65}$$

If we know the PVT relationships for a gas, then μ_{JT} is directly calculable. For an ideal gas,

$$\left(\frac{\partial V}{\partial T}\right)_P = \frac{R}{P} = \frac{V}{T} \qquad \mu_{JT} = -\frac{1}{C_P}\left[V - T\left(\frac{V}{T}\right)\right] = 0$$

Again it is useful to approach real gas behavior via a van der Waals gas. To work out the Joule–Thomson coefficient for this gas, we must differentiate volume with respect to temperature. The pure van der Waals equation is clumsy for this purpose, but we can derive an approximate relationship which will work for moderate pressures. In the expression $P\overline{V} = RT + Pb - a[(\overline{V} - b)/\overline{V}^2]$, it is only this last term which gives trouble. We can make the approximation that the volume of the molecules themselves is small compared to $\overline{V}$, that is, $\overline{V} - b \approx \overline{V}$, and that $1/\overline{V} \approx P/RT$. This then produces the modified van der Waals equation

$$P\overline{V} = RT + Pb - Pa/RT \tag{4-66}$$

$$\overline{V} = \frac{RT}{P} + b - \frac{a}{RT} \tag{4-67}$$

$$\left(\frac{\partial V}{\partial T}\right)_P = \frac{R}{P} + \frac{a}{RT^2}$$

$$\left(\frac{\partial H}{\partial P}\right)_T = \overline{V} - \frac{RT}{P} - \frac{a}{RT} = \left(\frac{RT}{P} + b - \frac{a}{RT}\right) - \frac{RT}{P} - \frac{a}{RT} = b - \frac{2a}{RT}$$

$$\mu_{JT} = \frac{1}{C_P}\left(\frac{2a}{RT} - b\right) \tag{4-68}$$

For very low temperatures the first term will dominate, and the coefficient will be positive, while for higher temperatures the coefficient will be negative, and the inversion temperature will be $T_{JT} = 2a/Rb$. This approximation to Joule–Thomson behavior of real gases is quite good, failing only at high pressures (compare Figures 4-13 and 3-14), where the approximations in the modified van der Waals equation fail. The full van der Waals equation predicts the fall-off of μ with lower T at high pressures as well.

Heat Capacity

In Chapter 3 we developed an expression for the difference between the heat capacity at constant volume and at constant pressure. This can now be simplified:

$$C_P - C_V = \left[\left(\frac{\partial E}{\partial V}\right)_T + P\right]\left(\frac{\partial V}{\partial T}\right)_P = T\left(\frac{\partial P}{\partial T}\right)_V\left(\frac{\partial V}{\partial T}\right)_P \tag{4-69}$$

This is true for any substance, including solids and liquids. In these condensed phases, C_P is very similar but not identical to C_V, since there is little volume change with temperature. It is common in condensed phases to express Eq. 4-69 in terms of the coefficient of thermal expansion α and the coefficient of compressibility β. α is the fractional volume change per degree Kelvin at constant pressure:

$$\alpha = \frac{1}{V}\left(\frac{\partial V}{\partial T}\right)_P$$

while β is the fractional volume decrease with increasing pressure at a constant temperature

$$\beta = -\frac{1}{V}\left(\frac{\partial V}{\partial P}\right)_T$$

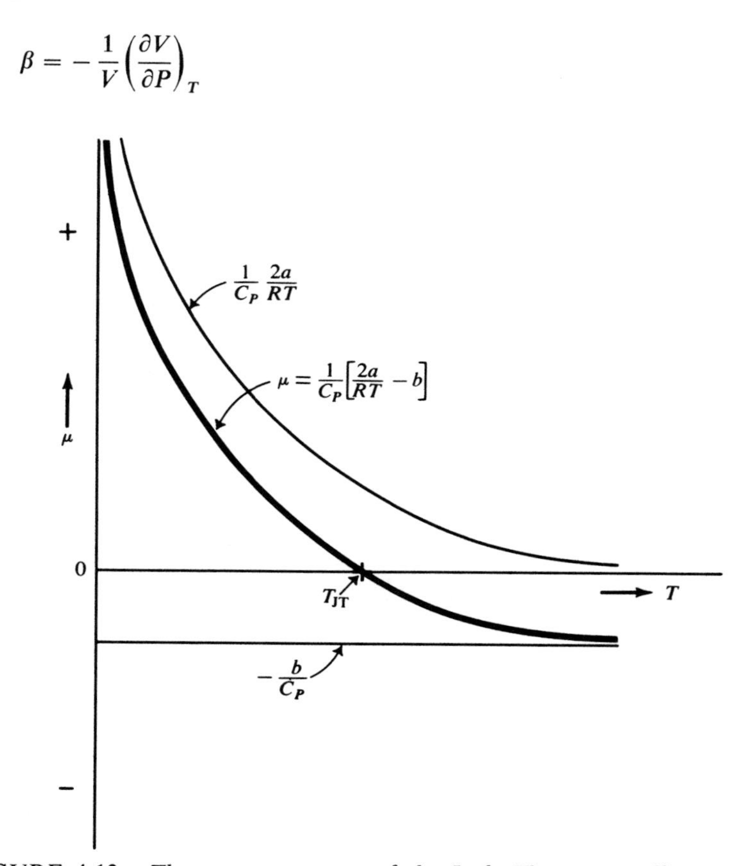

FIGURE 4-13. *The two components of the Joule–Thomson coefficient μ_{JT} for a van der Waals gas. At low temperatures, intermolecular attractions dominate, and at high temperatures, the finite size of the molecules becomes more important.*

Equation 4-69 can be initially simplified by eliminating $(\partial V/\partial T)_P$.

$$C_P - C_V = TV\alpha\left(\frac{\partial P}{\partial T}\right)_V \tag{4-70}$$

How do we express $(\partial P/\partial T)_V$ in terms of α and β? This requires a little mathematical sleight of hand. Since both of the ultimately desirable variables α and β involve partials of V in terms of P and T, we might begin by treating V as a function of P and T:

$$dV = \left(\frac{\partial V}{\partial P}\right)_T dP + \left(\frac{\partial V}{\partial T}\right)_P dT$$

Differentiating with respect to T at constant V,

$$\left(\frac{\partial V}{\partial T}\right)_V = 0 = \left(\frac{\partial V}{\partial P}\right)_T\left(\frac{\partial P}{\partial T}\right)_V + \left(\frac{\partial V}{\partial T}\right)_P$$

This now yields

$$\left(\frac{\partial P}{\partial T}\right)_V = -\frac{(\partial V/\partial T)_P}{(\partial V/\partial P)_T} = \frac{\alpha}{\beta}$$

which can be substituted in Eq. 4-70. The final expression for difference in heat capacities is

$$C_P - C_V = VT\frac{\alpha^2}{\beta} \tag{4-71}$$

This expression is equally true for solids, liquids, and gases, although its most common application is to condensed phases.

Entropy of Mixing—Ideal Gases

The last property of entropy which we want to consider is the entropy of mixing, in order to demonstrate that the thermodynamic entropy of mixing obeys the same rules as the statistical entropy of mixing as developed in Section 2-17.

Assume at the start that there are n_a moles of ideal gas A and n_b moles of ideal gas B, occupying volumes V_a and V_b. If the gases are initially separated by a partition and the partition is then removed, the particles will mix in a total volume $V_{a+b} = V_a + V_b$. If the process is isothermal, then ΔE will be zero. With no PV work done, w will be zero, and q as well. If the process were reversible, then $q = 0$ would require that $d\mathscr{S} = dq/T = 0$ as well. But the mixing process is *not* reversible, and $dq \neq Td\mathscr{S}$. What is $\Delta\mathscr{S}$ for the mixing process?

The way to solve this problem is to imagine some *equivalent* reversible process and calculate its entropy changes. This reversible process may be divided into two steps: the reversible expansion of each of the gases individually to their final volume V_{a+b}, and the *reversible adiabatic* mixing of the contents of these expanded volumes. Which process is more important in the entropy change, expansion or mixing? We will see that, oddly enough, the entire entropy change comes from the expansion; the reversible adiabatic mixing makes no contribution.

Max Planck was the one who first proposed the hypothetical experiment that we now carry out. The first step is the reversible expansion of the two gases each to a volume V_{a+b}. The entropy changes in these two isothermal expansions are

$$\Delta\mathscr{S}_a = n_a R \ln\left(\frac{V_{a+b}}{V_a}\right)$$

$$\Delta\mathscr{S}_b = n_b R \ln\left(\frac{V_{a+b}}{V_b}\right)$$

The final reversible adiabatic mixing, like all reversible adiabatic processes, produces no entropy change. Then the total entropy change is

$$\Delta\mathscr{S} = n_a R \ln\left(\frac{V_{a+b}}{V_a}\right) + n_b R \ln\left(\frac{V_{a+b}}{V_b}\right)$$

$$\Delta\mathscr{S} = -n_a R \ln\left(\frac{V_a}{V_{a+b}}\right) - n_b R \ln\left(\frac{V_b}{V_{a+b}}\right)$$

$$\Delta\mathscr{S} = -n_a R \ln X_a - n_b R \ln X_b$$

and, *per mole* of gas,

$$\Delta\overline{\mathscr{S}} = -R(X_a \ln X_a + X_b \ln X_b) \tag{4-72}$$

But how can we carry out such a reversible adiabatic mixing? This is Planck's contribution; he came up with the idea of using selectively permeable membranes. Planck assumed two membranes, one of which passes A molecules freely but blocks B (membrane α), and the other of which passes B, but not A (membrane β). He then imagined the gases each in one of two telescoping cylinders of volume V_{A+B} each, as shown in Figure 4-14. The right end of the A cylinder is sealed with membrane β and the left end cylinder of B with membrane α. Planck does not have to show that these membranes can be found for any particular real substances; if they are conceivable, they can be used. As long as the laws of thermodynamics are not violated by their existence, then whether or not such membranes exist for particular molecules is irrelevant.

Now slide one cylinder slowly into the other. At the intermediate stage of mixing of Figure 4-14, A molecules pass through membrane α and bounce off β, so that the volume available to them remains V_{a+b} throughout the mixing process. A similar statement holds for molecules of type B. The pressure on the left side of cylinder A, P_a, arises only from collisions of A molecules, since no B molecules can get past membrane α. But interestingly enough, the pressure on the *right* end of the A cylinder, which is closed by membrane β, is also caused only by the A molecules, since the B molecules go through membrane β without hindrance and contribute nothing to the pressure. Similar argument shows that the pressure on both ends of cylinder B is the same as well. Since the pressure at both ends of each cylinder is the same, the system is in a state of mechanical equilibrium at all stages of the mixing. It takes an infinitesimal amount of mechanical work to push the cylinders together or pull them apart. Hence the mixing is reversible.

For many components, Eq. 4-72 can be generalized.

$$\Delta \mathscr{S}_{\text{mix}} = -R \sum_j X_j \ln X_j \tag{4-73}$$

Note that this is the same result that was derived for the statistical entropy of mixing, Eq. 2-81. But from Planck's imaginary mixing process we see that the entropy change really comes from the expansion. There is no mixing interaction between the A and B molecules; as ideal gases they are unaware of each other's existence except on collision along lines of centers. After expansion they have more room in which to move, increasing the disorder of the system. It makes no difference to the entropy whether they are then mixed in a final volume of $V_a + V_b$ or not.

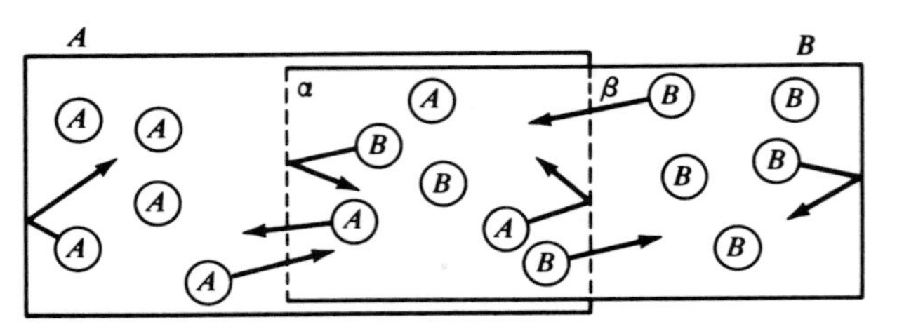

FIGURE 4-14. *Planck's hypothetical apparatus for reversible, adiabatic mixing of gases, using selectively permeable membranes, α permeable only to A molecules, and β permeable only to B.*

4-7 *WORK FUNCTION A*

Our original goal was to find a potential function which would be minimized at equilibrium. It should be a state function, perhaps some combination of the state functions we already know about: pressure, temperature, volume, energy, enthalpy, and entropy. We also know some of the properties this function must have. Under constant energy conditions, a system will tend spontaneously towards a state of maximum entropy, while under conditions of constant entropy the system will tend to occupy the lowest energy level available to it. A simple function with these properties immediately suggests itself: $E - \mathscr{S}$. For the sake of consistency in units, this must be modified to something of the form of $E - T\mathscr{S}$.

Definition of the Work Function A

Let us tentatively define a function which will be called the work function, or the Helmholtz free energy function, by

$$A \equiv E - T\mathscr{S} \tag{4-74}$$

and see if it has the properties of our desired potential.

Differentiating,

$$dA = dE - d(T\mathscr{S}) = dq + dw - Td\mathscr{S} - \mathscr{S}dT \tag{4-75}$$

Until now, dw has represented only the work of mechanical expansion, or PV work, but other types of work are possible. Some of these are the following.

(1) PV expansion $\quad dw = -PdV$
(2) surface expansion: (surface tension $= \gamma$) $\quad dw = \gamma dA$
(3) stretching a rod: [tensile force $= K(l)$] $\quad dw = K(l)\,dl$
(4) magnetization: ($\mathbf{H}$ field produces magnetic moment $\mathbf{M}_m$) $\quad dw = \mathbf{H} \cdot d\mathbf{M}_m$
(5) polarization of dielectric: ($\mathbf{E}$ field produces dipole moment $\mathbf{M}_e$) $\quad dw = \mathbf{E} \cdot d\mathbf{M}_e$
(6) electrochemical work: (charge dq moved through potential drop $\mathscr{E}$) $\quad dw = -\mathscr{E}dq$

The final example of electrochemical work is important in electrochemistry. If the unit of charge is one electron then the work is $w = -e\mathscr{E}$, where e is the electronic charge. For 1 mole of electrons, $w = -Ne\mathscr{E} = -\mathscr{F}\mathscr{E}$, where $\mathscr{F}$ is Faraday's constant or 1 mole of electronic charge. Thus if one chemical equivalent of reaction is done on 1 mole of substance, the amount of work done is $w = -\mathscr{F}\mathscr{E}$. If n chemical equivalents of work per mole are done in a cell, then the work done is $w = -n\mathscr{F}\mathscr{E}$.

We can now group all these different forms of work into two classes: PV work, and other kinds of work which we call the external work dw_{ext}. Then $dw = dw_{ext} - PdV$.

Reversible Isothermal Processes

The work function can be regrouped in a slightly more useful form.

$$dA = (dq - Td\mathscr{S}) + dw - \mathscr{S}dT \qquad (4\text{-}76)$$

In a reversible process, $dq = Td\mathscr{S}$, and the term in parentheses is zero. Furthermore, for isothermal processes, $dT = 0$, leaving only

$$dA = dw \qquad \text{(reversible, isothermal)} \qquad (4\text{-}77)$$

$$\Delta A = w$$

The change in the work function of a substance in any reversible isothermal process is just equal to the work done on it from the outside. Moreover, any work done by the system is paid for by a drop in the work function; hence its name.

As an example, take a system and do work on it in a reversible isothermal manner. If the working substance is an ideal gas, then this work will be lost as heat; if the gas is not ideal, then some of the work will go to changing the internal energy. The change in the work function in this process is equal to the *total* work done, but the internal energy change is less, by an amount equal to the heat loss since $\Delta E = w + q$ (remember that q is negative in this process). The same argument holds if heat is fed in. Suppose that 5 cal of net heat are fed in and 8 cal of useful work are done on the surroundings. The change in A will be -8 cal, although the internal energy change will be only -3 cal. The drop in A represents the work done, while the drop in E represents that part of the work done which was not compensated for by the addition of heat. Hence the Helmholtz free energy represents the *available* work of the system, as opposed to its total energy.

Reversible Isothermal Processes—PV Work Only

Let us put one more restriction on the work function and assume now that only PV work is involved. Then

$$dA = -PdV \qquad (4\text{-}78)$$

If the process were being carried out under constant volume conditions, such as in a bomb calorimeter, dV and hence dA would be zero, and A would be constant. *Thus at constant volume and temperature, the work function is unchanged for all reversible processes.* But remember that a reversible process is defined as one which is at equilibrium for all intermediate steps. Therefore, under conditions of constant volume and temperature $dA = 0$ at equilibrium. In a plot of A against extent of reaction, A must be at a maximum, minimum, or inflection point at equilibrium, and the only question is which (see Figure 4-15).

The customary way of finding whether a zero-slope point is a maximum, minimum, or inflection point is to look at the sign of the second derivative. But an easier way here is to look at points slightly away from equilibrium and see if they have a higher A, lower A, or either, depending upon how one leaves the equilibrium point. For this we need an irreversible process.

Irreversible Isothermal Processes

For a spontaneous, irreversible process, the work done by the system is less than that for an equivalent reversible one, and the heat absorbed will be less than the reversible $Td\mathscr{S}$. Recall the expression for dA:

$$dA = (dq - Td\mathscr{S}) + dw - \mathscr{S}dT \tag{4-79}$$

The last term is zero because the process is isothermal, and the term in parentheses is negative. Hence

$$dA < dw \qquad \Delta A < w \tag{4-80}$$

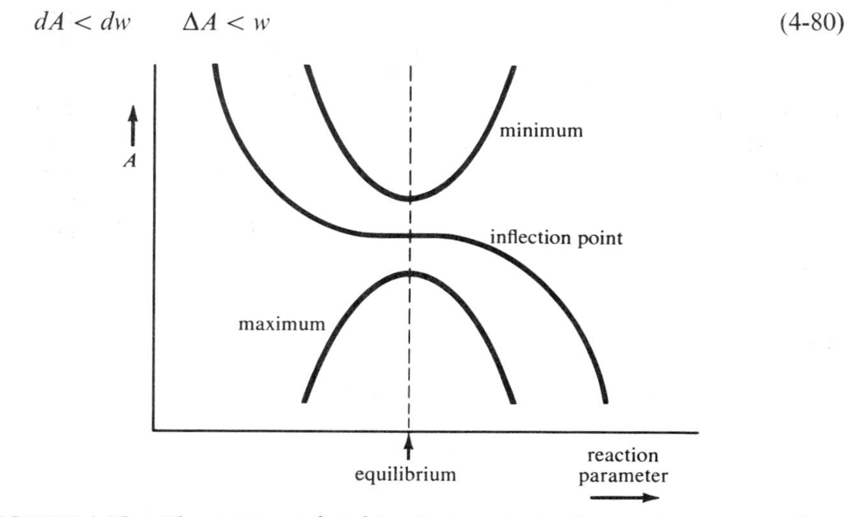

FIGURE 4-15. *The statement that $dA = 0$ at constant volume and temperature for a reversible (that is, equilibrium) process means that A is at a minimum, a maximum, or an inflection point at equilibrium. The fact that $dA < 0$ for an irreversible process, which is driving* toward *equilibrium, means that A is minimized at equilibrium and is therefore a useful potential function for chemical reactions at constant V and T.*

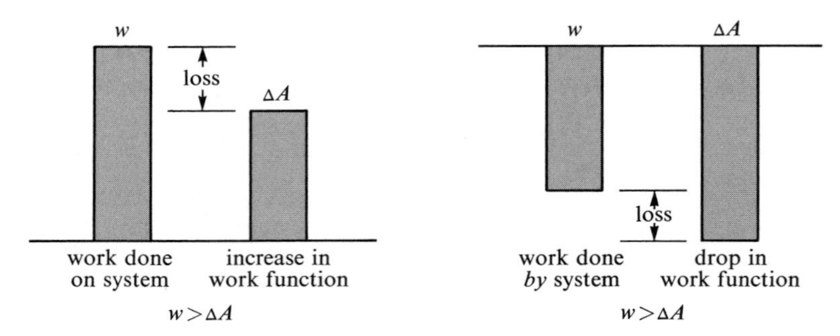

FIGURE 4-16. *An illustration of Murphy's law for the work function. The easiest way to remember the significance of the statement $w > \Delta A$ for irreversible processes is to remember that whether you are doing work* on *the system or extracting useful work* from *the system, you lose.*

This says that the *increase* in the work function produced inside the system is not as great as the work fed into the system. If a certain amount of work is done on a system irreversibly, its work function does not increase by this entire amount; some of the work is wasted in some nonconservative process such as friction (Figure 4-16). If work is done *by* the system, then both ΔA and w will be negative, but the fall in A is greater than the work done, and Eq. 4-80 still holds.

If there is no external work involved—electrical, gravitational, magnetic—and if the volume is held constant, then $w = 0$, and dA is negative for any spontaneous, irreversible process. Such a process is one which is driving *toward* equilibrium. Thus in any spontaneous isothermal process in which no work is done, the work function falls. Under these conditions the position of equilibrium is found by minimizing the work function A with respect to the parameters by which the path of the reaction is measured. It is apparent that A is here the chemical potential for which we have been seeking (Figure 4-17).

Summary

For any reversible process not involving external work,

$$dA = dw - \mathscr{S}dT = -PdV - \mathscr{S}\,dT \tag{4-81}$$

This is the state function whose existence was inferred from Eq. 4-61. It is the chemical potential function under certain restricted conditions. For *any* reversible isothermal process, the change in A is a measure of work done on or by the system. In any real process, A rises less than the work done on the system, and falls more than the work done by the system.

4-8 GIBBS FREE ENERGY G

A function more useful to the chemist is one involving constant pressure and temperature, since most chemical processes are carried out at constant pressure. (If they do not stay at constant pressure at all intermediate points of the reaction, at least at the *end* of the processes the pressure is brought back to its original value. For state functions, this is good enough.) This will turn out to be the function known as the Gibbs free energy, or commonly just free energy.

Definition of Free Energy G

The energy of a reaction was found earlier to be equal to the heat of reaction at constant volume, and the enthalpy was defined in such a way as to play a similar role at constant pressure. The enthalpy was built from the energy by adding a PV term. This suggests that a new free energy function, useful at constant P and T, might be

$$G \equiv A + PV \tag{4-82}$$

Let us see if this is so:

$$dG = dA + d(PV) = dE - Td\mathscr{S} - \mathscr{S}dT + PdV + VdP$$
$$dG = (dq - Td\mathscr{S}) + (dw + PdV) - \mathscr{S}dT + VdP$$

At constant temperature and pressure, the last two terms drop out.

$$dG = (dq - Td\mathscr{S}) + (dw + PdV) \tag{4-83}$$

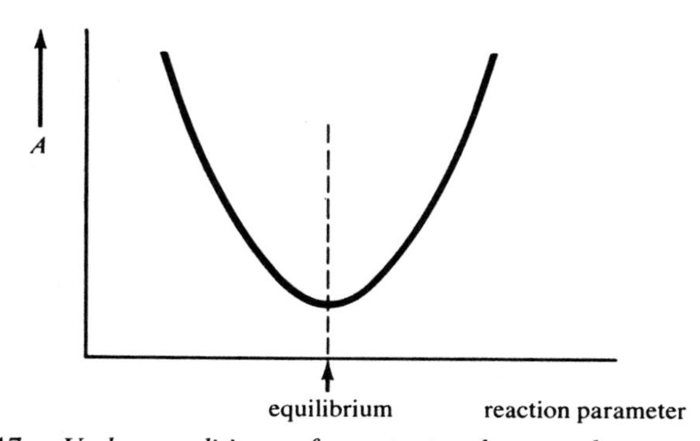

FIGURE 4-17. *Under conditions of constant volume and temperature, with no external work (electrical, magnetic, gravitational, and so on), the work function A is the proper potential function to minimize in order to find the conditions of chemical equilibrium.*

Furthermore, note that $dw = dw_{\text{ext}} - PdV$, so that the term in the second parentheses is dw_{ext}.

$$dG = (dq - T\,d\mathscr{S}) + dw_{\text{ext}} \tag{4-84}$$

Reversible Processes—Constant P and T

For reversible processes, the first term in parentheses is zero, leaving

$$dG = dw_{\text{ext}} \qquad \text{or} \qquad \Delta G = w_{\text{ext}} \tag{4-85}$$

The free energy increases in a reversible process by the amount of work done on the system *other than PV* work. Note that even under these conditions, the *work function A* still gives the total work done in the reversible process, including PV work. Under isothermal and constant pressure conditions,

$$dA = dG - PdV$$

If, for example, we did 5 cal of electrical work and 6 cal of compression work on a system, the free energy increase ΔG would be 5 cal, while the work function would rise by the full 11 cal.

If no external work is involved, then G is unchanged in a reversible process. For isothermal, isobaric processes, G is at a minimum, maximum, or saddle point at equilibrium; in order to decide among these alternatives we must look at the change in G in a spontaneous process.

Irreversible Processes

In an irreversible process, $dq < Td\mathscr{S}$ and

$$dG < dw_{\text{ext}} \qquad \Delta G < w_{\text{ext}} \tag{4-86}$$

The rise in free energy is less than the *external* work done on a system, and the fall in free energy is greater than the external work done by a system on its surroundings. If we were to do 5 cal of magnetization work on a system, the free energy might only rise by 4 cal, and in the act of using these 4 cal of free energy to do gravitational work, only 3 cal of work might be obtained.

In any irreversible process where external work is not involved, $dG < 0$ and the free energy falls. Hence, since spontaneous, irreversible processes are processes which are moving toward equilibrium, G must be a minimum at equilibrium. Thus for isothermal, constant *pressure* processes, the free energy G is the proper chemical potential function and not A.

Very few chemical reactions are truly isothermal and isobaric, with both temperature and pressure unchanged throughout the course of the reaction. But most reactions end at or can be brought back to the initial P and T, and the changes in the thermodynamic state functions such as E, H, S, A, and G are the same no matter what the details of intermediate pressures and temperatures. This, of course, is the prime virtue (and definition) of a state function. Hence in future discussions, unless special comments are made, such phrases as "isothermal," "isobaric," or "constant volume" are to be interpreted as meaning only a return to the initial temperature, or pressure, or volume, at the end of the reaction.

For any reversible process not involving external work,

$$dG = -\mathscr{S}dT + VdP \tag{4-87}$$

Note that G is the other new state function suggested earlier by Eq. 4-63.

EXAMPLE 3. Electrochemical Cell. Consider the chemical reaction of hydrogen and chlorine gases to give hydrochloric acid. The reaction is

$$\tfrac{1}{2}H_{2(g)} + \tfrac{1}{2}Cl_{2(g)} \rightarrow HCl_{aq}$$

We are going to carry this reaction out first irreversibly and then reversibly, and compare the free energy changes in the two processes.

(1) *Irreversible process.* First mix the gases in a flask, and then trigger the reaction with light (it will go photochemically). The two gases are in a highly metastable condition, and, when triggered, react explosively. Now dissolve the HCl produced in water, and finally cool the system down to the initial temperature before the explosion. It is quite obvious that this reaction is irreversible; violently so, in fact. Only PV work has been done, so that the free energy of the system must have fallen.

(2) *Alternate irreversible process.* The same reaction can also be carried out irreversibly in an electrochemical cell (Figure 4-18), with hydrogen gas bubbled over one electrode and chlorine gas over the other. The half-reactions occurring at each electrode are shown. Electrons produced at the hydrogen electrode (a) will flow to the chlorine electrode (b). If we assume that the wire connecting the electrodes has no resistance, then the flow of electrons along the wire does no work. The external work is zero, and the free energy change of the cell as a whole is negative. Note that this process, free flow of electrons against no opposing potential, is analogous to the free expansion of a gas into a vacuum, against no opposing pressure.

(3) *Reversible process.* Just as a free expansion of a gas can be made reversible by providing a suitable opposing back pressure, so this electron-flow reaction can be made reversible by supplying a back potential. Such a cell is shown in Figure 4-19. The potentiometer can be adjusted until the

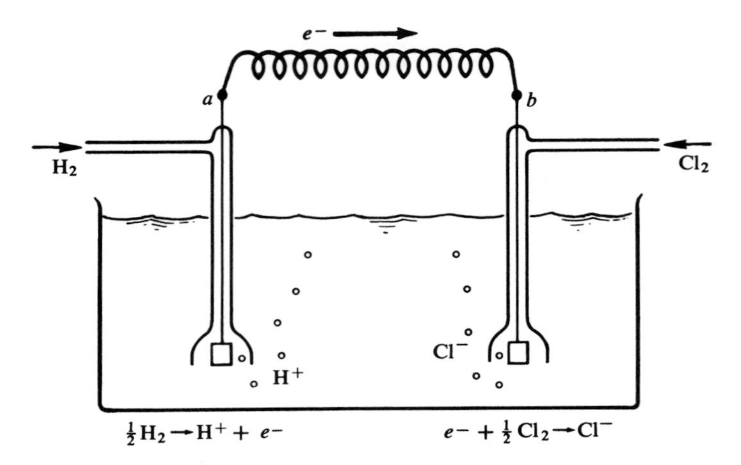

FIGURE 4-18. *An irreversible electrochemical reaction, with electrons flowing from one terminal to another freely, with no opposing back emf. This is analogous to the free expansion of a gas into a vacuum. No electrical work is done.*

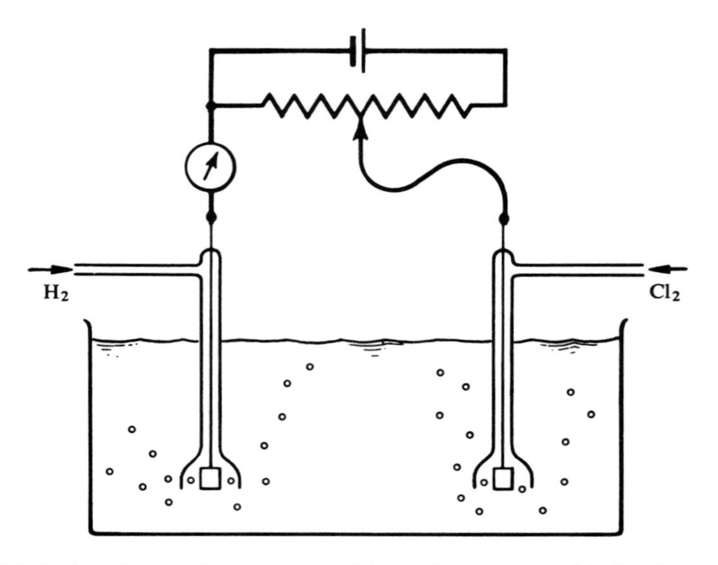

FIGURE 4-19. *A reversible way of conducting the reaction, with the electrons doing work against the maximum possible back emf at each instant. This is analogous to a reversible expansion of a gas, where $P_{ext} = P_{int}$. Like the reversible expansion, this reversible cell reaction would take an infinite length of time to complete, for with the back emf balancing the emf of the cell, the galvanometer reading will be zero and no current will flow. Still, this, like all reversible processes, is a convenient if unattainable limit for the behavior of real and irreversible processes.*

back emf is exactly balancing the potential of the cell itself. Electrons then will not flow, and a galvanometer in the circuit will show no current. This is then a reversible process, and like all reversible processes requires an infinite amount of time, being only the limiting case of real, irreversible processes. The free energy change of the cell-plus-potentiometer system must be zero. However, the cell goes from the same initial to the same final states as in the irreversible process of example (1). Since G is a state function, the free energy change of the cell itself must still be negative, but since the whole system does not change in free energy, the potentiometer must increase in G by the same amount by which the cell decreases. Work is done on the potentiometer, and in this case the battery is charged. The work done on the potentiometer is $n\mathscr{F}\mathscr{E}$, and the free energy change of the cell is $\Delta G_{\text{cell}} = -n\mathscr{F}\mathscr{E}$.

4-9 PROPERTIES OF A AND G

Nomenclature

Unfortunately, there are many overlapping names and symbols for the Gibbs and Helmholtz free energies. Table 4-2 lists the most important. Be sure,

TABLE 4-2 FREE ENERGY NOMENCLATURE

Modern notation	A = work function or Helmholtz free energy	G = free energy or Gibbs free energy
G. N. Lewis and the past two generations of American chemists	A = work function	F = free energy
Willard Gibbs, who started it all	ψ (no name)	ζ = free energy
A particularly pernicious practice of Europeans and of American physicists	F	G
A particularly lucid set of names used in Europe but unfortunately not in this country	free energy	free enthalpy

especially if reading European journals, to know what the symbols stand for. Also note that Europeans (and physicists) tend to use the symbol U for the energy E. There are undoubtedly other variants of those in Table 4-2. We shall use only the first variant with A and G.

The remainder of this section is concerned with relationships involving A and G in reversible processes with no external work. Then changes in A or G are given by

$$dA = -PdV - \mathscr{S}dT \tag{4-81}$$

$$dG = +VdP - \mathscr{S}dT \tag{4-87}$$

Isothermal Processes

At constant temperature, the equations become

$$\begin{aligned} dA &= -PdV \\ dG &= +VdP \end{aligned} \tag{4-88}$$

Notice that these are parts of the total differential of PV; in fact, $d(PV) = -dA + dG$.

For ideal gases, the above expressions produce a result which at first looks paradoxical. In straightforward integration, $\Delta A = -RT\ln(V_2/V_1) = RT\ln(P_2/P_1)$, and $\Delta G = RT\ln(P_2/P_1)$. Thus $\Delta A = \Delta G$. Is there no difference between A and G for an ideal gas? This explanation is simple: $G = A + PV$ and for the ideal gas $PV = RT$, so that $G = A + RT$. For an isothermal process $dA = dG$, and hence $\Delta G = \Delta A$. The free energy and work content are not the same, but they *change* in tandem for an ideal gas.

Temperature Dependence of Isothermal ΔA and ΔG

From earlier expressions for dA and dG,

$$\left(\frac{\partial A}{\partial T}\right)_V = -\mathscr{S} = \left(\frac{\partial G}{\partial T}\right)_P \tag{4-89}$$

At constant volume, for each component of a reaction

$$\left(\frac{\partial A_j}{\partial T}\right)_V = -\mathscr{S}_j \tag{4-90}$$

and, for the entire reaction,

$$\left(\frac{\partial \Delta A}{\partial T}\right)_V = -\Delta\mathscr{S} \tag{4-91}$$

where $\Delta\mathscr{S} = \sum_j \nu_j \mathscr{S}_j$ and $\Delta A = \sum_j \nu_j A_j$. But at constant temperature, $\Delta\mathscr{S}$

can be eliminated

$$\Delta A = \Delta E - T\,\Delta\mathscr{S}$$

$$\Delta\mathscr{S} = \frac{\Delta E - \Delta A}{T}$$

$$\left(\frac{\partial \Delta A}{\partial T}\right)_V = \frac{\Delta A - \Delta E}{T}$$

Now observe what happens if $\Delta A/T$ is differentiated with respect to temperature:

$$\left(\frac{\partial(\Delta A/T)}{\partial T}\right)_V = \frac{1}{T}\left(\frac{\partial \Delta A}{\partial T}\right)_V - \frac{1}{T^2}\,\Delta A = \frac{1}{T}\left[\left(\frac{\partial \Delta A}{\partial T}\right)_V - \frac{\Delta A}{T}\right]$$

Substituting back the expression for $(\partial\Delta A/\partial T)_V$, the following expression results, known as the Gibbs–Helmholtz equation:

$$\left(\frac{\partial(\Delta A/T)}{\partial T}\right)_V = -\frac{\Delta E}{T^2} \tag{4-92}$$

The importance of this equation is that it provides a method of calculating the work function change in a chemical reaction as a function of the reaction temperature.

A slightly different form of this equation, more convenient for some purposes, is

$$\left(\frac{\partial(\Delta A/T)}{\partial(1/T)}\right)_V = \Delta E \tag{4-93}$$

If $\Delta A/T$ is plotted against $1/T$, the slope at any point of the curve is the energy of the reaction at that temperature (see Figure 4-20).

Under constant pressure conditions, the same derivation can be made using H for E and G for A, giving another form of the Gibbs–Helmholtz equation.

$$\left(\frac{\partial(\Delta G/T)}{\partial T}\right)_P = -\frac{\Delta H}{T^2} \tag{4-94}$$

$$\left(\frac{\partial(\Delta G/T)}{\partial(1/T)}\right)_P = \Delta H \tag{4-95}$$

If the heat of the reaction is known as a function of temperature, the variation of the free energy with temperature may be calculated. Or conversely, free

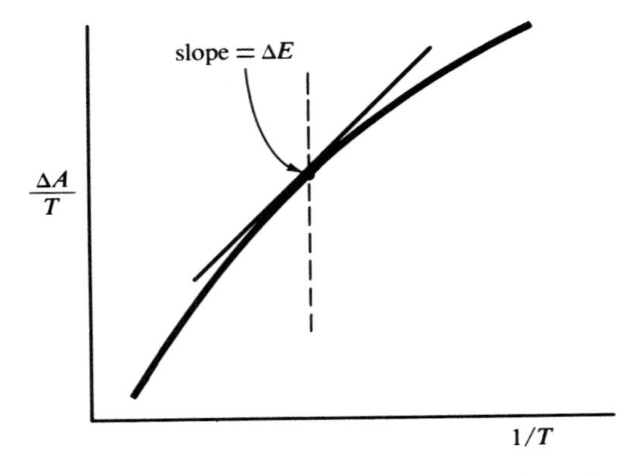

FIGURE 4-20. *From the Gibbs–Helmholtz equation with A, the slope of a plot of $\Delta A/T$ against $1/T$ is the internal energy change of the reaction ΔE.*

energy of reaction can be measured very conveniently in an electrochemical cell from the relationship $\Delta G(T) = -n\mathscr{F}\mathscr{E}(T)$. A plot of $\mathscr{E}(T)/T$ versus $1/T$ will have a slope of $-\Delta H/n\mathscr{F}$. This is a very convenient and common method of getting standard heats of reactions. (See Figure 4-21.)

Free Energy of Reactions

Recalling the standard convention for chemical reaction notation presented in Section 3-14, a generalized chemical reaction may be written.

$$0 = \sum_j \nu_j A_j \tag{3-84}$$

where the A's are chemical components, not work functions. Then using the

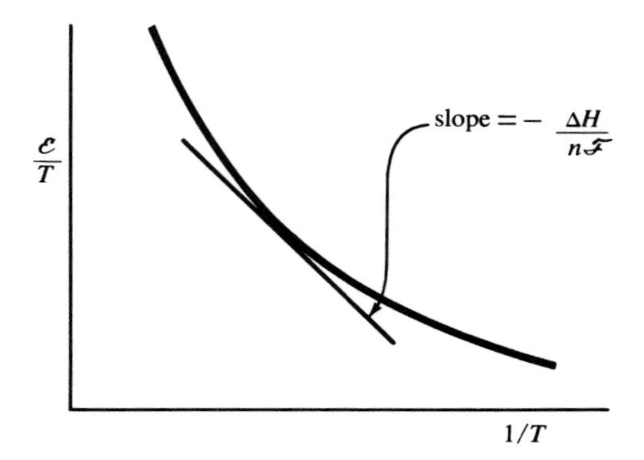

FIGURE 4-21. *Measurements of the variation of cell emf with temperature provide a way of obtaining the enthalpy of the cell reaction, using the Gibbs–Helmholtz equation in G.*

convention developed for the standard heats of reaction

$$\Delta H^0_{298} = \sum_j \nu_j \, \Delta H^0_{f(298)} \tag{3-85}$$

(in which ΔH^0_f was the standard enthalpy of formation of a compound from its elements), by analogy one can write

$$\Delta G^0_{298} = \sum_j \nu_j \, \Delta G^0_f \tag{4-96}$$

representing the standard free energy of a reaction. The ΔG^0_f's have yet to be defined, however, and this must wait until after the third law.

Practical Calculation of $\Delta G(T)$

The Gibbs–Helmholtz equation ties the change of free energy with temperature to the heat of reaction, and we already know how to find $\Delta H(T)$ from heat capacity considerations (Section 3-16). If heat capacity data are in the usual form of $C_{P,j} = a_j + b_j T + c_j T^2$, then the form of the dependence of H with temperature is

$$\Delta H(T) = \Delta H_0 + \Delta a T + \frac{\Delta b}{2} T^2 + \frac{\Delta c}{3} T^2 \tag{3-94}$$

The integration constant ΔH_0 is obtained by knowing the heat of reaction for any one temperature.

Now consider the Gibbs–Helmholtz equation in differential form.

$$d\left(\frac{\Delta G}{T}\right) = -\left(\frac{\Delta H}{T^2}\right) dT$$

$$= -\left(\frac{\Delta H_0}{T^2} + \frac{\Delta a}{T} + \frac{\Delta b}{2} + \frac{\Delta c}{3} T\right) dT$$

and integrating

$$\frac{\Delta G}{T} = I + \frac{\Delta H_0}{T} - \Delta a \ln T - \frac{\Delta b}{2} T - \frac{\Delta c}{6} T^2$$

$$\Delta G(T) = IT + \Delta H_0 - \Delta a T \ln T - \frac{\Delta b}{2} T^2 - \frac{\Delta c}{6} T^3 \tag{4-97}$$

The integration constant I can be evaluated, like H_0, by knowing the *free energy* at any given temperature. If the power-series approximation to C_P were valid all the way to absolute zero, then I would have physical significance;

$I = (\Delta G_0 - \Delta H_0)/T = -\Delta\mathscr{S}_0$. The negative of the integration constant would be the entropy of reaction at absolute zero. But in practice, I has no such real significance and is only a constant of integration.

In summary, the heat of reaction as a function of temperature requires heat capacity data in the relevant temperature range, and the heat of reaction at one temperature within this range; the free energy as a function of temperature requires the above information plus the free energy at any one temperature.

There is a theoretical flaw in the above expression which comes from the presence of the integration constant $I = -\Delta\mathscr{S}_0$. It is not like the constant ΔH_0, because it does not vanish when computing the changes in free energy between two temperatures. Recall that we had

$$\Delta H_2 - \Delta H_1 = \int_{T_1}^{T_2} C_P\, dT$$

and the constant ΔH_0 drops out. But when we try to do the same thing with free energy, we find

$$\Delta G = IT + \Delta H_0 - T \int \frac{1}{T^2}\left(\int C_P\, dT\right) dT$$

$$\Delta G_2 - \Delta G_1 = I(T_2 - T_1) - T \int_{T_1}^{T_2} \frac{1}{T^2}\left(\int_{T_1}^{T_2} C_P\, dT\right) dT$$

This says that the change in free energy between two temperatures is somehow a function of the integration constant, which has been shown to be $-\Delta\mathscr{S}_0$. Remember that so far we have not put any restrictions on $\Delta\mathscr{S}_0$. Yet $\Delta\mathscr{S}_0$ cannot be arbitrary; one can measure the emf of a cell at two different temperatures and get an absolute number which is this free energy difference. There is still a flaw in the theory, and it is going to take the third law to remove it.

4-10 *THIRD LAW—IDENTITY OF STATISTICAL AND THERMODYNAMIC ENTROPY*

Obviously $\mathscr{S}_0$, or $\Delta\mathscr{S}_0$ for a reaction, is not an arbitrary quantity; arguments like those used above demonstrate this. What kind of experiments can be devised to find out what the physical value of a particular $\mathscr{S}_0$ is?

About the turn of the century T. W. Richards was working with galvanic cells, making exactly the kind of measurements mentioned in the previous section. Recall that for such cells, $\Delta G = -n\mathscr{F}\mathscr{E}$, and that one can write the Gibbs–Helmholtz equation as $\Delta H = -n\mathscr{F}[\partial(\mathscr{E}/T)/\partial(1/T)]_P$. Thus a series of measurements of the emf of a cell can give both the free energy and heat

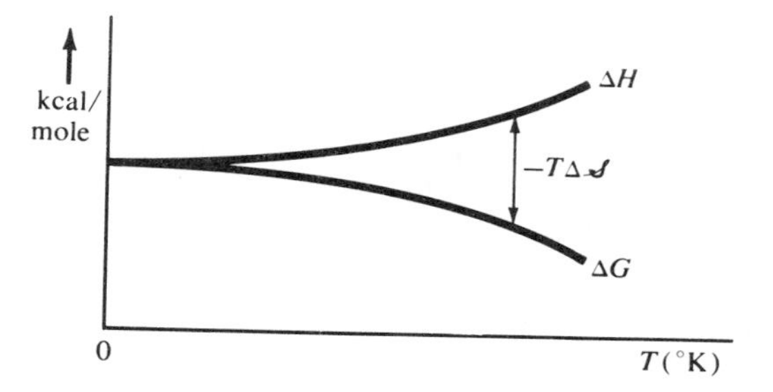

FIGURE 4-22. *As absolute zero is approached, both ΔH and ΔG for a reaction (obtained from emf measurements in an electrochemical cell) approach one another.*

of reaction involved. Richards did this for many different types of galvanic cells, and took the temperature as close as he could to absolute zero. His typical curves looked like Figure 4-22. The difference between these curves at any given temperature is just $\Delta H - \Delta G = T\Delta\mathscr{S}$; it thus appeared that at absolute zero, the free energy and heat of reaction approach the same value, or $T\Delta\mathscr{S}$ approaches zero. Note that this does not say that $\Delta\mathscr{S}$ goes to zero, only that it does not go to infinity faster than T goes to zero. Richards was, of course, never able to actually approach 0°K, but for every system he tested, the graph had the appearance of Figure 4-22.

Nernst Postulate

All that Richards was able to conclude was that $\mathscr{S}_0$ was not infinite, but from a purely statistical consideration, this would not be expected anyway. What Nernst did was to look more closely at Richards' plots. Not only did it appear that the curves approached one another, but their slopes seemed to go to zero, like Figure 4-22 and unlike Figure 4-23. But the slope of the plot of ΔG versus T is just the partial $(\partial \Delta G/\partial T)_P = -\Delta\mathscr{S}$. Therefore, if the *slope* approaches zero, so does $\Delta\mathscr{S}$, which implies that $\Delta\mathscr{S}_0 = 0$ for any reaction.

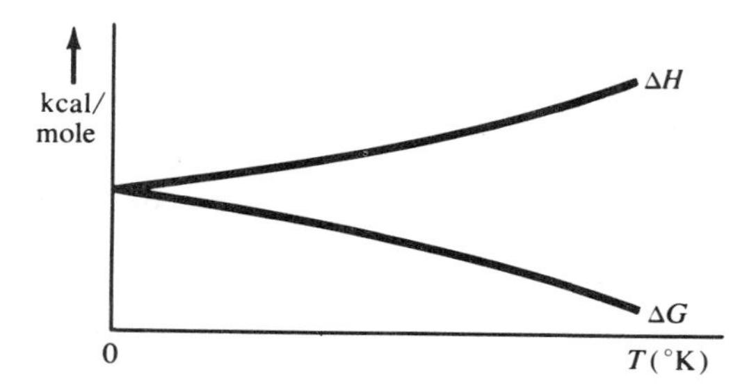

FIGURE 4-23. *As absolute zero is approached, the curves of ΔH and ΔG appear to level off to zero slope, as in Figure 4-22, and* not *like the curves in this diagram.*

Planck's Contribution

This does not mean, however, that the entropy per mole must remain fixed in a reaction; the number of moles will usually change, as in the reaction $C + O_2 \rightarrow CO_2$. But because of the chemical indestructibility of atoms, it would be reasonable to assert that the Nernst postulate is equivalent to saying that the entropy *per atom* for all atoms at absolute zero is the same arbitrary value. This is almost exactly what Planck did; he said that the simplest method for making the entropy *change* for any reaction at 0°K vanish would be to assign to all substances a *zero* entropy at that temperature. The whole business of entropy and absolute zero was finally cleaned up by G. N. Lewis, who formulated as the third law the statement that *the entropy for all pure, crystalline, perfectly ordered substances at absolute zero is zero, and the entropy of all other substances is positive.* This finally takes care of the question of statistical versus thermodynamic entropy by making the difference $\mathscr{S}_0$ vanish. Accordingly, since statistical and thermodynamic entropy are now seen to be identical, the script $\mathscr{S}$ will be abandoned and the italic S used from now on.

Why Pure? Why Perfectly Crystalline?

The reasons for Lewis' restrictions on substances whose entropy at absolute zero is zero are seen from the statistical interpretation of entropy as a measure of disorder. The reason for insisting on purity is obvious. Consider a mixture of two components, in a solid solution or alloy. Even at absolute zero the entropy of the alloy would not be zero. There would still be an entropy of mixing relative to the entropies of the pure unalloyed components.

This makes it difficult for the chemist, because there are very few pure substances available to him. Even elements cannot ordinarily be considered pure, because of the presence of isotopes. If, for example, a sample of chlorine gas were cooled to absolute zero, there would still be the residual entropy which comes from having Cl^{35} and Cl^{37} present. However, the chemist fortunately can usually ignore such isotopic factors because chemical reactions do not ordinarily separate isotopes, and the entropy factors are the same before and after reaction. In computing an entropy change for a reaction, this nonzero term representing the isotopic composition of reactants and products would appear on both sides of the equation, and cancel. Only if the reaction produced a separation of isotopes would the isotope contribution to entropy have to be included.

Another form of configurational entropy to consider is that of crystalline imperfections; any imperfection is an obvious sign of disorder, so we would naturally expect it to produce a positive entropy. An example of this disorder is the freezing in of crystal mispacking in linear molecules such as carbon monoxide. As was discussed in Section 2-17, the *energy* difference between a perfectly ordered crystal lattice and one with a few molecules lined up reversed could be very small, but the entropy effects would be noticeable. Measurements of this residual entropy have been made on CO, NO, N_2O, and a great many other molecules, and have agreed with the entropy expected from $S = k \ln W$.

The requirement of crystallinity is demonstrated by considering the metastable class of amorphous solids called glasses. Glycerol is a good example of a substance which can crystallize and can also form a glass. Is there an entropy change in going from crystalline glycerol to glass at 0°K? To find out, consider the cyclic process of Figure 4-24. The entropy change marked ΔS_0 cannot be measured directly. But from heat capacity measurements on crystal and glass, the two entropy changes in raising the temperature to the melting point can be calculated, and that of phase change at T_m can be calculated from T_m and the heat of fusion. The experimental result is that $\Delta S_0 = +4.6$ e.u./mole. This entropy rise comes from the more disordered state of the glass.

A test of the third law that is rather elegant involves the use of two crystalline forms of a single substance. If such a substance does exist in either of two *perfect crystalline* forms, then the entropy change in going from one to the other should be zero. Sulfur can exist in either the monoclinic or rhombic form. The monoclinic form is metastable under normal conditions, however, and will slowly change to the rhombic. Exactly the same sort of cycle can be carried out for sulfur as for glycerol, as in Figure 4-25. The entropy change we are seeking is the sum of the changes in the three steps of the cycle. Thus $\Delta S_0 = 8.81 + 0.26 - 9.04 = 0.03$ e.u. However, the errors in these measurements are around 0.15 e.u. Within the limits of accuracy of the data, the entropy change is indeed zero as predicted.

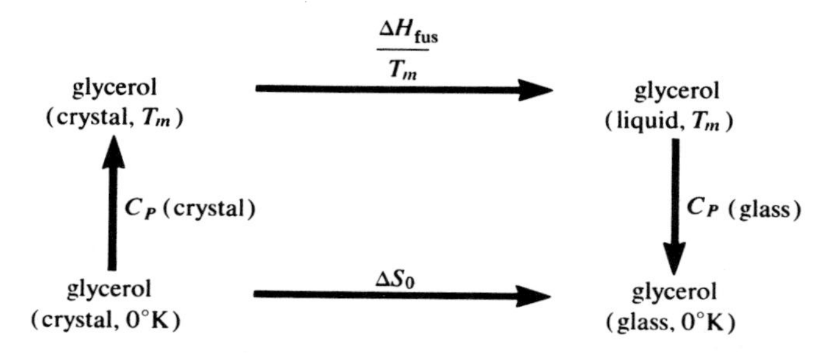

FIGURE 4-24. *An indirect way of calculating the entropy change when glycerol melts at absolute zero, using thermal measurements at higher temperatures and heat capacity extrapolations to* 0°K. *Such an alternative-path cycle is valid only because entropy is a state function.*

Alternate Form of the Third Law

Another very common version of the third law is that it is impossible to attain absolute zero in a finite number of steps. It is easy to show that this is equivalent to Lewis' statement. How does one attain very low temperatures? The simplest method is to take the desired object and put it against a colder object until thermal equilibrium is reached. This is fine, but then the limit of cooling is the coldest natural object, and this is no way to produce a new record for low temperatures.

A general method for making a substance colder than anything with which it could be placed in thermal equilibrium is the process of adiabatic relaxation. The general idea behind such a process is to take a substance, precool it as much as possible, and then let it do work on its environment adiabatically. This adiabatic work is paid for by a drop in internal energy and temperature. Adiabatic expansion of a precooled gas is at the heart of many refrigeration systems. (If the gas is also below its Joule–Thomson inversion temperature and gains an additional temperature drop from work against intermolecular attractive forces, so much the better.)

It is easy to understand adiabatic expansion from a statistical model. Suppose that initially there are 11 molecules in energy levels of 0, 2, and 4 units as in Figure. 4-26, such that the total energy is 18 units. (These are the *translational* energy levels, and of course are so closely spaced at room temperature that we seldom think of translational motion as being quantized.) Now cool the substance to the temperature of the coldest existing object, and let this lower the energy to 8 units. Since no mechanical work has been done on the system the energy levels are unchanged, but the distribution of molecules among levels is different. If the gas is now expanded adiabatically, it does work on the environment. At the same time, the volume expansion lowers the energy levels, and molecules in these levels now have only 4 units of energy. If the object used to precool the system of 11 molecules had been

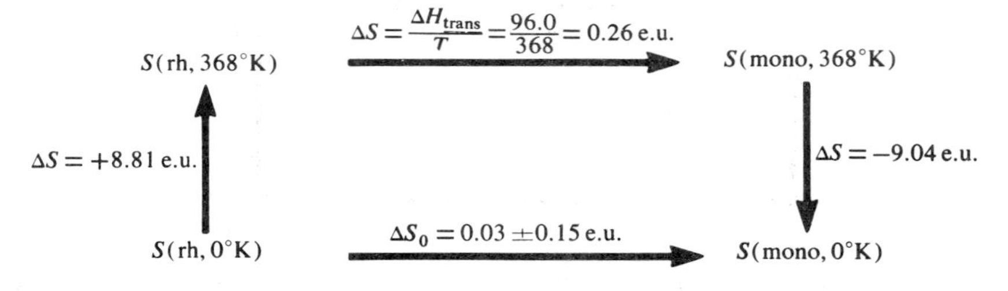

FIGURE 4-25. *The entropy of change from rhombic to monoclinic sulfur at absolute zero, obtained by the methods of Figure 4-24, is zero within the limits of accuracy of the thermal data. This is reasonable, for the two crystal packing schemes, although different, are equally ordered.*

the coldest object on earth, then the system would hold the new record for the coldest object after adiabatic expansion. Is there any reason why this process could not be repeated until absolute zero is reached?

The solid state analog of adiabatic expansion, which can be used to produce temperatures as low as 10^{-5} °K, is adiabatic demagnetization. The principle is the same but the experimental details are different. The substance, usually a paramagnetic salt, is placed in a strong magnetic field which aligns all of the vectors of the magnetic domains. As the entropy falls, heat $q = T\Delta S$ is given off. The salt is cooled as much as possible by placing it next to a colder object. It is then insulated from its surroundings and the magnetic field is turned off. The magnetic domains become misaligned again, and the configurational entropy having to do with the degree of order of the magnetic vectors rises. Yet the salt is sealed off from its environment adiabatically, so that $q = 0$, and the *total* entropy must be unchanged. The only other source of entropy is thermal, and the thermal entropy falls because the temperature falls; $\Delta S_{\text{therm}} + \Delta S_{\text{mag}} = 0$.

A possible continuous cooling process might be as follows. Take a large amount of a suitable paramagnetic salt, magnetize it, cool it as low as possible, isolate it, and demagnetize it. It is now a new coldest substance. Take half of it, remagnetize it, and use the other half to cool it and absorb some of the heat produced by the magnetization. Isolate the magnetized half and lower its temperature still more by removing the magnetic field. Divide this new coldest substance in half and repeat the cycle as many times as desired. Can absolute zero ever be reached by such a process?

It is easy to demonstrate that if Lewis' form of the third law is accepted, the answer is no. Assume that in some cooling reaction, substance A at temperature T_2 goes to substance B at temperature T_1, where $T_1 < T_2$. A might be the magnetized salt, and B the unmagnetized form. The expressions for

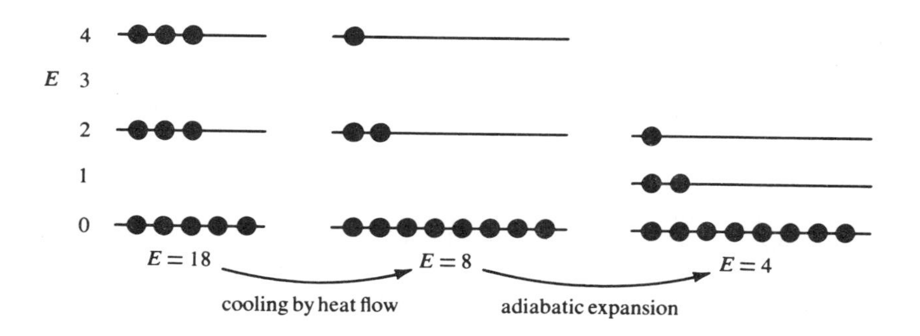

FIGURE 4-26. *After a substance has been cooled as low as possible by placing it next to colder objects and letting heat flow away, it can be cooled still further by adiabatic expansion. See Figure 4-12.*

the entropies of these substances are

$$S_A(T_2) = S_{0,A} + \int_0^{T_2} \frac{C_A}{T}\,dT$$

$$S_B(T_1) = S_{0,B} + \int_0^{T_1} \frac{C_B}{T}\,dT$$

If the reaction is to go, the entropy change must be greater than or equal to zero, or $S_B(T_1) \geqq S_A(T_2)$. Hence substituting for the entropy expressions,

$$S_{0,B} + \int_0^{T_1} \frac{C_B}{T}\,dT \geqq S_{0,A} + \int_0^{T_2} \frac{C_A}{T}\,dT$$

But by Lewis' statement of the third law, $S_{0,B} = S_{0,A}$, which leads to

$$\int_0^{T_1} \frac{C_B}{T}\,dT \geqq \int_0^{T_2} \frac{C_A}{T}\,dT$$

Remembering that the heat capacities are positive numbers, the arguments of both integrals are positive. Now T_2 was the starting temperature, so it was greater than zero, which makes the right side of the inequality greater than zero. But this in turn requires the left side to also be greater than zero.

$$\int_0^{T_1} \frac{C_B}{T}\,dT \geqq 0$$

But this can only occur if the final temperature T_1 is greater than zero. The final temperature of the process can never reach absolute zero.

4-11 *THIRD LAW ENTROPIES AND PHYSICAL PROPERTIES*

Now that S_0 has been shown to be zero for all pure crystalline solids, we are in a position to calculate *the* entropy of a substance at a standard temperature such as 298°K. This standard "third law" entropy S^0 can be tabulated as were heats of formation from elements, $\Delta H_f{}^0$. Furthermore, from the expression $\Delta G = \Delta H + T\Delta S$, a set of standard free energies of formation from elements, $\Delta G_f{}^0$, can also be tabulated. A table of ΔH^0, ΔG^0, and S^0 for various substances is given in Appendix 4. The tabulated third law entropies, being identical to statistical entropies, provide an idea of the relative disorder of various substances.

Examples of Third Law Entropy

One property of substances that correlates well with third law entropies is hardness. The molar entropy of diamond is 0.58 e.u. at 298°K, that of the hard metal tungsten is 8.0, that of the soft metal lead is 15.5, that of liquid mercury is 18.5 and of mercury vapor is 41.8. Hardness and low entropy are two manifestations of the presence of strong, directional bonds between atoms. Diamond has strong, tetrahedrally oriented covalent C—C bonds which maintain a rigid and ordered structure. In metals the interatomic forces are present but nondirectional. In liquids and gases the order and resistance to mechanical deformation are even less.

A second property which correlates with entropy is the chemical complexity of the substance. Thus the molar entropy of magnesium is 7.8, that of NaCl is 17.3, that of $MgCl_2$ is 21.4, and that of $AlCl_3$ is 40. As another example, the molar entropy of $CuSO_4$ is 27.1, that of the monohydrate is 35.8, that of the trihydrate is 53.8, and of the pentahydrate is 73.0. The extra entropy per hydrate water molecule in all of these examples, and in fact in most simple hydrated salts, is about 9.0 e.u., close to the molar entropy of ice at 298°K. This illustrates the fact that the degree of order in a hydrated salt crystal is similar to that in a crystal of pure ice. This is hardly a startling conclusion, but one which gives us confidence in the meaningfulness of third law entropy comparisons.

The expected increase in entropy upon melting and vaporization is found, with entropies of vaporization being larger in general than those of melting. In general, for substances which do not exhibit significant hydrogen bonding, the entropy rise with melting is 2–3 e.u. per mole, and 20–22 e.u. for vaporization. This is the origin of an empirical rule stated by F. Trouton in 1884, that the heat of vaporization in cal/mole divided by the normal boiling point in °K is approximately 21 for most liquids. For hydrogen-bonded substances such as water, the entropies of vaporization are closer to 25 or 26 e.u. because the hydrogen bonds must also be broken during vaporization. Acetic acid and formic acid have anomalously *low* entropies of vaporization, 14.9 and 15.4 e.u., respectively. This is because these molecules exist to an appreciable extent as dimers even in the vapor phase, and the vapor phase is less disordered relative to the liquid phase than in normal situations.

As a final example of the influence of complexity, consider the series of saturated hydrocarbons. For methane gas the standard third law entropy is 44 e.u., for ethane gas it is 54.9 e.u., and for propane gas it is 64.5 e.u. Notice, however, that even though the entropy per mole is rising, the entropy per atom is decreasing. A given number of atoms is more ordered when formed into a few large molecules than many small ones. Structural isomers, as expected, show small entropy differences. The standard entropy of *normal* butane is 74.1, but that of isobutane is 70.4. The isobutane molecule is more ordered than the long flexible straight chain. A more dramatic example occurs with pentane—the standard entropies are 83.3 for *n*-pentane, 82.0 for isopentane, and 73.2 for neopentane. The reasons are obvious from the carbon skeletons:

n-pentane:

C—C—C—C—C

isopentane:

C
|
C—C—C—C

and neopentane:

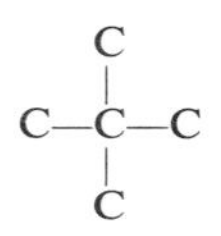

But in the liquid phase the molar entropies of isopentane and neopentane are much more alike: 62.8 and 62.4, respectively. Why is there a much smaller difference in the liquid? The molecules in a liquid are packed close together, and the packing of neo-type molecules is not significantly more ordered than the packing of the iso-molecules.

In summary, hard compounds generally have low entropies, and soft compounds, high. Light compounds have low entropies, heavy compounds have high entropies. Liquids and solids have low entropies compared with gases. Simple compounds have low entropies, complex compounds have higher entropies.

4-12 *STANDARD FREE ENERGIES*

Standard free energy values are similar to standard enthalpies in the sense that there is no absolute reference point of measurement of free energy. All that we can do is measure free energy *changes* in chemical processes. In place of absolute free energies, we use the free energies of formation of substances from their elements in agreed-upon standard states in a way exactly analogous to that for enthalpy. The standard free energy of a reaction is given by

$$\Delta G^0 = \Delta H^0 - T\Delta S^0 \tag{4-98}$$

This equation assumes by convention that ΔG^0 for an element in its standard state is zero; however, entropies are on an absolute scale, and the entropies of the elements in their standard states are not themselves zero.

EXAMPLE 4. The formation of water from its elements illustrates the application of Eq. 4-98

$$H_{2(g)} + \tfrac{1}{2}O_{2(g)} \rightarrow H_2O_{(l)}$$

The data for this reaction are tabulated below.

	$H_{2(g)}$	$O_{2(g)}$	$H_2O_{(l)}$
ΔH^0	0.0	0.0	−68.32 kcal/mole
ΔG^0	0.0	0.0	−56.69 kcal/mole
S^0	31.21	49.00	16.72 e.u./mole

The entropy of the reaction is

$$\Delta S^0 = +16.72 - 31.21 - \tfrac{1}{2}(49.00) = -39.00 \text{ e.u./mole}$$

Hence the standard free energy of formation of $H_2O_{(l)}$ is

$$\Delta G^0 = -68.32 - \frac{298(-39.00)}{1000} = -68.32 + 11.63$$

$$= -56.69 \text{ kcal/mole}$$

(Note that the $T\Delta S^0$ term was divided by 1000 because entropy units are in cal/deg mole.)

The large drop in entropy for this reaction is not surprising since the liquid product is obviously a more ordered system. The negative ΔH^0 term indicates that heat is given off as the reaction progresses, so that this reaction is favored from an energy standpoint. Since the total contribution to the free energy from the increased order (the entropy term) is so small compared to the big thermal drive given to the reaction, the net free energy is negative and the reaction is spontaneous. In the majority of cases the enthalpy term dominates the entropy term, and spontaneous reactions are quite likely to be exothermic. Of course, this assumption is based on the premise that the $T\Delta S^0$ term is small and unimportant, and the next example illustrates a case for which this assumption is not true.

Exception

The following reaction is an example of a chemical process in which the entropy term dominates.

$$Ag_{(s)} + \tfrac{1}{2}Hg_2Cl_{2(s)} \rightarrow AgCl_{(s)} + Hg_{(l)}$$

Thermodynamic data for the participating species are the following.

	$Ag_{(s)}$	$Hg_2Cl_{2(s)}$	$AgCl_{(s)}$	$Hg_{(l)}$
ΔH^0	0.0	−63.32	−30.36	0 kcal/mole
ΔG^0	0.0	−50.35	−26.22	0 kcal/mole
S^0	10.21	46.8	22.97	18.5 e.u./mole

$$\Delta H^0 = -30.36 - \frac{(-63.32)}{2} = +1.30 \text{ kcal/mole}$$

$$\Delta S^0 = 18.5 + 22.97 - \frac{46.8}{2} - 10.21 = +7.86 \text{ e.u./mole}$$

$$T\Delta S^0 = \frac{(7.86)(298)}{1000} = +2.34 \text{ kcal/mole}$$

$$\Delta G^0 = \Delta H^0 - T\Delta S^0 = +1.30 - 2.34 = -1.04 \text{ kcal/mole}$$

Here is an example of a spontaneous endothermic reaction. Even though ΔH^0 is positive, the presence of a disordered liquid product gives sufficient magnitude to the entropy term to make the over-all reaction spontaneous.

Hardboiling an Egg

Another example of a reaction in which the entropy dominates the energy in determining the direction of reaction is the hardboiling of an egg. Egg albumen is a protein whose structure is maintained in part by an immense number of hydrogen bonds. In the process of boiling an egg, the hydrogen bonds are broken and much of the order of the protein molecule is lost. If the temperature is sufficiently high, then the disordering of the system produced by the collapse of the structure of the protein molecule, and expressed as $T\Delta S$, is more than enough to counteract the large ΔH required to break all of the hydrogen bonds. The result is that ΔG is negative and reaction at or above such a temperature is spontaneous.

Boiling of Water

As an example of the comparative effects of ΔH and $T\Delta S$ on the spontaneity of a reaction, consider the reaction

$$H_2O_{(l,\ 1\text{ atm})} \rightarrow H_2O_{(g,\ 1\text{ atm})}$$

The heat of reaction ΔH is simply the heat of vaporization, and can be calculated at any temperature from heat capacity data and the standard heats of formation of liquid and gaseous water, using

$$\Delta H_{(T)} = \Delta H_0 + \int \Delta C_P \, dT = \Delta H_0 + \Delta a T + \frac{\Delta b}{2} T^2 + \frac{\Delta c}{3} T^3$$

Similarly, the entropy of reaction can be calculated from heat capacity data and standard third law entropies, using

$$\Delta S_{(T)} = \Delta S_0 + \int \frac{\Delta C_P}{T} \, dT = \Delta S_0 + \Delta a \ln T + \Delta b T + \frac{\Delta c}{2} T^2$$

$$T\Delta S_{(T)} = T\Delta S_0 + \Delta a T \ln T + \Delta b T^2 + \frac{\Delta c}{2} T^3$$

A plot of ΔH and $T\Delta S$ for the vaporization of water (with trends in the curves slightly exaggerated) is shown in Figure 4-27. The reaction is endothermic—heat of vaporization is absorbed—but *less* so at higher temperatures. At the critical point, of course, the heat of vaporization is zero, because the distinction between liquid and vapor phase loses meaning. At lower temperature, the increase in entropy in going from liquid to gas phase is essentially constant, and $T\Delta S$ is nearly linear in T. But gradually, as the critical point is approached, the entropy of vaporization must go to zero, and the $T\Delta S$ curve must level off and fall to zero. A little of this tendency is visible in

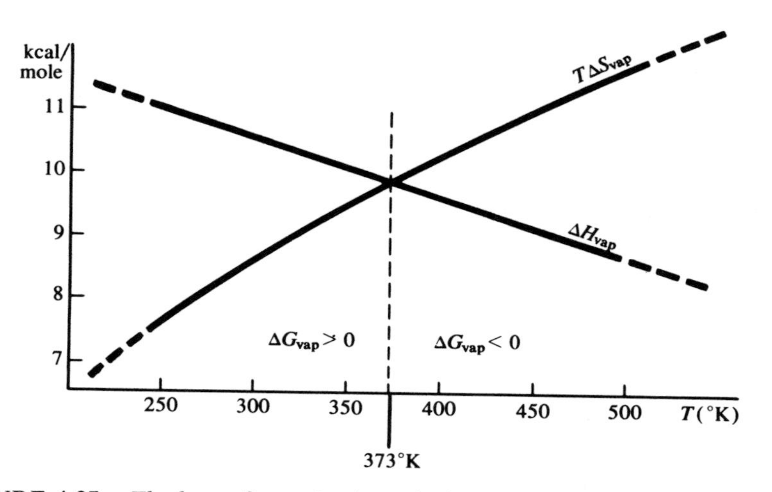

FIGURE 4-27. *The heat of vaporization, which opposes a liquid-to-gas transition, falls with temperature rise. The $T\Delta S$ term, which favors the production of the more disordered gas, rises at T rises. The point at which the entropy term first balances the enthalpy term is the boiling point of the liquid.*

the figure. At low temperatures, the great increase in enthalpy upon vaporization overbalances the effect of the entropy increase, $\Delta H > T\Delta S$, ΔG is positive, and the liquid phase is more stable. At higher temperatures, the $T\Delta S$ term arising from disordering upon vaporization dominates the ΔH term, ΔG is negative, and the vapor phase (at 1 atm partial pressure) is more stable. At 373°K, the enthalpy and entropy factors are exactly balanced and the two phases are in equilibrium. This temperature is the boiling point. It is generally true that at low temperatures enthalpy changes dominate the ΔG expression, while at high temperatures entropy changes are more important.

4-13 GENERAL METHOD OF FINDING THERMODYNAMIC RELATIONSHIPS

Before proceeding to the applications of thermodynamics, it is appropriate at this time to review the thermodynamic relationships covered thus far. A very concise summary, in which everything that we have covered so far can be summarized in six master equations, I–VI, has been proposed by A. Tobolsky, *J. Chem. Phys.* **10**, 644 (1942).

Fundamental Relationships

(a) *First and second laws.* The first and second laws can be summarized by

$$dE = TdS - PdV \tag{I}$$

Equation I is only true for certain conditions. It applies only to *reversible* processes in which all work is *mechanical.* Since the majority of the problems you will ever work involve equilibrium situations, these are not serious limitations as long as they are kept in mind.

(b) *Other energy functions.* Three more master equations serve as definitions of H, A, and G.

$$H \equiv E + PV \tag{II}$$

$$A \equiv E - TS \tag{III}$$

$$G \equiv H - TS \tag{IV}$$

From these four equations and the fact that E, H, A, and G are state functions, come the following useful relationships, which should be derivable upon need and not memorized:

$$dH = TdS + VdP \tag{IIa}$$

$$dA = -SdT - PdV \tag{IIIa}$$

$$dG = -SdT + VdP \tag{IVa}$$

Only two more master equations remain, but before turning to them, let us first work out the implications of Eqs. I–IV.

(*c*) *Consequences of I–IV.*

From Eq. IIIa,

$$\left(\frac{\partial A}{\partial T}\right)_V = -S \qquad \text{(IIIb)}$$

From Eq. IVa,

$$\left(\frac{\partial G}{\partial T}\right)_P = -S \qquad \text{(IVb)}$$

These show the temperature dependence of A and G. Substituting for S in Eqs. IIIb and IVb, and going through a little manipulation, we have the two forms of the Gibbs–Helmholtz equation.

$$\frac{\partial(A/T)}{\partial(1/T)} = \Delta E \qquad \text{(IIIc)}$$

$$\frac{\partial(G/T)}{\partial(1/T)} = \Delta H \qquad \text{(IVc)}$$

From Eq. IIIa,

$$\left(\frac{\partial A}{\partial V}\right)_T = -P \qquad \text{(IIId)}$$

From Eq. IVa,

$$\left(\frac{\partial G}{\partial P}\right)_T = +V \qquad \text{(IVd)}$$

These show the isothermal behavior of A and G. Applying the Euler reciprocity principle to Eqs. IIIa and IVa, respectively,

$$\left(\frac{\partial S}{\partial V}\right)_T = +\left(\frac{\partial P}{\partial T}\right)_V \qquad \text{(IIIe)}$$

$$\left(\frac{\partial S}{\partial P}\right)_T = -\left(\frac{\partial V}{\partial T}\right)_P \qquad \text{(IVe)}$$

where S is in terms of P, V, and T.

(d) *Heat capacities.* The last two master equations are

$$C_V \equiv \left(\frac{\partial q}{\partial T}\right)_V \tag{V}$$

$$C_P \equiv \left(\frac{\partial q}{\partial T}\right)_P \tag{VI}$$

and they bring with them the following consequences.

For reversible processes, $dq = TdS$. Hence Eqs. V and VI provide us with the following relationships:

$$\left(\frac{\partial S}{\partial T}\right)_V = \frac{C_V}{T} \tag{Va}$$

$$\left(\frac{\partial S}{\partial T}\right)_P = \frac{C_P}{T} \tag{VIa}$$

These show the temperature dependence of S.

With these six master equations, indicated by Roman numerals, and the expressions immediately derivable from them, one can obtain any thermodynamic relationship desired.

Applications

Tobolsky's second contribution is a systematic method of deriving thermodynamic relationships. The power of Tobolsky's method is best illustrated by applying it to several sample cases.

EXAMPLE 5. What is the volume dependence of energy under isothermal conditions?

$$\left(\frac{\partial E}{\partial V}\right)_T = ?$$

The first step in determining this relationship is to express the energy as a total differential in terms of two of the three variables P, V, or T. Since we want $(\partial E/\partial V)_T$, the logical step is to express dE as

$$dE = XdV + YdT \qquad X = \left(\frac{\partial E}{\partial V}\right)_T \tag{4-99}$$

where X is the term we want to find. The next step is to reduce the left side of the equation to terms having dV and dT, and equate coefficients. Substituting Eq. I for dE,

$$T\,dS - PdV = XdV + Y\,dT \tag{4-100}$$

We can rid this equation of dS by expressing S in terms of V and T.

$$T\left[\left(\frac{\partial S}{\partial V}\right)_T dV + \left(\frac{\partial S}{\partial T}\right)_V dT\right] - PdV = XdV + YdT \tag{4-101}$$

Equation 4-101 still contains S's, but these can be replaced by substituting their equivalent expressions, Eqs. IIIe and Va, respectively:

$$T\left[\left(\frac{\partial P}{\partial T}\right)_V dV + \frac{C_V}{T}\,dT\right] - PdV = XdV + YdT \tag{4-102}$$

Since only dV and dT terms are now present, we can solve for X by equating it with the dV coefficients on the left side of Eq. 4-102.

$$X = \left(\frac{\partial E}{\partial V}\right)_T = T\left(\frac{\partial P}{\partial T}\right)_V - P \tag{4-103}$$

This is the thermodynamic equation of state which was first derived as Eq. 4-59. As a dividend, the dT coefficient provides $(\partial E/\partial T)_V$.

$$Y = \left(\frac{\partial E}{\partial T}\right)_V = C_V \tag{4-104}$$

EXAMPLE 6. What is the pressure dependence of energy under isothermal conditions?

$$\left(\frac{\partial E}{\partial P}\right)_T = ?$$

Beginning as we did in the first example, but expressing E in terms of the variables P and T which are involved in the sought-for expression,

$$dE = XdP + YdT \qquad X = \left(\frac{\partial E}{\partial P}\right)_T \tag{4-105}$$

$$T\,dS - PdV = XdP + YdT \tag{4-106}$$

Equation 4-106 contains both dS and dV terms, and we want only dP and dT terms. We can get rid of entropy by expressing it in terms of P and T. Since P and T were chosen as the independent variables in this problem, volume is a dependent variable, so that we can expand dV as an exact differential in terms of P and T as well.

$$T\left[\left(\frac{\partial S}{\partial T}\right)_P dT + \left(\frac{\partial S}{\partial P}\right)_T dP\right] - P\left\{\left(\frac{\partial V}{\partial T}\right)_P dT + \left(\frac{\partial V}{\partial P}\right)_T dP\right\} = XdP + YdT \tag{4-107}$$

The S-containing terms can be replaced by their equivalent expressions, Eqs. VIa and IVe, respectively:

$$T\left[\frac{C_P}{T} dT - \left(\frac{\partial V}{\partial T}\right)_P dP\right] - P\left[\left(\frac{\partial V}{\partial T}\right)_P dT + \left(\frac{\partial V}{\partial P}\right)_T dP\right] = XdP + YdT \tag{4-108}$$

Collecting dT and dP terms,

$$\left[C_P - P\left(\frac{\partial V}{\partial T}\right)_P\right] dT + \left[-T\left(\frac{\partial V}{\partial T}\right)_P - P\left(\frac{\partial V}{\partial P}\right)_T\right] dP = XdP + YdT \tag{4-109}$$

Equating dP coefficients, we obtain

$$X = \left(\frac{\partial E}{\partial P}\right)_T = -T\left(\frac{\partial V}{\partial T}\right)_P - P\left(\frac{\partial V}{\partial P}\right)_T \tag{4-110}$$

Solving for Y as an added dividend yields

$$Y = \left(\frac{\partial E}{\partial T}\right)_P = C_P - P\left(\frac{\partial V}{\partial T}\right)_P \tag{4-111}$$

Once the logic of attack is understood, Tobolsky's method provides a means of obtaining what would otherwise be very difficult relationships. But whether used for this purpose or not, it is a convenient and compact six-equation summary of the basic relationships of thermodynamics.

REFERENCES AND FURTHER READING

H. A. Bent, *The Second Law* (Oxford University Press, Oxford, 1965). A beautifully written book with a strong physical feel for what the equations of thermodynamics and statistical mechanics *mean*. Unconventional in its derivations. Strong historical and anecdotal flavor.

J. D. Fast, *Entropy* (McGraw-Hill, New York, 1962). A good treatment of entropy as a measure of disorder.

EXERCISES

4-1. A homeowner has the bright idea of using an extra refrigerator to cool his living room during the summer. He therefore sets up the machine in the middle of the room, leaving the refrigerator door open to get the benefit of its cooling coils. Room temperature is initially 25°C, and it may be assumed that the refrigerator is operating between 25 and 0°C. The machine ordinarily is capable of freezing 1 kg of ice per hour with these operating temperatures. Calculate the temperature change in the living room, with the new temperature, after 1 h of operation of the refrigerator. Assume ideal thermodynamic efficiency, and that the heat capacity of the room is 100 kcal/deg. The heat of fusion of water is 80 kcal/g.

4-2. A brick of heat capacity C_1, initially at temperature T_1, is placed on another brick of heat capacity C_2, initially at temperature T_2.

(a) What is the total entropy change in the bricks after thermal equilibrium has been reached?

(b) If $C_1 = 2000$ cal/deg and $C_2 = 1000$ cal, and if $T_1 = 0$°C and $T_2 = 100$°C, what will the final temperature be, and what will the entropy change be in cal/deg?

4-3. Suppose that the two bricks of the previous problem can be brought to some common temperature T_c, so determined that the total entropy of the two bricks remains unchanged from what it was before they were brought together.

(a) Derive an expression for T_c.

(b) Calculate a numerical value for T_c and compare it with the result of the previous problem.

4-4. One mole of an ideal gas in contact with a heat reservoir at 25°C expands isothermally from 100 to 1 atm pressure. Make a table showing the ΔS for the gas, for the heat reservoir, and for the two systems combined, if in the expansion

(a) 2730 cal of work is done

(b) 1000 cal of work is done

(c) no work is done.

4-5. Consider the following nonisothermal cycle, in which P is the vapor pressure of water at a temperature T:

$$\begin{array}{ccccc}
H_2O_{(l)} & \rightarrow & H_2O_{(l)} & \rightarrow & H_2O_{(l)} \\
100°C,\ 1\ \text{atm} & & 100°C,\ P\ \text{atm} & & T,\ P \\
\uparrow & & & & \downarrow \\
H_2O_{(g)} & \leftarrow & H_2O_{(g)} & \leftarrow & H_2O_{(g)} \\
100°C,\ 1\ \text{atm} & & T,\ 1\ \text{atm} & & T,\ P
\end{array}$$

(a) Set up an expression for the entropy change during the entire cycle by adding together the terms for each of the steps. Neglect the entropy change attending the compression of the liquid, and assume the vapor to be an ideal gas.
(b) Neglecting the difference between the heat capacities of liquid and vapor, and hence assuming that ΔH_{vap} is independent of temperature, derive an expression for the vapor pressure of water at temperature T.
(c) If $\Delta H_{vap} = 9.720$ kcal/mole, calculate the vapor pressure of water at 105°C.
Note that this problem illustrates the use of the state function properties of entropy to obtain a relationship between quantities *other* than entropy.

4-6. In the diagram below are drawn a series of adiabatic reversible paths, $1 \rightarrow 2$, $1' \rightarrow 2'$, and $1'' \rightarrow 2''$, each one starting at the temperature T_2 and ending at the temperature T_1. The points 1, 1′, and 1″ are labeled in an order such that for any process proceeding to the right along the isothermal T_2, heat is absorbed by the system. Prove that the adiabatics a and b cannot intersect. (*Hint*: Assume that they *do* intersect at the temperature T_1, and then show this assumption forces you to violate something heretofore assumed to be correct.)

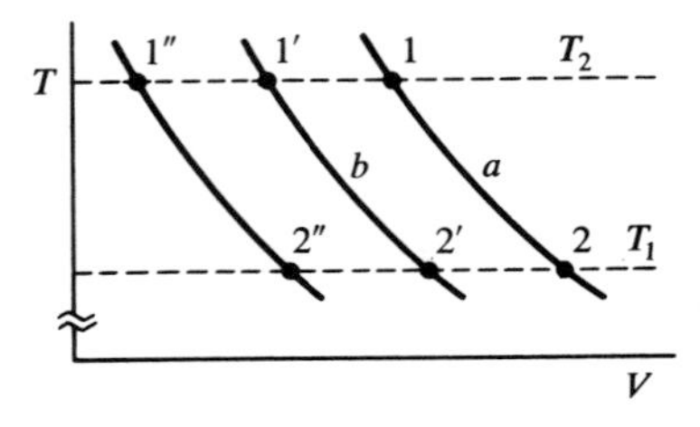

4-7. Consider a chamber containing a gaseous mixture of H_2, N_2, and NH_3, pressurized by a movable piston, and suppose that the mixture is at equilibrium with respect to the reaction

$$3H_{2(g)} + N_{2(g)} = 2NH_{3(g)}$$

Suppose, furthermore, that the introduction of a suitable catalyst shifts the equilibrium to the left by increasing the rate of the back reaction more than it does the forward reaction, with a consequent increase in the number of moles of gas in the chamber and a rise in the piston's equilibrium position, and that upon removing the catalyst the system return to its original state. The cycle could then be repeated:

catalyst in, piston up; catalyst out, piston down; and the piston could be connected so as to do useful work. To keep the engine from cooling off—presumably this might interfere with its efficient operation, and by the first law the energy that appears as work must come from somewhere—the engine is allowed to remain in thermal contact with its surroundings. This costs nothing. Would not this be a good way to operate an oceangoing vessel, particularly in the tropics?

(a) Show the way in which this violates thermodynamic principles.

(b) Point out the flaw in the initial assumptions which makes this impossible result appear.

4-8. An ideal gas is carried through a Carnot cycle. Draw diagrams of this cycle using each of the following as coordinates:

P and V	E and S
T and P	S and V
T and S	T and H

In each case make the first coördinate the vertical axis. Number the successive legs of the curve 1, 2, 3, 4 in the same way in each plot.

4-9. A cylinder with a frictionless piston contains 1 mole of an ideal gas. In addition, the closed end of the cylinder is joined to the piston by means of a spring that obeys Hooke's law with a potential energy of $\frac{1}{2}k(L-L_0)^2$, where k is the Hooke's law constant, L is the length of the spring, and L_0 is its equilibrium length. An observer studies the PVT relationships of the gas without knowing about the presence of the spring.

(a) Assuming that k is independent of temperature, what will the observer find for the apparent equation of state of the gas? Let V and V_0 be the volumes corresponding to L and L_0, and let A be the cross-sectional area of the cylinder.

(b) From the apparent equation of state, in conjunction with one of the thermodynamic equations of state, show how the energy of this "gas" depends upon its volume.

Note: This problem shows how hidden forces operating between molecules might also be measured to some extent by a thermodynamic analysis of the equation of state of the gas.

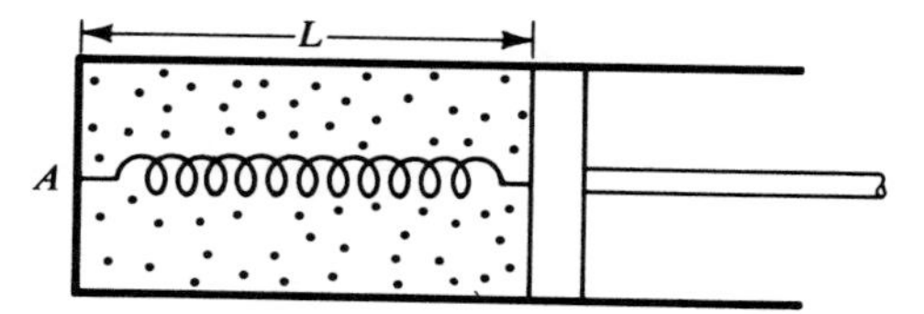

4-10. A possible equation of state for a liquid is

$$V = V_0[1 + \alpha T - \beta(P-1)]$$

in which P is the pressure in atmospheres, V_0 is the molar volume at 0°C and 1 atm, and α and β are constants. Under these conditions, prove that

$$(\partial S/\partial P)_T = -V_0\alpha$$
$$(\partial S/\partial V)_T = \alpha/\beta$$
$$(\partial H/\partial P)_T = V_0(1 - 273\alpha - \beta(P-1))$$

Calculate ΔS and ΔH in cal/deg and cal when 1 mole of water is compressed at 20°C from 1 to 25 atm. (For water near 20°C and 1 atm, $\alpha = 2.1 \times 10^{-4}$ deg^{-1} and $\beta = 49 \times 10^{-6}$ atm^{-1}.)

4-11. (a) Evaluate $(\partial E/\partial V)_T$ for a van der Waals gas, making use of one of the thermodynamic equations of state.

(b) From the purely mathematical properties of an exact differential, show that if $(\partial E/\partial V)_T$ is a function only of volume, then C_V is a function only of temperature.

(c) Derive an expression for the variation of entropy with volume for a van der Waals gas during an isothermal process.

(d) In one experiment, a van der Waals gas is heated to twice its initial volume, then expanded isothermally to twice this new volume, then cooled to its original temperature, and to a final volume V_f. In a second experiment, the same gas at the original starting conditions is expanded to V_f in a one-step isothermal process. Calculate the difference in entropy in the two final states. How can you account for this difference in terms of the molecular picture of entropy?

4-12. A strip of rubber is subjected to reversible elongations and retractions with the work involved given by $dw = -f dL$, where L is the length and f is the force. Assuming the energy E to be a function of length L and temperature T, we can write

$$dq_{\text{rev}} = \left(\frac{\partial E}{\partial T}\right)_L dT + \left[\left(\frac{\partial E}{\partial L}\right)_T - f\right] dL$$

By setting $dS = dq_{\text{rev}}/T$ and evaluating the second partials $(\partial^2 S/\partial L \partial T)$ in two different ways, prove that

$$\left(\frac{\partial E}{\partial L}\right)_T = f - T\left(\frac{\partial f}{\partial T}\right)_L$$

4-13. Calculate the entropy change per liter of solution when pure N_2, H_2, and NH_3 gases are mixed to form a solution having the final composition of 15 mole % N_2, 55% H_2, and 30% NH_3 at 273°K and 1 atm total pressure. Assume ideal gas behavior.

4-14. Prove that $(\partial A/\partial T)_V = -S$.

4-15. (a) Derive $(\partial S/\partial V)_E = P/T$. (*Hint:* dE is an exact differential.)

(b) Verify the above for the special case of an ideal gas, that is, evaluate the partial directly.

4-16. From the first and second laws and related definitions, prove

$$\left(\frac{\partial S}{\partial V}\right)_T = \left(\frac{\partial P}{\partial T}\right)_V$$

$$\left(\frac{\partial T}{\partial P}\right)_S = \left(\frac{\partial V}{\partial S}\right)_P$$

4-17. (a) Show that, if $(\partial E/\partial V)_T = 0$, then the equation of state of the substance must be of the form: $P = Tf(V)$, where $f(V)$ is *any* arbitrary function of V but of V only.

(b) Show that, if $(\partial H/\partial P)_T = 0$, then the equation of state of the substance must be of the form $V = Tg(P)$, where $g(P)$ is *any* arbitrary function of P but of P only.

(c) Show, in view of the previous results, that if both the E and H relationships are true, then PV/T must be a constant.

Note: It was proven in the text that $PV = RT$ implied the E and H relationships given above. You have now proved that the reverse is true as well.

4-18. Under what conditions is the enthalpy the potential function whose minimization marks the point of chemical equilibrium? Prove this rigorously by showing the following.

(a) The enthalpy change is zero for a process at equilibrium under these conditions.

(b) This is a true minimum point and not a maximum or an inflection point.

Draw a plot of enthalpy versus some reaction coordinate under these conditions as the reaction approaches, reaches, and passes the equilibrium point.

4-19. One mole of an ideal, monatomic gas initially at 1 atm and 25°C is expanded isothermally and irreversibly to 44.8 liters, under conditions such that the work done by the gas is 100 cal. Calculate ΔS and ΔG.

4-20. Consider an electrolytic cell

$$Ag_{(s)},\ AgCl_{(s)},\ HCl_{(aq)},\ Cl_{2(g)},\ Pt_{(s)}$$

(a) At 25°C and 1 atm the electromotive force of the cell is 1.1372 V. What is the free energy change in calories with the passage of one faraday of electricity from left to right within the cell?

(b) The temperature coefficient of the emf is -0.000595 V/deg. What are ΔS in cal/deg and ΔH in cal for the change in state which accompanies the passage of one faraday?

Note that parts (a) and (b) can be worked without any knowledge of the cell reaction.

(c) Show that, when one faraday of electricity passes through the cell, the following change in state occurs:

$$Ag_{(s)} + \tfrac{1}{2}Cl_{2(g)} \rightarrow AgCl_{(s)}$$

The heat of formation of $AgCl_{(s)}$ is tabulated as -30.36 kcal. How does this compare with the ΔH of part (b)?

(d) Assuming chlorine to be an ideal gas, and neglecting the volumes of the solids compared with those of the gas, show that the total reversible work involved in the operation of the cell is

$$w = \mathscr{E}\mathscr{F} - \tfrac{1}{2}RT$$

From this and the results of part (a), calculate A for the reaction.

4-21. (a) Show that, in a plot of energy against temperature, the enthalpy and the free energy always slope in opposite directions.

(b) What is the separation between these two curves at any temperature?

(c) What important observations about the behavior of these curves near absolute zero lead to the third law?

4-22. Methylammonium chloride exists in a number of crystalline forms, and the thermodynamic properties of the β and γ forms have been investigated down to temperatures near absolute zero. Below 20° or so it begins to become difficult to measure heat capacities with accuracy, but Debye has shown that heat capacities in this region can be approximated very well by

$$C_P \cong C_V = \frac{464.5\,T^3}{\theta^3}\ \frac{\text{cal}}{\text{deg mole}}$$

The constant θ can be determined for a given substance by measuring just one heat capacity in this range of 0–20°K. From the information given below, find the heat of transition from the β form to the γ form at 220.4°K

C_P for β at 12.0°K = 0.202 cal/deg mole.

$\int C_P\, d \ln T$ from 12.0 to 220.4°K = 22.326 cal/deg mole for β.

C_P for γ at 19.5°K = 1.426 cal/deg mole.

$\int C_P\, d \ln T$ for γ from 19.5 to 220.4°K = 23.881 cal/deg mole.

The normal transition temperature for 1 atm pressure is 220.4°K.

4-23. Assume that the limiting slope, as T approaches zero, of a plot of ΔG against T is finite instead of zero. Prove that ΔC_P for the reaction will *still* approach zero at absolute zero.

4-24. Assume that

$$\lim_{T\to 0}\left(\frac{\partial \Delta G}{\partial T}\right)_P = 0$$

for reactions involving perfect crystalline solids, prove that

$$\lim_{T\to 0}\left(\frac{\partial \Delta A}{\partial T}\right)_V = 0$$

4-25. It has been suggested that α-cyanopyridine might be prepared from cyanogen and butadiene by the reaction

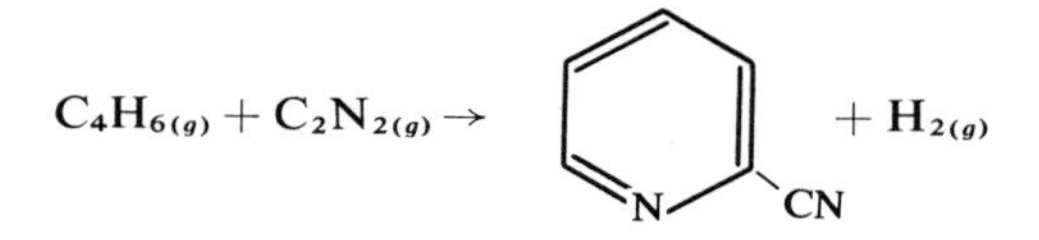

In view of the thermodynamic data given below, would you consider it worthwhile to attempt to work out this reaction? Why, or why not?

	ΔH_f^0 kcal/mole	S^0 cal/deg mole
butadiene$_{(g)}$	26.748	66.42
cyanogen$_{(g)}$	71.820	57.64
α-cyanopyridine$_{(s)}$	62.000	77.09
$H_{2(g)}$	0.0	31.21

4-26. Show that, during an *isotropic*, reversible, adiabatic expansion of an ideal gas, the populations of the energy levels are unchanged. Assume a monatomic gas in a rectangular container of dimensions a, b and c cm, with energy states

$$\varepsilon_{jkl} = \frac{h^2}{8m}\left(\frac{j^2}{a^2} + \frac{k^2}{b^2} + \frac{l^2}{c^2}\right)$$

Assume also that the most probable distribution of energy exists at all times:

$$n_{jkl} = \frac{n}{q} \exp\left(\frac{-\varepsilon_{jkl}}{kT}\right)$$

where $q = (2\pi mkT)^{3/2} V/h^3$.

By an isotropic expansion is meant one in which each dimension increases by the same proportion, so that

$$\frac{da}{a} = \frac{db}{b} = \frac{dc}{c} = \frac{1}{3}\frac{dV}{V}$$

Since the expansion is to be carried out adiabatically,

$$VT^{C_V/R} = \text{const}$$

The goal is the calculation of the change in population of a level with volume change under adiabatic conditions:

$$\left(\frac{dn_{jkl}}{dV}\right)_{q'} = ?$$

where q' is the heat (q is the molecular partition function).

This problem is best approached in several steps.
(a) Show that

$$\left(\frac{dn_{jkl}}{dV}\right)_{q'} = \left(\frac{\partial n_{jkl}}{\partial T}\right)_V \left(\frac{dT}{dV}\right)_{q'} + \left(\frac{\partial n_{jkl}}{\partial V}\right)_T$$

(b) Show that

$$\left(\frac{\partial n_{jkl}}{\partial T}\right)_{V'} = \frac{n_{jkl}}{kT^2}\,(\varepsilon_{jkl} - \tfrac{3}{2}kT)$$

Remember that q is also a function of T and V.
(c) Check the results of part (b) by showing that they satisfy the requirement that molecules are neither created nor destroyed during the temperature change:

$$\sum_{jkl} \left(\frac{\partial n_{jkl}}{\partial T}\right)_V = 0$$

(d) Derive an expression for

$$\left(\frac{\partial n_{jkl}}{\partial V}\right)_T = ?$$

The expression given above for isotropic expansions will be useful in evaluating

$$\left(\frac{\partial \varepsilon_{jkl}}{\partial V}\right)_T$$

(e) Check the results of part (d) by showing that molecules are neither created nor destroyed during the volume change:

$$\sum_{jkl} \left(\frac{\partial n_{jkl}}{\partial V}\right)_T = 0$$

(f) Using your results in parts (a)–(e), show that

$$\left(\frac{dn_{jkl}}{dV}\right)_{q'} = 0$$

or that, although the levels change in energy during the expansion, their populations are unchanged.

4-27. (a) Following the pattern of solution of Exercise 4-26, show that for an expansion of dimension a, with b and c held stationary, there *is* a redistribution of populations, as given by

$$\left(\frac{dn_{jkl}}{da}\right)_{q'} = \frac{n_{jkl}}{kTa}\,\frac{2}{3}\left(\frac{h^2}{8m}\right)\left(2\frac{j^2}{a^2} - \frac{k^2}{b^2} - \frac{l^2}{c^2}\right)$$

$$= \frac{n_{jkl}}{kTa}\,\frac{2}{3}\left(\frac{3h^2}{8m}\,\frac{j^2}{a^2} - \varepsilon_{jkl}\right)$$

(b) Imagine now that populations are plotted in a three-dimensional "quantum space" of coordinates $x = j/a$, $y = k/b$, and $z = l/c$, x, y, and z ranging from $-\infty$ to $+\infty$. On what kind of a surface through this plot will there be *no* population change? What kind of surfaces in this plot will contain those quantum states that have the *same* increase or decrease in population? Sketch an xy cross section through quantum space, with contours of equal $(\partial n_{jkl}/\partial a)_{q'}$. Note that your result indicates an increase in the populations of high quantum states in the direction in which the expansion took place, a, and a compensating decrease in quantum population in the other two directions. During an adiabatic expansion, the quantum levels drop in energy in the direction of the expansion. In the isotropic expansion, the molecules drop in energy along with the levels, and follow their quantum levels down. In the anisotropic expansion, in contrast, the molecules lag behind their levels and end in slightly higher quantum states in the direction of the expansion, with a compensating flow inward to lower states in y and z.

(c) The energy can be written

$$\varepsilon_r = \frac{h^2}{8m}(x^2 + y^2 + z^2) = \frac{h^2}{8m} r^2$$

All states with the same $r^2 = x^2 + y^2 + z^2$ will have the same energy, and will be the degenerate states of energy level ε_r. (Or, more properly, all energy states between r and $r + dr$ can be grouped together as the degenerate states of energy ε_r.) Show that, even though the populations of individual x, y, z quantum states have shifted, the populations of the *energy levels* ε_r have *not*.

This can be done by integrating over a spherical shell in quantum space, and by showing that the populations of the shells are constant during the adiabatic expansion

$$\left(\frac{dn_r}{da}\right)_{q'} = \int_{\phi=0}^{2\pi} \int_{\theta=0}^{\pi} \left(\frac{dn_{xyz}}{da}\right)_{q'} \sin\theta \, d\theta d\phi = 0$$

where the conversion to spherical polar coordinates is

$$\frac{j}{a} = x = r \sin\theta \cos\phi$$

$$\frac{k}{b} = y = r \sin\theta \sin\phi$$

$$\frac{l}{c} = z = r \cos\theta$$

The conclusion from this problem is that, once again, in an adiabatic reversible expansion, the energy levels themselves have shifted but their populations have remained unchanged.

Chapter 5

THERMODYNAMICS OF PHASE CHANGES AND CHEMICAL REACTIONS

5-1 FREE ENERGY AND THE MEANING OF PHASE CHANGES

WITH THE ground rules of thermodynamics in hand, we are ready to apply them to chemical systems. The simplest of all possible systems are those which consist of only one component. For such systems, the only "chemical change" which can occur is a phase change.

A pure substance can exist as a gas, a liquid, or one or more crystalline solids. Each of these phases will have associated with it a molar free energy, and any change from phase A to phase B will be accompanied by a free energy of phase change

$$A \rightarrow B \qquad \Delta G = ? \tag{5-1}$$

If the free energy of this "reaction" as written is negative, then phase A will spontaneously convert to phase B. If positive, then B will convert to A. Only under conditions where A and B have the *same* molar free energy can both exist in equilibrium.

The principal variables in inducing phase changes are temperature and pressure. How does free energy vary with P and T? At 0°K, the entropy of a system is zero and the enthalpy and free energy are the same. Figure 5-1a shows the behavior of G and H as the temperature is increased at constant pressure. Since $(\partial H/\partial T)_P = C_P > 0$, the enthalpy increases with increasing temperature. On the other hand, $(\partial G/\partial T)_P = -S < 0$, and free energy decreases as the temperature is raised. There is another special relationship between the two curves in Figure 5-1a. Since

$$H - G = TS \tag{5-2}$$

the distance between the two curves at any point is T times the entropy. Furthermore, the free energy curve falls off more and more rapidly with T, for the slope $-S$ becomes more and more negative.

$$-\left(\frac{\partial^2 G}{\partial T^2}\right)_P = \left(\frac{\partial S}{\partial T}\right)_P = \frac{C_P}{T} > 0 \tag{5-3}$$

This is shown in Figure 5-1b.

Solid, Liquid, and Gas Free Energy Curves

Diagrams of the type of Figure 5-1 apply to all phases, but their slopes and intercepts vary from one phase to another. In general, the enthalpy of a gas is much greater than that of the liquid, which in turn is somewhat greater than the enthalpy of the solid. At absolute zero, enthalpies and free energies are identical. This means that the molar free energy of a gas is greater than that of a liquid, which is greater than that of the solid. A typical plot of the free energy behavior of solid, liquid, and gas phases is shown in Figure 5-2. At any given temperature, the downward slope of the gas curve is greatest, liquid intermediate, and solid least. This is intuitively obvious from the entropies of the three phases.

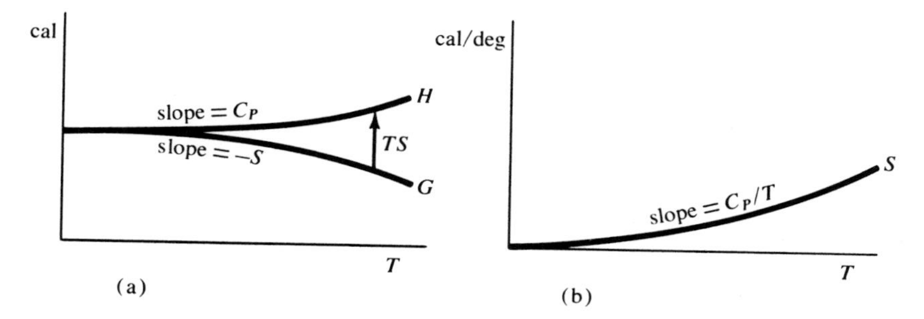

FIGURE 5-1. *The slopes of the free energy and enthalpy curves are the entropy and heat capacity at constant pressure. Both approach zero at absolute zero. The slope of the curve of entropy itself is C_P/T, and this, too, approaches zero at absolute zero.*

In the absence of a reversal of these trends, it is obvious that the liquid and solid G curves will cross at a sufficiently high temperature. This is the melting point of the solid. At all temperatures below the crossing temperature, the solid has the lower molar free energy and hence is the thermodynamically stable state. Above this point it is the liquid which is the stable phase. As the curves are drawn here, the gas curve does not cross the liquid curve until a higher temperature yet, the boiling point. Above this point the molar free energy of the gas is less than that of the other two phases and the gas phase is the stable state. Except for a supercooled liquid, these metastable states in which a different phase has a lower free energy are very difficult or impossible to obtain experimentally. But we can calculate the free energy

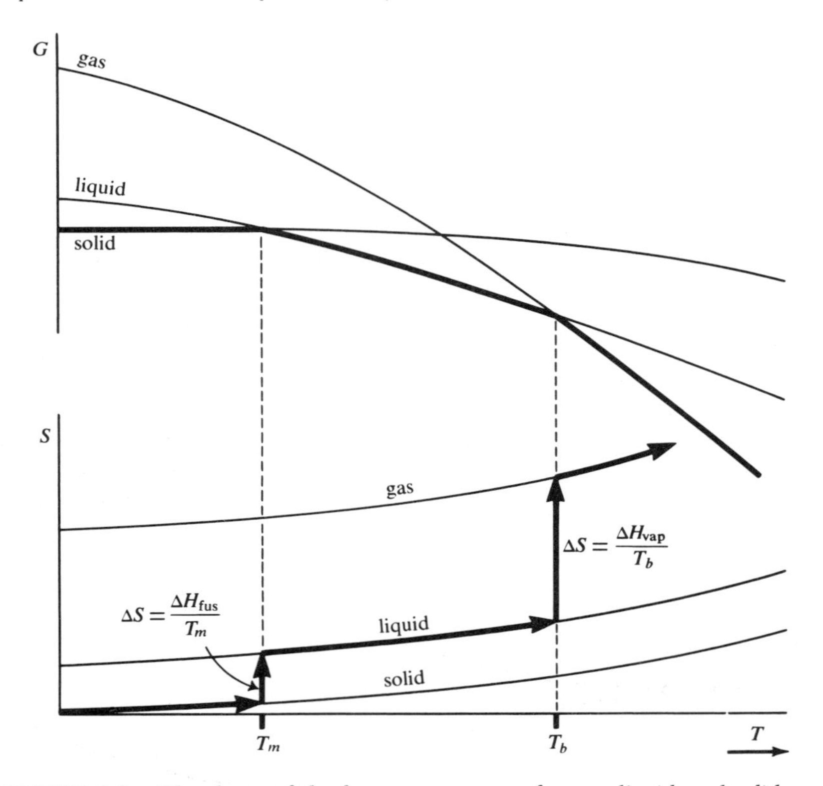

FIGURE 5-2. *The slopes of the free energy curves for gas, liquid, and solid are all negative, since S must be positive. The downward slope is greatest at any temperature for the gas, as its molar entropy is greatest. The stable form of a substance at any point is that form with the lowest molar free energy. The point where the liquid curve falls below the solid curve is the melting point, and that where the gas curve falls below the liquid is the boiling point. Compare Figure 4-27. The entropy changes involved in the transitions are shown below.*

which liquid water would have well above its boiling point even though we cannot produce it.

The entropies of the three phases are shown in the bottom of Figure 5-2, along with the two entropies of phase transition. A plot of the entropy per mole of a substance, in its most stable state at each temperature, would follow the heavy arrows on the S diagram. Compare this explanation of the occurrence of a boiling point with that of Section 4-12. The two are entirely equivalent, but emphasize different aspects of the transition.

In all of these transitions the free energy is continuous but the entropy is discontinuous. Such behavior is characteristic of first order transitions. A first order transition is defined as one in which the free energy itself is continuous, but all of its first derivatives are discontinuous. Second order transitions will be considered in Section 5-6.

Effect of Pressure

At a given temperature the molar free energy of a phase will change with pressure at a rate proportional to its molar volume.

$$\left(\frac{\partial \overline{G}}{\partial P}\right)_T = +\,\overline{V} \tag{5-4}$$

Gases have high molar volumes; liquids and solids have relatively small molar volumes. The molar volumes of ice, water, and water vapor at 0°C, for example, are 19.65, 18.00, and 22,400 cm^3. This means that the gas curve will be most sensitive to changes in pressure. The liquid curve will be slightly more sensitive than the solid curve if the liquid is less dense than the solid, as is usually the case (recall that water is an exception). Figure 5-3 shows the effect of pressure upon molar free energies. In the lower half of the figure are plotted the variations of melting and boiling points with pressure, of the conditions for two-phase equilibrium. Both boiling point and freezing point rise with pressure (if the liquid is less dense than the solid). By Eq. 5-4, the molar free energy of the gas is increased much more than that of the liquid by an increase in pressure, and equilibrium is shifted in favor of the condensed phase. Only an increase in temperature can restore equilibrium between phases. Le Chatelier's principle, that an equilibrium system subjected to stress shifts its equilibrium point in such a way as to relieve the stress, is valid here. The liquid-vapor equilibrium curve of Figure 5-3 can be interpreted either as showing the way in which boiling point varies with pressure, or the way in which equilibrium vapor pressure of a liquid varies with temperature, since the boiling point is just the temperature at which the vapor pressure becomes equal to the total external pressure.

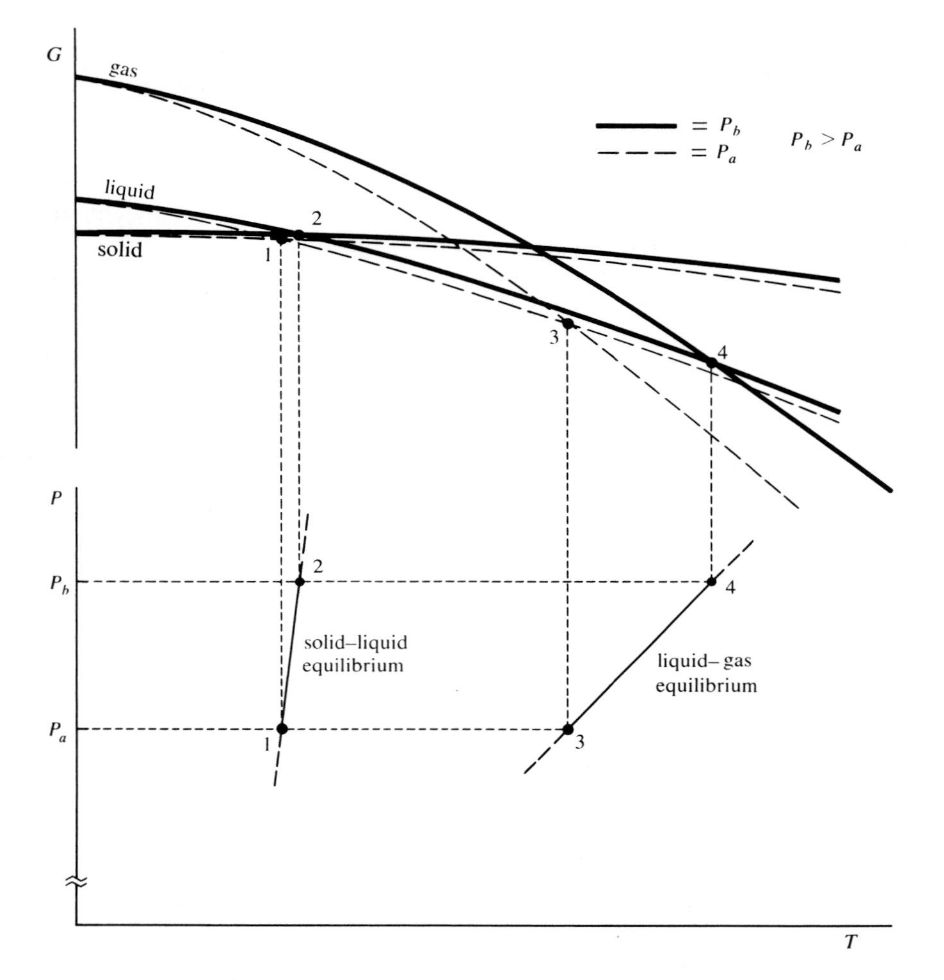

FIGURE 5-3. *The shift in intersection points of the free energy curves produced by pressure (top), cause the shifts in melting point and boiling point recorded in the P-versus-T diagram at the bottom. As the free energy of the gas is most sensitive to pressure, the boiling point is more affected than the melting point.*

At sufficiently low pressures, the gas phase free energy curve can be made to intercept the liquid and solid phase curves at a temperature *below* the melting point temperature, as shown in Figure 5-4. The first phase transition from the solid as the temperature is raised is a sublimation to the gas phase rather than a melting, at T_{subl}. Solid carbon dioxide, or dry ice, shows this sublimation behavior at normal atmospheric pressure. Only at higher pressures will dry ice melt to liquid CO_2.

Normal melting behavior results at pressures such as P_c in Figure 5-5, and sublimation at lower pressures such as P_a. By appropriately adjusting the pressure to P_b, one can create a situation in which all three free energy curves intersect at one point (point 3). For this given pressure and temperature, the solid, liquid, and gas all have the same free energy and can exist in thermodynamic equilibrium. This unique point is called the triple point. In the *PT* diagram at the bottom of Figure 5-5, the curve through points 1, 3, and 5 marks the points of intersection of the free energy curves above for liquid and gas at various pressures. The curve through 2, 3, and 4 similarly marks the intersection of gas and solid *G* curves, and line 3-6 marks the much smaller variation of melting point with pressure. (Only the gas phase free energy curve is shown varying with pressure here for simplicity, and to be consistent line 3-6 should therefore be vertical.) The intersection of all three *PT* curves at point 3 is the three-phase triple point.

As has already been mentioned, curves 1-3-5 and 2-3-4 represent the vapor pressure curves for liquid and solid, respectively. Note that the condensed phase with the lower molar free energy, and hence the greater thermodynamic stability, also has the lower vapor pressure. Vapor pressure itself can be used as a measure of phase stability.

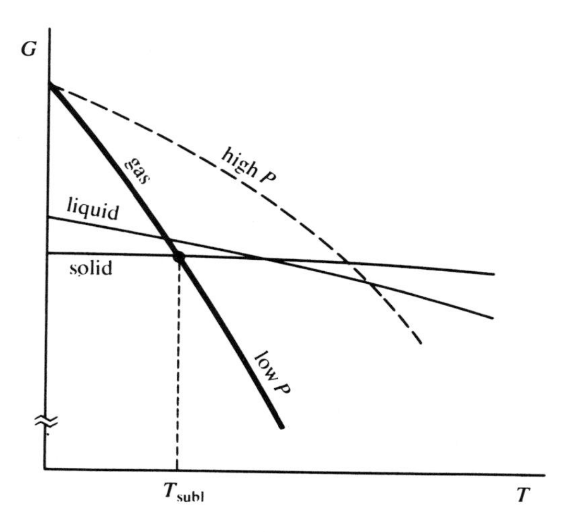

FIGURE 5-4. *At sufficiently low pressures, the gas free energy curve crosses the other two at a lower temperature than the solid–liquid intersection. As temperature rises, the gas becomes the most stable phase before the liquid does, and the solid sublimes.*

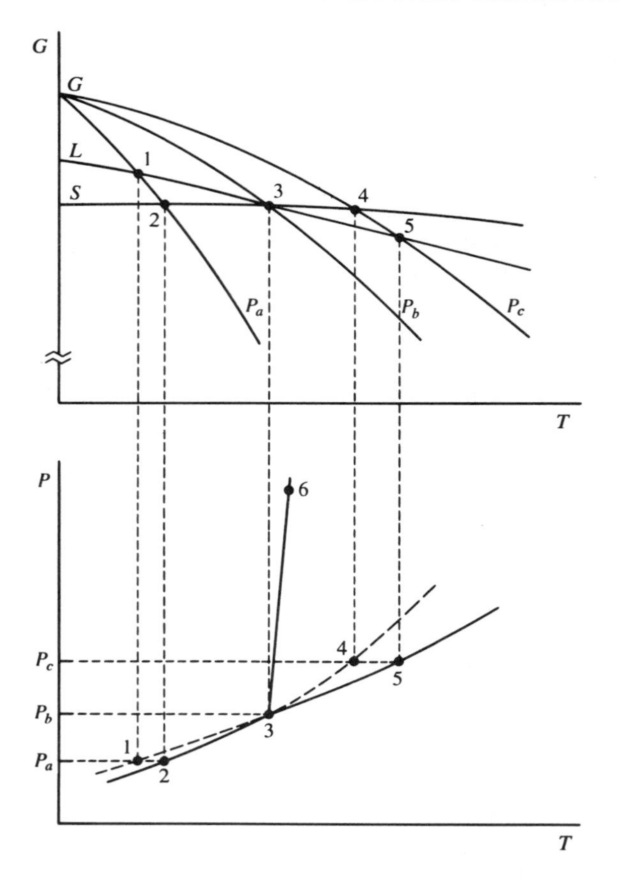

FIGURE 5-5. *A summary plot correlating free energy-temperature diagrams with pressure-temperature diagrams. At high pressure P_c the solid melts, while at low pressure P_a it sublimes. At pressure P_b and the temperature of point* 3, *the three phases can exist in equilibrium. This is called the triple point.*

5-2 CALCULATION OF PHASE EQUILIBRIUM CURVES

The curves of Figure 5-5 were obtained from the condition of equal free energy in phases at equilibrium. The mathematical expression for these curves is known as the Clapeyron equation.

Pure Substance, No External Atmosphere

If two pure phases, gas and liquid, for example, are in equilibrium, then the molar free energy is the same in each phase.

$$\bar{G}_{(l)} = \bar{G}_{(g)} \tag{5-5}$$

If the temperature or pressure are changed, but equilibrium is maintained, then the changes in free energy of the two phases must be the same.

$$d\bar{G}_{(l)} = d\bar{G}_{(g)} \tag{5-6}$$

Replacing dG by its value from Eq. IVa, (Section 4-13)

$$-\bar{S}_{(l)}\,dT + \bar{V}_{(l)}\,dP = -\bar{S}_{(g)}\,dT + \bar{V}_{(g)}\,dP \tag{5-7}$$

This expression links pressure and temperature. They both cannot be varied at will if phase equilibrium is to be maintained. Rearranging the latter equation gives the dependence of equilibrium pressure on temperature which is known as the Clapeyron equation. It shows that the ratio of pressure change to temperature change, with the free energy of phase change held at zero, is just the ratio of the changes in the first derivatives of this free energy of phase change with respect to temperature and pressure, $\Delta\bar{S}$ and $\Delta\bar{V}$.

$$\frac{dP}{dT} = \frac{\bar{S}_{(g)} - \bar{S}_{(l)}}{\bar{V}_{(g)} - \bar{V}_{(l)}} = \frac{\Delta\bar{S}_{\text{vap}}}{\Delta\bar{V}_{\text{vap}}} \tag{5-8}$$

$$\frac{dP}{dT} = \frac{\Delta\bar{H}_{\text{vap}}}{T\Delta\bar{V}_{\text{vap}}} \tag{5-9}$$

Presence of an External Atmosphere

What is the effect of an external atmospheric pressure upon the vapor pressure of a solid or liquid? Any change in free energy of the given substance in the gas phase is given by

$$d\bar{G}_{(g)} = -\bar{S}_{(g)}\,dT + \bar{V}_{(g)}\,dP_{(g)} \tag{5-10}$$

where $P_{(g)}$ is the partial vapor pressure of the given substance. Free energy changes of the liquid phase are given by

$$d\bar{G}_{(l)} = -\bar{S}_{(l)}\,dT + \bar{V}_{(l)}\,dP_{(l)} \tag{5-11}$$

where $P_{(l)}$ is the *total* pressure on the liquid phase. In the previous section this was equal to the liquid's own vapor pressure $P_{(g)}$. Now it is the *total* atmospheric pressure P_T. If equilibrium is maintained during an alteration of the three variables T, $P_{(g)}$, and P_T, then the molar free energy changes of liquid and vapor must be the same.

$$-\bar{S}_{(l)}\,dT + \bar{V}_{(l)}\,dP_T = -\bar{S}_{(g)}\,dT + \bar{V}_{(g)}\,dP_{(g)} \tag{5-12}$$

The Clapeyron equation, giving the change in vapor pressure with temperature, now at a constant total atmospheric pressure, is

$$\left(\frac{\partial P_{(g)}}{\partial T}\right)_{P_T} = \frac{\Delta \bar{S}}{\bar{V}_{(g)}} = \frac{\Delta \bar{H}_{\text{vap}}}{T\bar{V}_{(g)}} \tag{5-13}$$

The only difference between this equation and Eq. 5-9 is the replacement of the molar volume change on evaporation by the molar volume of the gas. Since the molar volume of the gas is so much greater than that of the liquid, the difference between these two equations is usually negligible.

The external atmospheric pressure has a very small effect upon vapor pressures of condensed phases. From Eq. 5-12 at constant temperature,

$$\bar{V}_{(l)}\, dP_T = \bar{V}_{(g)}\, dP_{(g)} \tag{5-14}$$

$$\left(\frac{\partial P_{(g)}}{\partial P_T}\right)_T = \frac{\bar{V}_{(l)}}{\bar{V}_{(g)}} \tag{5-15}$$

For water, the molar volume of the liquid is 18.0 cm^3 and that of the gas is around 30,000 cm^3 at 373°K. The change in vapor pressure is therefore smaller than the change in atmospheric pressure by a factor of 18/30,000, and can be disregarded except under very unusual pressure conditions.

Other Phase Changes

Thus far, comments made about liquid-vapor transitions also apply to any other phase changes, and the Clapeyron equation is perfectly general. Ignoring the minor corrections for external atmosphere, one can write the Clapeyron equation for a general phase change as

$$\frac{dP}{dT} = \frac{\Delta \bar{H}_\phi}{T\Delta \bar{V}_\phi} \tag{5-16}$$

where ϕ is the phase change. This is the equation for each of the curves in the PT plot of Figure 5-5. $\Delta \bar{H}_{\text{vap}}$ and $\Delta \bar{V}_{\text{vap}}$ are always positive, so the slope of the liquid-gas equilibrium curve is always positive. The same is true of the solid–gas equilibrium curve. But although $\Delta \bar{H}_{\text{fus}}$ is always positive, $\Delta \bar{V}_{\text{fus}}$ is not necessarily positive. Ice, for example, contracts when it melts. Therefore the slope of the solid–liquid equilibrium curve may be either positive or negative, depending on the substance in question.

The crystal structure of ice is a large, open cage network of hydrogen bonds. When ice melts, the hydrogen bonded cage collapses in upon itself so that at 0°C water is more dense than ice. As a result, line 3-6 of Figure 5-5 for

H_2O has a negative slope. The melting point of ice decreases as pressure increases. Increased pressure favors the more condensed phase, as Le Chatelier's principle would predict, and ice can be melted by pressing on it, a fact of some importance for the smooth sliding of ice skaters. (Of course, if ice were not less dense than water, all of the ice would be at the bottom of the lake and there would be no skating anyway.)

It is a simple matter to calculate the effect of pressure on the melting point of ice. The densities and molar volumes of ice and water at 0°C are

$$\rho_{\text{ice}} = 0.917 \text{ g/cm}^3 \qquad \overline{V}_{\text{ice}} = 19.65 \text{ cm}^3\text{/mole}$$

$$\rho_{\text{water}} = 1.000 \text{ g/cm}^3 \qquad \overline{V}_{\text{water}} = 18.00 \text{ cm}^3\text{/mole}$$

$$\Delta\overline{H}_{\text{fus}} = 1440 \text{ cal/mole}$$

The energy units of $\Delta\overline{H}$ must be converted from calories to liter atm or cm^3 atm by 1 cal = 41 cm^3 atm. Then,

$$\frac{dT}{dP} = \frac{T\Delta\overline{V}}{\Delta\overline{H}}$$

$$\frac{dT}{T} = \frac{\Delta\overline{V}}{\Delta\overline{H}}\,dP = \frac{-1.65}{1440 \times 41}\,dP = -2.8 \times 10^{-5}\,dP$$

Assuming a melting point of 273.2°K at 1 atm, the melting point at 100 atm would be

$$\ln\left(\frac{T}{273.2}\right) = -2.8 \times 10^{-5} \times 99$$

$$T = 272.5°\text{K}$$

$$\Delta T = -0.7°\text{C}$$

Another example of a solid which is less dense than its melt is type metal alloy, which has been carefully chosen so that the melt expands as it solidifies and forms a sharp image of the type face of the mold into which it is poured.

Clausius–Clapeyron Equation: One Phase a Gas

Equation 5-16 can be simplified if one of the phases is a gas. The change in volume on evaporation or sublimation is almost equal to the molar volume of the gas:

$$\Delta\overline{V} = \overline{V}_{(g)} - \overline{V}_{(s,\,l)} \cong \overline{V}_{(g)}$$

Furthermore, it is usually a good approximation to consider the vapor to be an ideal gas.

$$\overline{V}_{(g)} \cong \frac{RT}{P}$$

Therefore, the Clapeyron equation can be written

$$\frac{dP}{dT} = \frac{\Delta\overline{H}_{\phi}}{T\Delta\overline{V}_{(g)}} \cong \frac{P\Delta\overline{H}_{\phi}}{RT^2} \tag{5-17}$$

or

$$\frac{d\ln P}{dT} = \frac{\Delta\overline{H}_{\phi}}{RT^2} \tag{5-18}$$

where

$$\Delta\overline{H}_{\phi} = \Delta\overline{H}_{\text{vap}} \quad \text{or} \quad \Delta\overline{H}_{\text{subl}}$$

This form is known as the Clausius–Clapeyron equation.

Remember that Eq. 5-18 applies only to solid–gas and liquid–gas equilibria. $\Delta\overline{H}_{\text{subl}}$, since it includes the heat of fusion and of vaporization, is greater than $\Delta\overline{H}_{\text{vap}}$. Therefore, the slope of the solid–gas curve is always greater than the slope of the liquid–gas curve at any given temperature, as can be seen in Figure 5-5.

5-3 PRESSURE AS A MEASURE OF ESCAPING TENDENCY

As was mentioned in Section 5-1, equilibrium vapor pressure can be used as a measure of the escaping tendency of molecules from a phase. At temperature T_b in Figure 5-6, the pressure of the vapor in equilibrium with the liquid is P_3 and the pressure of the vapor in equilibrium with the solid is P_4. At this temperature, the vapor pressure of the solid is greater than that of the liquid. If solid and liquid were placed in an enclosed evacuated container together, the solid would not be in equilibrium with its vapor until enough solid had sublimed to bring the vapor pressure up to P_4. On the other hand, vapor would *condense* into liquid as long as the vapor pressure were greater than P_3 This means that more solid would sublime until the solid phase disappeared, leaving liquid and vapor with pressure P_3. However, at temperature T_a, the vapor pressure of the solid (P_1) is *less* than that of the liquid (P_2). If any liquid were present, it would evaporate and condense on the surface of the solid until no liquid remained. In general, the phase that is stable in equilibrium with the vapor is the phase that has the *lower* equilibrium vapor pressure.

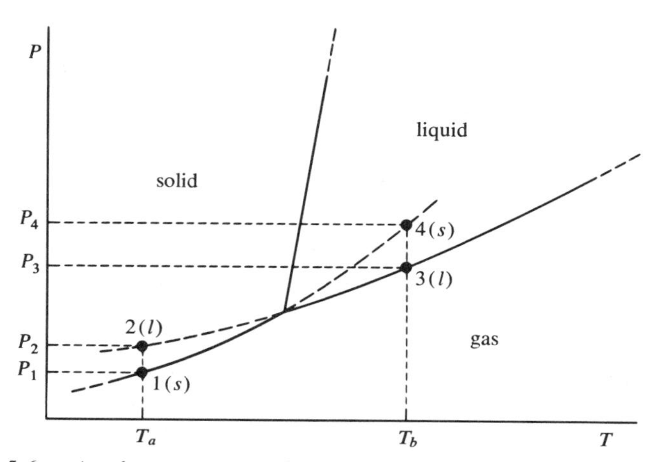

FIGURE 5-6. *An alternative way of viewing the relative stability of solid and liquid phases is to note that the more stable phase at a given temperature is the one with the* lower *vapor pressure. Vapor pressure, in this view, is a measure of the escaping tendency from a condensed phase.*

5-4 SINGLE COMPONENT PHASE DIAGRAMS

A phase diagram for water, including pertinent numerical values, is given in Figure 5-7. The diagram is not drawn to scale, but is exaggerated to call attention to certain characteristics of water. This plot includes one phenomenon that we have not previously considered—the critical point. As the pressure on a gas is increased, the molecules are pushed closer together and

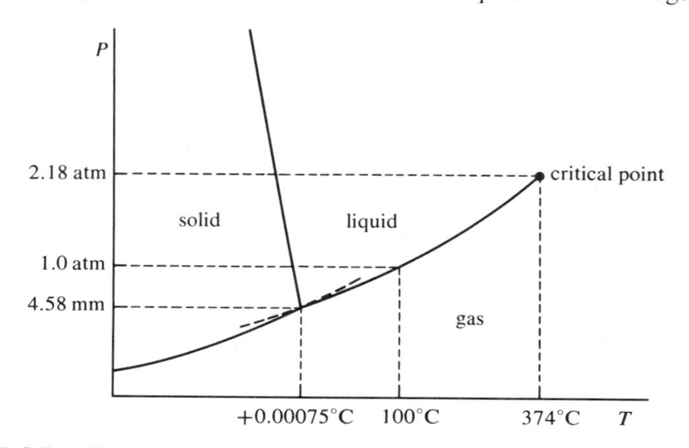

FIGURE 5-7. *The PT phase diagram for water. Since the ice structure collapses on melting and the molar volume decreases, the solid-liquid equilibrium line has a reverse slope.*

the density rises. As the temperature of a liquid is increased, the molecules move more rapidly, the molecular volume increases, and the density falls. At sufficiently high temperature and pressure, the densities of the two phases become equal and the difference between the phases, detectable typically by a liquid meniscus, disappears. One phase results, and its characterization as a very dense gas or a very open liquid is a matter of taste. The critical point specifies the highest pressure and temperature at which the phenomenon of a meniscus separating two phases can be observed. By circumventing the critical point, it is possible to go from a system that is clearly liquid to a system that is clearly vapor without ever having seen two phases present at the same time.

The surprising thing, in a sense, is not that "conventional" two-phase behavior should disappear above a critical point, but that there should ever be a liquid phase at all. The normal thing to expect might be that the more a gas is compressed the closer the molecules come together, in a smooth and continuous manner. This sudden shrinkage, emission of heat, and drop in entropy might well be considered abnormal behavior to Fred Hoyle's Black Cloud. In a sense it is we who live in the freakish corner of the universe below the critical points of gases.

Figure 5-7 actually shows only a very small part of the complete phase diagram for water. The complete diagram includes six different forms of ice and is shown in Figure 5-8. The dotted lines indicate regions that have not yet been thoroughly investigated. The diagram in Figure 5-7 is collapsed into nothingness on the abscissa of Figure 5-8. Ice I is the common crystalline form that exists under normal conditions. By a historical accident, there

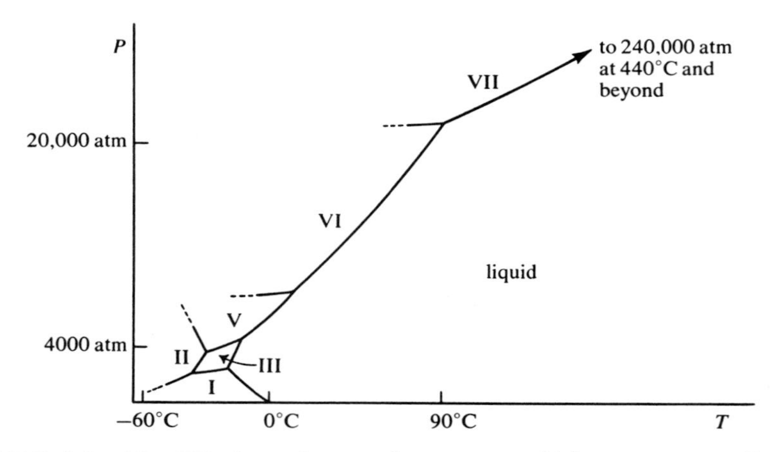

FIGURE 5-8. *The PT phase diagram for water, at higher pressures. Roman numerals mark the different crystal forms of ice. Figure 5-7 in its entirety is flattened out along the horizontal axis in this diagram.*

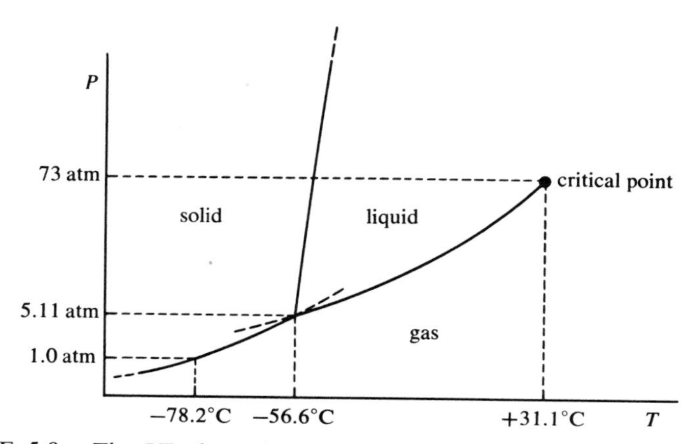

FIGURE 5-9. *The PT phase diagram for carbon dioxide. Dry ice sublimes at normal pressures because this pressure is less than that of the triple point for* CO_2. *At pressures above 5.11 atm dry ice will melt in what we usually think of as a "normal" manner.*

is no ice IV. Interestingly, ice VII melts above 100°C at 27,000 atm, higher than the boiling point of water at 1 atm. Scientists looking for a new phase have followed the ice VII/water equilibrium line out to 240,000 atm at 440°C with no new developments.[1]

Carbon dioxide presents an interesting contrast to water because it has the same basic phase diagram as water, but seen under different conditions (Figure 5-9). The one major difference is in the slope of the solid-liquid equilibrium curve; the CO_2 curve has a positive slope, whereas that for H_2O is negative. Because the triple point of CO_2 is above 1 atm, solid CO_2 sublimes rather than melting under normal conditions. However, above 5.11 atm, CO_2 behaves just like water, except that it follows the example of the majority of liquids and expands upon melting. (See Figure 5-9.)

5-5 *FREE ENERGY DIAGRAMS AND THE CRITICAL POINT*

Is it possible to account for the occurrence of a critical point between liquid and gas phases with a free energy diagram of the type of Figure 5-3 or 5-5? Figure 5-10 shows how this can be done. Below the critical point, the free energy curve of the gas phase slopes down more steeply than that of the liquid, for the molar entropy of the gas (and its degree of disorder) is always greater. On the other hand, the downward slope of the liquid curve is

[1] For ice IX, melting point 114.4°F, relative to which all other forms of H_2O are unstable at room temperature, see Kurt Vonnegut, *Cat's Cradle* (Dell, New York, 1963). What is thermodynamically unsound about "ice IX"?

increasing at a faster *rate*, because

$$\left(\frac{\partial^2 \bar{G}}{\partial T^2}\right)_P = -\left(\frac{\partial \bar{S}}{\partial T}\right)_P = -\frac{C_P}{T} \tag{5-3}$$

and the heat capacity C_P of a liquid is *greater* than that of a gas with which it is in equilibrium. (See Figure 4-11, for example.) This is reasonable, for a liquid, when heated, can dissipate some of this heat by doing work against the attractive intermolecular forces of its closely packed neighboring molecules in a way in which the widely dispersed gas cannot. The liquid can absorb more heat before its temperature rises by a specified amount. This "intermolecular work" effect in a liquid is analogous in a sense to the storage of energy in the internal degrees of freedom of a gas of polyatomic molecules.

If the liquid free energy curve has a less negative slope than the gas, but one which is dropping faster, then there will eventually be some high temperature at which the slope of the liquid curve will catch up with that of the gas. This will be the point at which the molar entropies are equal, or at which the degree of molecular disorder is the same in the two phases. This is the situation at T_c in Figure 5-10a.

Now, if the pressure is raised, the free energy of the gas will be increased, and at a more rapid rate than that of the liquid, because

$$\left(\frac{\partial \bar{G}}{\partial P}\right)_T = +\bar{V} \tag{5-4}$$

and below the critical point, the molar volume of the gas is greater than that of the liquid. At the critical pressure P_c the molar volumes will be equal and the two curves will coincide in both ordinate and slope (Figure 5-10b). At this point, the free energies, molar volumes, molar entropies, and molar enthalpies will all be identical in the two phases, and there will be no physical way of telling where one phase ends and another begins. This is the critical point.

Critical point behavior would not be expected in a solid-liquid phase equilibrium line such as the ice VII-water line of Figure 5-8. The reason can be seen by looking at Figure 5-10a with the word "gas" replaced by

---

FIGURE 5-10. *Free energy explanation of the occurrence of a critical point.* (*a*) *At sufficiently high temperatures, the* slope *of the liquid free energy curve will match that of the gas, since the second derivative of G with respect to T is greater for the liquid. At this temperature, the molar entropies of the two phases will be equal.* (*b*) *If now the pressure is raised, at some higher pressure P_c and temperature T_c both the slopes and absolute values of the G curves will coincide. The liquid and gas phases will have identical molar entropies, free energies, and enthalpies, and will no longer be distinguishable as two separate phases. This is the* critical point.

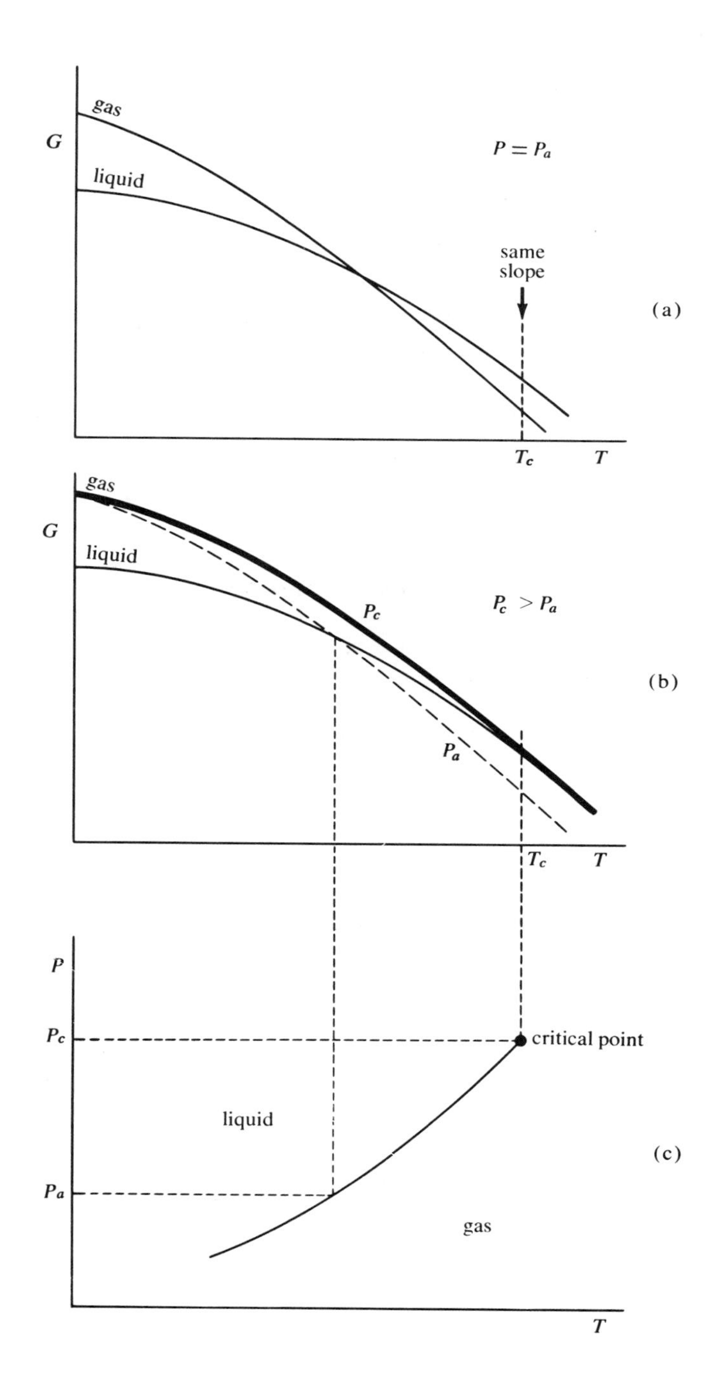
gas
G
liquid
P = Pa
same slope
Tc
T
(a)
gas
G
liquid
Pc
Pc > Pa
Pa
Tc
T
(b)
P
Pc
critical point
liquid
Pa
gas
T
(c)

"liquid," and "liquid" by "solid." The solid curve will *not* accelerate its downward slope in such a way as to become parallel to the liquid curve, as the liquid curve did earlier for the gas. For the heat capacity of a solid is *less* than that of a liquid with which it is in equilibrium (see Figure 4-11). The extra heat storage ability of a liquid comes from the ability of its molecules to increase their thermal motion and do work against the attractive forces of near-neighbor molecules. The molecules of a gas are free to increase their thermal motion but too far from one another for the work contribution to be significant; molecules of a solid are close enough to exert strong forces on one another but too tightly held in the crystal to do much work against intermolecular forces when heat is fed in. Vibration about a mean position is increased, but the mean position remains the same until the crystal breaks up and melts. Another way of stating the same conclusions is that the molar entropy of a liquid will rise with temperature faster than that of a solid, whose order is maintained as long as melting does not occur. So although ice VIII may exist somewhere beyond 240,000 atm and 440°C, a critical point is not expected.

5-6 FIRST AND SECOND ORDER TRANSITIONS

Earlier in this chapter we introduced the idea of first order transitions. For such transitions the free energy is a continuous function, but its first derivatives are not. A graphical interpretation of a typical first order transition is given in Figure 5-11. The figure includes a free energy plot as well as plots of the two first derivatives of free energy, entropy, and volume:

$$\left(\frac{\partial G}{\partial T}\right)_P = -S \qquad \left(\frac{\partial G}{\partial P}\right)_T = +V$$

In a second order transition both the free energy and its first derivatives are continuous, but its second derivatives are discontinuous. As depicted in Figure 5-12a, the entropy changes in a continuous manner with temperature, but the *rate of change in disorder* alters at the phase transition point. This plot

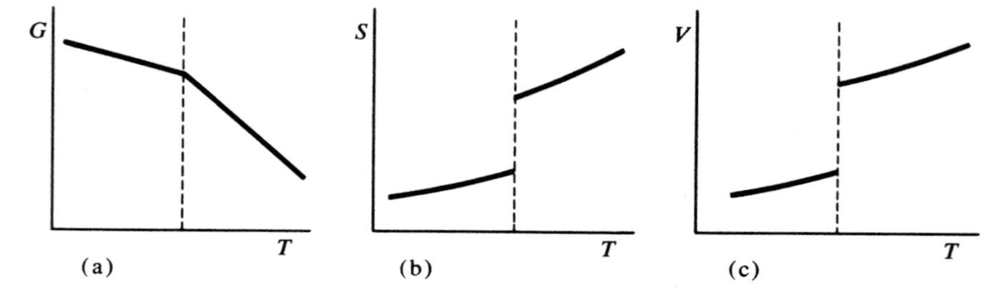

FIGURE 5-11. *The behavior of free energy, entropy, and molar volume at a first order phase transition.*

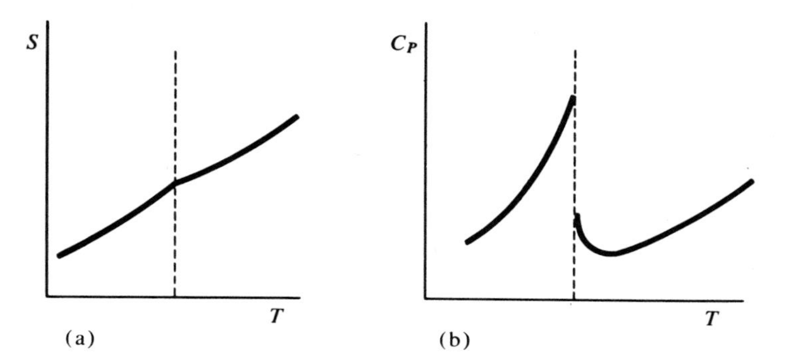

FIGURE 5-12. *The behavior of entropy and heat capacity at a second order phase transition.*

can be thought of as the limit of the normal first order transition, in which the entropy change at transition goes to zero. The break in this ST curve means that a plot of the first derivative of entropy with respect to temperature, C_P/T or C_P, will be discontinuous, as shown in Figure 5-12b. Because the heat capacity curve often has the shape of a Greek λ, this second order transition is frequently referred to as a λ-point transition. In general, for a second order transition, all second derivatives of free energy are discontinuous.

$$-\left(\frac{\partial^2 G}{\partial T^2}\right)_P = \left(\frac{\partial S}{\partial T}\right)_P = \frac{C_P}{T} \tag{5-3}$$

$$-\left(\frac{\partial^2 G}{\partial P^2}\right)_T = -\left(\frac{\partial V}{\partial P}\right)_T = \beta V \tag{5-19}$$

$$\left(\frac{\partial^2 G}{\partial P\,\partial T}\right) = \left(\frac{\partial V}{\partial T}\right)_P = \alpha V \tag{5-20}$$

where β is the coefficient of compressibility and α is the coefficient of thermal expansion.

In this type of transition, there is no abrupt change in free energy nor in the extent of disorder of the system. But there is an abrupt change in the ability of the system to store heat. This behavior is characteristic of processes involving order–disorder transitions. Crystalline ammonium chloride is a good example of such a substance. The crystal structure of NH_4Cl (Figure 5-13) is pseudo-body-centered in the sense that if chlorine and nitrogen were identical and no hydrogen were present, the crystal would be body centered. Every nitrogen is surrounded by eight chlorines; and every chlorine is surrounded by eight nitrogens.

As shown in Figure 5-13, there are two different ways of orienting the ammonium ion tetrahedron. These two arrangements are not energetically equivalent, because there is interaction from one unit cell to the next. At low temperatures, one would expect to find an ordered NH_4Cl crystal in which all hydrogen atoms were in the same choice of alternates throughout the entire crystal. As the temperature is increased and the crystal begins to absorb heat, the orientations of the ammonium ion tetrahedra from one unit cell to the next will become increasingly random. The heat capacity at constant pressure is in part a function of the randomizing of the tetrahedra, because misaligned tetrahedra in neighboring cells are in an energetically unfavorable, or excited state. Eventually, a point is reached where there is a 50–50 mixture, and complete randomness in the orientation of the tetrahedra. When this state is reached, the crystal can no longer use this particular mode of energy absorption, and the heat storing ability will suddenly fall, as shown in Figure 5-12b. For NH_4Cl this point of heat capacity discontinuity occurs at 243°K.

Another example of a second order transition is β-brass. In the most ordered form of an equimolar alloy, copper atoms occupy the corners of a cubic lattice and zinc atoms the centers of the cubes. (The Cu and Zn positions are entirely equivalent, however. If the Zn atoms are taken as establishing the corners of the cubes, then the Cu atoms occupy the centers.) This again is a pseudo-body-centered lattice.

At low temperatures, Cu and Zn atoms are perfectly ordered in their respective sites. But as the temperature rises, mistakes and disorder begin to appear as Cu and Zn atoms exchange positions. Entropy rises, and heat can be absorbed by creating disorder. But at 742°K the disordering is complete and Cu and Zn atoms occupy sites in a random manner. No *more* disorder can be created, and this storehouse for heat is thereafter closed. The heat

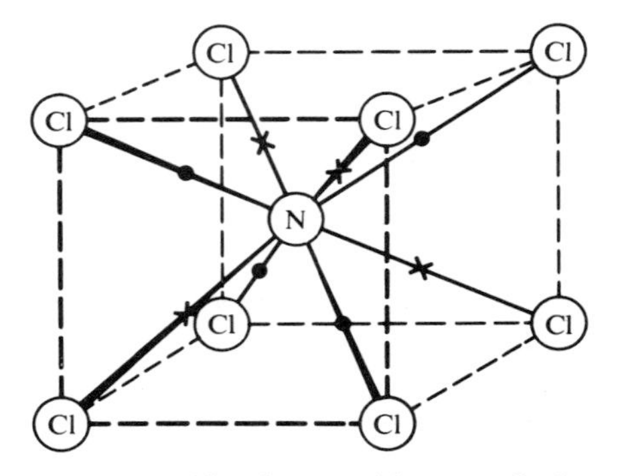

FIGURE 5-13. *The ammonium chloride crystal lattice is built up from a cubic array of* Cl^- *ions with a* NH_4^+ *ion at the center of each cube [Why, then, is the chemical formula not* $(NH_4)Cl_8$*?]. There are two possible sets of positions for the four protons, marked by dots and by crosses. The degree of regularity of arrangement of these protons on adjacent ammonium ions will affect the entropy of the crystal.*

capacity falls suddenly and the entropy begins to rise with T at a much slower rate. The curves, again, are of the type of Figure 5-12.

5-7 *PHASE RULE—FIRST EXPOSURE*

One-component systems provide a simple introduction to the phase rule and the concept of degrees of freedom. Figure 5-7 presents the now familiar phase diagram of H_2O. In the open areas of this diagram there is only one phase present. Whenever there is only a single phase, pressure and temperature can be varied independently at will. On the other hand, if two phases are to remain in equilibrium, then the only possible choices of P and T are those along one of the two-phase equilibrium lines. If either P or T is chosen, then the other is fixed and there is only one degree of freedom. If either P or T is changed, the change in the other is determined by the Clapeyron equation. Moreover, *three* phases can be maintained in equilibrium only at the triple point. There are *no* degrees of freedom. Any alteration in either P or T will result in the destruction of at least one of the three phases.

All of the preceding observations can be summarized by a simple formula:

$$f = 3 - p \tag{5-21}$$

where f represents the degrees of freedom and p is the number of phases. This is the most elementary statement of the phase rule, about which we will have more to say later.

5-8 *PARTIAL MOLAR VOLUME*

There can be no chemistry without at least two chemical species. As soon as we leave pure substances for mixtures, we find that the molar values of extensive thermodynamic variables such as volume, enthalpy, entropy, or free energy of each component depend upon the kind and amount of other components present. The volume is the easiest variable to illustrate.

The volume per mole of pure water is 18 cm^3, and that of pure ethanol is 58 cm^3. If half a mole of each are mixed in a beaker, we might expect the final volume to be 9 plus 29 or 38 cm^3. Instead, the volume is observed to be only 37.1 cm^3. This volume shrinkage upon mixing is found with many easily miscible liquid pairs in which the association of unlike molecules is easier than that between like molecules. A plot of volume per mole versus mole fraction for the ethanol-water system is shown in Figure 5-14. If there were no shrinkage, the volume would follow the dashed line given by the expression

$$\overline{V}_{\text{ideal}} = X_A \overline{V}_A^{\bullet} + X_B \overline{V}_B^{\bullet} \tag{5-22}$$

(Here A represents water and B represents ethanol, and the superscript filled circles indicate pure components.) The true volume is given by the solid curve.

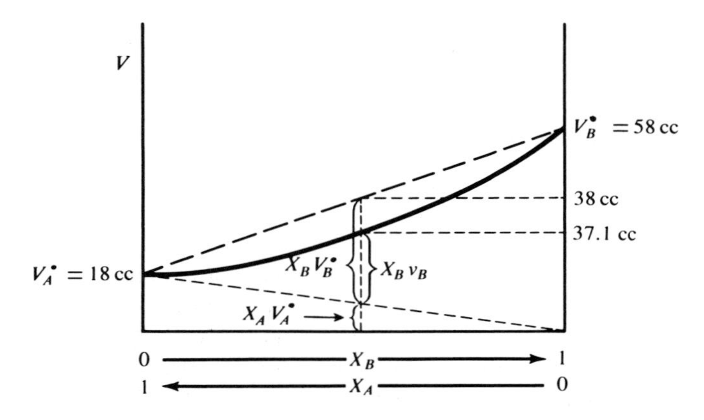

FIGURE 5-14. *Total volume of a binary liquid mixture of water (A) and ethanol (B) as a function of the mole fraction composition of the solution. If all of the observed shrinkage upon mixing were to be ascribed to a component B, then it would have an* apparent *molar volume under these conditions of v_B. Volumes for an ideal solution, with no shrinkage or expansion, are shown by a dashed line.*

If ethanol is considered to be the solute and water the solvent, then we can think of all of the shrinkage as arising from some deficiency in the ethanol. Then the "apparent molar volume" of ethanol in a solution with water at a given mole fraction would be v_B in the expression

$$\overline{V}_{\text{true}} = X_A \overline{V}_A^{\bullet} + X_B v_B \tag{5-23}$$

For the equimolar solution of the preceding paragraph,

$$v_B = \frac{37.10 - \frac{1}{2}(18.0)}{\frac{1}{2}} = 56.20 \text{ mole cm}^3$$

Although the molar volume of pure ethanol is 58 cm^3, in an equimolar aqueous solution it appears to be 56.2 cm^3.

But it is as wrong to assign all of the shrinkage effect to one component as it is to ignore the shrinkage. The molar volume of each component is affected by the presence of the other. The proper approach is indicated in Figure 5-15. In this figure, the total volume is plotted against the number of moles of B in one mole of A. At the equimolar point, the volume is not 18 plus 58 cm^3, but is only 74.2 cm^3. The *slope* of this curve at any concentration is the rate of change of volume per mole of ethanol, the amount of water being held constant. With the additional restrictions of constant pressure and temperature, this is the definition of the *partial molar volume*.[2]

[2] The partial molar volume of a pure substance is simply its volume per mole. This section represents an extension of the meaning of the bar symbol to components of a mixture.

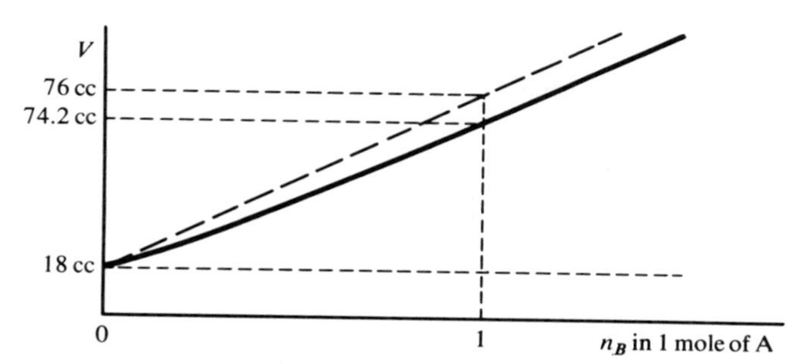

FIGURE 5-15. *If the total volume of a mixture is plotted against the number of moles of component B in 1 mole of A, then the slope of the curve at any point is the partial molar volume of component B. In this diagram, A is water and B is ethanol.*

$$\bar{V}_A \equiv \left(\frac{\partial V}{\partial n_A}\right)_{P,\,T,\,n_B} \qquad \bar{V}_B \equiv \left(\frac{\partial V}{\partial n_B}\right)_{P,\,T,\,n_A} \tag{5-24}$$

The partial molar volume of one component is the *rate* of change in volume with number of moles of that component, all other components being held at a fixed level, or, alternatively, it is the *increase* in volume obtained by adding 1 mole of a component to such a large volume of solution that the change in solution concentration is negligible.

The partial molar volumes of ethanol and of water in an ethanol-water solution are shown in Figure 5-16. Note that whenever the partial molar volume of one component is rising, that of the other is falling. This is one expression of the Gibbs–Duhem equation, which will be derived later.

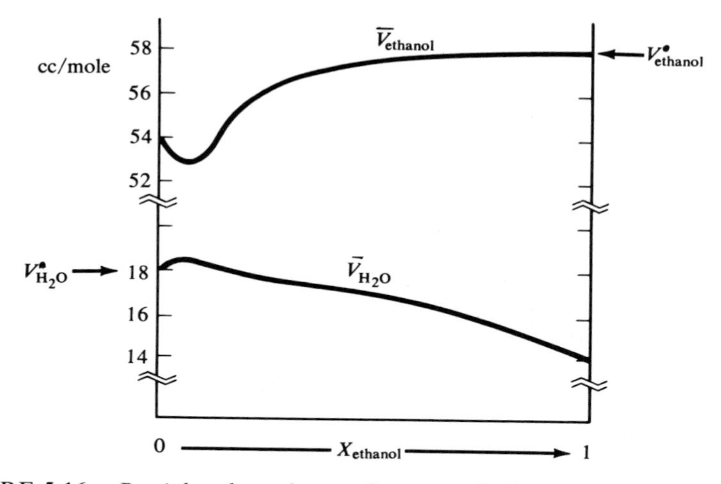

FIGURE 5-16. *Partial molar volumes of water and ethanol in solutions of varying concentrations.*

Partial molar volumes are well-defined experimental properties of components of a solution. If the forces of attraction between unlike molecules are greater than those between like molecules, as was the case for ethanol and water, then shrinkage will occur upon mixing, and the partial molar volumes will usually be less than the molar volumes of the pure components. If the intermolecular interactions are weaker, then expansion will occur and the partial molar volumes will usually be greater than the pure molar volumes. Only if the interactions between like and unlike molecules are the *same* will the partial molar volumes be equal to the pure molar volumes at all concentrations, so that the total volume will be given by

$$V = X_A \overline{V}_A^{\bullet} = X_B \overline{V}_B^{\bullet} \tag{5-25}$$

Such solutions are called *ideal* solutions.

How is the true total volume related to the partial molar volumes in *nonideal* solutions? For the answe' to this we must turn to the properties of simple homogeneous functions.

5-9 HOMOGENEOUS FUNCTIONS

Partial molar volume is only one member of a whole class of partial molar functions, all of which are related to simple homogeneous functions.

Definitions

Volume is a simple homogeneous function of the first degree with respect to the masses of all components:

$$V = V(m_1, m_2, m_3, \cdots)$$

This means that if the masses of all components are doubled, the volume is doubled. Or, in general,

$$V(\lambda m_1, \lambda m_2, \ldots) = \lambda V(m_1, m_2, \ldots) \tag{5-26}$$

where λ is any positive number. Volume is also a homogeneous function of the first degree with respect to the number of moles, since mole numbers are related to masses through a set of constants, the molecular weights. In general, a homogeneous function of degree α can be defined as

$$F(\lambda n_1, \lambda n_2, \ldots) = \lambda^{\alpha} F(n_1, n_2, \ldots) \tag{5-27}$$

An increase in all variables by a factor λ produces an increase in the function by a factor of λ^{α}.

Volume, energy, entropy, enthalpy, free energy, and the work function are all homogeneous functions of the first degree in masses or moles. Such quantities are also called *extensive* variables. On the other hand, pressure and temperature are homogeneous functions of the zeroth degree. For example,

$$P(\lambda n_1, \lambda n_2, \ldots) = \lambda^0 P(n_1, n_2, \ldots) \tag{5-28}$$

Properties of this kind are said to be *intensive*.

As a rule, we are only concerned with homogeneous functions of the zeroth and first degree. Is there such a thing as a "hyperextensive" physical function of degree greater than one? Yes—volume itself is a simple homogeneous function of the third degree in linear dimensions:

$$V(\lambda x, \lambda y, \lambda z) = \lambda^3 V(x, y, z) \tag{5-29}$$

The energy of a particle in a box is a homogeneous function of the second degree in the quantum numbers:

$$\varepsilon_{(j,k,l)} = \frac{h^2}{8m}\left(\frac{j^2}{a^2} + \frac{k^2}{b^2} + \frac{l^2}{c^2}\right) \tag{5-30}$$

$$\varepsilon_{(\lambda j, \lambda k, \lambda l)} = \lambda^2 \varepsilon_{(j,k,l)} \tag{5-31}$$

The energy of a rigid rotor is *not* a homogeneous function at all with respect to the quantum number J:

$$E_{\lambda J} = hB(\lambda^2 J^2 + \lambda J) \neq \lambda^\alpha E_J \tag{5-32}$$

For an ideal gas, volume is a homogeneous function of the zeroth degree in pressure and temperature:

$$V_{(P,T)} = \frac{RT}{P}$$

$$V_{(\lambda P, \lambda T)} = \lambda^0 V_{(P,T)} \tag{5-33}$$

However, for a van der Waals gas, the volume is *not* a homogeneous function of pressure and temperature:

$$V_{(P,T)} \cong \frac{RT}{P} + \left(b - \frac{a}{RT}\right) \tag{5-34}$$

Clearly, when talking about a property as a homogeneous function, one must specify the variables, because the choice of variables determines the degree of the function, or whether or not it is homogeneous at all. The variables with which we will usually be concerned will be pressure, temperature, and mole numbers.

Extensive properties are defined as homogeneous functions of the first degree with respect to pressure, temperature, and mole number:

$$F = F(T, P, n_1, n_2, \ldots)$$
$$F(T, P, \lambda n_1, \lambda n_2, \ldots) = \lambda F(T, P, n_1, n_2, \ldots) \tag{5-35}$$

Why go to all the trouble of invoking homogeneous functions? We do so because, like state functions, homogeneous functions have particularly useful properties.

Euler's Theorem

Euler's theorem for homogeneous functions leads to some very important relationships. For a completely general homogeneous function of degree α in the variables $x, y, z, \ldots,$

$$f(\lambda x, \lambda y, \lambda z, \ldots) = \lambda^{\alpha} f(x, y, z, \ldots) \tag{5-36}$$

Multiplying all of the extensive dependent variables by a given constant λ gives the function a new value λ^{α} times the old. Differentiating Eq. 5-36 with respect to λ,

$$\begin{aligned}\frac{df(\lambda x, \lambda y, \ldots)}{d\lambda} &= \frac{\partial f}{\partial(\lambda x)}\frac{\partial(\lambda x)}{\partial \lambda} + \frac{\partial f}{\partial(\lambda y)}\frac{\partial(\lambda y)}{\partial \lambda} + \cdots \\ &= \alpha\lambda^{\alpha-1} f(x, y, \ldots)\end{aligned} \tag{5-37}$$

Simplifying,

$$x\frac{\partial f}{\partial(\lambda x)} + y\frac{\partial f}{\partial(\lambda y)} + z\frac{\partial f}{\partial(\lambda x)} + \cdots = \alpha\lambda^{\alpha-1} f \tag{5-38}$$

For the special case of $\lambda = 1$,

$$x\frac{\partial f}{\partial x} + y\frac{\partial f}{\partial y} + z\frac{\partial f}{\partial x} + \cdots = \alpha f \tag{5-39}$$

Since we are primarily interested in first degree functions, we can set $\alpha = 1$ as well:

$$f = x\frac{\partial f}{\partial x} + y\frac{\partial f}{\partial y} + z\frac{\partial f}{\partial z} + \cdots \tag{5-40}$$

If f is *volume*, then we have a relationship between total volume and the partial molar volumes:

$$V = n_1 \overline{V}_1 + n_2 \overline{V}_2 + n_3 \overline{V}_3 + \cdots \tag{5-41}$$

The total volume of a mixture is the *sum* of the products of the number of moles of each component times the slope of the volume-versus-concentration curve for that component. This is not an immediately obvious property. For example, this is analogous to saying that one can calculate his altitude on a mountain by taking the sum of the product of his longitude and latitude, respectively, with the respective north-south and east-west slopes at that point. This approach is manifest nonsense, but only because altitude is not usually a homogeneous function of first degree in geographical location. This property is a special property of *extensive* variables:

$$V = \sum_j n_j \overline{V}_j \tag{5-42}$$

Gibbs–Duhem Equation

Since volume is a simple homogeneous function of the first degree, Eq. 5-42 is true. But volume is also a state function, and dV is an exact differential. Therefore,

$$dV = \left(\frac{\partial V}{\partial n_1}\right) dn_1 + \left(\frac{\partial V}{\partial n_2}\right) dn_2 + \cdots = \sum_j \overline{V}_j\, dn_j \tag{5-43}$$

We now know that both the mole numbers n_j and the partial molar volumes $\overline{V}_j$ can change in Eq. 5-42. The total differential is then

$$dV = \sum_j \overline{V}_j\, dn_j + \sum_j n_j\, d\overline{V}_j \tag{5-44}$$

Subtracting Eq. 5-43 from Eq. 5-44 yields an important relationship between partial molar quantities:

$$\sum_j n_j\, d\overline{V}_j = 0 \tag{5-45}$$

Partial molar volumes are not independent quantities. Whenever changes are made in the partial molar volumes of the components, the weighted sum of all the changes is zero, if the weighting factors are the mole numbers of the components.

For a two-component system, Eq. 5-45 becomes

$$n_A d\overline{V}_A + n_B d\overline{V}_B = 0 \qquad (5\text{-}46)$$

The two partial molar volumes are linked, and the changes in one can be predicted by a knowledge of the changes in the other:

$$d\overline{V}_B = -\frac{n_A}{n_B}\, d\overline{V}_A \qquad (5\text{-}47)$$

As the partial molar volume of one component increases, that of the other has to decrease, as is seen in Figure 5-16.

The Gibbs–Duhem equation, a particularization of Eq. 5-46 for the case in which changes are produced by changing the amount of *one* of the components, is

$$n_A\left(\frac{\partial \overline{V}_A}{\partial n_A}\right)_{n_B} + n_B\left(\frac{\partial \overline{V}_B}{\partial n_A}\right)_{n_B} = 0 \qquad (5\text{-}48)$$

A similar equation exists for the effect of changes in n_B.

Other Partial Molar Quantities

There are many partial molar quantities in addition to partial molar volume. For example, the partial molar enthalpy of the jth component is designated by $\overline{H}_j$. The partial molar heat of solution can be found if we know the heat of solution

$$\overline{\Delta H}_{S_j} = \left(\frac{\partial \Delta H}{\partial n_j}\right)_{n_i} \qquad (5\text{-}49)$$

This quantity is the rate of change of the heat of solution per mole of solute added, and in fact is the differential heat of solution discussed in Chapter 3.

If the variable is mass, then the partial molar mass of the jth component is simply the molecular weight

$$\overline{m}_j = \left(\frac{\partial M}{\partial n_j}\right)_{n_i} = M_j \qquad (5\text{-}50)$$

The total mass is the number of moles times the molecular weight:

$$m = \sum_j n_j \bar{m}_j = \sum_j n_j M_j \tag{5-51}$$

There is never any shrinkage or expansion of mass upon mixing (barring nuclear reactions).

In general, if J represents any extensive property, the partial molar expression for the jth component is

$$\bar{J}_j = \left(\frac{\partial J}{\partial n_j}\right)_{T,P,n_i} \tag{5-52}$$

where the subscript n_i signifies a constant number of moles of all components *except* for the jth.

5-10 *PARTIAL MOLAR FREE ENERGY OR CHEMICAL POTENTIAL*

The most useful partial molar quantity is the partial molar free energy $\bar{G}_j$. It is so useful that it is given the name of chemical potential and a separate symbol μ_j to emphasize its generality.

Basic Properties

The chemical potential is defined as

$$\mu_j \equiv \bar{G}_j = \left(\frac{\partial G}{\partial n_j}\right)_{T,P,n_i}$$

As we shall see, the concept of chemical potential is not uniquely bound to free energy. But since μ_j is defined under conditions of constant pressure and temperature, it is perhaps reasonable to begin by associating μ_j with free energy. Just as pressure and temperature are the "natural" variables for free energy, so the other important functions are also usually associated with two particular variables:

free energy	$G = G(T, P)$	$dG = -SdT + VdP$	(5-53)
work function	$A = A(T, V)$	$dA = -SdT - PdV$	(5-54)
enthalpy	$H = H(S, P)$	$dH = TdS + VdP$	(5-55)
energy	$E = E(S, V)$	$dE = TdS - PdV$	(5-56)

Imagine now a box enclosing a collection of different chemical species at a certain over-all pressure and temperature. The free energy of the entire system is now not only a function of P and T, but of the amount of each chemical component present:

$$G = G(T, P, n_1, n_2, \ldots)$$

The total change in free energy as all of the independent variables change is

$$dG = \left(\frac{\partial G}{\partial T}\right)_{P, n_i} dT + \left(\frac{\partial G}{\partial P}\right)_{T, n_i} dP + \sum_j \bar{G}_j \, dn_j \tag{5-57}$$

From our knowledge of the derivatives of free energy, we can rewrite this equation as

$$dG = -SdT + VdP + \sum_j \mu_j \, dn_j \tag{5-58}$$

The difference between this equation and Eq. 5-53, which did not involve the chemical potentials, is that now we are considering variations in the free energy of a system with changes in the amounts of chemical components and not just with pressure and temperature. This expanded treatment is necessary for multicomponent systems.

From the definition of the work function,

$$A = G - PV$$
$$dA = dG - VdP - PdV$$

This plus Eq. 5-58 yields

$$dA = -SdT - PdV + \sum_j \mu_j \, dn_j \tag{5-59}$$

Similarly, for enthalpy,

$$H = G + TS$$
$$dH = dG + S\,dT + T\,dS$$
$$dH = TdS + VdP + \sum_j \mu_j \, dn_j \tag{5-60}$$

and for energy,

$$E = H - PV$$
$$dE = dH - VdP - PdV$$
$$dE = TdS - PdV + \sum_j \mu_j \, dn_j \tag{5-61}$$

All of these new expressions reduce to the older ones of Eqs. 5-53 through 5-56 if the amounts of all components are held constant.

By introducing the chemical potential term into these equations, we allow for the fact that these functions do change when the amounts of participating species are altered. At this point, it makes no difference how the change in components is brought about; it could be caused by reaction or by physical addition or removal of components from the system.

In each of Eqs. 5-58 through 5-61, μ plays a different role, even though it is the same quantity in all cases. All of the following equalities are equally true:

$$\mu_j = \left(\frac{\partial G}{\partial n_j}\right)_{T,P,n_i} = \left(\frac{\partial A}{\partial n_j}\right)_{T,V,n_i} = \left(\frac{\partial H}{\partial n_j}\right)_{S,P,n_i} = \left(\frac{\partial E}{\partial n_j}\right)_{S,V,n_i}$$

It is important to realize that while the chemical potential is defined by all of these expressions, only the first one is a partial molar quantity. As defined, a partial molar quantity is the rate of change of a function with respect to the concentration of one component under conditions of *constant pressure and temperature*. The chemical potential *is* the rate of change of enthalpy with amount of one component, when entropy, pressure, and amounts of all other components are held constant, but it is *not* the partial molar enthalpy.

Relationships between extensive variables also apply to their corresponding partial molar expressions; for example, in

$$dG = -SdT + VdP + \mu_1 dn_1 + \mu_2 dn_2 + \cdots \tag{5-62}$$

Euler cross partials between n_1 and P or T yield

$$\left(\frac{\partial \mu_1}{\partial T}\right)_{P,n_1,\ldots} = -\left(\frac{\partial S}{\partial n_1}\right)_{T,P,n_2,\ldots} = -\bar{S}_1 \tag{5-63}$$

$$\left(\frac{\partial \mu_1}{\partial P}\right)_{T,n_1,\ldots} = +\left(\frac{\partial V}{\partial n_1}\right)_{T,P,n_2,\ldots} = +\bar{V}_1 \tag{5-64}$$

The first of these equations involves the partial molar entropy of component 1, and the second, its partial molar volume. These expressions are analogous to

$$\left(\frac{\partial G}{\partial T}\right)_P = -S \qquad \left(\frac{\partial G}{\partial P}\right)_T = +V$$

which apply to single-component systems. Other expressions involving G, V, H, E, and so on, also have their analogues with μ_j, $\bar{V}_j$, $\bar{H}_j$, $\bar{E}_j$, and so on.

But the cross partials between concentration terms yield a new relationship

$$\left(\frac{\partial \mu_1}{\partial n_2}\right)_{P,\,T,\,n_1} = \left(\frac{\partial \mu_2}{\partial n_1}\right)_{P,\,T,\,n_2}$$

This says that the sensitivity of the chemical potential of any one component to changes in the amount of a second is the same as the sensitivity of the second to changes in the first. There is reciprocity in the effects of two components on each other's chemical potential.

Dependence of μ on Concentration in Gas Mixtures

For an ideal gas,

$$\left(\frac{\partial G}{\partial P}\right)_T = V = \frac{RT}{P}$$

Or, at constant temperature,

$$dG = VdP = RT\, d \ln P \tag{5-65}$$

The equivalent expression for partial molar quantities in an ideal gas mixture is

$$d\mu_j = \overline{V}_j\, dP = RT\, d \ln P_j \tag{5-66}$$

where P_j is the partial pressure of the jth component.

From Dalton's law we know that in such a mixture each of the components acts as though it were alone with a pressure equal to its partial pressure and a volume equal to the total volume of the system:

$$P_j = X_j P_T \qquad P_T = \sum_j P_j \tag{5-67}$$

Under these conditions one can find the *change* in the chemical potential of the gas from the change in its partial pressure:

$$\mu_{j(2)} = \mu_{j(1)} + RT \ln \frac{P_{j(2)}}{P_{j(1)}} \tag{5-68}$$

Just as we found it useful to have a reference level for energy measurements, it is also helpful to have a reference level from which to measure chemical potentials. The standard state that is commonly employed for this purpose is that for which the *partial pressure* of the component gas is 1 atm. With this definition, Eq. 5-68 can be written as

$$\mu_{j(P_j, T)} = \mu^0_{j(T)} + RT \ln \frac{P_{j(\text{atm})}}{1 \text{ atm}}$$

or simply

$$\mu_{j(P_j, T)} = \mu^0_{j(T)} + RT \ln P_j \tag{5-69}$$

where it is implicit that the argument of the logarithm is really the unitless ratio of P_j in atmospheres to 1 atm.

Chemical Potential and Escaping Tendency

In Section 5-3 we saw that for systems of one component the free energy served as a measure of the escaping tendency of a substance from a condensed phase, liquid or solid. In multicomponent systems, the chemical potential plays the same role in determining the escaping tendency of a component from one phase into another.

If iodine is added to the two-phase system water–carbon tetrachloride, it can be seen to have a different solubility in each solvent. The iodine will establish an equilibrium between the two phases, and at equilibrium the chemical potential of the iodine must be the same in both phases. For if *dn* moles of iodine are taken from water (w) to carbon tetrachloride (c) (at constant temperature and pressure), the total change in free energy is

$$dn^c_{I_2} = dn \qquad dn^w_{I_2} = -dn$$
$$dG = \mu^c_{I_2}\, dn - \mu^w_{I_2}\, dn$$
$$dG = (\mu^c_{I_2} - \mu^w_{I_2})\, dn$$

At equilibrium, $dG = 0$ and the chemical potential of both phases has to be the same:

$$\mu^c_{I_2} = \mu^w_{I_2}$$

As long as dG is less than zero there will be a spontaneous tendency for the iodine to go from the water phase to the carbon tetrachloride phase:

$$dG < 0 \qquad \mu^c_{I_2} < \mu^w_{I_2}$$

In general, whenever several immiscible phases have a component in common, this component will diffuse out of phases in which it has a high chemical potential and into phases in which it has a low chemical potential. This process will continue until its chemical potential is the same in all phases. This condition is the criterion for equilibrium distribution of the component in the system.

5-11 THERMODYNAMICS OF CHEMICAL REACTIONS

Following the notation introduced in Section 3-14, a general chemical reaction can be written as

$$n_a A + n_b B + \cdots \rightarrow n_c C + n_d D + \cdots \tag{3-80}$$

or, by condensing terms,

$$\sum_j v_j A_j = 0 \qquad v_j \begin{cases} + n_j \text{ for products} \\ - n_j \text{ for reactants} \end{cases} \tag{5-70}$$

As used in this context, v_j is a stoichiometric coefficient or stoichiometry number. It is not necessarily the actual number of moles involved in a process. However, if the only way in which the amounts of components change is through reaction (that is to say, there is no physical addition or removal of components), then the change in the mole number of each component is related to its stoichiometry number by

$$\frac{dn_1}{v_1} = \frac{dn_2}{v_2} = \frac{dn_3}{v_3} = \cdots = d\xi \tag{5-71}$$

where ξ is a reaction parameter commonly called the "extent of reaction." It indicates how many stoichiometric units of the reaction, as written, actually have occurred. The actual number of moles of a given component which are consumed or produced would be related to ξ by

$$dn_j = v_j \, d\xi \tag{5-72}$$

Employing this new notation, the free energy change in a chemical system which arises from reaction or from changes in temperature and pressure can be written as

$$dG = -SdT + VdP + \left(\sum_j \mu_j v_j\right) d\xi \tag{5-73}$$

where the term $\sum_j u_j v_j$ is given the special name of the reaction potential:

$$\sum_j \mu_j v_j = \widetilde{\Delta G} \tag{5-74}$$

If changes in amounts of components occur only through the chemical reaction whose progress is measured by ξ, then the reaction potential is the rate of change of total free energy per unit of reaction at constant temperature and pressure.

$$\widetilde{\Delta G} = \left(\frac{\partial G}{\partial \xi}\right)_{P,\,T} \tag{5-75}$$

The reaction potential is identical to the free energy change of a reaction as defined in Section 4-12. The standard reaction potential is the standard free energy change of a reaction as calculated from the tabulated values of the free energies of formation.

Universality of Reaction Potential

Having introduced the reaction potential, we can now rewrite Eqs. 5-58 through 5-61 as

$$dG = -SdT + VdP + \widetilde{\Delta G}\, d\xi \tag{5-76}$$

$$dA = -SdT - PdV + \widetilde{\Delta G}\, d\xi \tag{5-77}$$

$$dH = TdS + VdP + \widetilde{\Delta G}\, d\xi \tag{5-78}$$

$$dE = TdS - PdV + \widetilde{\Delta G}\, d\xi \tag{5-79}$$

These four equations are less general, inasmuch as they limit changes in components to one chemical reaction, while the earlier set included physical addition or removal as well. In each case, $\widetilde{\Delta G}$ gives the rate of change of the function per unit of chemical reaction under appropriate reaction conditions.

$$\widetilde{\Delta G} = \left(\frac{\partial G}{\partial \xi}\right)_{T,\,P} = \left(\frac{\partial A}{\partial \xi}\right)_{T,\,V} = \left(\frac{\partial H}{\partial \xi}\right)_{S,\,P} = \left(\frac{\partial E}{\partial \xi}\right)_{S,\,V} \tag{5-80}$$

Under conditions of constant P and T, the reaction potential calculated from tables of standard free energies of formation, $\Delta G_f{}^0$, is the *free energy* change per unit of reaction. But under constant V and T conditions, the *same* numbers give the *work function* change. If the reaction could be carried out at constant S and P, the reaction potential value would measure the change in enthalpy, and, at constant S and V, the energy. When working with a bomb calorimeter, one can use tables of $\Delta G_j{}^0$ data at constant P and T, relabeling them as $\Delta A_j{}^0$'s.

The spontaneity of a reaction under various experimental conditions depends upon the value of $\widetilde{\Delta G}\, d\xi$. If this term is less than zero, the reaction will be spontaneous.

$$\begin{array}{ll} \text{const } T \text{ and } P & dG = \widetilde{\Delta G}\, d\xi < 0 \\ \text{const } T \text{ and } V & dA = \widetilde{\Delta G}\, d\xi < 0 \\ \text{const } S \text{ and } P & dH = \widetilde{\Delta G}\, d\xi < 0 \\ \text{const } S \text{ and } V & dE = \widetilde{\Delta G}\, d\xi < 0 \end{array}$$

Instead of four different potential functions for use in four different circumstances, we now have only one, the reaction potential. A reaction is spontaneous if $\widetilde{\Delta G} < 0$.

5-12 HOMOGENEOUS IDEAL GAS REACTIONS—MATHEMATICAL FRAMEWORK

From Eq. 5-69, we can find the chemical potential of a gas at other than standard conditions:

$$\mu_j = \mu_j^{\,0} + RT \ln P_j \tag{5-81}$$

Then the reaction potential is

$$\widetilde{\Delta G} = \sum_j \nu_j \mu_j = \widetilde{\Delta G}^0 + RT \sum_j \nu_j \ln P_j \tag{5-82}$$

$$\widetilde{\Delta G} = \widetilde{\Delta G}^0 + RT \ln \prod_j P_j^{\nu_j} \tag{5-83}$$

As an example, let us apply Eq. 5-83 to the following gas phase reaction:

$$2HCl + \tfrac{1}{2}O_2 \rightarrow H_2O + Cl_2$$

$$\widetilde{\Delta G} = \widetilde{\Delta G}^0 + RT \ln\left(\frac{P_{H_2O} P_{Cl_2}}{P_{HCl}^2 P_{O_2}^{\frac{1}{2}}}\right) \tag{5-84}$$

Here $\widetilde{\Delta G}$ is the free energy change per stoichiometric unit of reaction at 298°K and constant P (or the work function change per stoichiometric unit of reaction at 298°K and constant V), for the reactant and product gases at specified partial pressures. $\widetilde{\Delta G}^0$ is the standard reaction potential with all reactants and products at 1 *atm partial pressure* and 298°K.

The relative magnitudes of the $\widetilde{\Delta G}^0$ term and the logarithm term in Eq. 5-84 determines the thermodynamic progress of the reaction. Starting with all reactants and no products, $\widetilde{\Delta G}$ will be negative and infinitely large no matter what value $\widetilde{\Delta G}^0$ might have, and the reaction will go spontaneously to the right. On the other hand, with all products and no reactants, the opposite situation will exist and the formation of at least some reactants is inevitable. At some intermediate mixture of reactants and products, the two terms will cancel, $\widetilde{\Delta G}$ will be zero, and the reaction will be at equilibrium. Therefore, if you know the reaction potential for any one set of conditions, for example, the standard state, and also know how the gas concentrations deviate from this standard state, then you can calculate the reaction potential under any other conditions.

At equilibrium, the partial pressure ratio which forms the argument of the log term is defined as the equilibrium constant:

$$K_P = \prod_j P_j^{\nu_j} = \frac{P_{H_2O}P_{Cl_2}}{P_{HCl}^2 P_{O_2}^{\frac{1}{2}}} \qquad (\widetilde{\Delta G} = 0) \tag{5-85}$$

Another example of the application of reaction potentials is the water–gas reaction

$$H_2 + CO_2 \rightarrow H_2O + CO \qquad (\text{gas phase, } T = 298°K)$$

$\widetilde{\Delta G}^0$ for this reaction, calculated from tables of $\Delta G_f{}^0$, is $+6.81$ kcal/mole. If all components are at one atmosphere partial pressure, then the reaction will go spontaneously in reverse.[3] From the principle of Le Chatelier, we know that the reaction can be driven forward by reducing the partial pressures of the products sufficiently or by increasing the partial pressures of the reactants. But how great must these changes be in order to favor formation of products? If

$$P_{H_2} = P_{CO_2} = P_{H_2O} = 1 \text{ atm} \qquad P_{CO} = 0.1 \text{ atm}$$

then

$$\widetilde{\Delta G} = 6.81 + 0.593 \ln \frac{1 \times 0.1}{1 \times 1} = +5.45 \frac{\text{kcal}}{\text{mole}}$$

[3] To the thermodynamicist, a "spontaneous" process is one which will go of its own accord if given enough time, and no suggestion of rapidity is involved. By this use of the term, the weathering away of the entire continent of North America is a spontaneous process. Only on this time scale is the water gas reaction at 298°K rapid as well as spontaneous.

Reactants are still favored over products. More drastic changes are called for. If

$$P_{H_2} = P_{CO_2} = 100 \text{ atm} \qquad P_{H_2O} = P_{CO} = 0.1 \text{ atm}$$

then

$$\widetilde{\Delta G} = 6.81 + 0.593 \ln 10^{-6} = -1.38 \frac{\text{kcal}}{\text{mole}}$$

At last the reaction conditions favor formation of products, though not by much. This is an example of how serious a barrier of 7 kcal/mole is to a reaction.

Equilibrium

At equilibrium $\widetilde{\Delta G} = 0$, $\prod_j P_j^{\nu_j} = K_P$, and Eq. 5-83 becomes

$$\widetilde{\Delta G}^0 = -RT \ln K_P = -2.303\ RT \log_{10} K_P \tag{5-86}$$

Solving for K_P,

$$K_P = e^{-\widetilde{\Delta G}^0/RT} = 10^{-\Delta G^0/2.303RT} \tag{5-87}$$

For the water–gas reaction at 298°K, the equilibrium constant is

$$K_P = e^{-6.81/298R} = e^{-11.5} \cong 10^{-5}$$

The negative exponent indicates that at equilibrium there is a preponderance of reactants over products. In general, as long as the actual $\prod_j P_j^{\nu_j}$ is less than K_P, the reaction is spontaneous in the direction of product formation.

Much of the material of this section is summarized in Figure 5-17. At constant P and T,

$$dG = \widetilde{\Delta G}\, d\xi \tag{5-88}$$

and the slope of the curve is $\widetilde{\Delta G}$. The reaction parameter is a function of the partial pressures, although not a simple one. If in the water–gas reaction there were initially 1 mole of each species, and if after ξ units of reaction there were $1 - \xi$ moles of each reactant and $1 + \xi$ moles of each product, then Eq. 5-83 would be

$$\widetilde{\Delta G} = \widetilde{\Delta G}^0 + RT \ln\left(\frac{1+\xi}{1-\xi}\right)^2 \tag{5-89}$$

For less simple reactions the expression in terms of reaction coordinate is clumsier yet. Still the principle remains. This free energy curve has the rather special property that a knowledge of the *slope* at any one point (such as the standard state) tell how *far away* from that point is the condition for zero slope (equilibrium). This is the meaning of Eq. 5-86.

Since $\widetilde{\Delta G}^0$ is a function of temperature but is independent of partial pressure, we have demonstrated incidentally the existence of an equilibrium constant which depends only upon temperature.

Chemical Potential and Equilibrium Constant in Terms of Mole Fraction and Molarity

The chemical potential of a gas is given by

$$\mu_j = \mu_j^{\,0} + RT \ln P_j \tag{5-69}$$

But in a gaseous mixture, the partial pressure is related to the total pressure through mole fraction.

$$P_j = X_j P \qquad P = \text{total pressure} \tag{5-90}$$

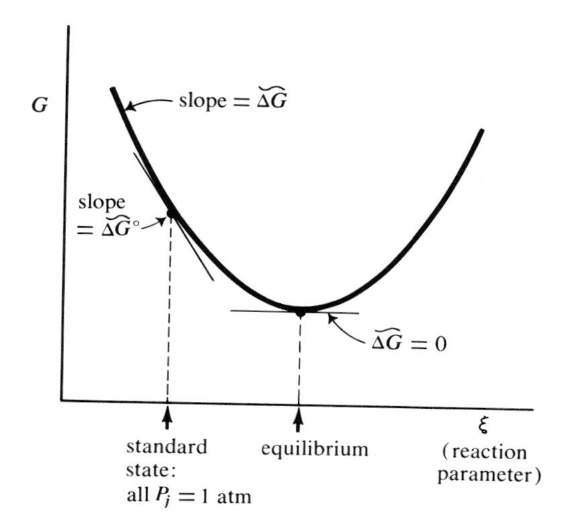

FIGURE 5-17. *The reaction potential for a chemical reaction, $\widetilde{\Delta G}$, is the slope in a plot of the total free energy of the system of reactants and products against the extent of reaction, or reaction parameter, ξ. The slope at such a point in the reaction that all gases are at a partial pressure of* 1 atm *is the standard reaction potential $\widetilde{\Delta G}^0$. The slope at equilibrium is zero.*

Hence

$$\mu_j = \mu_j{}^0 + RT \ln P + RT \ln X_j \tag{5-91}$$

If the reaction is being run at constant total pressure, then the first two terms on the right can be combined into a new constant:

$$\mu_j = \mu_j{}^\bullet + RT \ln X_j \tag{5-92}$$

where $\mu_j{}^\bullet = \mu_j{}^0 + RT \ln P$ is the chemical potential of the *pure substance* at a total pressure P.

The equilibrium constant can be expressed equally well in terms of pressure, concentration, or mole fraction:

$$K_P = \prod_j P_j^{\nu_j} \qquad P_j = \text{atmospheres} \tag{5-93}$$

$$K_c = \prod_j c_j^{\nu_j} \qquad c_j = \frac{\text{moles}}{\text{liter}} \tag{5-94}$$

$$K_N = \prod_j N_j^{\nu_j} \qquad N_j = \frac{\text{molecules}}{\text{cm}^3} \tag{5-95}$$

$$K_X = \prod_j X_j^{\nu_j} \qquad X_j = \text{mole fraction} \tag{5-96}$$

Since

$$P_j = \frac{n_j RT}{V} = c_j RT = N_j\left(\frac{1000RT}{N}\right)$$

K_P and K_c are related by

$$K_P = K_c(RT)^{\Delta\nu} \tag{5-97}$$

$$\Delta\nu = \sum_j \nu_j$$

where $\Delta\nu$ is the difference in the number of moles of products and reactants. In a similar way

$$K_P = K_N\left(\frac{1000RT}{N}\right)^{\Delta\nu} \tag{5-98}$$

$$P_j = X_j P$$

$$K_P = K_X P^{\Delta\nu} \tag{5-99}$$

In the previous section we saw that both $\widetilde{\Delta G^0}$ and K_P were functions only of temperature and not pressure. From Eqs. 5-97, and 5-98 we can see that K_c and K_N are also functions only of temperature. However, K_X (Eq. 5-99) is a function of both temperature and pressure.

5-13 HOMOGENEOUS IDEAL GAS REACTIONS—EXAMPLES

Dissociation of Water

Consider a system consisting of water vapor in a glass bulb at 298°K with a total pressure equal to its normal vapor pressure of 27 mm. At equilibrium, what percent of the water will be dissociated?

The most convenient way to solve this problem is to set up a table listing the mole numbers of each component at initial and equilibrium conditions along with equilibrium mole fraction and pressure data, as Table 5-1. Here α represents the fraction of water dissociated at equilibrium. Defining α in this way has the advantage that absolute mole numbers drop out of the mole fraction expressions.

TABLE 5-1 DISSOCIATION OF WATER

	$H_2O_{(g)}$	$\rightarrow$	$H_{2(g)}$	+	$\frac{1}{2}O_{2(g)}$	*Total moles*
Initial moles	n		0		0	n
Moles at equilibrium	$n(1-\alpha)$		$n\alpha$		$\frac{1}{2}n\alpha$	$n(1+\frac{1}{2}\alpha)$
Mole fraction	$\dfrac{1-\alpha}{1+\alpha/2}$		$\dfrac{\alpha}{1+\alpha/2}$		$\dfrac{\alpha/2}{1+\alpha/2}$	
Partial pressure	$\dfrac{1-\alpha}{1+\alpha/2}P$		$\dfrac{\alpha}{1+\alpha/2}P$		$\dfrac{\alpha/2}{1+\alpha/2}P$	

The equilibrium constant in terms of degree of dissociation is

$$K_P = \frac{P_{H_2}P_{O_2}^{\frac{1}{2}}}{P_{H_2O}} = \left[\left(\frac{\alpha}{1+\alpha/2}\right)\left(\frac{\alpha/2}{1+\alpha/2}\right)^{\frac{1}{2}} \Big/ \left(\frac{1-\alpha}{1+\alpha/2}\right)\right]P^{\frac{1}{2}} \tag{5-100}$$

Simplifying,

$$K_P = \frac{\alpha^{\frac{3}{2}}P^{\frac{1}{2}}}{\sqrt{2}(1-\alpha)(1+\alpha/2)^{\frac{1}{2}}} \tag{5-101}$$

This as it stands would be extremely cumbersome to solve for α. But it can be simplified for this particular reaction. K_P can be calculated from handbook values of $\widetilde{\Delta G}^0$:

$$\widetilde{\Delta G}^0_{298} = +56.690 \text{ kcal/mole}$$

$$K_P = 10^{-(\widetilde{\Delta G}^0/2.303RT)} = 10^{-[56,690/(2.303)(1.987)(298)]} = 2 \times 10^{-42}$$

Such a small value for K_P indicates that we can expect α to be quite small. Therefore reasonable simplifying assumptions about α are

$$\alpha \ll 1$$

thus

$$1 - \alpha \cong 1 \qquad 1 + \frac{\alpha}{2} \cong 1$$

The total pressure is 27/760 atm = 0.035 atm. Then

$$2 \times 10^{-42} = \alpha^{\frac{3}{2}} P^{\frac{1}{2}}$$

$$\alpha = \left(\frac{(1.414)(2 \times 10^{-42})}{(0.035)^{\frac{1}{2}}} \right)^{\frac{2}{3}} = 6.11 \times 10^{-28}$$

It is obvious that there is very little dissociation of water vapor at room temperature, and our simplifying assumptions about α were proper. The reason for this is the large free energy barrier for dissociation; $\widetilde{\Delta G}^0 \cong +56.7$ kcal/mole.

Dissociation of Dinitrogen Tetroxide

The dissociation of N_2O_4 can be followed by measuring the density of the vapor in the system. At room temperature and 1 atm total pressure, one N_2O_4 molecule in six is dissociated. This is all the information needed to calculate the standard free energy of dissociation of N_2O_4. Again, the most systematic method of procedure is to set up a table such as Table 5-2. This shows initial conditions before dissociation and conditions at an arbitrary time t when a certain fraction α of N_2O_4 has dissociated. If D represents the dimer and M the monomer,

$$K_P = \frac{P_M{}^2}{P_D} = \frac{4\alpha^2 P^2/(1 + \alpha)^2}{P(1 - \alpha)/(1 + \alpha)} = \frac{4\alpha^2 P}{1 - \alpha^2} \tag{5-102}$$

TABLE 5-2 DISSOCIATION OF DINITROGEN TETROXIDE

	$N_2O_{4(g)}$ →	$2NO_{2(g)}$	*Total moles*
Initial moles	n	0	n
Moles at time t	$n(1-\alpha)$	$2n\alpha$	$n(1+\alpha)$
Mole fraction	$\frac{1-\alpha}{1+\alpha}$	$\frac{2\alpha}{1+\alpha}$	
Partial pressure	$\frac{1-\alpha}{1+\alpha}P$	$\frac{2\alpha}{1+\alpha}P$	

The original statement of the problem contained the information that at room temperature and 1 atm total pressure, one molecule of N_2O_4 in six is dissociated, or that $\alpha = 0.167$. Substituting this value of α into Eq. 5-102,

$$K_P = \frac{4(0.167)^2}{1-(0.167)^2} = 0.1143$$

With K_P, $\widetilde{\Delta G}^0$ can be calculated

$$\widetilde{\Delta G}^0_{298} = -2.303RT\log_{10} K_P = -(2.303)(1.987)(298)\log_{10} 0.1143$$

$$\widetilde{\Delta G}^0_{298} = +1.288 \text{ kcal/mole of } N_2O_4$$

By measuring the percent dissociation as a function of temperature, one can calculate $\widetilde{\Delta G}^0_{298}$ as a function of temperature, and hence can obtain the enthalpy of reaction $\widetilde{\Delta H}^0_{298}$.

From Eq. 5-102, the degree of dissociation depends on pressure in the following way:

$$\alpha = \left(\frac{K_P}{K_P + 4P}\right)^{\frac{1}{2}} \tag{5-103}$$

As the pressure rises toward infinity, α approaches zero, and as the pressure approaches zero, α approaches 1. These changes are exactly what would be expected from Le Chatelier's principle.

This reaction is so simple that it is easy to make a quantitative plot of free energy versus reaction coordinate analogous to Figure 5-17. In the process, we shall learn something about what makes chemical reactions go. The dissociation of N_2O_4 requires energy to accomplish. Why does the reaction occur at all? Why do we not find all of the material in the form of the more stable dimer?

If there are n_D moles of dimer and n_M moles of monomer in a closed container, then the total free energy of the system is

$$G = n_D \mu_D + n_M \mu_M \tag{5-104}$$

For each component,

$$\mu_j = \mu_j^0 + RT \ln P + RT \ln X_j \tag{5-91}$$

and

$$G = n_D \mu_D^0 + n_M \mu_M^0 + (n_D + n_M)RT \ln P + RT(n_D \ln X_D + n_M \ln X_M) \tag{5-105}$$

The free energy per mole of gas mixture is

$$\bar{G} = (X_D \mu_D^0 + X_M \mu_M^0) + RT \ln P + RT(X_D \ln X_D + X_M \ln X_M) \tag{5-106}$$

This can be divided into two terms: the free energy of the pure gases taken separately at total pressure P, and the free energy of mixing

$$\bar{G}_{\text{pure}} = X_D \mu_D^0 + X_M \mu_M^0 + RT \ln P \tag{5-107}$$

$$\bar{G}_{\text{mix}} = RT(X_D \ln X_D + X_M \ln X_M) \tag{5-108}$$

$$\bar{G} = \bar{G}_{\text{pure}} + \bar{G}_{\text{mix}} \tag{5-109}$$

Equation 5-109 provides the key to the driving force behind the reaction. While it is true that dissociating the dimer is an energetically unfavorable process, nevertheless by dissociating some dimer into monomer a mixture is formed. This mixture has a higher entropy than pure N_2O_4, and hence a lower free energy. The equilibrium point is a balance between the unfavorable free energy of the dissociation process and the favorable free energy of mixing once some N_2O_4 has dissociated.

For this simple system, the reaction parameter ξ is just equal to the degree of dissociation α. Then the mole fractions and total number of moles (assuming 1 mole of N_2O_4 in the absence of dissociation), from Table 5-2, are

$$X_D = \frac{1 - \xi}{1 + \xi} \qquad X_M = \frac{2\xi}{1 + \xi} \qquad n = 1 + \xi \tag{5-110}$$

The free energy components for $1 + \xi$ moles are

$$G_{\text{pure}} = (1+\xi)\left[\left(\frac{1-\xi}{1+\xi}\right)\mu_D{}^0 + \left(\frac{2\xi}{1+\xi}\right)\mu_M{}^0\right] + (1+\xi)RT\ln P \qquad (5\text{-}111)$$

$$G_{\text{pure}} = \mu_D{}^0 + \widetilde{\Delta G}{}^0\xi + (1+\xi)RT\ln P$$

$$G_{\text{mix}} = (1+\xi)RT\left[\left(\frac{1-\xi}{1+\xi}\right)\ln\left(\frac{1-\xi}{1+\xi}\right) + \left(\frac{2\xi}{1+\xi}\right)\ln\left(\frac{2\xi}{1+\xi}\right)\right]$$

$$G_{\text{mix}} = RT[(1-\xi)\ln(1-\xi) + 2\xi\ln 2\xi - (1+\xi)\ln(1+\xi)] \qquad (5\text{-}112)$$

where

$$\widetilde{\Delta G}{}^0 = 2\mu_M{}^0 - \mu_D{}^0$$

For simplicity, assume a total pressure of 1 atm, so that the last term in Eq. 5-111 drops out.

From standard free energy tables,

$$\mu_D{}^0 = +23.491 \text{ kcal/mole}$$
$$\mu_M{}^0 = +12.390 \text{ kcal/mole}$$
$$\widetilde{\Delta G}{}^0 = +1.288 \text{ kcal}$$

And

$$G_{\text{pure}} = 23.491 + 1.288\,\xi \text{ kcal} \qquad (5\text{-}113)$$

Values of G_{pure} and G_{mix} are tabulated in Table 5-3 and plotted in Figure 5-18.

TABLE 5-3

ξ	0.0	0.1	0.15	0.2	0.4	0.6	0.8	1.0
G_{pure}	23.491	23.620	23.684	23.749	24.006	24.263	24.521	24.780
G_{mix}	0	−0.310	−0.392	−0.453	−0.566	−0.534	−0.373	0
G	23.491	23.310	23.292	23.296	23.440	23.729	24.148	24.780

The intrinsic molar free energy of unmixed components favors pure dimer; the free energy of mixing favors a 50-50 mixture. The true equilibrium point, at 16.7% dissociation, represents a balance between the two trends.

Free energy diagrams of the type of Figures 5-17 or 5-18 in general are not easy to calculate for even mildly complex reactions. Nevertheless, the principles which they illustrate should be kept in mind when equations such as Eqs. 5-83 or 5-86 are used.

Ammonia Equilibrium

Ammonia in equilibrium with its elements provides a more complex system than the N_2O_4/NO_2 system.

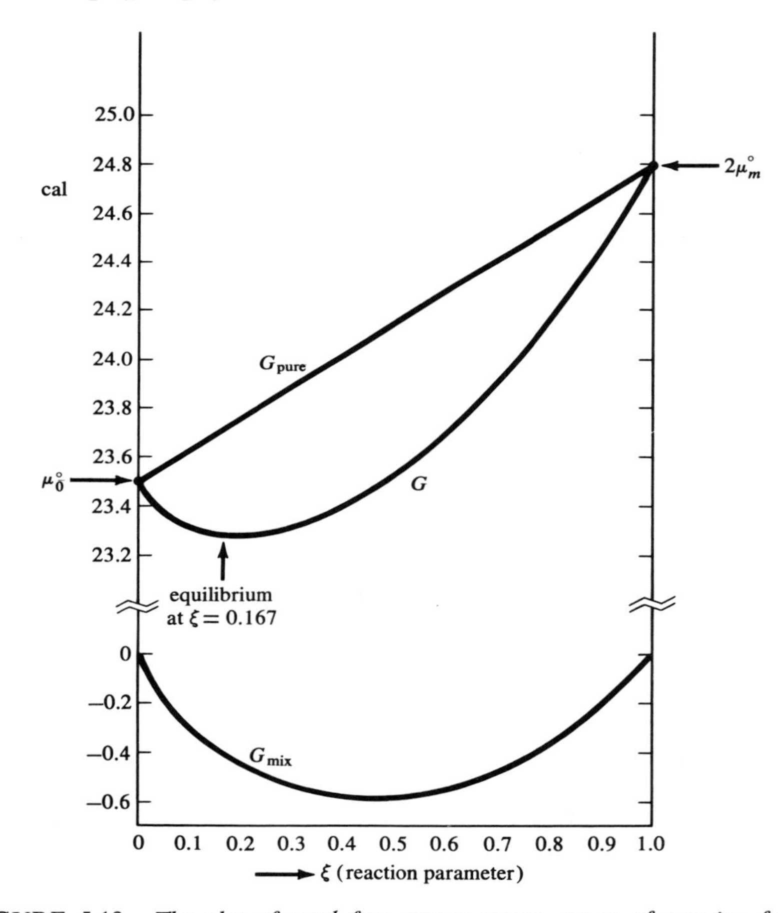

FIGURE 5-18. *The plot of total free energy versus extent of reaction for the dissociation of* N_2O_4. *Molar free energy considerations favor the undissociated state* (G_{pure}), *whereas the free energy of mixing favors a 50/50 mixture* (G_{mix}). *The true equilibrium point, at the minimum of* total *free energy, is a compromise between these two opposing tendencies.*

TABLE 5-4 SYNTHESIS OF AMMONIA

	$\frac{3}{2}H_2$	+ $\frac{1}{2}N_2$	$\rightleftarrows$ NH_3	*Total moles*
Initial moles	0	0	n	n
Time t moles	$\frac{3\alpha n}{2}$	$\frac{\alpha n}{2}$	$n(1-\alpha)$	$n(1+\alpha)$
Mole fraction	$\frac{3\alpha}{2(1+\alpha)}$	$\frac{\alpha}{2(1+\alpha)}$	$\frac{1-\alpha}{1+\alpha}$	
Partial pressure	$\frac{3\alpha P}{2(1+\alpha)}$	$\frac{\alpha P}{2(1+\alpha)}$	$\frac{(1-\alpha)P}{1+\alpha}$	

$$\tfrac{3}{2}H_2 + \tfrac{1}{2}N_2 \rightleftarrows NH_3 \tag{5-114}$$

The steps leading to partial pressures are given in Table 5-4. Note that, although the reaction is written as an association, α can still be defined as a degree of dissociation. This is a matter of convenience. The framework of Table 5-4 is also for the special case where all H_2 and N_2 have come from the dissociation of NH_3, or where the H_2 to N_2 mole ratio is 3 : 1. The equilibrium constant for the reaction as written is

$$K_P = \frac{P_{NH_3}}{P_{H_2}^{\frac{3}{2}} P_{N_2}^{\frac{1}{2}}} = \frac{1-\alpha}{(3\alpha/2)^{\frac{3}{2}}(\alpha/2)^{\frac{1}{2}}} \frac{1+\alpha}{P}$$

$$K_P = \frac{4(1-\alpha^2)}{3^{\frac{3}{2}}\alpha^2 P}$$

(a) At 10 atm and 400°C, it was observed that 3.85% of the gas mixture was NH_3, starting from pure NH_3. What are K_X, K_P, and α? Since $X_{NH_3} =$ 0.0385,

$$X_{N_2} + X_{H_2} = 0.9615$$

$$X_{N_2} = \frac{0.9615}{4} = 0.2404$$

$$X_{H_2} = 0.7212$$

$$K_X = \frac{0.0385}{(0.7212)^{\frac{3}{2}}(0.2404)^{\frac{1}{2}}} = 0.128$$

From Dalton's law at 10 atm total pressure,

$$P_{NH_3} = 10(0.0385) = 0.385 \text{ atm}$$
$$P_{H_2} = 10(0.7212) = 7.212 \text{ atm}$$
$$P_{N_2} = 10(0.2404) = 2.404 \text{ atm}$$
$$K_P = \frac{0.385}{(7.212)\ (2.404)^{\frac{1}{2}}} = 0.0128$$

K_P could have been obtained directly from K_X by using Eq. 5-99 with $\Delta\nu = -1$.

The extent of dissociation α can be found from any of the mole fraction expressions. For example,

$$X_{NH_3} = \frac{1-\alpha}{1+\alpha} = 0.0385$$
$$\alpha = 0.926$$

Or, at equilibrium, 92.6% of the NH_3 is dissociated into H_2 and N_2.

How will these numerical data be altered if we change the experimental conditions? What happens if the total pressure is 50 atm instead of 10 atm?

(b) At 50 atm and 400°C, 15.11% of the mixture is observed to be NH_3. Proceeding as before,

$$\begin{array}{rlrl} X_{NH_3} &= 0.1511 & P_{NH_3} &= 50 \times 0.1511 \text{ atm} = 7.55 \text{ atm} \\ X_{N_3} &= 0.2122 & P_{N_3} &= 10.61 \text{ atm} \\ X_{H_2} &= 0.6367 & P_{H_2} &= 31.84 \text{ atm} \\ \hline X_{total} &= 1.0000 & P_{total} &= 50.00 \text{ atm} \end{array}$$

$$K_X = \frac{0.1511}{(0.6367)^{\frac{3}{2}}(0.2122)^{\frac{1}{2}}} = 0.643$$

The equilibrium constant based upon mole fractions is quite different at 10 and 50 atm total pressure.

Calculating the degree of dissociation from the mole fraction of N_2 (remember that any one of the mole fractions equations can be used),

$$X_{N_2} = \frac{\alpha}{2(1+\alpha)} = 0.2122$$
$$\alpha = 0.736$$

Therefore, NH_3 is 73.6% dissociated at 50 atm and 400°C. The degree of dissociation has decreased as the pressure was increased, exactly as the principle of Le Chatelier predicts.

But in spite of the changes in K_X and α, K_P is the same at both pressures:

$$K_P = K_X P^{-1} = \frac{0.643}{50} = 0.0128$$

The position of equilibrium does shift with pressure, but it does so in such a way that K_P remains constant.

Phosgene Equilibrium—Effect of an Inert Gas

Table 5-5 is set up in terms not only of reactants and products, but of a certain amount of inert gas which plays no direct role in the reaction. What effect will this inert gas have upon equilibrium?

From Eq. 5-99, the general relationship between K_X and K_P is

$$K_X = K_P P^{-\Delta\nu} \tag{5-99}$$

In this case, $\Delta\nu = -1$:

$$K_X = K_P P = \frac{X_{COCl_2}}{X_{CO}X_{Cl_2}} \tag{5-115}$$

The effect of the inert gas on the equilibrium can be followed by observing the change in the mole ratio of $COCl_2$ and Cl_2:

$$\frac{n_{COCl_2}}{n_{Cl_2}} = \frac{X_{COCl_2}}{X_{Cl_2}} = X_{CO}K_P P = n_{CO}K_P\left(\frac{P}{n}\right) \tag{5-116}$$

TABLE 5-5 PHOSGENE EQUILIBRIUM

	CO	+	Cl_2	⇄	$COCl_2$	*Inert gas*	*Total moles*
Start	a		b		0	c	$a+b+c$
Time t	$a-x$		$b-x$		x	c	$a+b+c-x=n$
Mole fraction	$\frac{a-x}{a+b+c-x}$		$\frac{b-x}{a+b+c-x}$		$\frac{x}{a+b+c-x}$	$\frac{c}{a+b+c-x}$	

(1) What is the effect on equilibrium of increasing the pressure? From Eq. 5-116, if P rises while n_{CO} and n remain constant, more $COCl_2$ is produced at the expense of Cl_2. Le Chatelier's principle would also predict that a rise in pressure would form a shift in the direction which produced a lesser number of moles of gas.

(2) What is the effect of adding inert gas while keeping the pressure constant? If the total number n of moles rises while P is constant, then Eq. 5-116 predicts a smaller ratio of products to reactants. This agrees with the idea that dilution favors a shift which produces more moles of gas.

(3) What is the effect of adding inert gas at constant volume? Here P and n rise proportionally, the ratio P/n is unchanged, and the equilibrium point is unchanged also. The diluting effect of more inert gas is balanced by the repressive effect of greater pressure and no net change occurs.

5-14 *TEMPERATURE AND THE EQUILIBRIUM CONSTANT*

In Section 5-13 we saw that, although the degree of dissociation of ammonia and K_X varied with pressure at 400°C, K_P was constant. If K_P is measured at several temperatures, however, data such as those in the first three lines of Table 5-6 are found. Dissociation is favored by higher temperatures.

TABLE 5-6 AMMONIA EQUILIBRIUM CONSTANTS

T	350°C	400°C	450°C
	623°K	673°K	723°K
K_P	0.0266	0.0129	0.00659
$\ln K_P$	−3.627	−4.350	−5.023
$1/T$	0.00161	0.00149	0.00138

The dependence of K_P on temperature can be found by modifying the Gibbs–Helmholtz equation:

$$\left(\frac{\partial(\widetilde{\Delta G^0}/T)}{\partial T}\right)_P = -\frac{\widetilde{\Delta H^0}}{T^2} \tag{5-117}$$

$$\widetilde{\Delta G^0} = -RT \ln K_P \tag{5-86}$$

$$\frac{d \ln K_P}{dT} = \frac{\widetilde{\Delta H^0}}{RT^2} \tag{5-118}$$

Note the formal resemblance between this last equation and the Clausius–Clapeyron equation, Eq. 5-18. Why should this occur?

Again, rearranging the Gibbs–Helmholtz equation produces a form which is particularly useful in making plots which lead to enthalpies of reaction:

$$\frac{d \ln K_P}{d(1/T)} = -\frac{\widetilde{\Delta H}^0}{R} \tag{5-119}$$

The slope of the curve of Figure 5-19 indicates an enthalpy of formation of ammonia from nitrogen and hydrogen of some −12.3 kcal/mole at 400°C. This exothermic reaction is favored by a drop in temperature, again in agreement with Le Chatelier's principle.

Measurement of the equilibrium constant at any one temperature will lead to a knowledge of the free energy of a reaction; measurement at several temperatures will lead to the enthalpy as well.

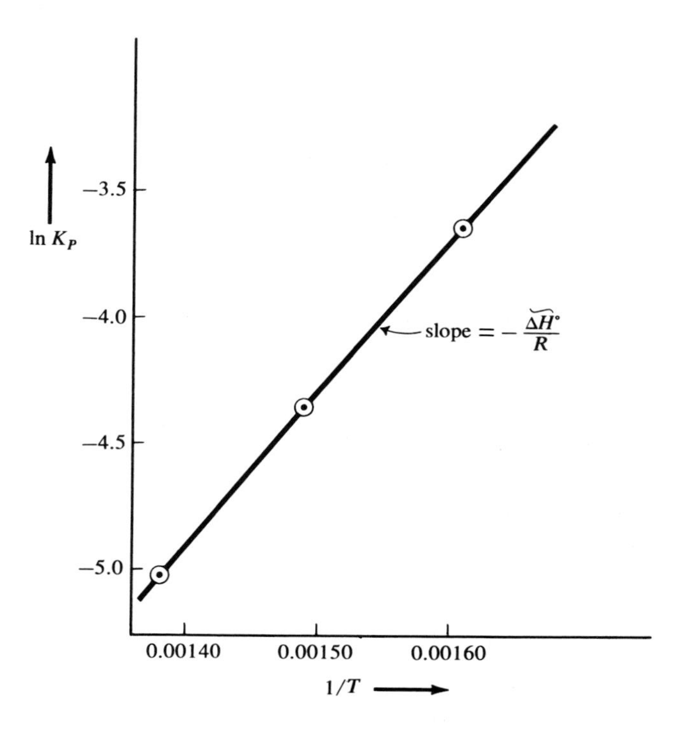

FIGURE 5-19. *The change in equilibrium constant K_P with temperature leads by way of a Gibbs–Helmholtz plot to the standard enthalpy of reaction $\widetilde{\Delta H}^0$.*

Practical Expression for $K_{P(T)}$

In Section 3-16 we found that the enthalpy of reaction at any temperature could be calculated from the enthalpy at some one known temperature and the heat capacity behavior of reactants and products between the two temperatures. If heat capacity is approximated by

$$\Delta C_P = \Delta a + \Delta bT + \Delta cT^2 \tag{3-92}$$

then enthalpy of reaction is given by

$$\widetilde{\Delta H}^0 = \widetilde{\Delta H}_0{}^0 + \Delta aT + \frac{\Delta b}{2}T^2 + \frac{\Delta c}{3}T^3 \tag{3-94a}$$

where $\widetilde{\Delta H}_0{}^0$ is the integration constant calculated from a knowledge of $\widetilde{\Delta H}^0$ at one temperature. Substituting the above expression for ΔH^0 into Eq. 5-119,

$$d \ln K_P = \left(\frac{\widetilde{\Delta H}_0{}^0}{RT^2} + \frac{\Delta a}{RT} + \frac{\Delta b}{2R} + \frac{\Delta cT}{3R}\right) dT \tag{5-120}$$

Integrating gives a practical expression for K_P:

$$\ln K_P = -\frac{I}{R} - \frac{\widetilde{\Delta H}_0{}^0}{RT} + \frac{\Delta a}{R}\ln T + \frac{\Delta bT}{2R} + \frac{\Delta cT^2}{6R} \tag{5-121}$$

where I is the same integration constant as in Eq. 4-97. I is evaluated from a knowledge of $\widetilde{\Delta G}^0$ or K_P at any one temperature, usually but not necessarily the same as that for the known $\widetilde{\Delta H}^0$.

For an example of the application of Eq. 5-121, let us return to the familiar water–gas reaction:

$$H_2 + CO_2 \rightleftarrows H_2O + CO$$

$$\widetilde{\Delta H}^0_{298} = +9.83 \text{ kcal/stoich. unit}$$

$$\widetilde{\Delta G}^0_{298} = +6.81 \text{ kcal/stoich. unit}$$

These and C_P data lead to

$$\ln K_P = 3.97 - \frac{4990}{T} + 0.259 \ln T - 1.56 \times 10^{-3}T + 2.53 \times 10^{-7}T^2 \tag{5-122}$$

$$\text{at } 298°\text{K}: \ln K_P = -11.7 \qquad \text{or } K_P \cong 10^{-5}$$

$$\text{at } 800°\text{K}: \ln K_P = -1.63 \qquad \text{or } K_P \cong 0.20$$

The extent of reaction can be calculated with the help of Table 5-7.

$$K_P = K_X = \frac{\alpha^2}{1-\alpha^2} \tag{5-123}$$

TABLE 5-7 WATER–GAS REACTION

	H_2 +	CO_2 ⇄	CO +	H_2O	*Total moles*
Start (moles)	n	n	0	0	$2n$
Equilibrium (moles)	$n(1-\alpha)$	$n(1-\alpha)$	$n\alpha$	$n\alpha$	$2n$
Mole fraction	$\frac{1-\alpha}{2}$	$\frac{1-\alpha}{2}$	$\frac{\alpha}{2}$	$\frac{\alpha}{2}$	

Solving for α,

$$\alpha = \frac{(K_P)^{\frac{1}{2}}}{1 + (K_P)^{\frac{1}{2}}} \tag{5-124}$$

$$\text{at } 298°\text{K}: K_P \cong 10^{-5} \qquad \alpha = 0.00317$$

$$\text{at } 800°\text{K}: K_P \cong 0.20 \qquad \alpha = 0.309$$

Hence, we see that the production of CO is favored by high temperatures, just as predicted for an endothermic reaction from Le Chatelier's principle.

5-15 NONIDEAL GASES AND FUGACITY

Everything in Sections 5-12 to 5-14 is wrong in the sense that it is written in terms of ideal gases and not real gases. Nevertheless, it is a starting point for the next task, dealing with homogeneous gas phase reactions of real gases.

Ideal Gases and Real Gases

The chemical potential of an ideal gas is given by

$$d\mu_j = \overline{V}_j\, dP = RT\, d \ln P_j \tag{5-66}$$

which, upon integration, gives

$$\mu_j = RT \ln P_j + B_{(T)} \tag{5-125}$$

where the integration constant B is a function only of temperature. The standard state is chosen such that $\mu_j = \mu_j^0$, when $P_j = 1$ atm. Then $B_{(T)}$ is $\mu_j^0(T)$, the standard chemical potential of the pure gas, and Eq. 5-125 becomes

$$\mu_j = \mu_j^0 + RT \ln P_j \tag{5-69}$$

For real gases, things are not so easy. In the first place, for real gases

$$\overline{V}_j \neq \frac{RT}{P_j}$$

and hence

$$d\mu_j \neq RT\, d \ln P_j$$

Since the ideal gas equations are not directly applicable to real gases, we are faced with a dilemma. We can either abandon the equations or abandon the variable P_j. Having put this much work into developing the ideal gas equations, we find it easier to keep them and to change the variable so that real gas behavior is predicted. We then define a new variable which has the dimensions and general properties of pressure, but which makes the *ideal* gas equations work for *real* gases. This new variable is called the fugacity f_j. It is a "corrected pressure" which applies to real gases just as the partial pressure applies to ideal gases. Since we know that as pressure goes to zero all gases approach ideal gas behavior, we define fugacity by

$$\mu_j \equiv RT \ln f_j + B_{(T)} \tag{5-126}$$

and

$$\lim_{P \to 0} \frac{f_j}{P_j} = 1 \tag{5-127}$$

Our problem is now one of relating fugacity, a variable which makes the equations work but which cannot be measured directly, to partial pressure, an easily measurable variable which unfortunately does not fit the equations. Once this relationship has been determined, it is a simple matter to use fugacities instead of pressures in all the equations.

Calculation of Fugacity from PVT Data

The simplest and most direct way to calculate the fugacity of a real gas is from *PVT* data, which, for a gas over a wide range of conditions, describe how it deviates from ideal behavior. Let us define α as the difference between the ideal gas molar volume under any conditions and the true measured molar volume:

$$\alpha \equiv \bar{V}_{\text{ideal}} - \bar{V}_{\text{real}} = \frac{RT}{P} - \bar{V} \tag{5-128}$$

Then

$$\bar{V} = \frac{RT}{P} - \alpha \tag{5-129}$$

Substituting Eq. 5-129 into the real gas version of Eq. 5-66 produces an expression for the change in the chemical potential which is valid for real gases:

$$d\mu = \bar{V}\,dP = RT\,d\ln f \tag{5-130}$$

$$\frac{RT}{P}\,dP - \alpha\,dP = RT\,d\ln f \tag{5-131}$$

$$d\ln f = d\ln P - \frac{\alpha}{RT}\,dP \tag{5-132}$$

Integrating from zero pressure,

$$\int_0^f d\ln f = \int_0^P d\ln P - \frac{1}{RT}\int_0^P \alpha\,dP \tag{5-133}$$

At $P = 0$, the gas is ideal and $f = P$ (by the original definition of fugacity). Thus Eq. 5-133 becomes

$$\ln f = \ln P - \frac{1}{RT}\int_0^P \alpha\,dP \tag{5-134}$$

If α is known in terms of previously tabulated PVT data, the integral can be evaluated and Eq. 5-134 can be used to go from pressure to fugacity. An alternative approach is to obtain an analytical expression for α from an equation of state such as the van der Waals equation.

As a matter of fact, α is very close to being constant for a given temperature within the range of pressures from zero to normal atmospheric pressure (Fig. 5-20). If we assume that α is constant, then Eq. 5-134 becomes

$$\ln f = \ln P - \frac{\alpha P}{RT} \tag{5-135}$$

or, in exponential form,

$$\frac{f}{P} = e^{-\alpha P/RT} \cong 1 - \frac{\alpha P}{RT} + \cdots \tag{5-136}$$

Rearranging Eq. 5-136 gives the relationship between fugacity and pressure that we have been seeking:

$$\frac{f}{P_{\text{true}}} = \frac{RT - \alpha P}{RT} = \frac{P\bar{V}}{RT} = \frac{P_{\text{true}}}{P_{\text{ideal}}} \qquad \bar{V} = \bar{V}_{\text{ideal}}$$

$$\frac{f}{P_{\text{true}}} = \frac{P_{\text{true}}}{P_{\text{ideal}}} \tag{5-137}$$

To a first approximation the true pressure is the geometric mean of the fugacity and the ideal pressure. The ratio of fugacity to real pressure is called the fugacity coefficient γ.

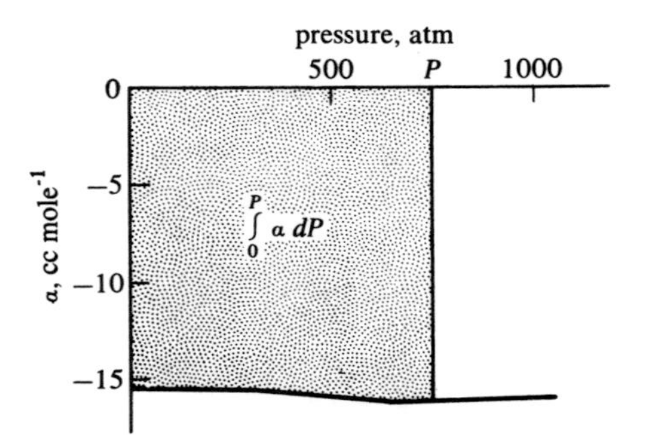

FIGURE 5-20. *Plot of the difference between ideal and real molar volumes of hydrogen gas,* $\alpha = \bar{V}_{\text{ideal}} - \bar{V}_{\text{real}}$, *as a function of pressure at* 298°K. *To a first approximation,* α *is independent of pressure.*

$$\frac{f}{P} \equiv \gamma \tag{5-138}$$

and is a convenient measure of the nonideality of a gas. To the extent that Eq. 5-137 is true, the fugacity coefficient is the same as a quantity often used to describe nonideality, the compressibility factor $Z = P\overline{V}/RT$.

$$\gamma = \frac{f}{P} = \frac{P}{P_{\text{ideal}}} = \frac{P\overline{V}}{RT} \tag{5-139}$$

A more accurate value for fugacity can be found by using the van der Waals equation to calculate α

$$P\overline{V} \cong RT + \left(b - \frac{a}{RT}\right)P$$

$$\alpha = \frac{RT}{P} - \overline{V} = \frac{a}{RT} - b \tag{5-140}$$

and using this α in Eq. 5-134.

Fugacity coefficients, like other useful thermodynamic quantities, are tabulated in reference books and handbooks. A typical plot of γ versus P at two different temperatures appears in Figure 5-21. At very low pressures, γ is less than unity, because the molecules are far enough apart that their bulk is not a major factor in determining their motion. However, the intermolecular forces of attraction are important, and intermolecular attraction causes γ to be less than unity. At high pressures the bulk of the molecules become more important. This phenomenon is related to the b term in the van der Waals equation, the excluded volume term. At high pressures, γ is greater than unity.

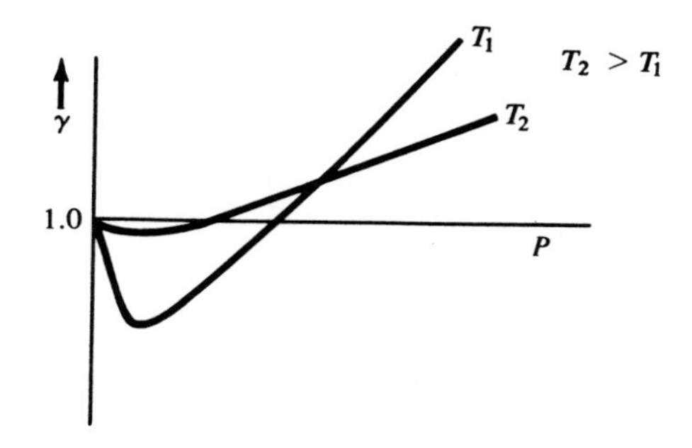

FIGURE 5-21. *Behavior of the fugacity coefficient γ at low temperature T_1 and at high temperature T_2.*

Principle of Corresponding States

A separate γ-versus-P plot could be made for every gas, but fortunately is unnecessary. If the plots are made in terms, not of P, V, and T, but of the so-called "reduced pressure" P_R, "reduced temperature" T_R, and "reduced volume" V_R, then the same plot will do for all gases. This is the principle of corresponding states, the principle that all gases seem to act in the same manner at the same fractional distances from zero pressure and temperature to their respective critical points. The reduced variables are defined as the ratio of the observed value to the critical point value.

$$P_R \equiv \frac{P}{P_c} \qquad T_R \equiv \frac{T}{T_c} \qquad V_R \equiv \frac{V}{V_c}$$

All gases exhibit much the same deviation from ideality at the same P_R, T_R, and V_R. Hence one plot of γ versus P_R (for a series of different T_R's) will serve for all gases (Figure 5-22). Since critical point data are readily available for most gases, we have only to measure the experimental pressure and temperature, calculate P_R and T_R, and obtain γ graphically. A representative sample of critical point data appears in Table 5-8.

TABLE 5-8 TYPICAL CRITICAL POINT DATA

Gas	P_c(atm)	T_c(°K)	$\bar{V}_c$(cm³/mole)
He	2.26	5.3	57.6
H_2	12.8	33.3	65.0
O_2	49.7	153.4	74.4
H_2O	217.7	647.2	45.0
Hg	200	1823.0	45.0

Applications of Fugacity to Equilibrium Calculations

For ideal gases, K_P is independent of pressure. But for real gases, we might expect K_P to show some deviation with pressure. Just how much K_P does change for the ammonia system is shown in Table 5-9. By any criteria, K_P is not very constant; but things can be much improved by changing to

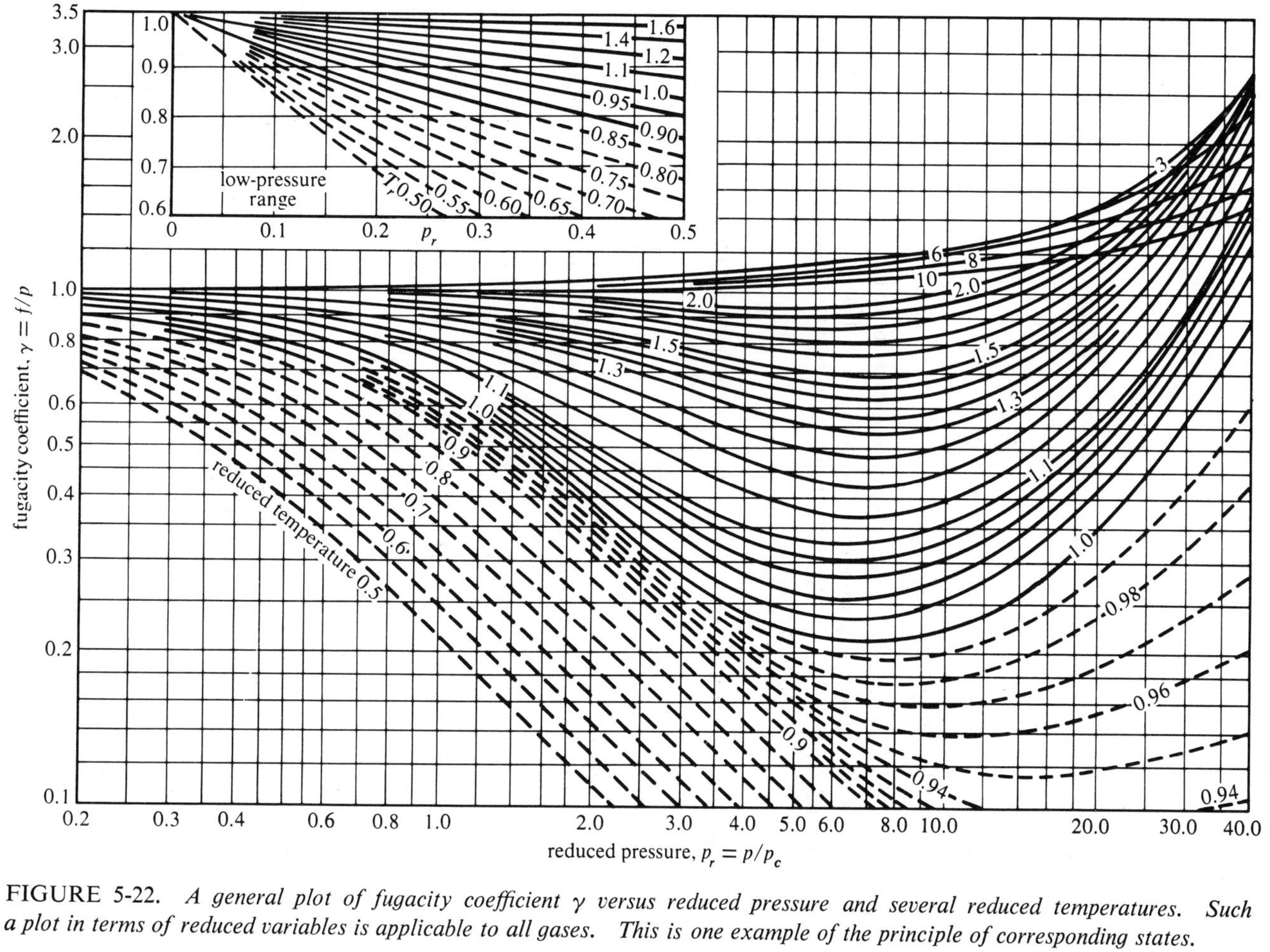
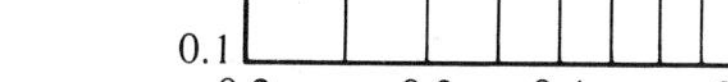

FIGURE 5-22. *A general plot of fugacity coefficient γ versus reduced pressure and several reduced temperatures. Such a plot in terms of reduced variables is applicable to all gases. This is one example of the principle of corresponding states.*

fugacities. Making use of

$$\frac{f}{P} = \gamma \tag{5-138}$$

we can write an equilibrium constant in terms of fugacity:

$$K_f = K_P K_\gamma \tag{5-141}$$

$$K_f = \frac{P_{NH_3}}{P_{H_2}^{\frac{3}{2}} P_{N_2}^{\frac{1}{2}}} \frac{\gamma_{NH_3}}{\gamma_{H_2}^{\frac{3}{2}} \gamma_{N_2}^{\frac{1}{2}}} \tag{5-142}$$

As Table 5-9 shows, this K_f equilibrium constant remains "constant" over a much wider range of pressures than does K_P. The failure at 1000 atm reflects the failures of the approximations used to arrive at γ, including the principle of corresponding states.

TABLE 5-9 K_P AND K_f AS A FUNCTION OF DEVIATION FROM IDEALITY
$\frac{3}{2}H_2 + \frac{1}{2}N_2 \rightleftarrows NH_3$, $T = 450°C$

$P_{(atm)}$	10	30	100	300	600	1000
K_P	0.00659	0.00676	0.00725	0.00884	0.01294	0.02328
K_γ	0.995	0.975	0.880	0.688	0.497	0.434
K_f	0.00655	0.00659	0.00636	0.00608	0.00642	0.01010

Fugacity of Condensed Phases

Fugacity was defined as a corrected pressure or an idealized pressure. Just as vapor pressure measures the escaping tendency of ideal gases from condensed phases, fugacity measures the escaping tendency of real gases from condensed phases.[4] At equilibrium, the chemical potential of a component in a liquid has to be equal to the chemical potential of that component in the gas phase. The fugacities of one component in two condensed phases must also be equal, for each must be in equilibrium with the same vapor phase. If a volatile liquid is only one component of a solution, then its escaping

[4] "Entropy" and "enthalpy" are made-up words created in that rash of late nineteenth-century etymology which also produced "anion," "cation," "electrode," and "scientist." But "fugacity," like "energy," was a perfectly good if obscure English word going back at least to 1656 (OED). Fugacity is the quality possessed by a fugitive.

tendency will be less than if it were present in pure form, or

$$f_j < f_j^{\bullet}$$

where $f_j^{\bullet}$ represents the fugacity of the pure liquid. We see in the coming sections that, for *ideal solutions*,

$$f_j^{\bullet} = X_j f_j^{\bullet} \tag{5-143}$$

Thus the fugacity of a component of a condensed phase must be defined equal to the fugacity of its vapor in equilibrium with it. We shall be able to use the framework developed for gas phase reactions to handle equilibria involving condensed phases as well.

Standard States and Activity

For *ideal gases*, we have the following general equation for the chemical potential:

$$\mu_j = RT \ln P_j + B_{(T)} \tag{5-125}$$

In the standard state ($P_j = 1$ atm), $B_{(T)} = \mu_j^0$. Hence the equation above becomes

$$\mu_j = \mu_j^0 + RT \ln P_j \tag{5-69}$$

For a *real gas*, these equations are fundamentally the same, except that up to this point we have not defined a standard state:

$$\mu_j = RT \ln f_j + B_{(T)} \tag{5-126}$$

$$\mu_j = \mu_j^0 + RT \ln \frac{f_j}{f_j^0} \tag{5-144}$$

By analogy with the ideal gas case, the most convenient standard state for a real gas would be the one for which the logarithmic term is zero. This standard state has two requirements. For mathematical simplicity, it should be the state for which $f_j^0 = 1$ atm. It should also be a state in which the gas is behaving ideally.

Figure 5-23 is a plot of fugacity versus pressure. The dotted line represents the situation if the gas were ideal. Real gas behavior is given by the solid line. As shown in the figure, the standard state is one in which $f = 1$ atm, but it is not the actual state of the real gas at unit fugacity. It is a hypothetical state in which the gas is behaving ideally at a fugacity of 1 atm. It is merely a reference state from which to measure fugacities, and is not a physically realizable state.

While such a standard state serves well for gases, it is not necessarily convenient for condensed phases. For example, at room temperature H_2O has a vapor pressure of 23.8 mm. There is no condition for which H_2O at 298°K has a vapor pressure of 1 atm. A convenient standard state for pure liquid H_2O is the pure component itself at a specified temperature and 1 atm total pressure. But this would mean that the fugacity of liquid water in its standard state would not be of unit magnitude; it would be 23.8/760 atm.

It is convenient to define the *activity* of a substance as the ratio of its fugacity to the fugacity in any convenient standard state.

$$a_j \equiv \frac{f_j}{f_j^0} \tag{5-145}$$

The concept of activity is then a generalization of the fugacity treatment to situations in which the most convenient standard state is not one of *unit* fugacity. Then

$$\mu_j = \mu_j^0 + RT \ln a_j \tag{5-146}$$

If the pure liquid is the standard state for solution components, then the activity of a pure liquid is 1.0, and that of a solution component is less than 1.0. Because of the way in which the standard state of a gas is chosen, its activity is always numerically equal to its fugacity, although a_j is a unitless quantity.

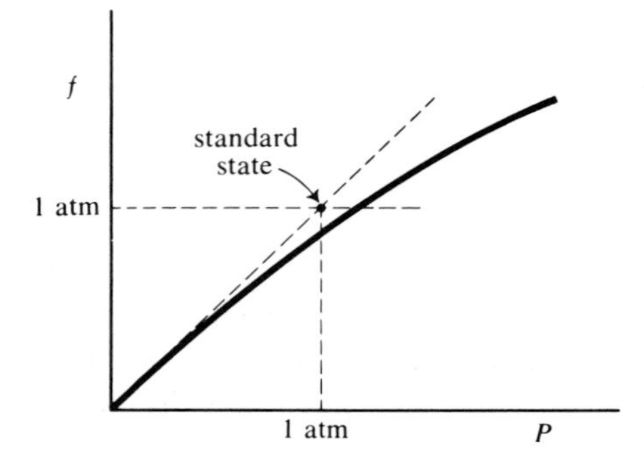

FIGURE 5-23. *The standard state for gases is a hypothetical state of unit fugacity in which the gas is still behaving ideally, or in which the partial pressure is also* 1 atm. *This state is only chosen for computational convenience, and it is immaterial that the standard state is not a physically attainable state.*

Selecting a different standard state will necessarily change the numerical value of the activity. However, this does not change the chemical potential. As one changes the standard state and hence the numerical value of the activity, $\mu_j{}^0$ is changed by a compensating amount, and μ_j itself remains unchanged. (See Section 6-3 for a different choice of standard states for solutes.)

5-16 MOLECULAR PICTURE OF THE HIGHER THERMODYNAMIC FUNCTIONS

The two fundamental expressions relating thermodynamic quantities to partition functions have already been obtained:

$$\bar{E} = kT^2\left(\frac{\partial \ln Q}{\partial T}\right)_V \tag{2-72}$$

$$\bar{S} = k \ln Q + \frac{\bar{E}}{T} \tag{2-88}$$

where Q is the molar sum over states or partition function, and where all energies are measured from the same level. It is often computationally convenient to calculate partition functions using energies measured from the ground state of each molecular species present, but using the condition of completely dissociated molecules with infinitely separated atoms as a zero for external energy measurements. Then the equations become

$$\bar{E} = kT^2\left(\frac{\partial \ln Q'}{\partial T}\right)_V + \bar{E}_0 \tag{5-147}$$

$$\bar{S} = k \ln Q' + \frac{\bar{E} - \bar{E}_0}{T} \tag{5-148}$$

where $\bar{E}$ is based on dissociated molecules, and $\bar{E}_0$ is the molar energy of the ground state of the molecules on this scale. The prime on Q' is a reminder of the altered zero level for energy measurements.

From these two relationships flow others for all of the thermodynamic functions:

$$\bar{A} = \bar{E} - T\bar{S} = -kT \ln Q \tag{5-149}$$

$$P = -\left(\frac{\partial A}{\partial V}\right)_T = kT\left(\frac{\partial \ln Q}{\partial V}\right)_T \tag{5-150}$$

$$\bar{H} = \bar{E} + P\bar{V}$$

$$\mu = \bar{G} = \bar{A} + P\bar{V} = P\bar{V} - kT \ln Q \tag{5-151}$$

For an ideal gas, the molecular partition function is related to the molar one by

$$Q = \frac{q^N}{N!} \cong \left(\frac{qe}{N}\right)^N \tag{2-74}$$

and the thermodynamic relationships become

$$\bar{E} = RT^2\left(\frac{\partial \ln q}{\partial T}\right)_V \tag{5-152}$$

$$\bar{S} = R \ln\left(\frac{qe}{N}\right) + \frac{\bar{E}}{T} \tag{5-153}$$

$$\bar{A} = -RT \ln\left(\frac{qe}{N}\right) \tag{5-154}$$

$$\mu = -RT \ln\left(\frac{q}{N}\right) \tag{5-155}$$

since $RT \ln e = RT = P\bar{V}$ for an ideal gas.

We can now identify the other Lagrangian multiplier α of Section 2-10. Recall from that derivation that $\beta = 1/kT$ and $e^{-\alpha} = n/q$. We now can see that

$$\alpha = \ln\frac{q}{n} = -\frac{\mu}{RT} \tag{5-156}$$

If $\mu' = \mu/N$ is thought of as a *molecular* chemical potential, then two symmetrical relationships result:

$$-\alpha = \frac{\mu'}{kT}$$

$$+\beta = \frac{1}{kT}$$

Ideal Monatomic Gas

For an ideal monatomic gas

$$q = q_t q_0 = \frac{(2\pi mkT)^{\frac{3}{2}}V}{h^3}\, e^{-\varepsilon_0/kT} \tag{5-157}$$

$$\bar{E} = \tfrac{3}{2}RT + \bar{E}_0 \tag{5-158}$$

$$\bar{S} = \tfrac{5}{2}R + R\ln\left(\frac{(2\pi mkT)^{\frac{3}{2}}V}{Nh^3}\right) \tag{5-159}$$

The molar entropy of argon gas at 273°K and 1 atm pressure, calculated from the mass of an Ar molecule and Eq. 5-159, is 36.67 cal/deg mole. The third law value from heat data is 36.98 cal/deg mole, in very good agreement.

$$\bar{A} = -RT \ln\left(\frac{(2\pi mkT)^{\frac{3}{2}} V}{Nh^3}\right) - RT + \bar{E}_0 \tag{5-160}$$

Note that the scale constant $\bar{E}_0$, absent in the entropy expression, has reappeared.

$$P = -\left(\frac{\partial A}{\partial V}\right)_T = +RT\left(\frac{1}{V}\right) = \frac{RT}{V} \tag{5-161}$$

This is only the verification of a relationship assumed in the derivations.

$$\mu = -RT \ln\left(\frac{q}{N}\right) = -RT \ln\left(\frac{q_t}{N}\right) + \bar{E}_0 \tag{5-162}$$

Again, $\bar{E}_0$ has appeared, since

$$-RT \ln(e^{-\varepsilon_0/kT}) = +RT \frac{\varepsilon_0}{kT} = \bar{E}_0$$

Diatomic Molecules—Rotation

To the monatomic gas results must be added internal modes of freedom for diatomic molecules. For rotation,

$$q_r = \frac{8\pi^2 IkT}{\sigma h^2} \tag{5-163}$$

$$\bar{E}_r = RT \tag{5-164}$$

$$\bar{S}_r = R \ln q_{\text{rot}} + R = R \ln\left(\frac{8\pi^2 IkTe}{\sigma h^2}\right) \tag{5-165}$$

The molar entropy of F_2 gas at 298°K and 1 atm is 48.48 e.u. from thermal measurements. The Sackur–Tetrode equation (Eq. 5-159) gives a translational component of 36.88 e.u., and Eq. 5-165 gives a rotational entropy of 11.50 e.u. Thus only 0.10 e.u. are left for vibrational and electronic disorder.

$$\bar{A}_r = -RT \ln\left(\frac{8\pi^2 IkT}{\sigma h^2}\right) \tag{5-166}$$

$$P_r = -\left(\frac{\partial A}{\partial V}\right)_T = 0 \tag{5-167}$$

Since the rotational contribution to the work function is independent of volume, there is no pressure effect from molecular rotation.

$$\mu_r = \bar{A}_r \quad \text{since} \quad P_r = 0 \tag{5-168}$$

The entire difference between A and G is bound up in translational motion for an ideal gas.

Diatomic Molecules—Vibration

For natural vibration frequency ν_0, and for $x = h\nu_0/kT$,

$$q_v = \frac{1}{1 - e^{-x}} \tag{5-169}$$

$$\bar{E}_v = RT\left(\frac{x}{e^x - 1}\right) = \bar{H}_v \tag{5-170}$$

$$\bar{A}_v = RT \ln (1 - e^{-x}) = \mu_V = \bar{G}_v \tag{5-171}$$

$$\bar{S}_v = \frac{\bar{E}_v - \bar{A}_v}{T} = R\left(\frac{x}{e^x - 1}\right) - R \ln (1 - e^{-x})$$

Diatomic Molecules—Electronic

If excited electronic states can be ignored, and if the degeneracy of the ground state is g_0, then

$$q = g_0$$
$$\bar{E}_e = 0$$
$$\bar{S}_e = R \ln g_0$$
$$\bar{A}_e = \bar{H}_e = \mu_e = 0$$

Equilibrium Constants

The expressions for μ permit us to calculate equilibrium constants from statistical molecular data alone.

$$\mu = -RT \ln \frac{q}{N} \tag{5-155}$$

It is convenient to separate translational and zero point contributions from the rest:

$$q = q_t \underbrace{q_r q_v q_e} q_0$$

$$q = q_t \quad q_i \quad q_0$$

$$(q_i = \text{internal energy terms})$$

And if q' is the partition function calculated from the lowest energy level of the molecule as zero,

$$q = q' q_0 = q' e^{-\varepsilon_0/kT}$$

$$(q' = q_t q_i = q_t q_r q_v q_e)$$

The translational partition function can be written

$$q_t = \frac{(2\pi mkT)^{\frac{3}{2}}}{h^3} \frac{RT}{P} \tag{5-172}$$

and in the standard state for an ideal gas, where $P = 1$ atm,

$$q^0 = [q']_{\text{std}} = [q_t q_i]_{\text{std}} = \frac{(2\pi mkT)^{\frac{3}{2}} RT}{h^3} q_i \tag{5-173}$$

Note that the energy units of k and R must differ because of their origin. Constant k is in *erg/deg mole*, whereas R is in cm^3 *atm/deg mole*. Then (using β for $1/kT$ for compactness)

$$q' = \frac{q^0}{P}$$

$$\mu = -RT \ln \frac{q}{N} = -RT \ln \left[\left(\frac{q'}{N} \right) e^{-\beta\varepsilon_0} \right]$$

$$\mu = -RT \ln \left[\left(\frac{q^0}{N} \right) e^{-\beta\varepsilon_0} \right] + RT \ln P \tag{5-174}$$

$$\mu = \mu^0 + RT \ln P$$

This result we have seen earlier. What we have just done is derive a statistical expression for the chemical potential in the *standard state*:

$$\mu^0 = -RT \ln \left[\left(\frac{q^0}{N} \right) e^{-\beta\varepsilon_0} \right] \tag{5-175}$$

The standard reaction potential can now be calculated using Eq. 5-175.

$$\widetilde{\Delta G}^0 = \sum_j \nu_j \mu_j$$

$$\widetilde{\Delta G}^0 = -RT \ln \prod_j \left[\left(\frac{q_j^{\ 0}}{N} \right)^{\nu_j} e^{-\beta \nu_j \varepsilon_{0j}} \right] \tag{5-176}$$

If $\sum_j \nu_i N\varepsilon_0 = \Delta \bar{E}_0$, then

$$\widetilde{\Delta G}^0 = -RT \ln \prod_j \left[\left(\frac{q_j^{\ 0}}{N} \right)^{\nu_j} \right] + \Delta \bar{E}_0 \tag{5-177}$$

This ΔE_0 is the molar heat of reaction at absolute zero, provided all molecular species have their energies measured from a common reference level. Consider as an example the reaction

$$2CO_{(g)} \rightarrow 2C_{(gr)} + O_{2(g)}$$

The energy situation at absolute zero before, during and after reaction is shown in Figure 5-24. For CO, $\varepsilon_0 = -9.15$ eV, and for O_2, -5.08 eV. For the entire reaction, involving the rupturing of two C—O bonds and the forming of one O—O, $\Delta\varepsilon_0 = +13.22$ eV, and $\Delta \bar{E}_0 = N\Delta\varepsilon_0 = 304$ kcal/mole.

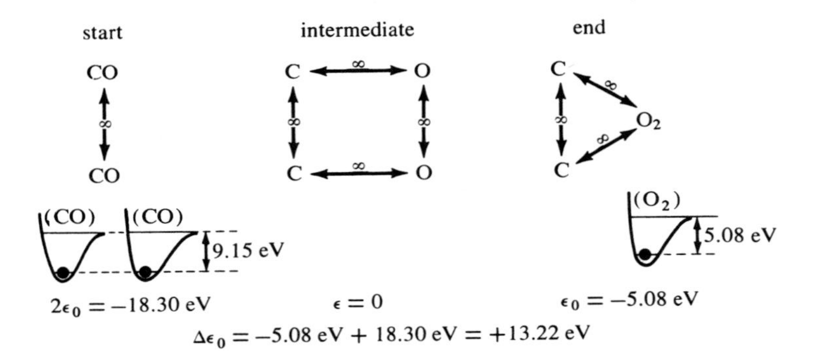

FIGURE 5-24. *If the zero reference level for measurement of energy in diatomic molecules is the state in which all atoms are infinitely far removed from one another* (*b*), *then the zero point energy of the reaction shown is the difference between the energy which must be expended to dissociate two* CO *molecules in their lowest energy state* (*a*) *and the energy which is regained when the* O_2 *molecule associates in its lowest energy state* (*c*).

The equilibrium constant for an ideal homogeneous gas reaction can now be written in molecular terms.

$$\widetilde{\Delta G}^0 = -RT \ln K_P = -RT \ln\left[\prod_j \left(\frac{q_j^0}{N}\right)^{\nu_j} e^{-\beta\Delta\varepsilon_0}\right]$$

$$K_P = \prod_j \left(\frac{q_j^0}{N}\right)^{\nu_j} e^{-\Delta \bar{E}_0/RT} \tag{5-178}$$

Note that, as expected, a large positive $\Delta\bar{E}_0$ such as that for the CO reaction will mean a small K_P and few products. Also, a large q_j^0 for products, or a great number of states in the sum over states of product molecules, will favor products. This is the entropy contribution.

We can now calculate the equilibrium constant for any reaction for whose participants we know how to write partition functions. This can be a tedious affair, however, and Eq. 5-178 is most often used in simple reactions or isotope reactions where many terms cancel.

A particularly important application of the statistical expression for equilibrium constants is in the calculation of reaction rate constants in chemical kinetics (see Gardiner, for example). Concentrations are commonly expressed in molecules per cm^3, and the equilibrium constant in these units is

$$K_N = \prod_j \left(\frac{q_j^0}{1000RT}\right)^{\nu_j} e^{-\Delta\bar{E}_0/RT}$$

But the quantity in parentheses is the partition function *per cm*^{3}:

$$\frac{q_j'}{V_{(\text{cm}^3)}} = \frac{q_j^0}{PV_{(cm^3)}} = \frac{q_j^0}{1000RT} = q_j^* \tag{5-179}$$

and the equilibrium constant expressed in terms of q_j^* is

$$K_N = \prod_j (q_j^*)^{\nu_j} e^{-\Delta\bar{E}_0/RT} \tag{5-180}$$

EXAMPLE 1. Dissociation of Na_2 dimer at 1000°K. A particularly simple reaction is

$$Na_{2(g)} \rightleftarrows 2Na_{(g)}$$

The three molecular parameters needed to calculate K_P are shown in Figure 5-25. Spectroscopic measurements show ε_0 to be -0.73 eV. The fundamental vibration frequency is $\tilde{\nu} = 159.23\ \text{cm}^{-1}$, and the equilibrium internuclear separation is 3.08 Å.

The full expression for K_P is

$$K_P = \underbrace{\left(\frac{(2\pi m_{\text{Na}} kT)^{\frac{3}{2}} RT}{h^3}\right)^2 \left(\frac{h^3}{(2\pi m_{\text{Na}_2} kT)^{\frac{3}{2}} RT}\right)}_{K_t}$$

$$\times \underbrace{\left(\frac{\sigma h^2}{8\pi^2 I_{\text{Na}_2} kT}\right)}_{K_r} \underbrace{\left(1 - e^{-x}\right)}_{K_e} \underbrace{e^{-\Delta \bar{E}_0/RT}}_{K_0} \qquad (5\text{-}181)$$

Let us look at each of the four major components of K_P to see their relative importance.

(a) *Translation*

$$K_t = \left(\frac{2\pi kT}{h^2}\right)^{\frac{3}{2}} RT \left(\frac{m_{\text{Na}}^2}{m_{\text{Na}_2}}\right)^{\frac{3}{2}}$$

$$K_t = (2.78 \times 10^{60})(82.05 \times 10^3)(83.1 \times 10^{-36})$$
$$(g^{-\frac{3}{2}}\ \text{cm}^{-3})(\text{cm}^3\ \text{atm})(\text{g}^{\frac{3}{2}})$$

$$K_t = 1.898 \times 10^{31}\ \text{atm}$$

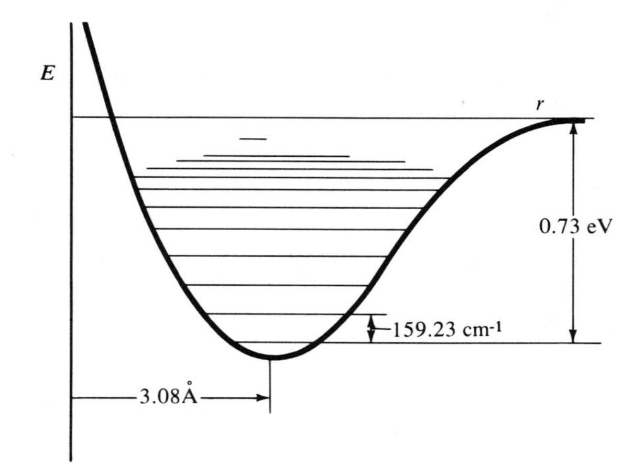

FIGURE 5-25. *The equilibrium constant for the dissociation of* Na_2 *dimers may be calculated statistically from a knowledge only of the zero point dissociation energy* (0.73 eV), *the equilibrium internuclear separation* (3.08 Å), *and the fundamental spacing of the vibrational energy levels* (159.23 cm^{-1}). *All of these are obtainable from spectroscopic measurements.*

The translational motion greatly favors dissociation because there are twice as many molecules moving in the monatomic form, and many more states in the sum over states.

(b) *Rotation*

$$I_{Na_2} = \mu r^2 = \left(\frac{m_1 m_2}{m_1 + m_2}\right) r^2 = \tfrac{1}{2} m_{Na} r^2$$

$$I_{Na_2} = 1.81 \times 10^{-38} \text{ g cm}^2$$

$$\frac{8\pi^2 k}{2h^2} = 1.24 \times 10^{38} \text{ erg}^{-1} \text{ deg}^{-1} \text{ sec}^{-2}$$

$$K_r = \frac{1}{1.24 \times 10^{38} \times 10^3 \times 1.81 \times 10^{-38}}$$

$$K_r = 4.46 \times 10^{-4} \text{(unitless)}$$

Because the diatomic molecule has rotational states and monatomic Na does not, the rotation contribution favors *association*.

(c) *Vibration*

$$x = \frac{h\nu_0}{kT} = \frac{hc\tilde{\nu}_0}{kT} = 0.229$$

$$K_V = 1 - e^{-x} = 0.205$$

This is a preference for *association*, but since the vibrational mode is only partially excited, the preference is a weak one. The electronic modes can be ignored completely.

(d) *Zero point*

$$\Delta \bar{E}_0 = 23.053 \text{ kcal/eV} \times 0.73 \text{ eV} = 16.83 \text{ kcal}$$

$$\frac{\Delta \bar{E}_0}{RT} = \frac{16.83}{1.98 \times 10^3} = 8.470$$

$$e^{-\Delta \bar{E}_0/RT} = 2.10 \times 10^{-4}$$

The moderately large dissociation energy of 0.73 eV means a moderate preference for *association*.

(e) *Total effect*

The over-all K_P is the product of the individual terms.

$$K_P = 3.64 \times 10^{23}$$

It is apparent that all other effects are dominated by the translational factor, and sodium exists in the gas phase at 1000°K mainly as the monomer.

Dissociation of I_2

As an example of the accuracy of the K_P calculations, Table 5-10 compares

TABLE 5-10 DISSOCIATION CONSTANTS FOR $I_2 \rightleftarrows 2I$

T	$K_C{}^*{}_{(obs)}$	$K_C{}^*{}_{(cal)}$
800°C	0.0113 moles/l	0.0114 moles/l
1000°C	0.165 moles/l	0.165 moles/l
1200°C	1.22 moles/l	1.23 moles/l

* $K_C = K_P/RT$ for this reaction.

calculated and measured values of K_C for the reaction

$$I_{2(g)} \rightleftarrows 2I_{(g)}$$

at three temperatures. The accuracy here is better than 1%.

Example of Isotope Reaction

In the isotope exchange reaction

$$O_2^{16} + O_2^{18} \rightleftarrows 2O^{16}O^{18}$$

most terms drop out. At first glance, it would seem that $K_P = 1$. Instead, K_P is closer to 4. Why should this be? The entire K_P expression is

$$K_P = \underbrace{\left(\frac{m_C{}^2}{m_A m_B}\right)^{\frac{3}{2}}}_{K_t} \underbrace{\frac{(\mu_C/\sigma_C)^2}{(\mu_A/\sigma_A)(\mu_B/\sigma_B)}}_{K_r} \underbrace{1}_{K_V} \underbrace{(0.971)}_{K_0}$$

$$K_P = (1.0045)(3.96)(1.00)(0.971)$$
$$K_P = 3.86$$

Here $A = O_2^{16}$, $B = O_2^{18}$, and $C = O^{16}O^{18}$. The fact that species C is intermediate to A and B in mass means that K_t is nearly 1. In a similar way, K_V and K_0 are near 1, and even the ratio of reduced masses is close to 1. Only the symmetry numbers are really important; $\sigma_A = \sigma_B = 2$, whereas $\sigma_C = 1$. The two ends of the mixed molecules are different. Alternatively, species C is more randomized than an equimolar mixture of A and B, and the driving factor favoring product is entropy.

REFERENCES AND FURTHER READING

Phase changes are covered well in Bent, and chemical equilibria in Wall and Klotz. For the statistical mechanics of the higher thermodynamic functions, see the references of Chapter 2.

EXERCISES

5-1. Draw a plot of free energy or chemical potential versus temperature to explain the phenomenon of melting.

(a) What is the slope of the free energy curve with temperature of a substance in a given state?

(b) How does this slope change with temperature?

(c) What criterion is used in deciding where a phase transition will take place?

5-2. (a) Sketch a plot of free energy G versus temperature for a solid, a liquid, and a gas on the same graph. Why do all three curves slope downhill with increasing temperature? Which curve slopes downhill most steeply, and why?

(b) Indicate the melting point of the solid on this plot. What thermodynamic criterion did you use to locate the melting point?

(c) On a second plot, sketch the behavior of G and H as the temperature approaches absolute zero. Does the H curve rise or fall with increasing temperature? Prove it.

(d) On a third plot, sketch the G-versus-T curves for a gas at two different pressures, with $P_1 < P_2$. At a given T, does G rise or fall as pressure increases? Prove that your answer is correct.

(e) What is physically unrealistic about the following two graphs? (Consider the system to be at constant volume if you like.)

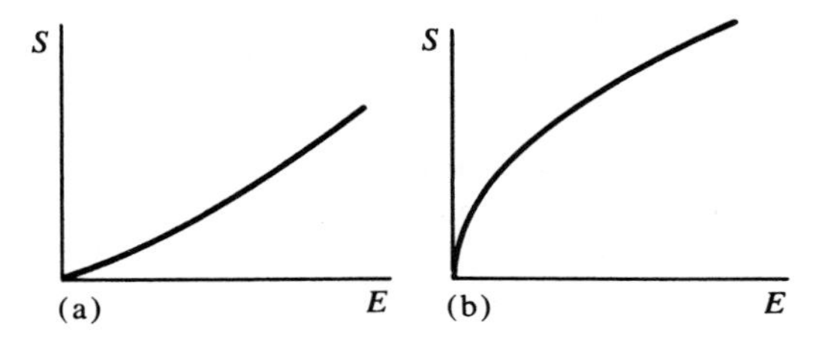

5-3. Draw rough plots of G, S, V, and C_P versus T at constant P, and versus P at constant T, for typical first and second order phase transitions.

5-4. In an article on machine guns in the 11th edition of the *Encyclopedia Britannica* there appears this statement: "The great difficulty which has to be met in all single-barrel machine guns is the heating of the barrel. The 7.5 pints of water in the water jacket of the Maxim gun are raised to the boiling point by 600 rounds of rapid fire—i.e., in about 1.5 minutes—and if firing be continued, about 1.5 pints of water are evaporated for every 1000 rounds."

From these data, calculate the entropy of vaporization of water.

5-5. The following data are found for Cl_2:

Temperature (°C)	Vapor pressure (atm)	Liquid density (g/cm³)	Vapor density (g/cm³)
0	3.65	1.468	0.0128
10	4.96	1.438	0.0175
20	6.57	1.408	0.0226

Calculate the enthalpy of vaporization of Cl_2 at 10°C. If Cl_2 were assumed to behave as an ideal gas, how would it change your result?

5-6. It is possible to supercool small drops of water to −40°C. Such drops are unstable, of course, and eventually nucleate to form ice crystals. Assume that such a drop is thermally isolated so that, when spontaneous ice formation occurs, it does so adiabatically and isobarically. C_P is 1 cal/deg g for liquid water and 0.5 cal/deg g for ice, and the heat of fusion of water is 80 cal/g at 0°C. Calculate the final temperature of the drop after the spontaneous process has occurred, and ΔH and ΔS for the process.

5-7. At 25°C the molar entropies of liquid and gaseous water at 1 atm are 16.716 and 45.106 cal/deg mole, respectively, and the heat of vaporization is 10.520 kcal/mole.

(a) Will the following change occur spontaneously at 25°C?
$H_2O_{(l)} = H_2O_{(g,\ 1\ \text{atm})}$

(b) Will the following change occur spontaneously at 25°C?
$H_2O_{(l)} = H_2O_{(g,\ 0.01\ \text{atm})}$

(c) How large must the partial pressure of water vapor be at 25°C before condensation to the pure liquid can occur?

(d) Estimate the normal boiling point of water from the data given above. State what assumptions or approximations you make.

5-8. Upon warming, rhombic sulfur changes reversibly to monoclinic sulfur at 95.5°C. At 120°C monoclinic sulfur melts.

(a) Show graphically approximately how the molar free energies of the two crystalline forms of sulfur stand with respect to each other in the neighborhood of the rhombic–monoclinic transition temperature.

(b) To the plot of part (a), add a curve for the molar free energy of liquid sulfur.

(c) Which has the higher melting point, rhombic or monoclinic sulfur?

(d) Which crystal form has the greater entropy? Show how you reached your conclusion.

(e) Which of the two solids do you suppose is the more soluble in a good solvent such as diethyl ether at 25°C? Justify your answer.

(f) Does the melting point of rhombic sulfur lie above or below the rhombic–monoclinic transition temperature?

5-9. From the following data, sketch the low temperature phase diagram of N_2. There are three crystal forms, α, β, and γ, which coexist at a triple point at 4650 atm and 44.5°K.

At the triple point, the volume increases upon phase transition are

α to γ, 0.165 cm^3/mole
β to γ, 0.208 cm^3/mole
β to α, 0.043 cm^3/mole

At 1 atm and 36°K, the β to α transition occurs with a volume increase of 0.22 cm^3/mole.

5-10. The following data are found for CCl_4:

Pressure	Melting point	Volume change on fusion
1 atm	−22.6°C	0.0258 cm^3/g
1000 atm	+15.3°C	0.0199 cm^3/g
2000 atm	+48.9°C	0.0163 cm^3/g

Estimate the latent heat of fusion of CCl_4 at 1000 atm pressure.

5-11. The following data are available for the two solid forms of carbon at 25°C:

Form	Heat of combustion (kcal/mole)	Molar entropy (cal/deg mole)	Density (g/cm^3)
diamond	94.484	0.5829	3.513
graphite	94.030	1.3609	2.260

(a) What is the free energy of the transition from graphite to diamond at 1 atm and 25°C? In which direction is the process spontaneous?
(b) Estimate the pressure at which the two forms would be in equilibrium at 25°C, and at 1000°C. Assume (incorrectly) that the densities are independent of pressure.

5-12. From the following experimental data, make a rough sketch of the phase diagram of acetic acid.
(1) Solid acetic acid can exist in two phases, α and β. The low pressure form α melts at 16.1°C under its own vapor pressure of 9.1 mm.
(2) The high pressure form β is more dense than α, and both solids are more dense than the liquid.
(3) The normal boiling point of the liquid is 118°C.
(4) Phases α and β and liquid are in equilibrium at 55°C and 2000 atm.

5-13. Which of the following quantities are equal to the chemical potential (more than one may be):

$$\left(\frac{\partial A}{\partial n_j}\right)_{T,P,n_i} \qquad \left(\frac{\partial H}{\partial n_j}\right)_{T,P,n_i}$$

$$\left(\frac{\partial E}{\partial n_j}\right)_{S,V,n_i} \qquad \left(\frac{\partial A}{\partial n_j}\right)_{T,V,n_i}$$

$$\left(\frac{\partial G}{\partial n_j}\right)_{T,V,n_i} \qquad \left(\frac{\partial H}{\partial n_j}\right)_{S,P,n_i}$$

n_i represents all other components except n_j.

5-14. Consider the oxidation of oxalic acid:

$$2(COOH)_{2(s)} + O_{2(g)} \rightleftarrows 4CO_{2(g)} + 2H_2O_{(g)}$$

(a) Calculate $\Delta H^0_{298°}$, $\Delta G^0_{298°}$, and K_P at 298°K from tabulated thermal data.
(b) *Assuming* that ΔH^0 is not a function of temperature in the range considered, calculate ΔG^0 at 398°K.
(c) Write an expression for K_P in terms of the fraction, α, of the original oxygen which has *not* reacted and the total pressure P_T, assuming that no CO_2 or H_2O was present at the beginning.
(d) Calculate the fraction of oxygen remaining at equilibrium, α, at 398°K and 1 atm total pressure. You will have to make an approximation, but an extremely good one under these conditions. State clearly what approximation was made.
(e) What effect on the equilibrium constant K_P will an increase in pressure have? Will such an increase favor the reaction to the right or to the left?
(f) Which reaction will be favored by the addition of an inert gas at constant pressure?
(g) In view of the results of parts (a)–(d), how do you account for the presence of *any* unreacted oxalic acid anywhere in the world as long as there is oxygen in the atmosphere?

5-15. Consider the following reaction:

$$N_{2(g)} + 3H_{2(g)} \rightarrow 2NH_{3(g)}$$

Using information available in tables in this chapter and the Appendices:
(a) Derive a general expression for ΔH as a function of temperature.
(b) Derive a similar expression for ΔG as a function of temperature.
(c) Calculate ΔG and the equilibrium constant K_P at 500°K.
(d) At some temperature, not necessarily that of part (c), the equilibrium constant K_P is 0.2 for the reaction as written above. Calculate the degree of dissociation of pure ammonia at 1 atm pressure at this temperature.

5-16. At 600°C, the enthalpy of the following reaction:

$$C_{gr} + 2H_{2(g)} \rightarrow CH_{4(g)}$$

is 21.045 kcal/mole. The third law entropies, in cal/deg mole, are 4.8 for C, 38.9 for H_2, and 56.6 for CH_4.

(a) What is the equilibrium constant K_P for this reaction at 600°C?
(b) What additional experimental information would be needed to calculate K_P at *any* temperature? Make an appropriate approximation for short temperature ranges and estimate K_P at 750°C.
(c) If the object of the process is to obtain the greatest yield of methane gas, is it better to use high or low temperatures? High or low pressures?

5-17. (a) At 900°K, the enthalpy of the following reaction:

$$C_2H_{6(g)} \rightarrow C_2H_{4(g)} + H_{2(g)}$$

is 24.42 kcal/mole and the free energy change is 5.35 kcal/mole. If pure ethane is passed over a dehydrogenation catalyst at this temperature and 1 atm pressure, what will be the mole percent of hydrogen present at equilibrium? Estimate the percent hydrogen at 1000°K.
(b) If a gas mixture of initial molar concentration 10% C_2H_4, 10% C_2H_6, and 80% N_2 is passed over the catalyst at 900°K and 1 atm, what will be the composition of the effluent gas at equilibrium? What will be the composition if the same initial mixture is used at 100 atm?

5-18. In the reaction

$$CO_{2(g)} = CO_{(g)} + \tfrac{1}{2}O_{2(g)}$$

at 1 atm total pressure, the degree of dissociation α was found to vary with temperature.

T(°K):	1000	1400	2000
α:	2×10^{-7}	1.27×10^{-4}	1.55×10^{-2}

What is ΔS^o for the reaction at 1400°K?

5-19. At high temperature and pressure, a quite good equation of state for gases is $P(V - b) = RT$. Calculate the fugacity of N_2 at 1000 atm and 1000°C by this equation, if $b = 39.1$ cm³/mole. What is the fugacity coefficient?

5-20. The Deacon process, once used for making chlorine, uses the reaction

$$2HCl_{(g)} + \tfrac{1}{2}O_{2(g)} = H_2O_{(g)} + Cl_{2(g)}$$

(a) Using heat capacity and heat of formation data at 25°C, derive an expression for the enthalpy of the reaction as a function of T.
(b) Use free energy data and derive an expression for $\log_{10} K_P$ as a function of temperature.

5-21. Phosphorus pentachloride dissociates according to the reaction

$$PCl_{5(g)} = PCl_{3(g)} + Cl_{2(g)}$$

(a) Show that, if one begins with pure PCl_5, and if a fraction α of this has dissociated at equilibrium, then the equilibrium constant will be given by

$$K_P = \frac{\alpha^2 P}{1 - \alpha^2}$$

where P is the total pressure.

(b) At 250°C the equilibrium constant is 1.78 atm. Calculate α for total pressures of 0.01, 0.10, and 1.00 atm.

(c) Calculate $\widetilde{\Delta G}$ for an *equimolar mixture* at a pressure of 1 atm and a temperature of 250°C.

(d) Calculate $\widetilde{\Delta G}$ for a 10% dissociation of pure PCl_5 at 1 atm and 250°C.

5-22. 4.4 grams of $CO_{2(g)}$ are introduced into a 1 liter flask containing excess solid carbon at 1000°C, so that the equilibrium $CO_{2(g)} + C_{gr} = 2CO_{(g)}$ is attained. The gas density at equilibrium corresponds to an average molecular weight of 36.0.

(a) Calculate the equilibrium pressure and K_P.

(b) If an additional amount of helium gas is introduced until the total pressure is doubled, will the equilibrium amount of CO be increased, decreased, or unaffected?

(c) If the volume of the flask were doubled instead, with helium introduced to maintain the same total pressure, would the equilibrium amount of CO increase, decrease, or remain unchanged?

(d) If the K_P for the equilibrium doubles with a 10-deg increase in temperature, what is the standard enthalpy for the reaction?

5-23. On the planet Nirgend IV, in another universe, scientists have discovered a curious correlation between size of a containing vessel and separation of nuclei in diatomic molecules. Specifically, the interatomic distance has been found to increase as the cube root of the container volume, or roughly as the enlargement of linear dimensions. Oddly enough, the potential curve must also be affected in a parallel manner, for the fundamental vibration frequency of the diatomic molecule is found to remain unchanged. This phenomenon is found to persist even at such low pressures that intermolecular interactions can be completely neglected. (The origin of this phenomenon is not clear, but need not concern us. Some have suggested that a molecule in some way "knows" how far apart the walls are and instinctively shrinks at the prospect of collision. Others have suggested that a universe might be imagined in which this effect did not exist, but they have been quite properly reproved by their colleagues for idle speculation.) Allowing for this peculiar behavior, derive the equivalent of the ideal gas law for Nirgend IV, as applied to diatomic molecules.

5-24. Using the statistical mechanical expressions for equilibrium constants, calculate the standard free energy change in kilocalories per mole for the reaction

$$Cl_{2(g)} + I_{2(g)} = 2ICl_{(g)}$$

The value of ΔE_0 can be approximated by the standard heat of reaction at room temperature for the purpose of this problem. Compare your result with that obtained by thermal measurements.

5-25. A certain system may exist in either of two and only two molecular energy states, of energies 0 and ε, respectively. Write expressions for E, C_V, S, and A using this information. You may take the entropy at absolute zero to be any value you wish, but explain your choice.

5-26. At 25°C the absolute entropies of $O_{2(g)}$ and $O_{3(g)}$ are 49.00 and 57.10 cal/deg mole, respectively. What is the equilibrium partial pressure of O_3 at 25°C if we start initially with pure O_2 at 1 atm pressure? The reaction is the conversion of oxygen to ozone, and the heat of formation of ozone is 34.80 kcal/mole at this temperature.

5-27. For the dissociation of iodine

$$I_{2(g)} \rightarrow 2I_{(g)}$$

$\Delta H^0 = 35.480$ kcal. The moment of inertia of the I_2 molecule is 743×10^{-40} g/cm^2, the fundamental vibration line occurs at 214 cm^{-1}, and the electronic partition function $q_e = 1$. For monatomic iodine, $q_e = 4$. Calculate the degree of dissociation of I_2 at 1000°C when the total pressure in the system is 1 atm.

5-28. (a) At 25°C, calculate the translational contribution to the free energy of the gases H_2, I_2, and HI. Assume I to be only I^{127}, and assume 1 atm partial pressures.

(b) Under the same conditions, calculate the rotational contributions to the free energies of the three molecules.

(c) Calculate the vibrational free energies of the three gases.

(d) Combine the results of parts (a)–(c) to obtain the total free energy of each of the three gases. Use this to calculate the free energy of formation of HI under standard conditions. Compare this with the thermal value in Appendix 4. How serious is the discrepancy?

Chapter 6

SOLUTIONS

6-1 IDEAL SOLUTIONS

Chemical Potential and Fugacity

THE SIMPLEST starting point for a discussion of solutions is the so-called "ideal solution." For ideal gases, it was assumed that there were *no* intermolecular forces. For ideal solutions, it will be assumed that all intermolecular forces are the *same*, whether between like or unlike pairs of molecules. But whether the solution is ideal or not, the following relations hold between chemical potential, fugacity and concentration.

For the chemical potential of gas j having fugacity f_j^g,

$$\mu_j^g = \mu_j^{g,0} + RT \ln f_j^g \tag{6-1}$$

If the gas behaves ideally, then the fugacity is just the partial pressure. Using Dalton's law of partial pressures, the above equation becomes

$$\mu_j^g = \mu_j^{g,0} + RT \ln f_j^{g,\bullet} + RT \ln X_j^g = \mu_j^{g,\bullet} + RT \ln X_j^g \tag{6-2}$$

where $\mu_j^{g,\bullet}$ is the chemical potential of the *pure* component j at the given temperature and total pressure and X_j^g is the mole fraction of j in the gas mixture.

Since any liquid j, whether pure or as a component of a solution, must be in equilibrium with gas j above the condensed phase, the chemical potentials and fugacities of j in the two phases must be the same:

$$\mu_j{}^l = \mu_j{}^g \qquad f_j{}^l = f_j{}^g$$

Raoult's Law

How does the fugacity of a condensed-phase component depend upon its concentration in the condensed phase, assuming ideal solution behavior? If intermolecular forces between like and unlike molecules are identical, then the escaping tendency of any one molecule is unaffected by the proportion of its neighbors which are like it or unlike it. The escaping tendency of a component is then proportional to the number of molecules of the substance which are present to escape, or

$$f_j{}^l = X_j{}^l f_j{}^\bullet \tag{6-3}$$

where $f_j{}^\bullet$ is the fugacity of the pure liquid (and also of the vapor in equilibrium with it). With the assumption of ideal gas behavior of the vapor, one can say that the vapor pressure of the jth component above the solution is the mole fraction of the j *in solution* times the vapor pressure of pure j

$$P_j{}^l = X_j{}^l P_j{}^\bullet \tag{6-4}$$

Either of the above equations is known as Raoult's law.

As an example, consider a mixture of benzene and toluene, which form a nearly ideal solution. At 25°C, the vapor pressure of benzene is 92 mm Hg, and that of pure toluene is 28 mm. For a 1 to 3 benzene-toluene solution, the partial pressure of benzene vapor, by Raoult's law, would be $\frac{1}{4} \times 92$ mm = 23 mm, and the partial pressure of toluene vapor above the mixture would be $\frac{3}{4} \times 28$ mm = 21 mm, giving a total equilibrium vapor pressure of 44 mm Hg. The two partial pressures and the total pressure are plotted in Figure 6-1 against the mole fraction of benzene or toluene, and all three are seen to be linear for an ideal solution.

Thermodynamics of Ideal Solutions

(a) *Chemical potential.* The chemical potential of a component, j, in the liquid phase is the same as that of its own vapor in equilibrium with it:

$$\mu_j{}^g = \mu_j{}^l = \mu_j^{l,\,0} + RT \ln\left(\frac{f_j{}^l}{f_j{}^0}\right) \tag{6-5}$$

where the standard state at the moment is that of unit fugacity of the gas phase ($f_j^0 = 1$ atm). We shall see later in connection with non-ideal solutions that other standard states are often more convenient. Now using Raoult's law:

$$\mu_j^l = \mu_j^{l,0} + RT \ln\left(\frac{f_j^{l,\bullet}}{f_j^0}\right) + RT \ln X_j^l$$

or combining the first two terms into the chemical potential of the pure liquid component, $\mu_l^{j\bullet}$:

$$\mu_j^l = \mu_j^{l,\bullet} + RT \ln X_j^l \tag{6-6}$$

The form of this expression is identical to that for gases.

(b) *Partial molar volume and the volume of mixing.* What happens to volumes and heat of mixing in an ideal solution? Remembering that the partial molar volume is the rate of change of chemical potential with pressure, at constant temperature,

$$\overline{V}_j = \left(\frac{\partial \mu_j}{\partial P}\right)_T = \left(\frac{\partial \mu_j^0}{\partial P}\right)_T + RT\left(\frac{\partial \ln f_j^\bullet}{\partial P}\right)_T + RT\left(\frac{\partial \ln X_j}{\partial P}\right)_T$$

The first term on the right is zero, since the standard chemical potential is defined at one pressure, and the last term is zero, since mole fraction is an

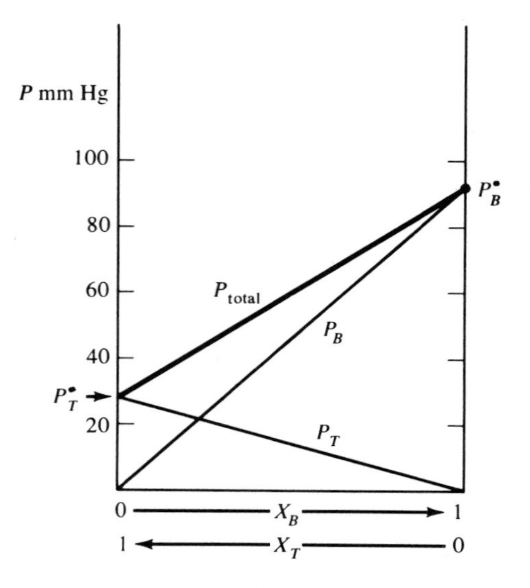

FIGURE 6-1. *Vapor pressure at 25°C in a benzene–toluene solution, which is effectively ideal. Raoult's law is obeyed, and the vapor pressure contribution of each component is proportional to its mole fraction concentration.*

independent variable with respect to pressure. The partial molar volume

$$\overline{V}_j = RT\left(\frac{\partial \ln f_j^\bullet}{\partial P}\right)_T \tag{6-7}$$

is the same as the molar volume of a pure liquid. Hence for ideal solutions

$$\overline{V}_j = \overline{V}_j^\bullet$$

$$V = \sum_j \overline{V}_j = \sum_j \overline{V}_j^\bullet \tag{6-8}$$

Thus there is no volume change on mixing if the components form an ideal solution.

(c) *Enthalpy.* By similar arguments involving the partial of μ with respect to T at constant pressure, the partial molar enthalpy of each component in the ideal solution can be shown to be the same as the molar enthalpy of the pure component. No heat is involved in preparing an ideal solution.

(d) *Free energy.* There *is* a free energy change, however. Before mixing, all the components were pure, and the total free energy was

$$G_1 = \sum_j n_j \mu_j^\bullet$$

After mixing, the chemical potential is given by Eq. 6-6. Then

$$G_2 = \sum_j n_j \mu_j = \sum_j n_j \mu_j^\bullet + RT \sum_j n_j \ln X_j$$

The free energy of mixing is therefore

$$\Delta G = G_2 - G_1 = RT \sum_j n_j \ln X_j \tag{6-9}$$

Since the mole fractions are all less than zero, the free energy change is always negative, that is, mixing is always spontaneous.

(e) *Entropy.* Since there is no heat of mixing, the entropy change is

$$\Delta S_{\text{mix}} = -\frac{\Delta G_{\text{mix}}}{T} = -R \sum_j n_j \ln X_j$$

or, per mole,

$$\frac{\Delta S_{\text{mix}}}{\text{mole}} = -R \sum_j X_j \ln X_j \tag{6-10}$$

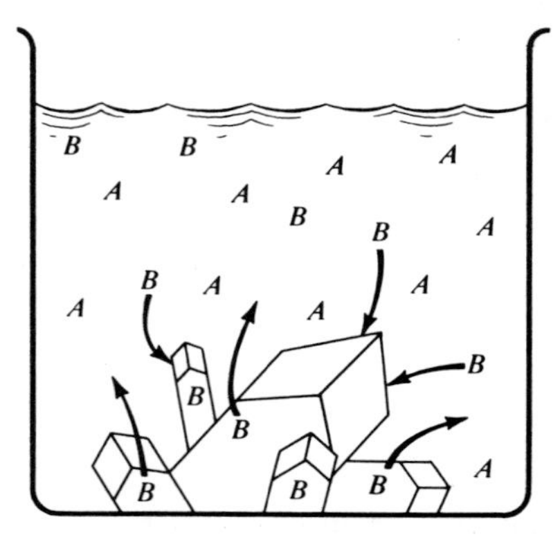

FIGURE 6-2. *A saturated solution of component B in A, in equilibrium with crystalline B. The chemical potential of B must be the same in both phases if equilibrium is to exist.*

Note that this is identical to the expression derived for the entropy of mixing ideal gases.

Solubility Behavior

Consider a system in which B is the solute, and A is the solvent, and the solution is saturated in B. There exists an equilibrium between crystalline B and molecules of B in solution in the solvent A (Figure 6-2). The condition for equilibrium, as always, is that the chemical potential of B in the crystal must equal the potential of B in solution:

$$\mu_B{}^S = \mu_B{}^l = \mu_B^{l,0} + RT \ln f_B{}^l \tag{6-11}$$

For an ideal solution, Raoult's law gives

$$f_B{}^l = X_B f_B{}^\bullet$$

and

$$\mu_B{}^S = \mu_B^{l,0} + RT \ln f_B^{l,\bullet} + RT \ln X_B{}^l$$
$$\mu_B{}^S = \mu_B^{l,\bullet} + RT \ln X_B{}^l \tag{6-12}$$

But what is this reference state of pure B? It is certainly not pure crystalline B. For if it were, then $\mu_B{}^S$ would be equal to this $\mu_B^{l,\bullet}$ at saturation and X_B would have to be 1. This would mean that equilibrium between solid B and solution could exist only when the solution was so concentrated that only B was present. The solid would be infinitely soluble. (This is true for liquids which form ideal solutions. One cannot have an equilibrium situation in which pure liquid B is present alongside a solution of liquid B in liquid A. Liquids forming ideal solutions are miscible in all proportions.)

A reasonable assumption to make about the pure reference state of Eq. 6-12 is that it is the pure, supercooled liquid at the specified temperature. Consider the vapor-pressure or fugacity versus temperature plot of Figure 6-3. If a liquid and vapor at equilibrium at point a are cooled carefully up to and past the triple point b, a supercooled liquid at point c can result, even though the thermodynamically stable system is that of solid and vapor at d. If the system is disturbed in any way, crystallization is usually spontaneous and immediate with liberation of the heat of fusion at that temperature. Now, remembering that at equilibrium the fugacities of pure solid B and B in solution must be equal,

$$f_B{}^S = f_B{}^l = X_B{}^l f_B^{l,\bullet} \quad \text{or} \quad X_B{}^l = \frac{f_B{}^S}{f_B^{l,\bullet}} \tag{6-13}$$

The mole fraction solubility of B is hence given by the ratio of fugacities of solid B and pure supercooled B at the given temperature. Note that no mention is made of any dependence upon solvent A. For ideal solutions, a given solid is equally soluble at the same temperature in *all* solvents. Common sense experience tells you that this is seldom true; sugar is not equally soluble in water, benzene, alcohol, and mercury. This simply points out the rarity of ideal solution behavior.

The assumption that the reference state of Eq. 6-12 is the supercooled liquid is reasonable, but is it right? What experimental evidence is there that this is the case? Entropy arguments would lead one to favor this assumption, for molecules in solution and in a liquid are disordered in much the

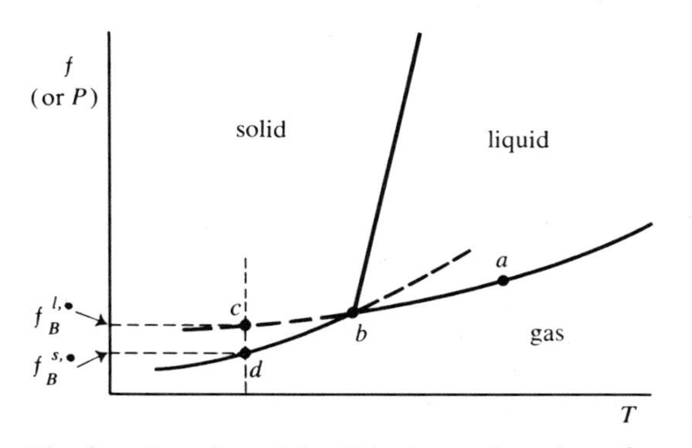

FIGURE 6-3. *The fugacity of a solid will be lower than that of a pure supercooled liquid, below its melting point. The mole fraction solubility of this substance in any solvent with which it forms an ideal solution will be equal to the ratio of these fugacities:* $X_B = f_B^{s,\bullet}/f_B^{l,\bullet}$.

same manner, while in a crystalline solid they are highly ordered. One can think of a liquid as the limiting case of a solution of the substance in a hypothetical solvent in which the solvent concentration has become vanishingly small. Again, the measured heats of solution of most solids are ordinarily very close to their heats of fusion at the same temperature, suggesting that it takes roughly the same amount of energy to break up a crystalline solid whether the end product is a pure liquid or molecules dispersed in a solution. Furthermore, the heats of solution of supercooled liquids are always very close to zero, suggesting that the difference between liquid and solution is small.

To discover the behavior of solubility with temperature, one need only rearrange the equilibrium condition equation slightly:

$$\mu_B{}^S = \mu_B{}^l = \mu_B^{l,\bullet} + RT \ln X_B{}^l \tag{6-12}$$

$$\ln X_B{}^l = \frac{1}{R}\left(\frac{\mu_B{}^S}{T} - \frac{\mu_B^{l,\bullet}}{T}\right) \tag{6-14}$$

The differential of this expression with respect to temperature contains expressions involving the rates of change of free energy (μ_B), which are related to molar enthalpies by the Gibbs–Helmholtz equation

$$\frac{d \ln X_B{}^l}{dT} = \frac{1}{R}\frac{d(\mu_B{}^S/T)}{dT} - \frac{1}{R}\frac{d(\mu_B^{l,\bullet}/T)}{dT}$$

$$= -\frac{\overline{H}_B{}^S}{RT^2} + \frac{\overline{H}_B^{l,\bullet}}{RT^2} \tag{6-15}$$

The difference between the partial molar enthalpy of the pure liquid and that of the solid is the partial molar heat of fusion of B:

$$\frac{d \ln X_B{}^l}{dT} = \frac{\Delta \overline{H}_{B.\,\mathrm{fus}}}{RT^2} \tag{6-16}$$

This equation provides a relationship between the dependence of solubility on T and the heat of fusion of the solid. It can be understood intuitively by considering the heat of fusion as arising primarily from the breaking up of the order of the crystalline form to obtain the liquid. Whether a solid is melted or dissolved, this same disordering process must be undergone. The greater the heat of fusion, and hence by implication the greater the barrier to the breaking of the crystalline order, the more the solubility is aided by raising the temperature. If there were no heat of fusion, or if no barrier had to be overcome, one would expect that solubility would be temperature independent. And indeed, this result follows immediately from Eq. 6-16.

This last expression can be integrated easily if one assumes that the heat of fusion is constant in the temperature range of interest. Then

$$\int_1^{X_B{}^l} d\ln X_B{}^l = \frac{\Delta \bar{H}_{B,\,\text{fus}}}{R}\int_{T_0}^{T}\frac{dT}{T^2}$$

$$\ln X_B{}^l = \frac{\Delta \bar{H}_{B,\,\text{fus}}}{R}\left(\frac{1}{T_0}-\frac{1}{T}\right) \tag{6-17}$$

But what is this temperature T_0? This is the temperature at which the equilibrium mole fraction of B is 1, or the temperature at which solid B is in equilibrium with a solution of B so concentrated that it is the only component there. This is just pure liquid B, so T_0 must be its melting point. Melting and solution are essentially the same phenomenon; a melt is the limit of a solution in which the amount of solvent goes to zero.

A plot of Eq. 6-17 for X_B near 1 is shown in Figure 6-4. If a solution of fixed concentration were being cooled along path 1, then at the intersection with the curve, crystalline B would begin to form. This would be described as the freezing point of a solution of A in solvent B. However, if instead, more and more B were being added to a solution at constant temperature along path 2, then at the same intersection point no more B would dissolve, and we would say that solvent A was now saturated with B. The description of this point as a freezing point or saturation point is a matter of viewpoint. In one description, B is the solvent, in the other, the solute.

For a more concrete example, consider water, which freezes at 0°C. As an antifreeze such as methanol is added, the freezing point falls. Now adding 12% by volume of methanol is observed to lower the freezing point

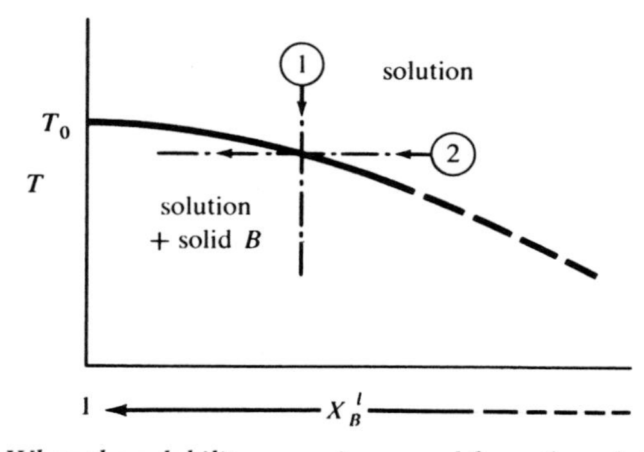

FIGURE 6-4. *When the solubility curve is crossed from the solution side, the experiment can either be described as the freezing of solvent B at a lower temperature because of the presence of solute (path 1), or as the saturation of the solvent by solute B and the precipitation of excess solute at the bottom (path 2). The assignment of the role of solvent or solute to component B is hence only a matter of convention.*

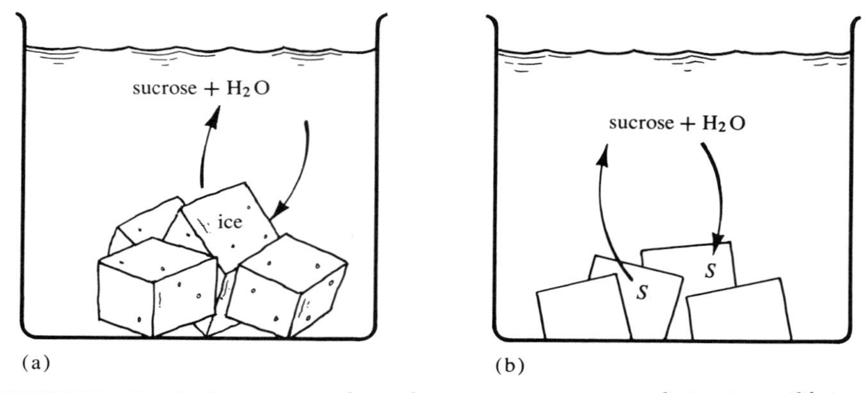

FIGURE 6-5. *Both situations above show a two-component solution in equilibrium with a pure solid phase of one of the components, yet in both cases it is water which is conventionally called the solvent. This is a reflection of the strong nonideality of sugar–water solutions.*

of the mixture to $-10°$. At lower temperatures, ice crystals begin forming. This is clearly a freezing-point depression phenomenon. Or is it? Suppose that instead you were working in a laboratory where the temperature was maintained at $-10°$, and were trying to dissolve ice crystals in methanol. As more ice was added, the methanol would become saturated until, when the water was 88% by volume, no more ice would dissolve. The solution would be saturated with ice. This would now be described as an experiment on the solubility of ice in methanol rather than the effect of methanol on the freezing point of water. Again, the labels "solvent" and "solute" are a matter of convenience.

So far, with two liquids, life is simple. But suppose you had a sucrose solution with ice in it (Figure 6-5a) at a temperature less than 0°C. Is this an experiment in lowering the freezing point of water by adding sucrose? Or are you finding the saturation point of ice in supercooled liquid sucrose? Now think of a sucrose solution with solid sucrose present (Figure 6-5b). Is this a saturated sucrose solution, or has the freezing point of liquid sucrose been lowered by adding water? Again, this is merely a question of semantics; the experimental situation is unchanged. The apparent one-sidedness of the descriptions is only a reflection of the nonideality of sucrose–water solutions.

6-2 COLLIGATIVE PROPERTIES

The traditional "colligative" or "collective" properties (vapor pressure lowering, freezing point lowering, boiling point elevation, and osmotic pressure) are so named because in ideal solutions the magnitude of the effect depends only upon the *number* of solute molecules present and not at all upon their kinds. These properties are most useful with solutes which are nonvolatile and which contribute nothing to the vapor phase.

Vapor Pressure Lowering

What happens to the vapor pressure of a volatile liquid when a nonvolatile solute is dissolved in it? Consider a system of solvent B and solute A (Figure 6-6). The vapor pressure of B is that for which there is an equilibrium exchange of B between liquid and vapor, and for which chemical potentials and fugacities of B in the two phases are equal. As nonvolatile solute A is added, the concentration of B falls. Not every molecule which reaches the surface is now of a type which can escape into the vapor, and the vapor pressure is lessened. Raoult's law gives

$$f_B{}^g = f_B{}^l = X_B f_B^{l,\bullet} \tag{6-18}$$

The *change* in fugacity produced by A is then

$$\Delta f_B = f_B{}^l - f_B^{l,\bullet} = (X_B - 1) f_B^{l,\bullet} = -X_A f_B^{l,\bullet} \tag{6-19}$$

For ideal vapor behavior,

$$\Delta P_B = -P_B{}^\bullet X_A \tag{6-20}$$

where $P_B{}^\bullet$ is the vapor pressure of pure B.

The vapor pressure lowering depends only upon how many solute molecules A are present, and not upon any other property of substance A (as long as it forms an ideal solution with B).

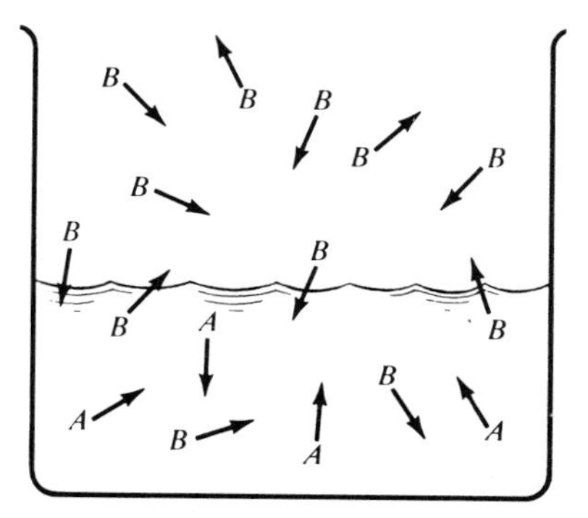

FIGURE 6-6. *Equilibrium between a solution of two components, only one of which is volatile, and the vapor of volatile component B.*

Freezing Point Depression

At the freezing point of B, equilibrium exists between solid B and liquid B, and the escaping tendencies in the two phases are identical. But as molecules of solute A are added to the liquid phase, then the proportion of molecules bouncing against the solid phase, which are B molecules and capable of being captured and added to the crystal lattice decreases. The escaping tendency of B from liquid to solid has diminished, whereas the reverse escaping tendency is unaffected. The crystals of B begin to dissolve. To prevent this and to restore equilibrium, one must favor the solid phase by lowering the temperature. At some lower temperature the two escaping tendencies again match and equilibrium is restored.

The quantitative requirement is that the two chemical potentials must again be equal at the new concentration of B:

$$\mu_B{}^S = \mu_B{}^l = \mu_B^{l,\bullet} + RT \ln X_B{}^l \tag{6-21}$$

As solute A is added, the ln X_B^l term falls. As T is lowered in compensation, $\mu_B{}^S$ and $\mu_B^{l,\bullet}$ both rise. [Recall that $(\partial \mu_B/\partial T)_P = -S_{B\cdot}$.] But $\mu_B^{l,\bullet}$ rises *faster* than $\mu_B{}^S$ because the molar entropy of the liquid is greater than that of the solid, and the ordering effect of the temperature drop is more pronounced. At some lower temperature, the large rise in $\mu_B^{l,\bullet}$ and the small drop in ln $X_B{}^l$ balance the small rise in $\mu_B{}^S$, and Eq. 6-21 again is balanced. This is illustrated in Figure 6-7.

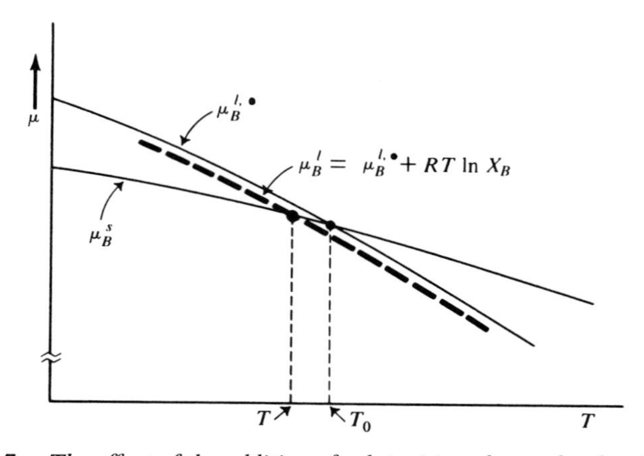

FIGURE 6-7. *The effect of the addition of solute A is to lower the chemical potential of solvent B from that of pure B. The dashed curve is a plot of chemical potential of B in solution as a function of temperature. This lowered escaping tendency curve of B in solution drops below the curve for pure crystalline B at a lower temperature, and the freezing point of the solution (or the melting point of crystalline B in the presence of solution) is lowered from T_0 to T.*

The relationship between the change in X_B and the temperature change needed to restore equilibrium is

$$\ln X_B = \frac{\mu_B{}^S - \mu_B^{l,\bullet}}{RT} \tag{6-22}$$

$$\frac{d \ln X_B}{dT} = -\frac{\overline{H_B{}^S} + \overline{H_B^{l,\bullet}}}{RT^2} = \frac{\Delta \overline{H}_{B,\,\text{fus}}}{RT^2} \tag{6-23}$$

This is the same expression as Eq. 6-16. Now consider solute A to be present in *low* concentrations:

$$d \ln X_B = d \ln (1 - X_A) \cong - dX_A$$

$$- \int_0^{X_A} dX_A = \int_{T_0}^{T} \frac{\Delta \overline{H}_{B,\,\text{fus}}}{RT^2}\, dT$$

$$- X_A = \frac{\Delta \overline{H}_{B,\,\text{fus}}}{R} \left(\frac{\Delta T}{TT_0}\right)$$

where again T_0 is the normal freezing point of B and $\Delta T = T - T_0$ is the freezing point depression.

For the small freezing point depressions which accompany small additions of A, T is close to T_0 and the approximation is valid:

$$X_A = - \frac{\Delta \overline{H}_{B,\,\text{fus}}}{R} \frac{\Delta T}{T_0{}^2}$$

$$\Delta T = T - T_0 = - \left(\frac{RT_0{}^2}{\Delta \overline{H}_{B,\,\text{fus}}}\right) X_A = - k_f{}' X_A$$

Again, as with vapor pressure lowering, the extent of a colligative effect is seen to depend only upon *how many* molecules of solute A are present and not on any other physical properties of A.

A more practical expression is in terms of molality:

$$X_A = \frac{n_A}{n_A + n_B} \cong \frac{n_A}{n_B} = \frac{m_A}{1000/M_B} = \frac{m_A M_B}{1000}$$

where M_B is the molecular weight of B. Then

$$\Delta T = - \left(\frac{RT_0{}^2 M_B}{1000\, \Delta \overline{H}_{B,\,\text{fus}}}\right) m_A = - k_f\, m_A \tag{6-24}$$

The molal freezing point depression constant k_f is a function only of the nature of the solvent B, and insofar as the solutions are ideal is the same for all solutes. Typical values of k_f are shown in Table 6-1.

TABLE 6-1 MOLAR FREEZING AND BOILING POINT CONSTANTS

Solvent	*Mol. wt.*	T_f (°C)	k_f (°/mole)	T_b (°C)	k_b (°/mole)
water	18.0	0	1.87	100	0.512
acetic acid	60.1	16.6	3.90	118.5	3.07
benzene	78.1	5.5	5.12	80.1	2.53
phenol	94.1	42	7.27	182	3.56
nitrobenzene	123.1	5.7	8.10	210.8	5.24
camphor	152.2	176	40		

Boiling Point Elevation

The logic and mathematics for the boiling point elevation derivation are almost identical with that of the previous section. There is an equilibrium exchange between the solvent B in the gaseous and solution phases, with the solute A molecules getting in the way and reducing the escaping tendency of the solvent. The boiling point is that temperature at which the vapor pressure of the system is equal to the external or atmospheric pressure. When a solute is added to a boiling liquid, the solvent vapor pressure falls and boiling ceases. Only if the temperature is raised and the innate escaping tendency of the solvent is increased sufficiently to compensate for the effect of dilution will boiling begin again.

As before, the chemical potentials must remain equal in the two phases, before and after, and

$$\mu_B{}^g = \mu_B{}^l = \mu_B^{l,\bullet} + RT \ln X_B{}^l$$

$$\frac{d \ln X_B{}^l}{dT} = -\frac{\Delta \overline{H}_{B,\,\text{vap}}}{RT^2} \tag{6-25}$$

This is illustrated in Figure 6-8.

Again, with the approximation of a small amount of solute A,

$$\Delta T_b = T - T_0 = +\left(\frac{RT_0{}^2}{\Delta \overline{H}_{B,\,\text{vap}}}\right) X_A \tag{6-26}$$

where T_0 is the normal boiling point. Changing from mole fraction to molality,

$$\Delta T_b = + k_b m_A \qquad k_b = \frac{RT_0{}^2 M_B}{1000\,\Delta\bar{H}_{B,\,\text{vap}}} \tag{6-27}$$

Boiling-point elevation constants are shown in Table 6-1 for the same substances whose freezing point depression constants were mentioned earlier.

Molecular Weight Determinations

These colligative properties provide a good means for determining molecular weights. A solution is made up with a known number of grams of the solute in 1000 cm^3 of the solvent. Then the vapor pressure lowering, freezing point depressing, or boiling point elevation produced by the solute is observed, leading via Eqs. 6-20, 6-24, or 6-27 to a knowledge of the number of *moles* of solute in the given amount of solvent. The ratio of grams solute to moles solute is then the molecular weight.

As an order-of-magnitude calculation of typical freezing point changes, consider a solution of glucose in water. The molecular weight of glucose is 180 g per mole. A solution of 100 g of glucose in 1000 g of water has a molality of 100/180 = 0.555. The freezing point depression of any solute of this molality in water is $-1.86°C \times 0.555 = -1.034°C$. This is a small temperature drop, but one which can be measured accurately with adequate care.

For most substances, one can get reasonable results with this method, but there are interesting anomalies. Sometimes the molecular weight seems to

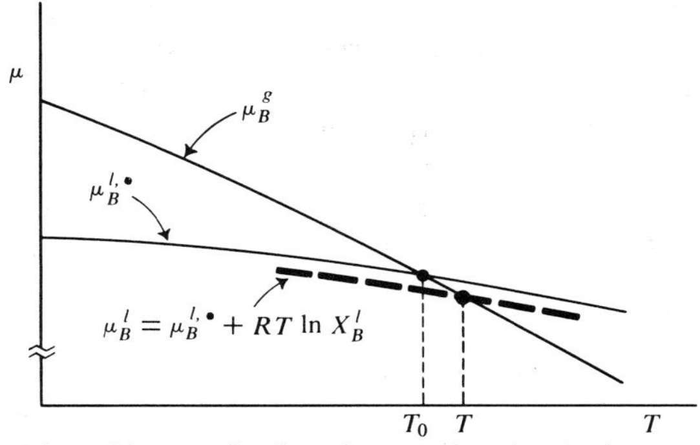

FIGURE 6-8. *The addition of solute lowers the chemical potential or escaping tendency of solvent B. As a consequence, the solution remains the phase with the lower μ_B until a higher temperature is reached, and the boiling point is raised from T_0 to T.*

depend on the solvent, which it should not do. An example of this behavior is benzoic acid, which in acetone gives a depression corresponding to its correct molecular weight of 122, but which in benzene shows an apparent molecular weight of 242. This is because benzoic acid dimerizes to a considerable extent in benzene. The reason the molecular weight does not come out to 244 is that the dimerization is not complete. With accurate measurements, one can find out precisely the extent of dimerization in systems of this kind. Electrolytes present a similar problem, the dissociation of molecular species to ions, producing molecular weights that are a fraction of their expected values, depending upon the number of ions formed. Each ion acts independently to inhibit the escape of solvent molecules. Again, by accurate measurements, one can ascertain the percent of ionization of a solute.

One of the most important classes of molecules whose molecular weights are to be determined are biological macromolecules such as proteins, and here the colligative properties discussed so far are of little help. As an illustration, consider a solution of 200 mg of the rather small protein cytochrome c in 10 cm^3 of water, equivalent to 20 g per thousand of solvent. The molecular weight of cytochrome c is 12,400, so the molality of such a solution is 20/12,400, or 1.6×10^{-3}. The freezing point depression observed will be $-1.80 \times 1.6 \times 10^{-3} = 0.003°C$. This small difference is entirely impractical to measure as a means of getting molecular weights. The methods discussed so far are pretty much restricted to low molecular weight solutes.

Osmotic Pressure

This last colligative property, osmotic pressure, is a standard method of measuring molecular weights of large molecules. The phenomenon used is that of selective diffusion through semipermeable membranes. Suppose one filled a sausage casing with a concentrated salt solution of a protein, fastened the ends, and put the casing into a large beaker of water. The membrane has small pores which will let the water and salt ions pass through freely, but which will block a sufficiently large molecule such as a protein. The water and salt will diffuse in and out until equilibrium is reached, but the protein will remain inside. This process is called *dialysis*, and is a convenient method of removing salts from protein solutions. Other membranes of different pore sizes can be used to separate molecules of different molecular weights, to purify water and to concentrate solutions.

In our sausage-bag experiment, the chemical potential of water inside the casing is less than that of pure water outside because of the presence of other molecules. More water will diffuse inside spontaneously, building up a net pressure difference across the membrane. This pressure opposes further flow, and when the pressure is high enough to balance the concentration difference, equilibrium exists. If the break point of the casing walls is reached first, then equilibrium of a different kind is established suddenly and the chemist ends up with gallons of a very dilute protein solution.

A more sophisticated version of this experiment is diagrammed in Figure 6-9. This consists of two chambers separated by a semipermeable membrane which will pass solvent B but not solute A. Two manometers measure the hydrostatic head and hence the pressure difference between the chambers. If solute A is added to the right-hand chamber, the chemical potential of B will fall, causing a spontaneous diffusion of B into the right compartment. This flow will continue until a net osmotic pressure difference is built up. Equilibrium is reestablished when the opposing effect of pressure in the right-hand compartment balances the effect of dilution and the chemical potentials are again equal.

At the beginning of the experiment, both compartments contain pure B, and their chemical potentials are both that of pure B:

$$\mu_B^{\bullet,L} = \mu_B^{\bullet,R} = \mu_B^{\bullet} \tag{6-28}$$

Adding A to the right side reduces the potential of B on that side:

$$\mu_B^{\bullet,L} > \mu_B^{R} = \mu_B^{\bullet} + RT \ln X_B \tag{6-29}$$

There is a diffusion of B from left to right in an attempt to equalize the potential, resulting in a pressure difference. But as the pressure in the right chamber increases, $\mu_B^{\bullet}$ in that chamber rises:

$$\left(\frac{\partial \mu_B}{\partial P}\right)_T = \overline{V}_B = \overline{V}_B^{\bullet} \tag{6-30}$$

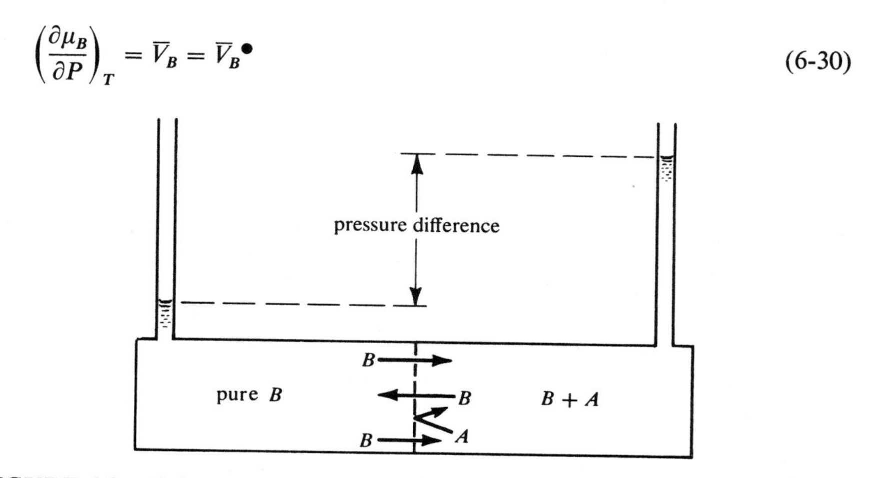

FIGURE 6-9. *Solute A, on the right side of a membrane which will pass solvent B but not A, lowers the chemical potential of B. As a result, more B diffuses into the right-hand chamber until the increased pressure at the right counterbalances the dilution effect of A and* μ_B *is once again equal in both chambers. This excess pressure is the* osmotic pressure.

For ideal solutions, the partial molar volume is the molar volume of pure substance. The change in chemical potential because of pressure is

$$\Delta\mu_B{}^R = \int_P^{P+\pi} \overline{V}_B{}^\bullet\, dP = \overline{V}_B{}^\bullet \pi \tag{6-31}$$

Adding this extra chemical potential to the right-hand compartment restores equilibrium:

$$\mu_B^{\bullet,L} = \mu_B{}^R = \mu_B{}^\bullet + RT \ln X_B + \pi \overline{V}_B{}^\bullet$$

$$\pi \overline{V}_B{}^\bullet = -RT \ln X_B \tag{6-32}$$

The chemical potential of B falls when A is added, but is restored by the osmotic pressure difference π.

In very dilute solutions, several simplifying approximations to the concentration terms can be made:

$$\ln X_B \cong -X_A$$

$$X_A = \frac{n_A}{n_A + n_B} \cong \frac{n_A}{n_B}$$

$$V = n_A \overline{V}_A{}^\bullet + n_B \overline{V}_B{}^\bullet \cong n_B \overline{V}_B{}^\bullet$$

Hence, for dilute solutions,

$$\pi \overline{V}_B{}^\bullet \cong X_A RT$$

$$\pi V \cong n_A RT \tag{6-33}$$

$$\pi \cong c_A RT \tag{6-34}$$

where c_A is the molar concentration in moles per liter of solution. Note the analogy between Eq. 6-33 and the ideal gas equation for n moles, $PV = nRT$, with solute molecules instead of gas molecules, solvent molecules instead of empty space, and with π and P being the excess pressure over that of pure solvent or perfect vacuum, respectively.

The cytochrome c molecular weight problem makes the usefulness of this technique apparent. In a dilute solution, molarity and molality will be approximately the same. If a solution of 200 mg of protein in 10 g of water is considered to be a 1.6×10^{-3} molar solution, then the osmotic pressure developed by such a solution is

$$\pi = (1.6 \times 10^{-3}\ \text{moles/liter}) \times (0.0821\ \text{liter atm/°mole}) \times 298° \times 760\ \text{mm Hg/atm} = 29\ \text{mm Hg}$$

A pressure of 29 mm of mercury in a manometer is easy to measure with accuracy, whereas a freezing point lowering of 0.003°C is scarcely measurable.

Freezing Point Diagrams in General

The linear approximation of freezing point behavior for very dilute solutions has been given in Eq. 6-24. Consider now the exact expression, valid over the entire mole fraction range. As an example, bismuth and cadmium form solutions in the liquid state, but are immiscible in the solid. The phases present then are the Bi–Cd solution, crystalline Bi, and crystalline Cd. The expression for change of solubility with temperature has already been derived:

$$\frac{d \ln X_B}{dT} = \frac{\Delta \overline{H}_{B,\text{fus}}}{RT^2} \tag{6-23}$$

Integrating this directly, from the melting temperature T_0 (assuming a constant heat of fusion),

$$\ln X_B = \frac{\Delta \overline{H}_{B,\text{fus}}}{R}\left(\frac{1}{T_0} - \frac{1}{T}\right) \tag{6-35}$$

First consider Bi as the solvent. As cadmium is added to lower the melting point of bismuth from its initial value of 546°K, the melting point curve follows the equation:

$$\ln X_{\text{Bi}} = \frac{\Delta \overline{H}_{\text{fus(Bi)}}}{R}\left(\frac{1}{546^\circ} - \frac{1}{T}\right)$$

which is plotted in Figure 6-10 as curve *a-b-c*. Similarly, the freezing point behavior of a solution of a little bismuth in much cadmium is given by

$$\ln X_{\text{Cd}} = \frac{\Delta \overline{H}_{\text{fus(Cd)}}}{R}\left(\frac{1}{596^\circ} - \frac{1}{T}\right)$$

and by curve *d-b-e*. At the temperature of 413°K, which corresponds to $X_{\text{Cd}} = 0.55$, the two curves cross. The area above both curves represents solution with no solid phase present, so this temperature is the lowest at which the solution can exist without something crystallizing out. This point is known as the *eutectic* point, or "easily melted" point.

As a solution (with a different mole-fraction composition) is cooled, its T-X point will eventually reach one of the curves. At this point, the solution

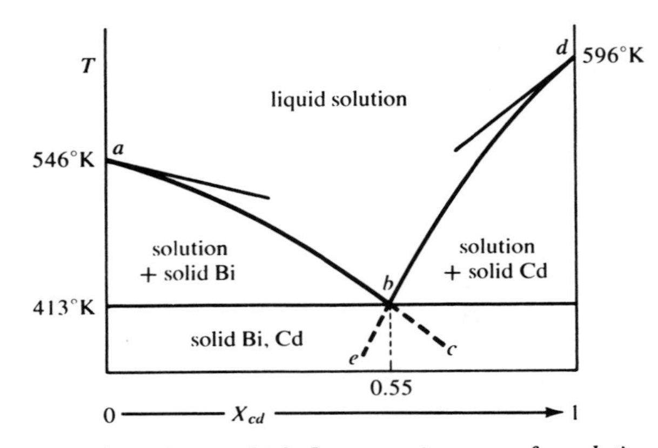

FIGURE 6-10. *At the point at which the saturation curve for solution and solid* Bi *crosses the saturation curve for solution and solid* Cd, *the liquid and both pure solid phases are in equilibrium. This point* (*b*) *is the eutectic* ("*easily melted*") *point.*

is saturated, and as more heat is removed some solid will begin to crystallize out. As solid is removed the composition of the solution will change to follow the solubility curve until the eutectic point is reached, at which both cadmium and bismuth will crystallize out. The portions *b-e* and *b-c* of the curves are only hypothetical. The curve *a-b* describes the solubility behavior of Bi, for example, but at point *b*, Cd becomes insoluble as well. The freezing point depression equation, Eq. 6-24, is just the linear asymptotic approximation to these curves at infinite dilution.

6-3 DEVIATIONS FROM IDEAL BEHAVIOR—HENRY'S LAW, ACTIVITIES, AND THE DEBYE–HÜCKEL MODEL

Positive and Negative Deviations

In an ideal solution, all intermolecular interactions are equivalent. The escaping tendency of a species is directly proportional to the number of molecules available to escape; $f_B = X_B f_B^{\bullet}$. A plot of fugacity (or vapor pressure) against mole fraction is linear as in Figure 6-11a.

Some unlike molecules deviate from ideality in that they associate with one another more easily than with molecules of their own kind. The ultimate limit of this type of behavior is compound formation. Since, in such a case, *B* molecules surrounded by *A* molecules are less likely to escape than if surrounded by other *B* molecules, the fugacity of *B* is less than that predicted by Raoult's law:

$$f_B < X_B f_B^{\bullet} \tag{6-36}$$

This situation is shown in Figure 6-11b. The partial molar volume of a component of such a solution will be less than that of the pure component because of closer intermolecular interactions, and there will be shrinkage upon mixing. Increased stability of the AB type associations means also that the partial molar enthalpy will be lower than the pure molar enthalpy, and that mixing will be exothermic. An increase in the temperature will tend to upset these associations and reduce the miscibility of the two substances, in agreement with Le Chatelier's principle. By the same principle, if shrinkage does occur upon mixing, then an increase in pressure should encourage the mixing process and enhance the solubilities of the two components in one another. This behavior is called *negative* deviation from ideality, and is shown by many substances which form close intermolecular associations, such as acetone and water, water and the halogen acids or the strong oxygen acids (sulfuric and nitric), pyridine and acetic acid, or chloroform and ethyl ether. A vapor pressure plot for a solution of acetone and chloroform is shown in Figure 6-12, and a fugacity plot is very similar.

The opposite type of behavior, positive deviation, is shown by very unlike molecules where a high price in free energy must be paid for mutual association. Mixtures of polar and nonpolar liquids such as hydrocarbons and water are good examples. The escaping tendency of a molecule in a solution is *greater* than in the pure component, and the fugacity is greater than expected

$$f_B > X_B f_B^\bullet \tag{6-37}$$

A typical fugacity curve is shown in Figure 6-11c, and Figure 6-13 is a vapor pressure plot of the acetone–carbon disulfide system. As might be expected, partial molar enthalpies are usually greater than those of the pure liquids. There is sometimes a volume increase on mixing, and heat will be absorbed in

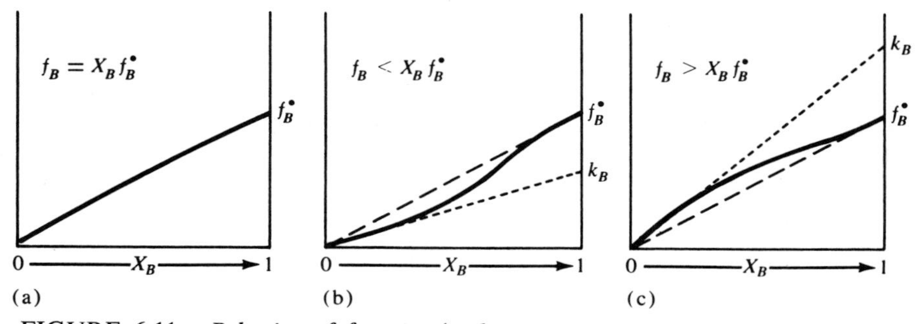

FIGURE 6-11. *Behavior of fugacity (and approximately, of vapor pressure) for (a) an ideal solution, (b) negative deviation, and (c) positive deviation.* $f_B^\bullet$ *is the fugacity of pure B, and* k_B *is the Henry's law constant discussed in Section 6-3.*

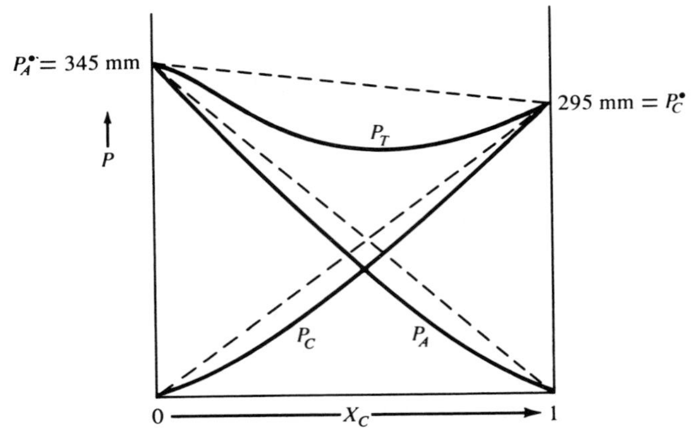

FIGURE 6-12. *Partial vapor pressures (P_C and P_A) and total vapor pressure (P_T) for a solution of acetone (A) and chloroform (C) of chloroform composition X_c. This solution shows negative deviation from ideality, and should be compared with Figure 6-1. A plot of fugacities would be virtually identical.*

the mixing process. The solubilities of the components will be increased as the temperature rises, in agreement with Le Chatelier's principle. Examples of such positive deviations include carbon disulfide and acetone, benzene and ethanol, n-propanol and water, and most nonpolar–polar pairs.

The framework for treating ideal solutions has been established:

$$\mu_j = \mu_j^{0} + RT \ln \frac{f_j}{f_j^{0}} = \mu_j^{\bullet} + RT \ln \frac{f_j}{f_j^{\bullet}}$$

$$\mu_j = \mu_j^{\bullet} + RT \ln X_j \tag{6-6}$$

This same working expression holds no matter whether the standard state of a solution component is taken to be that of the pure liquid component or that of unit fugacity of its vapor.

For nonideal solutions the same choice exists as with nonideal gases: to sacrifice either the equations or the variable. Again it is easier to keep the equations and to define a new variable for which they hold. The *activity*, defined as the ratio of fugacity to that in some standard state, then can play the role of a "corrected mole fraction" much as fugacity plays the role of "corrected pressure":

$$a_j \equiv \frac{f_j}{f_j^{0}} \tag{5-145}$$

$$\mu_j = \mu_j^{0} + RT \ln a_j \tag{5-146}$$

If the standard state is taken as that of pure liquid component, then this

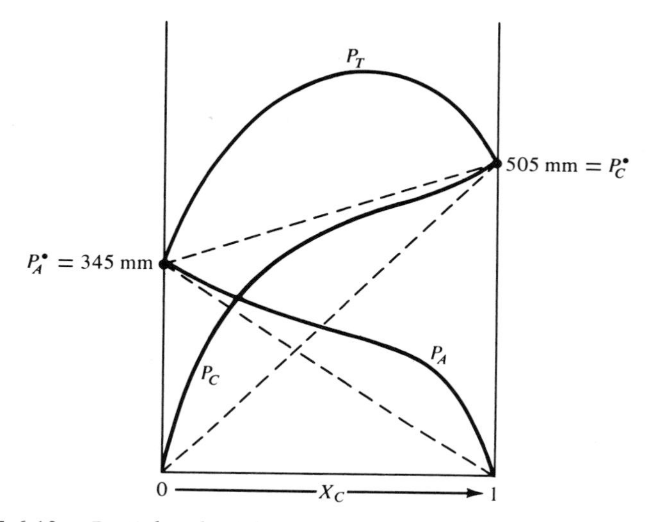

FIGURE 6-13. *Partial and total vapor pressures at 35°C for a mixture of acetone (A) and carbon disulfide (C), which shows positive deviation from ideality.*

becomes

$$\mu_j = \mu_j^{\bullet} + RT \ln a_j \tag{6-38}$$

and a "pseudo-Raoult's law" results:

$$f_j = a_j f_j^{\bullet} \neq X_j f_j^{\bullet} \tag{6-39}$$

Just as the ratio between the fugacity and pressure was called the fugacity coefficient, so the ratio between the activity and mole fraction is the activity coefficient γ; $\gamma_j \equiv a_j/x_j$. For ideal solutions, the activity coefficient is 1, for positive deviations it is greater than 1, and for negative deviations it is less than 1. These deviations are shown in Figure 6-14. For both positive and

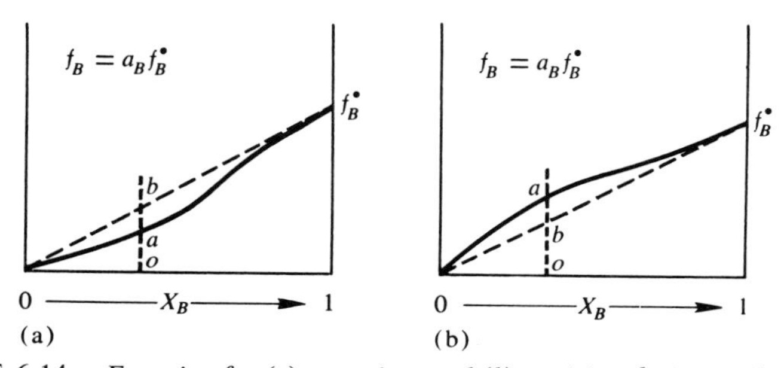

FIGURE 6-14. *Fugacity for (a) negative, and (b) positive deviation from ideality. In both cases, the ideal behavior is shown by a dashed straight line, and the fugacity coefficient is $\gamma = oa/ob$.*

negative deviations it is true that

$$\gamma_B = \frac{\overline{oa}}{\overline{ob}} = \frac{f_B}{X_B f_B^{\bullet}} = \frac{a_B}{X_B} \tag{6-40}$$

Note that γ_B is a function of concentration, and approaches unity at pure B.

This establishes a framework for treating equilibria in nonideal solutions. In both gas and condensed phases the relationship $\mu_j = \mu_j^0 + RT \ln a_j$ defines the activity of component j. The standard state μ_j^0 for gases is the state of *unit fugacity*, making the activity (which is unitless) numerically equal to the fugacity. For solutions, the standard state of each component is the pure component, and the activity is simply the ratio of the fugacity of the solution component to that of the pure component.

Now all of the previous expressions in equilibrium thermodynamics can be expressed in terms of activities, and then applied to condensed phases as well as gases:

$$\widetilde{\Delta G} = \widetilde{\Delta G^0} + RT \ln \prod_j a_j^{\nu_j} \tag{6-41}$$

$$\widetilde{\Delta G^0} = -RT \ln K_a \tag{6-42}$$

where $K_a = \prod_j a_j^{\nu_j}$ at equilibrium. The activities of the components of a solution can be determined by measuring such properties as freezing point depression, boiling point elevation, or osmotic pressure.[1] If the activity behavior as a function of concentration is found for one component of a binary solution, then the activity of the other component can be calculated using the Gibbs–Duhem equation. It is one of the tests of the validity of the concept of activity that the same numerical values for activity coefficients are found by several such methods, and that the use of these coefficients with measured concentrations (molarity, mole fraction) then leads to equilibrium constants which are truly constant.

Henry's Law

Raoult's law holds for the solvent in very dilute solutions because virtually the only molecules which a solvent molecule encounters around itself are other solvent molecules. But in such dilute solutions the solute, too, sees only one kind of neighboring molecule, that of solvent. The environment of a solute molecule will not be appreciably affected by addition of more solute until the solution becomes sufficiently concentrated that solute–solute encounters

[1] See Klotz, Lewis and Randall, or Wall for details on the practical calculation of activity coefficients.

become frequent. Under such dilute conditions, the escaping tendency or fugacity of solute should be proportional to the number of molecules there are to escape

$$f_A = X_A k_A \tag{6-43}$$

although the proportionality constant k will not be the fugacity of the pure solute. Equation 6-43 is Henry's law.

Henry's law behavior is shown in Figures 6-11b and 6-11c by dotted straight lines. The Henry's law constant k_B is the intercept of this line with the $x_B = 1$ axis. It is easy to show, using the Gibbs–Duhem equation, that if Raoult's law holds for one component of a two-component system, then Henry's law must hold in the same region for the other component. Assume that Raoult's law holds in a certain region for component A; $f_A = X_A f_A^{\bullet}$. The Gibbs–Duhem equation tells us that

$$X_A \frac{\partial \mu_A}{\partial X_A} + X_B \frac{\partial \mu_B}{\partial X_A} = 0$$

But the individual chemical potentials can be written

$$\mu_A = \mu_B{}^0 + RT \ln f_A \qquad \mu_B = \mu_B{}^0 + RT \ln f_B$$

Since the standard potentials are independent of changes in mole fraction of A, the Gibbs–Duhem equation becomes

$$X_A \frac{\partial \ln f_A}{\partial X_A} + X_B \frac{\partial \ln f_B}{\partial X_A} = 0$$

Using Raoult's law for component A,

$$\frac{\partial \ln f_A}{\partial X_A} = \frac{\partial \ln X_A}{\partial X_A} = \frac{1}{X_A}$$

$$\frac{\partial \ln f_B}{\partial X_A} = \frac{\partial \ln f_B}{\partial(1 - X_B)} = \frac{\partial \ln f_B}{\partial X_B}$$

Substituting in the Gibbs–Duhem equation and rearranging,

$$\frac{\partial \ln f_B}{\partial X_B} = \frac{X_A}{X_B} \frac{\partial \ln f_A}{\partial X_A} = \frac{1}{X_B}$$

$$d \ln f_B = \frac{dX_B}{X_B} = d \ln X_B$$

therefore,

$$f_B \propto X_B \quad \text{or} \quad f_B = k_B X_B$$

In regions where the fugacity of one component is proportional to mole fraction, with the fugacity of the pure component as a proportionality constant, the *other* component's fugacity is likewise proportional to its mole fraction, although with a different proportionality constant.

This Henry's law behavior suggests another convenient choice for standard states of solutes in dilute solutions. If the activity is chosen as the ratio of the fugacity to the fugacity of pure solute (that is, if the standard state is chosen to be pure solute), then the activity will not become identical to the mole fraction at great dilution. For while the activity expression is based on Raoult's law $f_B = a_B f_B^\bullet$, the true mole fraction expression is Henry's law $f_B = X_B k_B$.

The limit of infinitely dilute solutions, where the solvent obeys Raoult's law and the solutes obey Henry's law, is called an "ideally dilute" solution. With such an ideally dilute solution as the norm, deviations from this norm arising from nonideal behavior can be handled by defining the standard states in such a way that Raoult's and Henry's laws hold for solvent and solutes in terms of activities rather than mole fractions:

$$f_A = a_A f_A^\bullet \quad \text{for solvent} \tag{6-44}$$
$$f_B = a_B k_B \quad \text{for solutes}$$

The standard state of the solvent is pure solvent, as before. But now the standard state of each solute is a hypothetical pure solute in which Henry's law holds all the way to $X_B = 1$. The "reference state" of the solute, the state in which the activity becomes the mole fraction, is then an infinitely dilute solution, and is not the same as the standard state. The fact that the standard state is physically unattainable is no more disturbing than when calculating fugacities of gases (Figure 5-23); the standard state is chosen solely to make the mathematical calculations easy.

With this choice of reference state, then

$$\mu_B = \mu_B{}^0 + RT \ln X_B \tag{6-45}$$

in ideally dilute solutions, and

$$\mu_B = \mu_B{}^0 + RT \ln a_B = \mu_B{}^0 + RT \ln X_B + RT \ln \gamma_B \tag{6-46}$$

in less dilute solutions. But it must be remembered that $\mu_B{}^0$ now is the chemical potential, not of pure B but of B in a hypothetical pure state where

Henry's and not Raoult's law holds. It does not, therefore, correspond to any physically real μ_B.

A Molecular Theory of Activity Coefficients: Debye–Hückel Theory

If activity coefficients are anything more than a mathematical convenience in dealing with real solutions, then they should be calculable from first principles from the proper molecular model. Statistical models of imperfect solutions have been developed which lead to deviations from Raoult's law similar to those which are observed,[2] but the model which has had the greatest and most long-lasting impact upon chemistry has been the Debye–Hückel model for activity coefficients of ions in electrolytic solutions. Although we shall not consider solutions of electrolytes in detail, the Debye–Hückel theory is worth a close examination because of the molecular picture which it gives of some of the reasons for nonideality.

If a solute dissociates into ions, the ions can be considered to a first approximation as separate molecular species. Dissociation is then similar to the addition of more nondissociating solute. This behavior is seen in the effect of dissociation on colligative properties such as freezing point lowering. The Debye–Hückel theory assumes that if the ions were not charged, the solution would be ideal, or at least "ideally dilute" in the sense of the previous section. All deviations from ideality are attributed to the charges on the ions. Each ion will attract ions of the opposite charge to its own neighborhood to form an ion cloud. A positive ion will move through a local atmosphere of negative charge, and a negative ion, through a positive atmosphere. These ionic atmospheres exert a drag upon their ions, impeding their mobility through the solution. As a consequence, the *activity* of each ion is lowered. (See Figure 6-15.)

If the solution were ideally dilute,

$$\mu_j^{\text{ideal}} = \mu_j{}^0 + RT \ln X_j \tag{6-45}$$

The subscript j refers to one of the many species of ion assumed to be present in the solution. But because of charge interactions between ions, the true chemical potential of the jth ionic species will be different:

$$\mu_j = \mu_j^{\text{ideal}} + \mu_j^{\text{el}} = \mu_j{}^0 + RT \ln X_j + RT \ln \gamma_j \tag{6-47}$$

where μ_j^{el} is the added electrical contribution. The activity coefficient term, by the assumptions of the theory, can be associated entirely with the electrical contribution to the chemical potential.

[2] See Wall, for example.

$$\mu_j^{el} = RT \ln \gamma_j \tag{6-48}$$

The problem is then reduced to that of finding $\mu_j^{el} = (\partial G^{el}/\partial n_j)_{T,P,n_i}$, the partial molar electrical free energy of the solution for ionic species j. This can be done if we can find an expression for the extra free energy of the solution because of the ionic charges, as a function of the numbers of ions of each species present, n_j (moles).

The problem of calculating $G^{el}(n_i)$ as a function of ion concentrations was solved by Debye and Hückel by an ingenious *Gedankenexperiment*. Each ion, because of its own charge, is surrounded by an atmosphere of the opposite charge supplied by other ions. Imagine that the charge of each ion can be turned up and down like the volume on a hi-fi set. The ionic atmospheres will vary with the charge of each central ion. Now turn the charge on all of the ions down to zero, and then slowly raise all charges from zero to full value. If the central ion being considered is positive, then this recharging process is one of bringing infinitesimal increments of positive charge into an increasingly great negative atmosphere. Because of charge attraction, work can be done by the system on its surroundings; $w < 0$, and the free energy of the system will drop by an amount equal to the work done in the most efficient (that is, reversible) manner: $\Delta G = w$. This ΔG is just the electrical contribution to the free energy, G^{el}.

Before we can calculate the work involved in bringing a charge into a potential field, we must know the potential field itself. Given the charges on the ions, what will be the electrical potential around each ion arising from its ionic atmosphere of oppositely charged species? The starting point for this

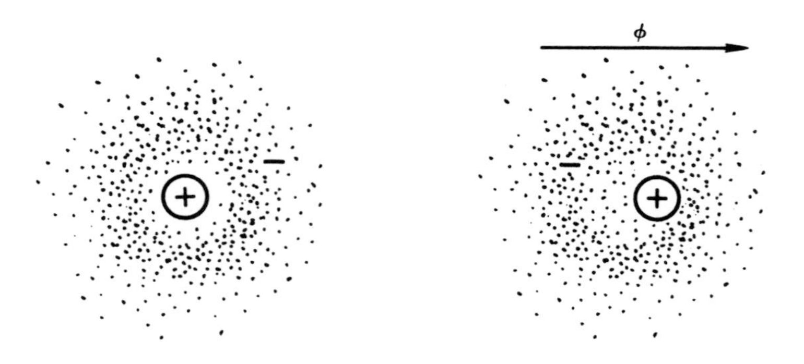

FIGURE 6-15. *An ion in solution will be surrounded by a cloud of opposite charge produced by neighboring ions (left). When the ion is moving (right), its motion will be impeded by the drag of this charge cloud, and its activity will be less than it would be in the absence of charges. This is the basis of the Debye–Hückel theory of activity coefficients.*

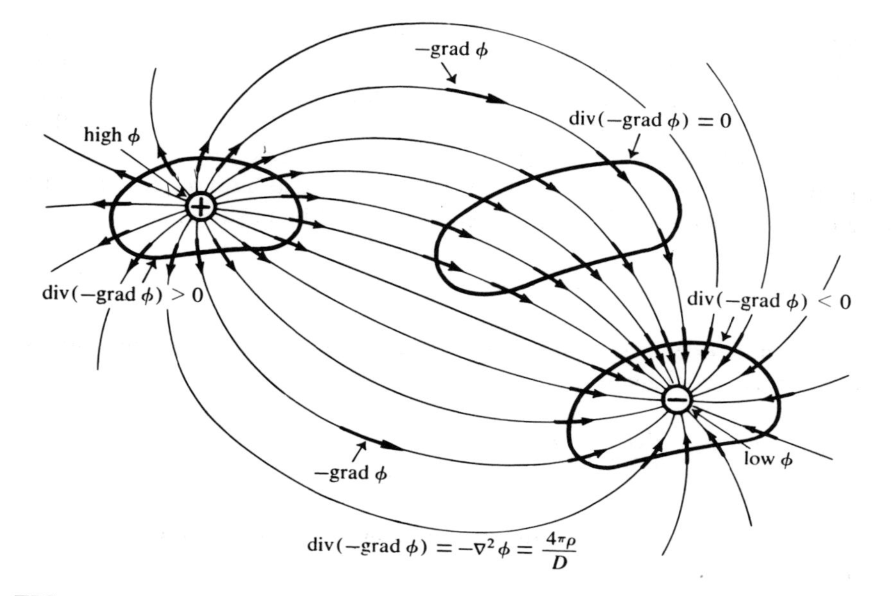

FIGURE 6-16. *According to the Poisson equation, the net outflow of electric field from a small unit of volume is proportional to the total charge within that volume. If the electric field is the negative of the* gradient *of the potential (the downhill slope), then the net outflow of field is the* divergence *of this field, or* −div(grad ϕ).

calculation is the Poisson equation

$$\nabla^2\phi = -\frac{4\pi\rho}{D} \tag{6-49}$$

a formidable looking equation which states an almost obvious fact: the net electrical field coming out of a small region of space is proportional to the total charge within this region. This is pictured in Figure 6-16. If ϕ is the electrical potential, then the electric field vector **E** points in the direction of the steepest downhill descent of ϕ. This is equivalent to saying that vector **E** is the negative of the *gradient* of ϕ, or in vector notation,[3]

$$\mathbf{E} = -\text{grad}\ \phi = -\nabla\phi \tag{6-50}$$

The *divergence* of a vector across a closed surface which encloses a region of space is the net vector flow through that surface. If a cheesecloth bag is

[3] For a good review of vectors and fields, see D. R. Corson and P. Lorrain, *Introduction to Electromagnetic Fields and Waves* (W. H. Freeman, San Francisco, 1962).

fastened to a water faucet, there will be a net flow of water out of the bag, and the divergence of the fluid flow vector will be positive. If the bag is placed over the drain, there will be a net flow inward, and the divergence will be negative. If the bag is placed in a running stream, as much water will flow in as out, and the divergence of the flow vector will be zero. In the same way, if a region contains positive charges (considered as a *source* of potential), then the divergence of the electric field vector will be positive:

$$\text{div}\,\mathbf{E} = -\text{div grad}\,\phi = -\nabla^2\phi > 0 \tag{6-51}$$

The div grad operator is the familiar Laplacean of quantum mechanics. For Cartesian coordinates,

$$\nabla^2\phi = \frac{\partial^2\phi}{\partial x^2} + \frac{\partial^2\phi}{\partial y^2} + \frac{\partial^2\phi}{\partial z^2}$$

and for spherical polar coordinates, assuming that ϕ itself is spherically symmetrical,

$$\nabla^2\phi = \frac{1}{r^2}\frac{d}{dr}\left(r^2\frac{d\phi}{dr}\right) \tag{6-52}$$

The Poisson equation simply says that the divergence of the electric field vector is proportional to the charge density within the element of volume ρ:

$$\text{div}\,\mathbf{E} = -\nabla^2\phi = \frac{4\pi\rho}{D} \tag{5-53}$$

The quantity D is the dielectric constant of the medium.

With this background, the derivation of the Debye–Hückel expression for activity coefficients can proceed in four steps.

(1) Use the Poisson equation to find the potential produced around an ion by its ionic atmosphere.

(2) Run all of the ionic charges up simultaneously from zero to their full values to get G^{el}.

(3) Take the partial of G^{el} to find μ_j^{el}.

(4) Equate this to $RT \ln \gamma_j$.

Let N_i be the mean concentration of ions of type i in molecules per cm^3, let n_i be the total number of *moles* of i, and let c_i be the concentration in moles per liter. Although the mean concentration will be N_i, the actual local concentration will vary from point to point through the solution because of the potential ϕ produced by the other ions. If an ion of charge z_i electrons or $z_i \varepsilon$ esu is at a point where the field is ϕ, it will have an electrostatic energy $z_i \varepsilon \phi$. Assuming a Boltzmann distribution, the concentration of ions i at a region of potential ϕ is

$$N_i' = N_i \exp\left(-\frac{z_i \varepsilon \phi}{kT}\right) \tag{6-54}$$

The total charge density at point in space where the potential is ϕ is the sum of the charge on each species of ion times the concentration of that species:

$$\rho = \sum_i N_i' z_i \varepsilon = \sum_i N_i z_i \varepsilon \exp\left(-\frac{z_i \varepsilon \phi}{kT}\right) \tag{6-55}$$

If $z_i \varepsilon \phi \ll kT$,

$$\rho = \sum_i N_i z_i \varepsilon \left(1 - \frac{z_i \varepsilon \phi}{kT}\right) \tag{6-56}$$

But $\sum_i N_i z_i = 0$ since the solution as a whole must be electrically neutral, and the charge density is

$$\rho = -\frac{\varepsilon^2 \phi}{kT} \sum_i N_i z_i^2 \tag{6-57}$$

This charge expression can now be used with the Poisson equation to calculate an expression for the potential ϕ:

$$\nabla^2 \phi = -\frac{4\pi\rho}{D} = +\left(\frac{4\pi\varepsilon^2}{DkT} \sum_i N_i z_i^2\right)\phi \tag{6-58}$$

If the constant K is defined by

$$K^2 \equiv \frac{4\pi\varepsilon^2}{DkT} \sum_i N_i z_i^2 \tag{6-59}$$

then the Poisson equation becomes

$$\nabla^2 \phi = \frac{1}{r^2} \frac{d}{dr}\left(r^2 \frac{d\phi}{dr}\right) = K^2 \phi \tag{6-60}$$

This is a differential equation for which the general solution, which can be verified by substitution, is

$$\phi = A \frac{e^{-Kr}}{r} + B \frac{e^{Kr}}{r} \tag{6-61}$$

But, as r goes to infinity, e^{Kr}/r also becomes infinite, whereas the potential ϕ does not. This is a physical boundary condition on the solution of the differential equation which requires that $B = 0$. If Kr is small, then

$$\phi = \frac{A}{r} e^{-Kr} \cong \frac{A}{r}(1 - Kr) = \frac{A}{r} - AK \tag{6-62}$$

As r goes to zero, the first term will dominate. But under these conditions, only the charge of the central ion will matter, and the effect of the ion cloud around it will be negligible. Hence, since for the central ion $\phi = z_i \varepsilon / Dr$, a second boundary condition requires that: $A = z_i \varepsilon / D$, and

$$\phi = \frac{z_i \varepsilon}{Dr} - \frac{z_i \varepsilon}{D(1/K)} \tag{6-63}$$

The first term is the potential produced by the central ion, and the second is the quantity which we are after, the potential produced by the atmosphere of other ions around the central one. The way in which it has been written emphasizes the fact that this ion cloud potential is equivalent to that which would be produced by a shell of charge, with a total charge equal but opposite in sign to that of the central ion, and with a radius of $1/K$. The potential from the ion cloud alone is

$$\phi' = -\frac{z_i \varepsilon}{D(1/K)} = -\frac{z_i \varepsilon K}{D} \tag{6-64}$$

Note that, in order to evaluate the constant A, we assumed that the ions were point charges, a step which will severely limit the applicability of our result. If instead, the ions were assumed to be hard spheres of diameter a_i, then the ion cloud potential would be

$$\phi' = -\frac{z_i \varepsilon}{D(\frac{1}{K} + a_i)} = -\frac{z_i \varepsilon K}{D(1 + Ka_i)} \tag{6-65}$$

Now for the charging experiment. If at any given instant, each ion has only a fraction f of its total charge, then $z_i^* = fz_i$ and $K^* = fK$. The potential from the ion cloud will be only

$$\phi^* = -\frac{z_i \varepsilon K}{D} f^2 \tag{6-66}$$

"Turning up" the charge by an amount df is equivalent to bringing in an increment of charge; $dq_i = z_i \varepsilon\, df$. The work per ion is

$$dw_i = \phi^*\, dq_i = -\left(\frac{\varepsilon^2 K}{D}\right) z_i^2 f^2\, df \tag{6-67}$$

and the total work involved with the charging of all of the ions is

$$dw = \sum_i VN_i\, dw_i = -\frac{V\varepsilon^2 K}{D}\left(\sum_i N_i z_i^2\right) f^2\, df \tag{6-68}$$

The electrical free energy is equal to the total work done:

$$G^{\text{el}} = w = -\frac{V\varepsilon^2 K}{D} \sum_i N_i z_i^2 \int_{f=0}^{1} f^2\, df$$

$$G^{\text{el}} = -\frac{V\varepsilon^2 K}{3D} \sum_i N_i z_i^2 \tag{6-69}$$

The free energy of the ions falls during the charging process, because they are in a more stable (and less mobile) situation.

Taking the partial of the free energy with respect to ionic species j, with $N_j = Nn_j/V$, where N is Avogadro's number,

$$\mu_j^{\text{el}} \equiv \left(\frac{\partial G^{\text{el}}}{\partial n_j}\right)_{P,\,T,n_i} = -\frac{NK\varepsilon^2 z_j^2}{2D} = RT \ln \gamma_j$$

(Remember that K is also a function of N_j.) The Debye–Hückel expression for activity coefficients in the limit of point ions is then

$$\ln \gamma_j = -\frac{\varepsilon^2 K z_j^2}{2DkT} \tag{6-70}$$

But we cannot observe individual activity coefficients γ_j, since we cannot vary the concentration of each ion at will. The solution as a whole must remain electrically neutral. We can observe a mean activity coefficient, and these observations can be compared with a modified form of Eq. 6-70 expressing mean activity coefficient. Assume that a solution contains a single electrolyte, which dissociates into ν_+ positive ions of charge z_+ and ν_- negative ions of charge z_- (z_- as used here is a negative number, not the magnitude of negative charge). The total number of ions per molecule is $\nu = \nu_+ + \nu_-$, and by electrical neutrality $\nu_+ z_+ + \nu_- z_- = 0$. In the absence of any method of varying one ion's concentration without simultaneously varying the other, we can only measure a mean chemical potential or activity of the ions. Since activity is related to chemical potential via a logarithm term, an arithmetic mean of chemical potentials implies a geometric mean of activity coefficients:

$$\begin{aligned} \nu \ln \gamma_{\pm} &= \nu_{+} \ln \gamma_{+} + \nu_{-} \ln \gamma_{-} \\ \gamma_{\pm}{}^{\nu} &= \gamma_{+}^{\nu_{+}} \gamma_{-}^{\nu_{-}} \end{aligned} \tag{6-71}$$

Using the values of the individual activity coefficients given by the Debye–Hückel model,

$$\begin{aligned} \nu \ln \gamma_{\pm} &= -\frac{\varepsilon^2 K}{2DkT} \nu_{+} z_{+}{}^2 - \frac{\varepsilon^2 K}{2DkT} \nu_{-} z_{-}{}^2 \\ \ln \gamma_{\pm} &= -\frac{\varepsilon^2 K}{2DkT} \left(\frac{\nu_{+} z_{+}{}^2 + \nu_{-} z_{-}{}^2}{\nu_{+} + \nu_{-}} \right) \\ \ln \gamma_{\pm} &= -\frac{\varepsilon^2 K}{DkT} |z_{+} z_{-}| \end{aligned} \tag{6-72}$$

But K itself is a function of ionic charges and concentrations:

$$K^2 = \frac{4\pi\varepsilon^2}{DkT} \sum_i N_i z_i{}^2$$

Replacing the concentration in ions/cm^3 by that in moles/liter, $N_i = c_i N/1000$, and recognizing that in dilute solutions the molarity (in moles/liter) is approximately equal to the product of the molarity (in moles/kg) and the pure solvent density ρ_0 (in kg/liter, or g/cm^3),

$$K^2 = \left(\frac{8\pi N^2 \varepsilon^2 \rho_0}{1000DRT} \right) \left(\frac{1}{2} \sum_i m_i z_i{}^2 \right) = B^2 I \tag{6-73}$$

The constant B is a function only of the solvent. The quantity I is the ionic strength, defined by

$$I = \frac{1}{2} \sum_i m_i z_i{}^2 \tag{6-74}$$

For water as the medium, $K = 0.3244 \times 10^8 I^{\frac{1}{2}}$ at 0°C, and $K = 0.3286 \times 10^8 I^{\frac{1}{2}}$ at 25°C. Then the Debye–Hückel limiting law for point ions becomes

$$\ln \gamma_{\pm} = -\frac{B\varepsilon^2}{2DkT}|z_+ z_-| I^{\frac{1}{2}} = -A|z_+ z_-| I^{\frac{1}{2}} \tag{6-75}$$

and in practical terms:

$$\begin{aligned} \log_{10} \gamma_{\pm} &= -0.4896 |z_+ z_-| I^{\frac{1}{2}} \text{ at } 0^\circ\text{C} \\ \log_{10} \gamma_{\pm} &= -0.509 |z_+ z_-| I^{\frac{1}{2}} \text{ at } 25^\circ\text{C} \end{aligned} \tag{6-76}$$

If the ions are assumed to be spheres of diameter a, then the Debye–Hückel equation is

$$\log_{10}\gamma_{\pm} = -\frac{A|z_+ z_-| I^{\frac{1}{2}}}{1 + Ka} \tag{6-77}$$

There is no *a priori* way of calculating the diameters of hydrated ions in solution, and they must be treated as empirical parameters. But it is at least satisfying that they fall in the range of 2.5–11 Å, and correlate well with the expected relative sizes of the actual ions (See Table 21-4 of Klotz).

How well does the Debye–Hückel model fit reality? After the care with which the theory has been expounded, you have a right to expect to be shown that it does fit, and such is the case in dilute solutions. The simple point-ion limiting law holds reasonably well for ionic strengths up to around 0.01 (0.01 molal KCl, or 0.0025 molal $CuSO_4$), and the spherical ion form (Eq. 6-77) holds to the neighborhood of 0.1 ionic strength. (See Figure 6-17.)

Equation 6-77 may seem like a trifling return on a considerable investment. Yet the fact remains that we have been able to account for the behavior of ionizing solutes using very simple physical assumptions, and to make this accounting quantitative under the conditions for which the assumptions are valid. There are no empirical adjustment constants in Eq. 6-75, and even the one in Eq. 6-77 is physically reasonable. Samuel Johnson once remarked that "A woman preaching is like a dog's walking on his hind legs. It is not done well; but you are surprised to find it done at all." The same might be said for the Debye–Hückel theory. The assumption that *all* deviation from ideality is to be ascribed to charge interactions is equivalent to saying that electrostatic forces have so long a range of influence that, long after neutral molecules would have reached the dilution necessary for ideal behavior, ions will still cause trouble. In this concentration range, the experimentally observed activity can be entirely accounted for by electrostatic drag between ions, and what might otherwise appear to be a device for making intractable equations come out properly, activity, acquires a real physical meaning.

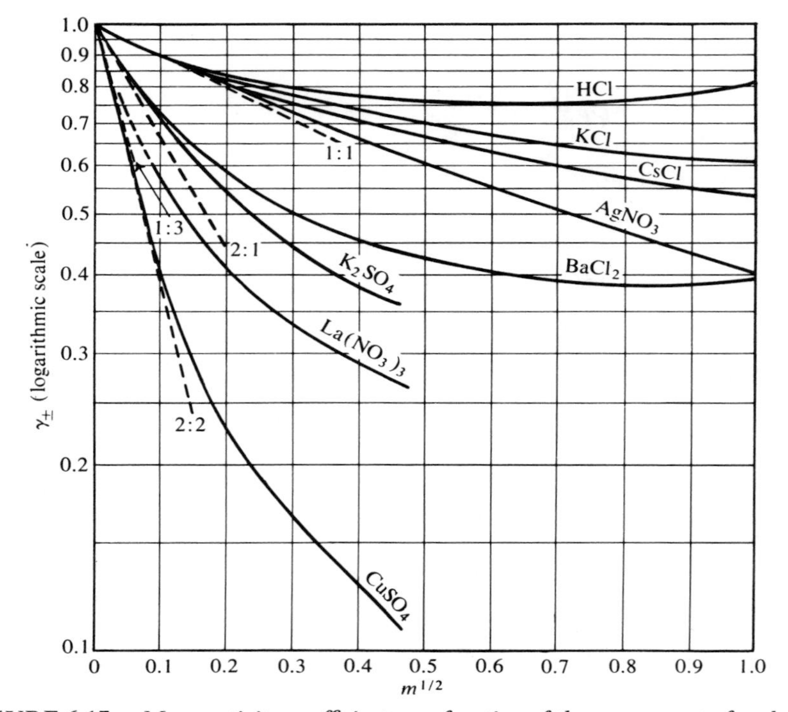

FIGURE 6-17. *Mean activity coefficient as a function of the square root of molarity for some typical electrolytes. Dashed lines show the behavior expected from the Debye–Hückel limiting law, which assumes point ions. The pairs of numbers (2 : 2, 2 : 1) indicate the number of cations and anions produced when the molecule dissociates.*

Distribution Coefficients

One useful consequence of Henry's law is the fact that, given a solute distributed between two immiscible solvents, such as iodine in benzene and water, the mole ratio of the solute in the two phases will be independent of the amount of solute present. At equilibrium, the fugacity in the two immiscible solvents, 1 and 2, must be the same; $f_A^1 = f_A^2$. Applying Henry's law for dilute solutions, $f_A^1 = k_A^1 X_A^1$ and the mole fraction ratio in the two solvents is the ratio of Henry's law constants. The distribution coefficient is independent of actual concentrations:

$$\frac{\alpha}{1-\alpha} = \frac{x_A^{\,1}}{x_A^{\,2}} = \frac{k_A^{\,2}}{k_A^{\,1}} \tag{6-78}$$

This, Nernst's distribution law, is the basis of a very powerful method of chemical separation called countercurrent distribution. The symbol α as used here is the fraction of the total moles of A to be found in solvent 1.

In countercurrent distribution, a solute is shaken and allowed to equilibrate between two immiscible phases, A and B. The two phases are then separated and fresh lots of A and B are added to their opposite partners. The solute

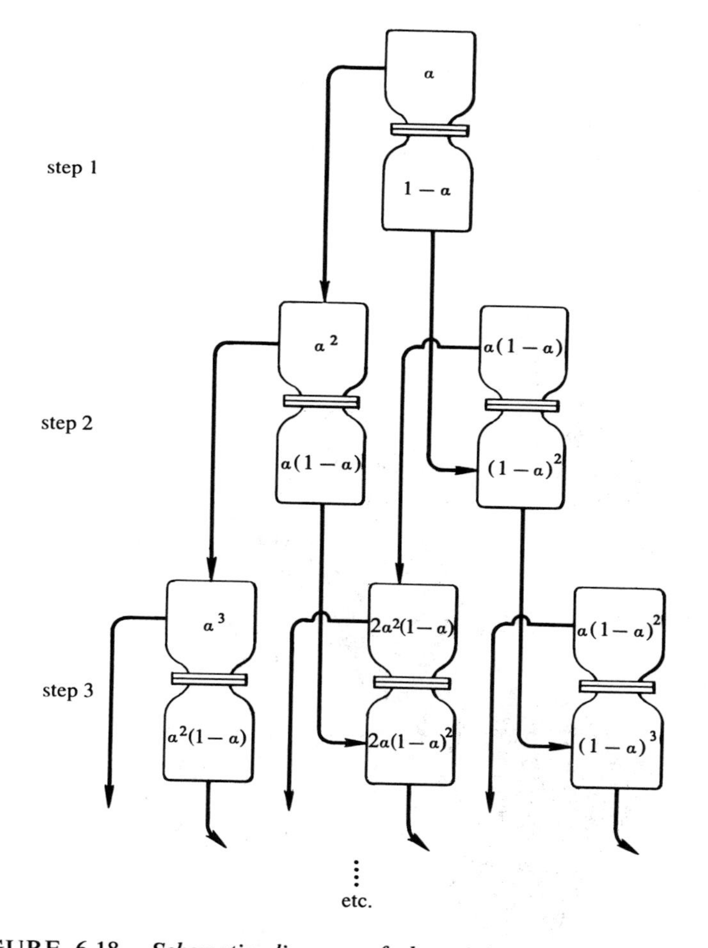

FIGURE 6-18. *Schematic diagram of the mixing processes during a countercurrent distribution experiment. At each step, after separation of immiscible solvents, the upper chamber is shifted one station to the left, and chambers with new solvents are added at each end of the chain. The distribution ratio between solvents* $\alpha/(1-\alpha)$ *equals the ratio of Henry's law constants.*

which had remained in solvent A is now redistributed between A and fresh B, and that which had gone into B is distributed between B and the new A. These two batches are allowed to settle, are separated, and the inner fractions are combined as shown in Figure 6-18 in a pyramiding fractionation scheme.

Countercurrent distribution machines with up to 2000 partition tubes have been built, connected in such a way that when the entire rack assembly is tipped forward, one of the component phases in each of the 2000 tubes flows automatically out of that tube and into the next neighboring tube. If a sample such as a mixture of two antibiotics is now introduced in the middle tube, and if the assembly is repeatedly shaken to attain equilibrium, allowed to rest while the solvents settle, and then tipped to make the transfer, with time the separation of even closely similar compounds is remarkably good. An example of a countercurrent distribution separation is shown in Figures 6-19 and 6-20.

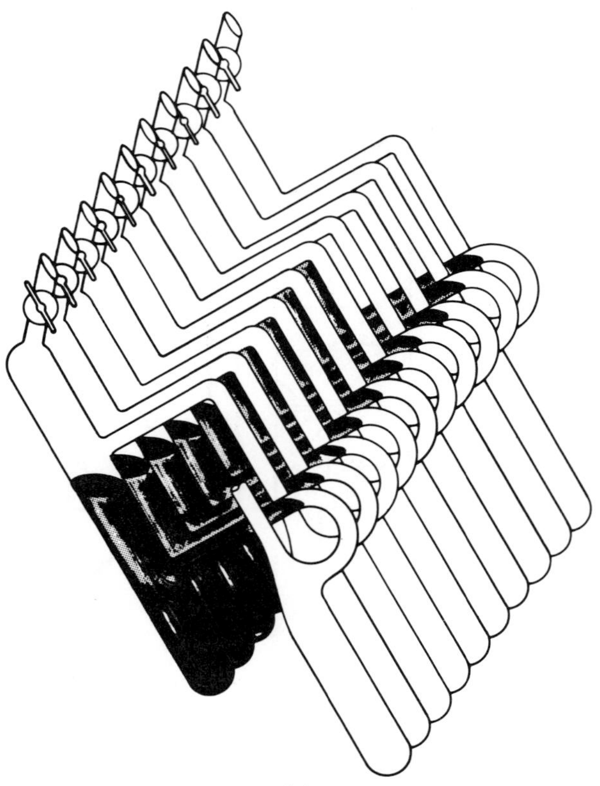

(a)

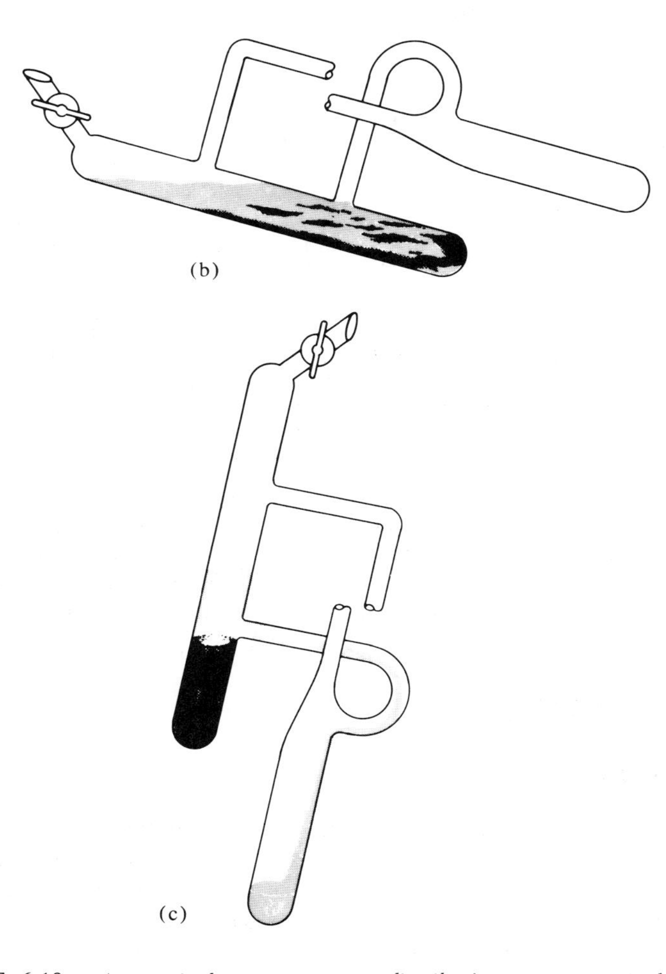

FIGURE 6-19. *A practical countercurrent distribution apparatus is built up of a bank from 10 to 2000 glass chambers arranged in parallel on a chassis as in* (a). *Two to four* cm^3 *of a heavier solvent and a similar amount of an immiscible lighter solvent are added to each tube, and the mixture to be separated is introduced at the right end of the bank of tubes or in the center. In operation, the entire chassis is tilted on its back* (b) *and agitated vigorously by a motor-driven cam. After a minute or so of mixing, the tubes are brought to the position of* (a) *and the immiscible phases are allowed to separate. When the rack is then tipped as in* (c), *all of the upper layers run off into the decant tubes. When the rack is tipped back to position* (b) *and beyond, the contents of each decant tube run into the next chamber to the left. Mixing is begun again and the cycle is repeated with fresh upper solvent being fed in at the right end. The entire process is carried out automatically with cams and timers, and produces the fractionation diagrammed in Figure 6-18, without the need for human manipulation.*

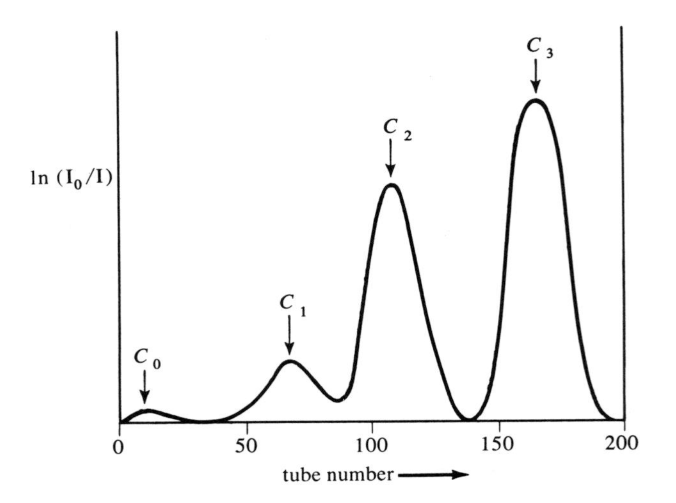

FIGURE 6-20. *The results of a countercurrent distribution separation of four closely related variants of the peptide antibiotic actinomycin* C. *Because of their slightly different distribution ratios in methyl butyl ether and* 1.75% *aqueous solution of sodium naphthaline-β-sulfonate, the four components are separated slowly into four regions on the apparatus. See H. Brockmann and H. Grone*, Chem. Ber. **87**, 1039 (1954).

6-4 PHASE RULE

The Basic Idea

In a system with a given number of components, the number of variables that can be altered independently without creating or destroying phases can be shown to be related to the number of phases present in the system. This relationship is known as the phase rule. Consider a general one-component phase diagram, like Figure 5-7. We saw in Section 5-7 that with only one phase present, both P and T could be varied independently without producing any phase changes. Two phases could coexist only at points along one of the lines linking P and T through the Clapeyron equation. Finally, *three* phases can coexist only at the triple point. Neither P nor T can be changed without first eliminating one phase. The empirical relationship between the number of phases p and the number of independent variables or degrees of freedom f is

$$f = 3 - p \qquad (5\text{-}21)$$

A two-component system is not as easy to diagram, since there are three independent variables: pressure, temperature, and concentration. The

Bi–Cd diagram of Figure 6-21 can be thought of as an isobaric slice through a full three-dimensional graph in P, T, and X_{Cd} at 1 atm pressure. The change of the diagram with pressure is extremely slight in this condensed-phase system, but it is there in principle.

At a P, T, and concentration represented by point a in the figure, P, T, and X_{Cd} can all be varied independently without altering the number of phases. But if the temperature is lowered to that of point b, then the solution is saturated in Bi, and crystallization of a second phase begins. As the temperature is lowered still further, and as more Bi comes out, the solution becomes richer in Cd and its concentration follows the solubility curve from b to e and finally to h. As long as there are two phases in equilibrium, there are only two degrees of freedom: P and *either* T or X. At 413°, the solution becomes saturated in Cd as well as Bi, and crystals of both metals begin to form. With three phases in equilibrium, the system is locked at the eutectic point h and has only pressure left as an independent variable. Any attempts to raise the temperature are frustrated by the utilization of the heat supplied to melt solid Bi and Cd, until all of the Cd is exhausted. Only then can the temperature rise, in what is now a two-phase situation. Any attempts to lower the temperature result only in hastening the solidification process of the liquid, and only after it has disappeared can the temperature of the two-phase mixture of crystalline Bi and Cd be cooled, along the line gj.

The empirical formula for the number of degrees of freedom in the two-component system is $f = 4 - p$. Comparing this to the one-component case, one might guess that the general phase relationship is $f = c + 2 - p$, where c is the number of components. This turns out to be correct, and is the usual expression for the phase rule.

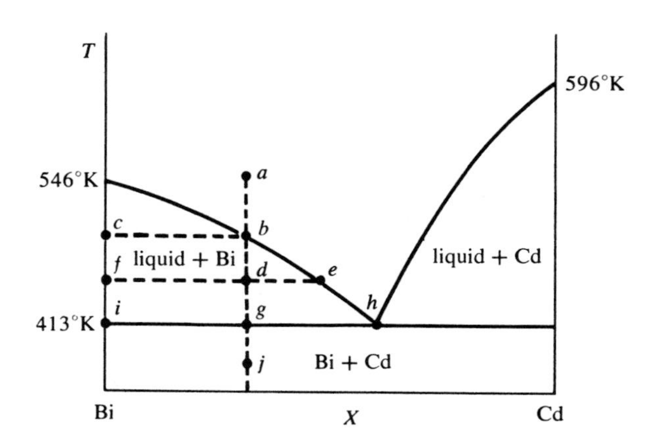

FIGURE 6-21. *Temperature-composition plot for* Bi *and* Cd. *For explanation of points (a)–(j), see text.*

Derivation of the Phase Rule

The number of degrees of freedom a system may possess is just the number of independent variables that it takes to describe the system. From a mathematical point of view, the number of independent variables is the total number of variables minus the number of restraining conditions. The expression for the total number of variables is easy. In each phase, the mole fraction of each component must be specified, so that for p phases and c components, there are pc mole-fraction variables. In addition, there are the two addition variables of temperature and pressure, giving a total of $pc + 2$ variables.

There are two basic types of restraining equations: the mole fractions in each phase sum to unity, and the chemical potential of each component must be the same in all phases. Thus there are p equations specifying that the mole-fraction sum in each phase is 1: $\sum_j X_j^\alpha = 1$. For each component, there are $p - 1$ equations establishing the equality of chemical potentials in all phases: $\mu_j^\alpha = \mu_j^\beta = \mu_j^\gamma = \cdots = \mu_j^p$. There are $c(p - 1)$ equations of this sort. The total number of restraining equations is therefore $p + pc - c$. Subtracting the number of equations of restraint from the number of total variables produces the number of *independent* variables:

$$f = pc + 2 - (p + pc - c) = c + 2 - p \tag{6-79}$$

[Parenthetically, this is of the same form as the expression relating the number of faces (f), edges (c), and vertices (p) in a polyhedron. Is this only coincidence?]

Number of Components

One important consideration when using the phase rule is to understand what is meant by the number of components in the system. The variable c is the number of chemically independent species necessary to describe the composition of *every* phase in the system. As in the derivation of the phase rule, the number of independent species is the total number of chemical species involved minus the number of restraining equations. In this case, the restraining equations turn out to be the equations of chemical equilibrium. To see what this means, consider the following examples.

In a system containing PCl_3, PCl_5, and Cl_2, there are three chemical species. They are related, however, by the equilibrium expression $PCl_5 \rightleftarrows PCl_3 + Cl_2$; given any two mole fractions, the third can be calculated from a knowledge of the equilibrium constant. In such a system, there are only two independent components. One could retain all three as components and add the equilibrium expression to the list of equations of restraint, but it is easier to use such equilibrium expressions to reduce the number of components. In the case of liquid water, it is known that water may associate with itself to form dimers, trimers, and large polymers. However, *each* of these associations is governed by an equilibrium equation, so that only one independent component is present. On the other hand, no known equilibrium exists between water and ethanol, so that a system with these two species would be a two-component system.

More than one phase may be involved in such considerations, as in the case of calcium carbonate and calcium oxide, both solids, with carbon dioxide, a gas. Even though each species is in its own phase, they are still linked by the equilibrium reaction $CaCO_3 \rightleftarrows CaO + CO_2$. Even though there are three separate phases involved, there are only two components in the system.

The case of water, ethanol, and ethylene is not as clear. There is a reaction linking the three; $C_2H_4 + H_2O \rightleftarrows C_2H_5OH$. However, the important point to be made is that equilibrium *must be attained* in order for the restraining equation to cancel out a variable. At room temperatures, this reaction proceeds extremely slowly, so that in most experiments equilibrium is never reached, and the dependence of the third species on the other two is never really established. Thus at room temperatures this is an effective three-component system. At elevated temperatures where equilibrium is more quickly achieved, the system behaves like a two-component system. Water offers another example of such a behavior; in the absence of a suitable catalyst, hydrogen and oxygen do not form water in a time comparable to that in which the experiment is conducted, nor does water break down to reach equilibrium with the gases. A system of hydrogen, oxygen, and water behaves like a three-component system, until a catalyst is added which permits equilibrium to be established rapidly.

6-5 TWO-COMPONENT SYSTEMS: LIQUID IMMISCIBILITY

The phase relationships in binary, or two-component systems, are easiest to illustrate and of the most general use. Aside from the intellectual challenge of plotting in four dimensions, three-component systems are the concern chiefly of the metallurgist. We shall look at phase equilibria involving two liquid phases, liquid and vapor, and liquid and solid. For two components there are three variables, P, T, and X_j, which can be plotted as the three spatial axes. Whenever a two-dimensional plot is given, its place as one slice through the full three-dimensional graph should be kept in mind.

In any two-component system, the phase rule is

$$f = 4 - p \tag{6-80}$$

In an ideal solution of two liquids, there is complete miscibility at all concentrations and never more than one phase (see Section 6-1). But in nonideal solutions, the positive deviation may be so great that the liquid separates into two phases: a solution of component A saturated in B, and another solution of component B saturated in A. It may not be possible to form one phase of intermediate composition.

An example of such behavior is the aniline–hexane system (Figure 6-22). At 25°C, only 10 mole % hexane can be added to aniline before the saturation point is reached and further hexane separates out as a second phase. If such a mixture is heated, more hexane will dissolve until finally the two liquids become completely miscible in all proportions at 60°C. This figure shows not only the compositions of the two immiscible solutions, but their relative mole amounts as well. Suppose that a solution of over-all composition X_c is brought to a temperature T_c. The composition of one of the two phases will be X_a and the other X_b, at the ends of a horizontal *tie line*. Moreover, the ratio of the total number of moles of the two solutions is given by the ratio of the *opposite* segments of the tie line:

$$\frac{n_a}{n_b} = \frac{CB}{AC} = \frac{X_b - X_c}{X_c - X_a} \tag{6-81}$$

This is the *lever principle*, and its proof is straightforward. Let n_a be the total number of moles in the phase of composition X_a, and n_b be that of phase X_b.

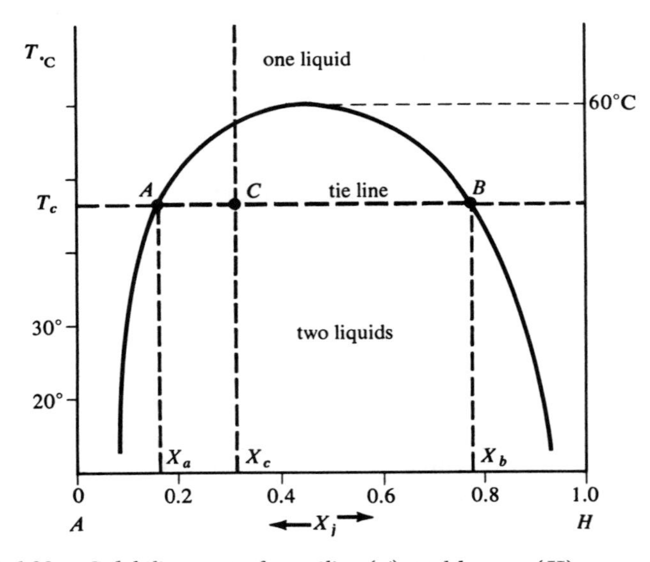

FIGURE 6-22. *Solubility curve for aniline* (A) *and hexane* (H) *at moderate pressures. Below* 60°C, *two immiscible solution phases exist, of compositions given by the end points of a horizontal tie line at the temperature concerned. At* 60°C *the compositions of the two phases coincide and only one phase exists. This is called the* upper consolute temperature.

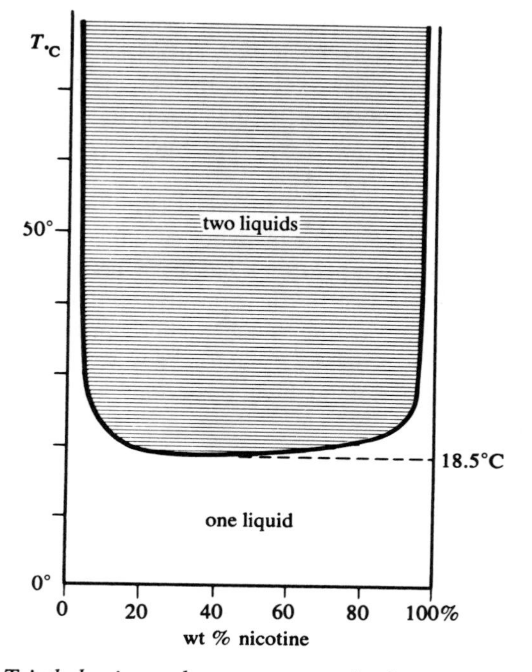

FIGURE 6-23. *Triethylamine and water are completely miscible* below *18.5°C, but separate into two immiscible solutions above this* lower consolute temperature. *The two-phase region is shaded with horizontal tie lines. What happens to the lever principle on a plot such as this of weight percent instead of mole fraction?*

Let n_h be the total number of moles of hexane. The total amount of hexane will be the sum of that in the two phases:

$$n_h = n_a X_a + n_b X_b = (n_a + n_b)X_c \tag{6-82}$$

from which Eq. 6-81 follows at once. The lever principle is of general validity in the phase diagrams that we shall encounter.

Why do the two phases become miscible at 60°? Hexane and aniline associate more readily with themselves than with each other. But as the temperature is raised, the kinetic energy of the molecules overcomes more and more of the energy barrier for association, and the solubility of each component in the other increases. At the so-called "upper consolute temperature" the saturation concentrations of the two phases become equal and the two phases become one. This behavior in a sense is like that of pure liquid and gas phases as the critical point is approached.

Triethylamine and water behave quite differently (Figure 6-23). At low temperatures they show negative deviation from ideality; the triethylamine–

water interaction is stronger than like-molecule interactions. Mixing is exothermic. But in turn, by Le Chatelier's principle, increasing the temperature should favor disassociation and the breakup of unlike-molecule associations. Above 18.5°C, this thermal disaggregation has proceeded to the point that the two components are no longer miscible in all proportions, and two phases can be obtained.

Nicotine and water show both phenomena (Figure 6-24), an upper and a lower consolute temperature. The tendency of unlike molecules to associate in all proportions at low temperatures has been broken up at 61°C, and a two-phase region begins. But by the time the temperature reaches 208°C, complete miscibility begins again. Le Chatelier's principle would lead one to predict that the mixing process for nicotine and water is exothermic at low temperatures and endothermic at high. Nicotine and water show negative deviation from ideality at low temperatures, and positive deviation at high.

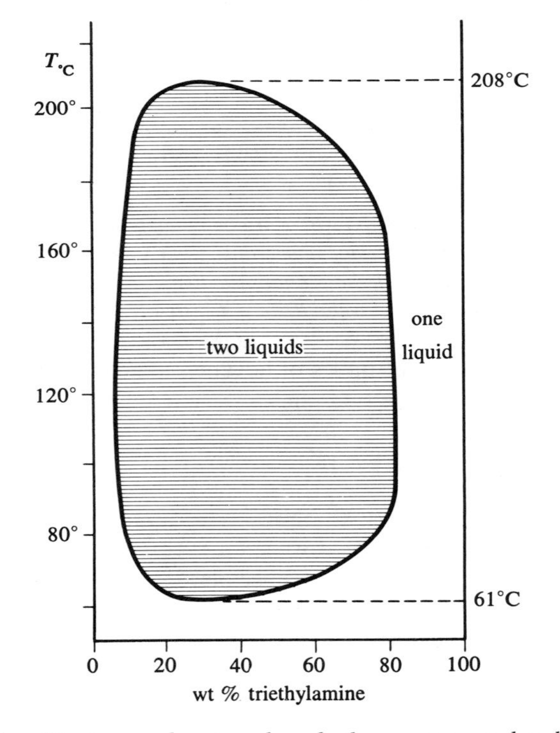

FIGURE 6-24. *Nicotine and water show both an upper and a lower consolute temperature, with limited miscibility at intermediate temperatures. These data were collected from mixtures sealed in capillaries, so that pressure in fact rises with temperature. For the true pressure when the upper consolute point was measured, see Exercise 6-14.*

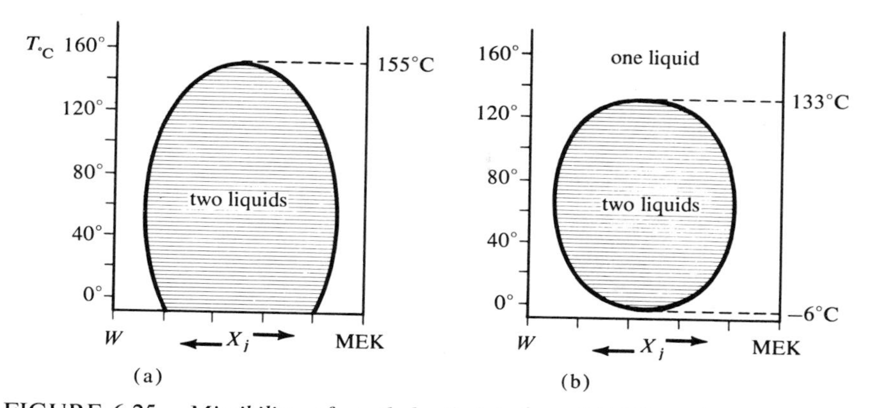

FIGURE 6-25. *Miscibility of methyl ethyl ketone* (MEK) *and water* (W) *at moderate pressures* (a) *and at* 150 atm *pressure* (b). *The vapor pressure of water at the upper consolute temperature of* 155°C *is approximately* 5 atm, *and the vapor pressure becomes* 150 atm *at* 343°C.

The effect of pressure on the methyl ethyl ketone–water system is shown in Figure 6-25. This pair of compounds, like water and nicotine, shows negative deviation at low temperatures and positive at high. But in both ranges there is shrinkage on mixing. An increase in pressure, accordingly, favors the association process and increases the mutual solubilities. The size of the two-phase region shrinks as pressure rises, until at a sufficiently high pressure it disappears and the liquids are soluble in all proportions. The full phase diagram for this type of behavior is shown in Figure 6-26.

6-6 *TWO-COMPONENT SYSTEMS: LIQUID–VAPOR EQUILIBRIUM*

The coordinates of the three-dimensional liquid–vapor phase diagram which we shall build up are shown in Figure 6-27. The two PT plots which form the end faces have already been encountered. They are just the liquid–vapor pressure plots of Chapter 5, calculated from the Clausius–Clapeyron equation.

Now consider an isothermal slice, or a PX plot at constant T. The vapor composition will be represented as $X_A{}^g$ and $X_B{}^g$, and the composition of the liquid by $X_A{}^l$ and $X_B{}^l$. From Raoult's law, for ideal solutions,

$$P_T = X_A{}^l P_A{}^\bullet + X_B{}^l P_B{}^\bullet = P_B{}^\bullet + (P_A{}^\bullet - P_B{}^\bullet) X_A{}^l \tag{6-83}$$

This is the straight line in Figure 6-28, and has been seen before in Figure 6-1. The vapor will be richer in the more volatile component than in the liquid, The partial pressure of a component in the vapor is proportional to the mole fraction *in the liquid*, by Raoult's law. But it is also related to the gas phase

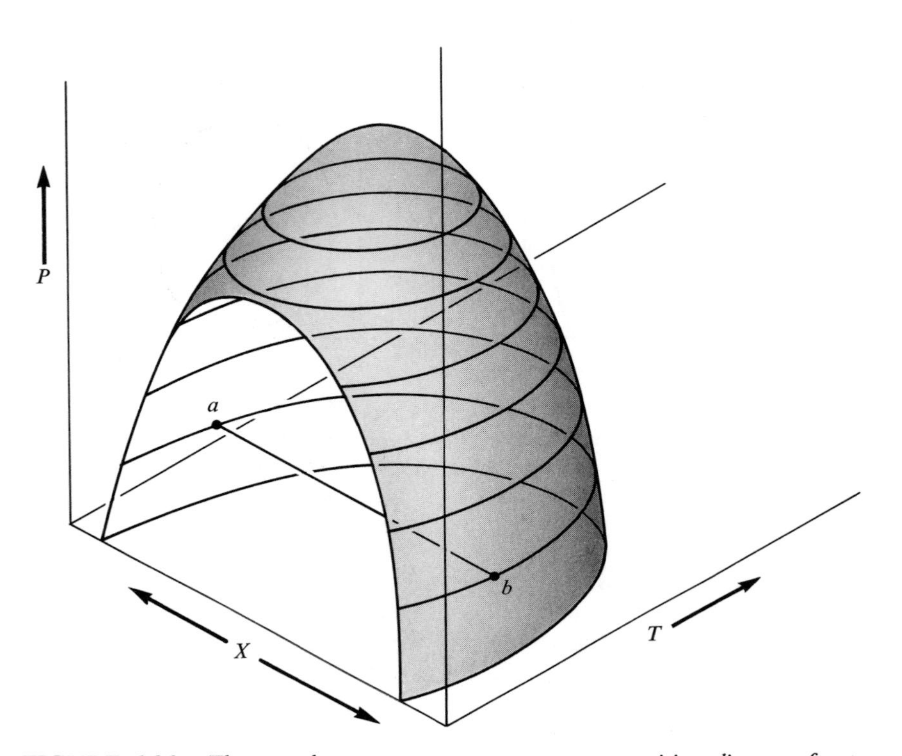

FIGURE 6-26. *The complete pressure–temperature–composition diagram for two partially miscible liquids. Figures 6-22 to 6-25 are sections normal to the P axis. The ends of the tie line ab give the compositions of the two immiscible phases which will be present for any P, T, and X_j point falling on the tie line. The relative molar amounts of the two phases will be given by the position of the PTX point on the tie line, in accordance with the lever principle.* (X_j *is the* over-all *composition of both phases.*)

composition by Dalton's law:

$$P_A = X_A{}^l P_A{}^\bullet = X_A{}^g P_T \tag{6-84}$$

From the previous two equations, the relationship between total pressure and *gas phase* composition can be found:

$$X_A{}^l = \frac{P_T - P_B{}^\bullet}{P_A{}^\bullet - P_B{}^\bullet} \tag{6-85}$$

$$X_A{}^g = \left(\frac{P_A{}^\bullet}{P_T}\right)\left(\frac{P_T - P_B{}^\bullet}{P_A{}^\bullet - P_B{}^\bullet}\right) \tag{6-86}$$

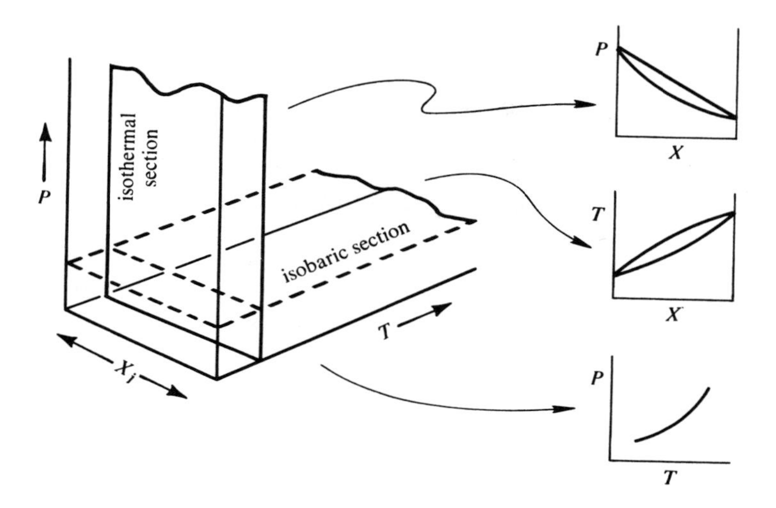

FIGURE 6-27. *The pressure, temperature, and over-all composition axes of a full three-dimensional two-component phase diagram. Isothermal, isobaric, and pure-component sections are shown to the right.*

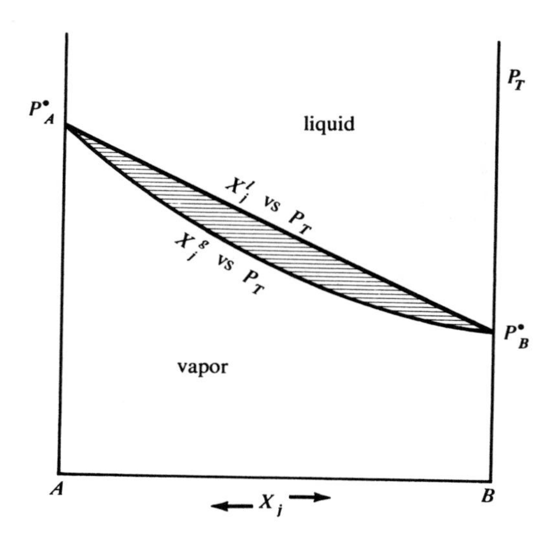

FIGURE 6-28. *The composition of liquid (X_j^l) and gas (X_j^g) phases in an ideal solution as a function of total pressure P_T. The two-phase region is shaded with horizontal tie lines.*

This latter expression is the curved line in Figure 6-28. The region above the liquid line is a one-phase liquid region, and that below the vapor curve is a one-phase vapor zone, in both of which pressure and composition can be varied independently without creating a second phase. But any point between the curves represents a pressure and composition for which two phases, solution and vapor, will be present. The compositions of these phases are given at the ends of a tie line parallel to the X_j axis, and the relative molar amounts of liquid and gas phase are given by the lever principle.

Benzene and toluene form a nearly ideal solution, and their vapor pressure diagram is shown in Figure 6-29. If a liquid of the composition of line *a-b-h* is subjected to the pressure at point *a* in a closed cylinder, then no vapor will be present. When the pressure falls to that at point *b*, a vapor phase will begin to appear, richer in benzene than the liquid (composition *c*). As the more volatile benzene predominates in the vapor, the liquid composition will fall along line *b-d-f* while the over-all gas phase composition goes from *c* to *e* to *g*. The last drop of liquid will have composition *f*, and below the pressure of tie line *g-f* only the gas phase is present. If a liquid of composition *a-h* is exposed to an external atmosphere, then it will have in equilibrium with it a vapor of partial pressure equal to that of the line *c-b*.

The phase rule is obeyed. For two components, $f = 4 - p$. But if variation is restricted to this isothermal plane then there is one less degree of freedom, or $f' = 3 - p$. When one phase is present, there are two degrees of freedom in this plane, P and X_j. But as long as both gas and liquid

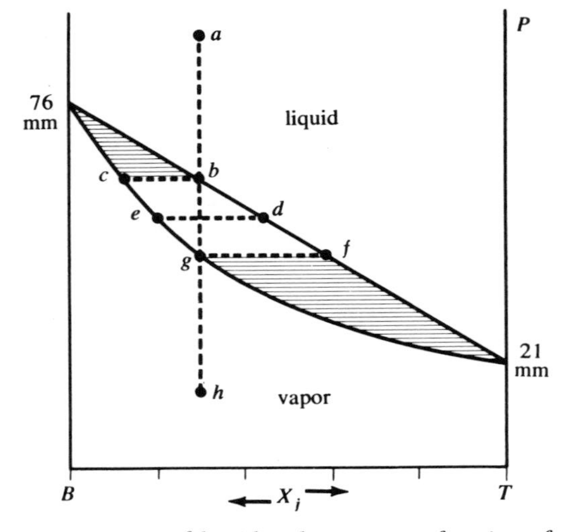

FIGURE 6-29. *Composition of liquid and vapor as a function of pressure at* 20°C *for benzene* (*B*) *and toluene* (*T*), *which form a nearly ideal solution. The straight liquid-composition line has been seen before in Figure 6-1, but at* 25°C.

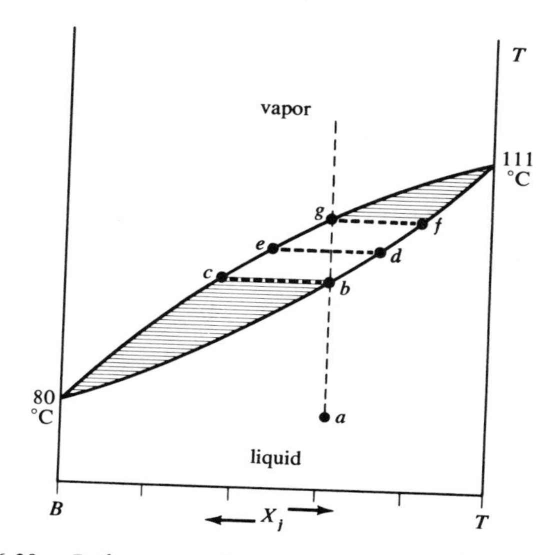

FIGURE 6-30. *Boiling point diagram for benzene (B) and toluene (T) at* 1 atm *pressure. At any temperature, the upper curve gives the composition of the vapor and the lower curve the composition of the liquid with which it is in equilibrium.*

coexist, there is only one degree of freedom, for the compositions of both phases are specified once P is given.

It is also possible from Eqs. 6-85 and 6-86 to calculate a TX plot of liquid and vapor composition versus temperature for a fixed total pressure. The vapor pressures of the pure components are known functions of temperature, $P^{\bullet}_{A(T)}$ and $P^{\bullet}_{B(T)}$, and can be varied while the total pressure is held constant. The resulting boiling point diagram is shown in Figure 6-30. The component with the higher vapor pressure, benzene, is naturally the one with the lower boiling point. If a liquid of composition *a-b-g* is heated in a closed piston at a constant external pressure, it will not show a vapor phase until the temperature rises to that of point *b*. The first vapor will have the composition of point *c*, and as the temperature is raised still higher (keeping the pressure constant), the vapor composition will rise from *c* to *e* to *g* while that of the liquid goes from *b* to *d* to *f*. At this point the composition of vapor equals that of the entire system and the last drop of liquid disappears. A liquid exposed to the atmosphere at the temperature of point *a* will have a vapor pressure less than the external atmospheric pressure, and point *b*, the boiling point, is that temperature at which the vapor pressure becomes equal to the atmospheric pressure. The lever principle again holds in calculating the relative molar amounts of liquid and vapor.

The full three-dimensional phase plot can be built up by studying a series of vapor pressure plots at different temperatures (Figure 6-31) and boiling

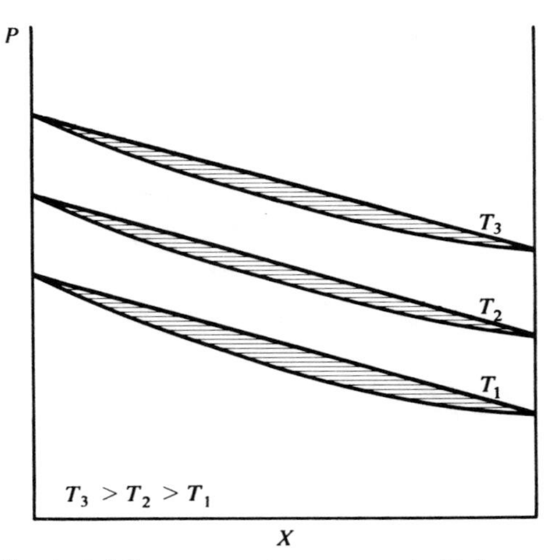

FIGURE 6-31. *The rise of the vapor pressure curves to higher pressures as temperature is increased.*

point plots at different pressures (Figure 6-32). As the temperature is raised, the vapor pressure is increased and the vapor pressure curves rise. As the pressure is increased, on the other hand, it becomes more difficult for molecules to escape to the gas phase, and boiling occurs only at higher temperatures. The complete diagram is shown in Figure 6-33. It consists of two surfaces: the upper giving liquid phase composition and the lower, gas

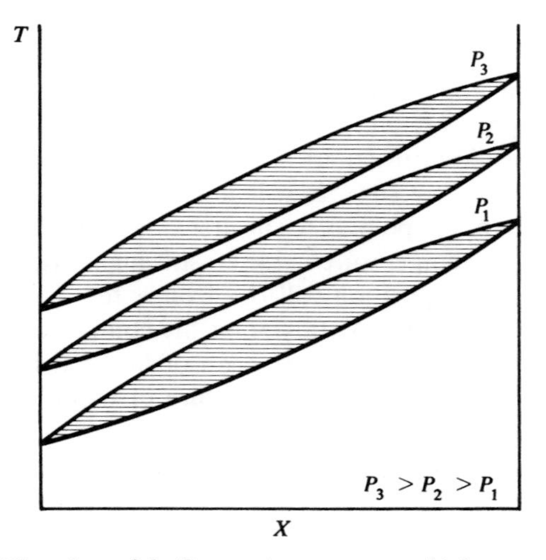

FIGURE 6-32. *The rise of boiling point curves to higher temperatures as the pressure is increased.*

phase. The regions above and below this pair of surfaces are one-phase zones with three degrees of freedom. But any point at a (P, T, X_j) *between* these surfaces represents a two-phase situation with gas and liquid compositions given by the intersections with the two surfaces of a line through this point and parallel to the X_j axis. The relative amounts of the two phases are inversely proportional to the distances from the over-all composition point to the intersections with the surfaces.

Two ideal liquids of different boiling points can be separated completely by the process of fractional distillation, illustrated in Figure 6-34. The process is simple, if tedious: if the initial mixture has a composition a, heat it to the boiling point at e and continue to heat it until the tie line is balanced (b-c) and there are equal molar amounts of liquid and vapor. Separate the vapor and

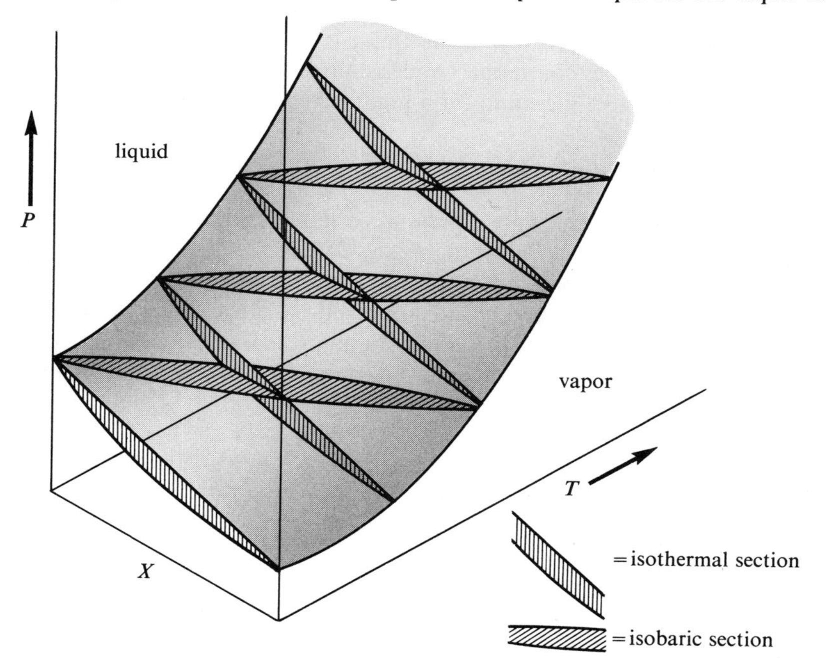

FIGURE 6-33. *The complete pressure–temperature–composition phase diagram for vapor and an ideal two-component solution. The upper surface represents solution compositions, and the lower, vapor compositions in two-phase situations. At any over-all composition point* between *these surfaces, the compositions of vapor and liquid are found by extending a line through this point and parallel to the X axis until it touches the surfaces. The relative amounts of liquid and vapor are found from this tie line by the lever principle. At any P, T, and over-all X point which does not lie between these two surfaces, there will be only one phase present, liquid or vapor. Sections corresponding to Figures 6-28 through 6-32 are marked. Shading in these sections does* not *represent tie lines.*

condense it. There will now be equal amounts of two liquids, one of concentration b and one of c. Heat the first again until half the material is vapor of composition d and half is liquid at e, then separate and cool. Treat the other half similarly to produce one fraction of composition $f(=e)$ and another of g. Combine the two central fractions. The end result of this two-stage process is one quarter of the original with composition d, one half with e, and one quarter with g. Now heat all three, separate, combine adjacent equal fractions, and repeat. The beauty of this process is that it can be carried out automatically in a bubble-cap fractionating still of the type in Figure 6-35. Vapor from one level bubbles up through the liquid of the level above and condenses, while liquid overflows back to the lower level. If mixture is fed in at the center, then the more volatile component flows out at the top (if the column is tall enough) while the less volatile liquid collects at the bottom. A simpler way to accomplish the same thing is to pack a column with glass beads, for which the continual condensation of vapor from below and reevaporation acts like a column of a great many smaller stages.

Negative Deviation from Ideality

For negative deviation the vapor pressure is less than that expected for ideal solutions, and the complete vapor pressure-versus-composition plot is like that of acetone and chloroform shown in Figure 6-36. At the minimum in vapor pressure, the vapor and liquid compositions must be the same, and the two curves touch. The boiling point diagram shows a maximum (Figure 6-37). Any attempts to fractionate a mixture of acetone and chloroform of

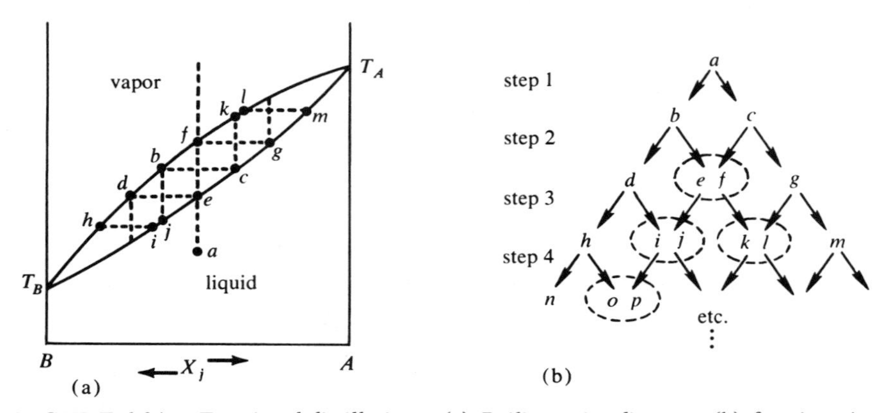

FIGURE 6-34. *Fractional distillation.* (a) *Boiling point diagram*; (b) *fractionation and recombination scheme. The liquid fraction from one sample and the vapor from the adjacent one (e and f, i and j, k and l, and so on) are of approximately the same composition and are recombined before the next heating step. This fractionation scheme is logically equivalent to the countercurrent distribution one in Figure 6-18.*

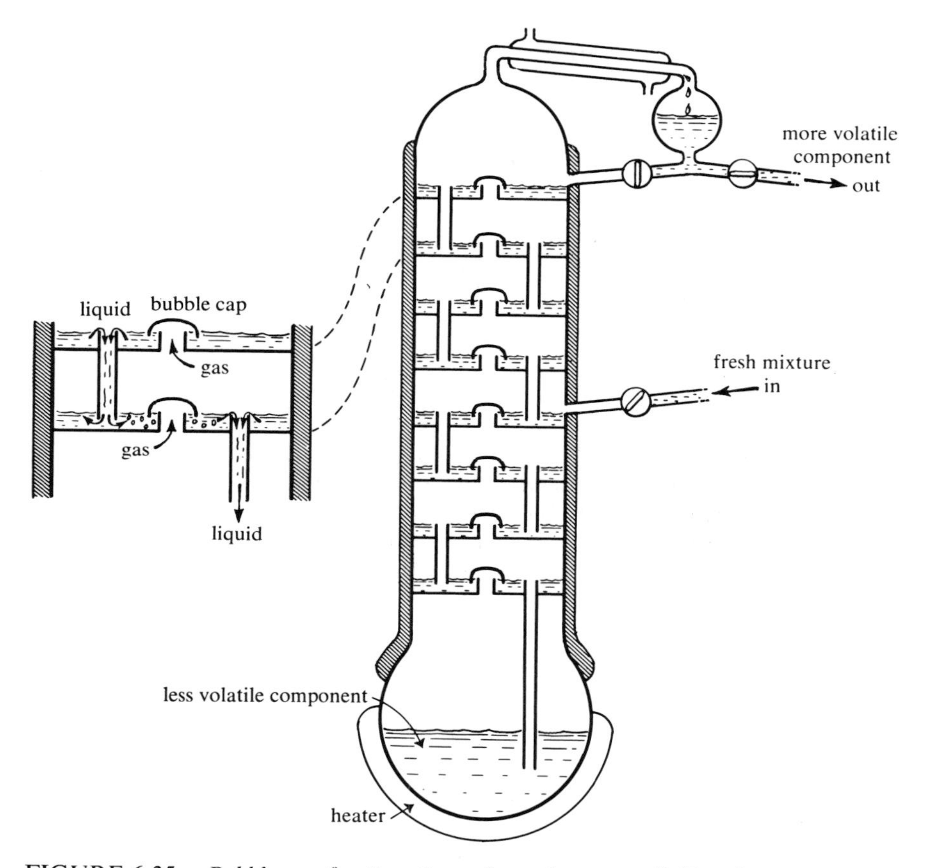

FIGURE 6-35. *Bubble-cap fractionating column for accomplishing the processes of Figure 6-34 automatically. Vapor from one stage equilibrates with liquid in the stage above, and excess liquid returns to the stage below.*

initial composition *a* by distillation will result in pure acetone and a constant boiling mixture, 0.65 mole fraction in chloroform, boiling at 64.5°C. Similarly, a starting mixture of composition *d* will yield only pure chloroform and the same "azeotropic" mixture (from "zein," to boil, and "atropos," unchanging).[4] Other systems which have such a maximum azeotrope are water and most of the inorganic acids, benzyl alcohol and *m*-cresol, and formic acid with either diethyl ketone or methyl *n*-propyl ketone. The three-dimensional *PTX* plot for negative deviation is like that for ideal solutions, but with a "trough" (Figure 6-38), along the bottom of which the two surfaces touch.

[4] This is the same Greek word that appears in "entropy," which was coined in the 1800's from "en-tropos" (involved in change), by analogy with "en-thalpein" (involved in heat) and the much older "en-ergon" (involved in work).

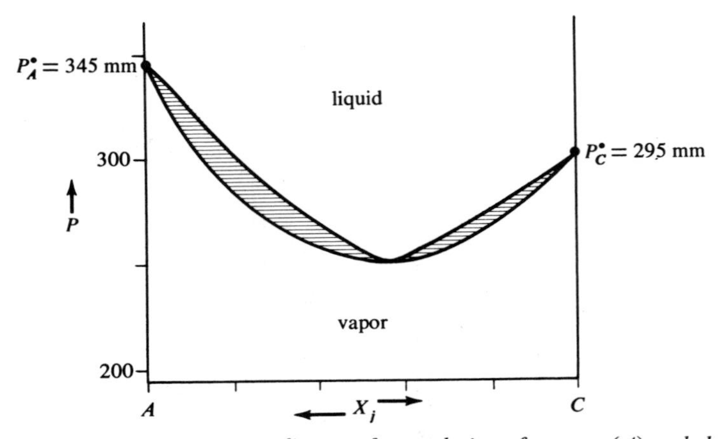

FIGURE 6-36. *Vapor pressure diagram for a solution of acetone* (A) *and chloroform* (C), *showing strong negative deviation from ideality. The upper, liquid composition curve has been seen already in Figure 6-12. Acetone and chloroform associate more strongly with one another than with themselves, and the escaping tendency of a molecule to the vapor phase is less for a solution than for a pure component. Two-phase regions are shaded with horizontal tie lines.*

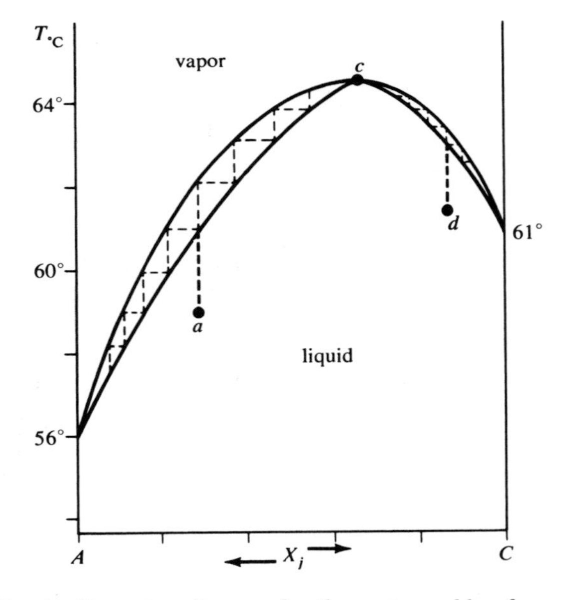

FIGURE 6-37. *Boiling point diagram for the acetone–chloroform system, showing negative deviation from ideality. Fractional distillation of a liquid of initial composition a will yield pure acetone in the distillate and an* azeotropic, *or constant boiling, mixture of composition c. If the initial concentration is that of point d, then the end result will be pure chloroform and the same azeotrope c.*

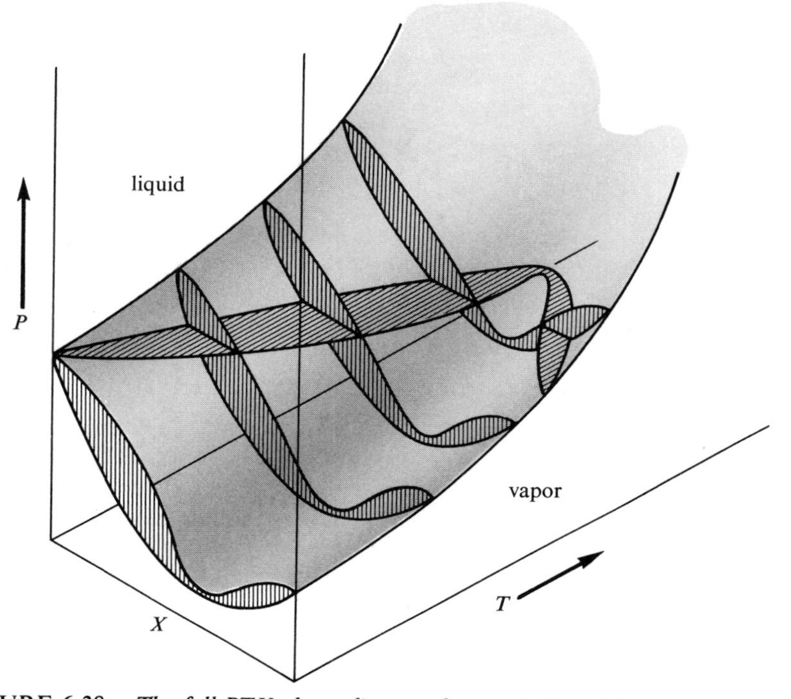

FIGURE 6-38. *The full PTX phase diagram for a solution with negative deviation from ideality, analogous to Figure 6-33 for ideal solutions. Isothermal and isobaric sections corresponding to Figures 6-36 and 6-37 are shown. Shading in these sections is for structural emphasis, and is normal to the tie lines. The dashed line follows the azeotropic composition points, where the liquid and vapor composition surfaces touch.*

Positive Deviation From Ideality

Positive deviation produces a vapor pressure curve with a maximum (Figure 6-39) and a boiling point diagram with a minimum boiling azeotrope (Figure 6-40). Ethanol, which normally boils at 78.4°C, also forms an azeotrope with water at 95.5% alcohol by weight, which boils at 78.1°C. It is impossible to prepare better than 95% ethanol starting from any more dilute solution of ethanol in water, and the common laboratory alcohol is this easily obtained azeotrope. A great many other liquid pairs form minimum boiling azeotropes because of molecular dissimilarity and reluctance to associate. These include water in combination with alcohols, hydrocarbons, ethers, esters, ketones, aldehydes, amines, and even organic acids. But alcohols themselves show negative deviation when mixed with hydrocarbons, esters, and alkyl halides.

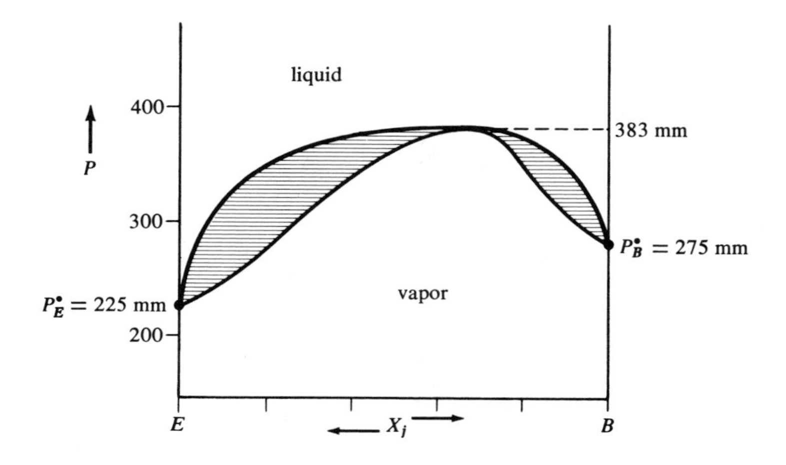

FIGURE 6-39. *Vapor pressure diagram of a solution of ethanol (E) and benzene (B), showing positive deviation. The upper, liquid composition curve is analogous to the acetone–carbon disulfide curve of Figure 6-13.*

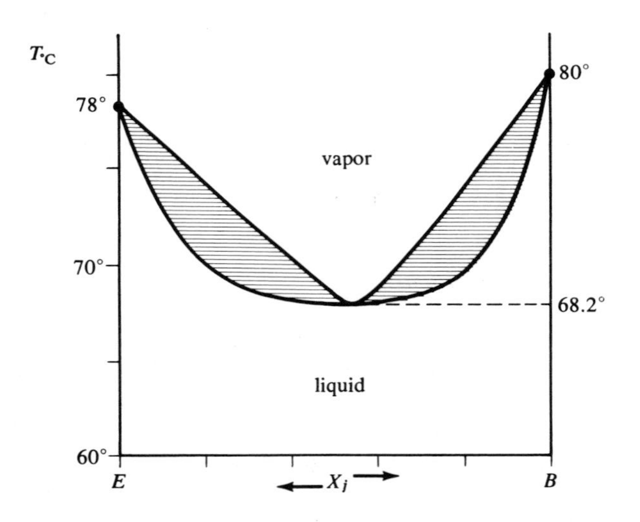

FIGURE 6-40. *Boiling point diagram at* 1 atm *for the ethanol–benzene system, showing a minimum-boiling azeotrope. Fractionation of a solution to the left of the azeotropic composition would yield pure ethanol and a distillate of the constant azeotropic composition. From this figure and Figure 6-39, try to imagine the full PTX diagram for positive deviation, analogous to Figures 6-33 and 6-38.*

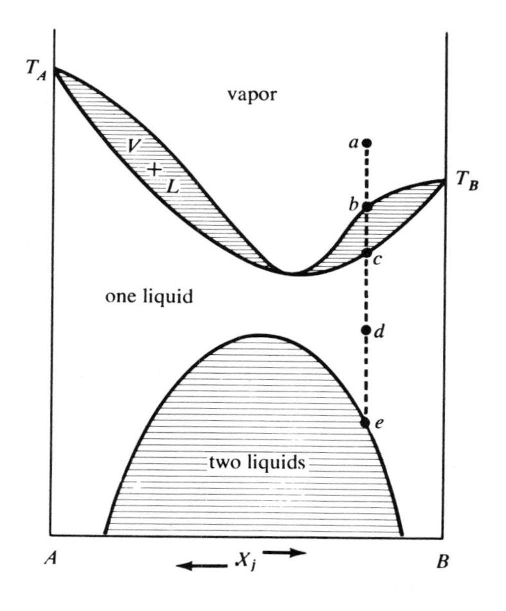

FIGURE 6-41. *Liquid–vapor phase diagram at constant pressure for a solution which shows both positive deviation from ideality and partial liquid miscibility at lower temperatures. Liquid–vapor and liquid–liquid two-phase regions are shaded with horizontal tie lines.*

If the positive deviation from ideality or reluctance of unlike molecules to associate is sufficiently great, then there may be immiscibility in the liquid phase like that of Figure 6-22, and an isobaric section may appear as in Figure 6-41. A vapor at *a* would begin to condense at *b* and form a two-phase system until *c*, at which point the last vapor would condense. There would be one liquid phase through *d* to *e*, where the temperature would have fallen so low that molecular motion was insufficient to keep the two components in association, and two immiscible phases would separate out. The effect of pressure on this diagram is shown in Figure 6-42. Decreased pressure lowers the boiling point curve and simultaneously encourages immiscibility, until just beyond Figure 6-42b the point is reached where the solutions remain immiscible all the way to the boiling point. In Figure 6-42c, vaporization occurs before the immiscible pair can be given enough thermal energy to make them mix. There will always be two liquid phases, α being almost pure A with a little B, and β the reverse.

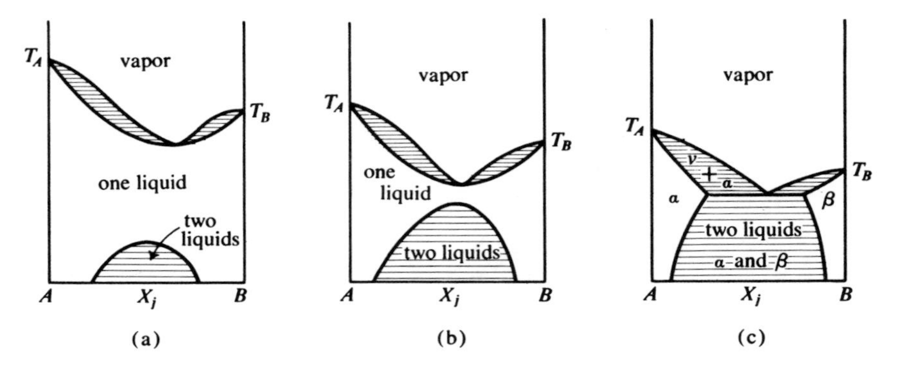

FIGURE 6-42. *The effect of pressure on the phase diagram of Figure 6-41:* (a) *high pressure,* (b) *intermediate pressure, and* (c) *low pressure.*

Water and butanol form such an extreme system (Figure 6-43). Much of this diagram should be obvious by now, but the three-phase point at h is new. As vapor at a is cooled, liquid condenses at b, and at temperature c, the vapor composition is at e and the liquid (phase α) at d. But at 92.4°C, a second liquid phase appears, immiscible with the first. At the instant that this second β phase first appears, the vapor is at composition h and the other α

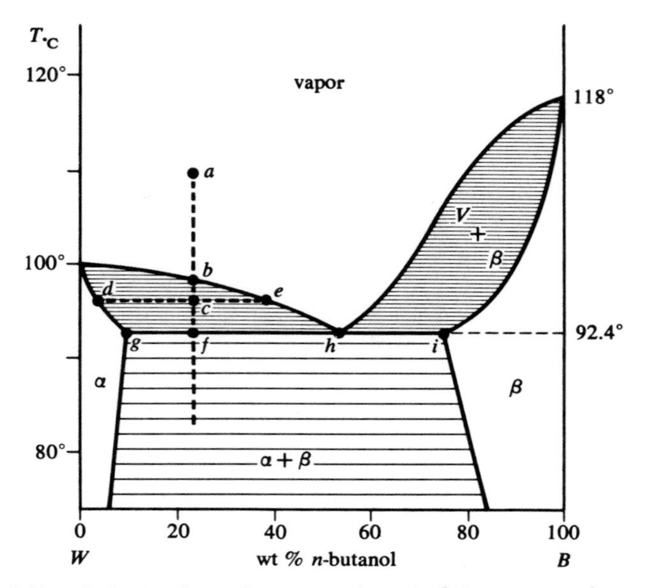

FIGURE 6-43. *Isobaric phase diagram at* 1 atm *of the system of water* (W) *and n-butanol* (B). *The immiscible solutions* α *and* β *vaporize before a high enough temperature can be reached to cause complete miscibility.*

phase at g, with the ratio of vapor to liquid being the ratio of distances gf to fh. (Note that since the composition axis is in weight percent and not mole fraction, the lever principle gives the weight ratio of the two phases and not the mole ratio. You should be able to prove this by analogy with Eq. 6-82.) As long as three phases are present and in equilibrium, the temperature cannot be lowered. Any withdrawal of heat results not in a temperature drop but in a condensation of more vapor, in the "reaction":

$$\text{vapor} = \alpha + \beta \tag{6-87}$$

This is consistent with the phase rule $f = 4 - p$, in that with three phases there is only one degree of freedom, and *none* within the isobaric section. As the last drop of vapor disappears, there remain two immiscible liquids for which the weight ratio of α to β is fi/gf. The butanol–water phase diagram at constant temperature looks like Figure 6-44. At very high pressures, the two immiscible liquids α and β can be made miscible and the solubility curve will close over as in Figure 6-22.

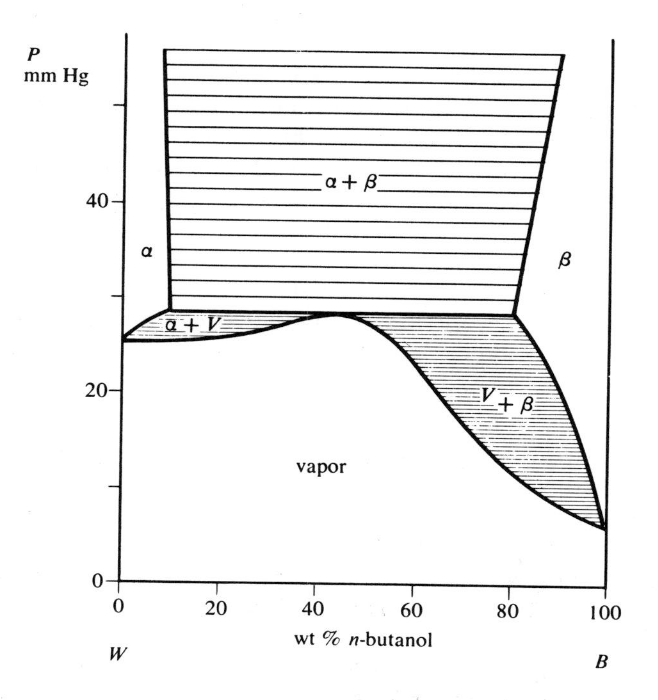

FIGURE 6-44. *Isothermal phase diagram at 25°C of the n-butanol–water system.*

Complete Vapor–Liquid Phase Diagram

All of the material in the last few sections can be summarized in a general three-dimensional *PTX* phase diagram for two components which show positive deviation from ideal solution behavior (Figure 6-45). The basic diagram shows two sets of surfaces. The first is the dome-shaped region of partial miscibility in the liquid phase (recall Figure 6-26). Outside this surface there is one liquid phase, $p = 1$ and $f = 4 - 1 = 3$. Here there are three degrees of freedom: P, T, and X_j. Within this surface there are two immiscible liquid phases whose composition is given by the termini of a tie line, α-β, and only two degrees of freedom exist. For once P and T are given, the compositions of the two phases are fixed.

The other surfaces in the diagram are the curved double-surface of liquid–vapor equilibrium analogous to Figures 6-33 and 6-38. A *PTX* point which falls *between* these surfaces marks a liquid–vapor two-phase situation. The compositions of the two phases are given by the intersections with the two surfaces of a tie line through the initial point and parallel to the X axis (see the line marked V-L in figure). The curves at which both surfaces come together at mole fractions of 0 and 1 (curves a_1-b_1-c_1-d_1-e_1 and a_2-b_2-c_2-d_2-e_2) are those obtained from the Clausius–Clapeyron equation for the respective pure components.

Five principal sections through the *PTX* plot are shown in Figure 6-46. The constant-pressure slice taken above the dome of immiscibility, (a), is analogous to the ethanol–benzene plot of Figure 6-40. The isobaric section taken below the point at which the immiscibility dome and liquid–vapor surfaces meet, (c), corresponds to the butanol–water system of Figure 6-43. The higher temperature isothermal section, (d), again is analogous to the ethanol–benzene diagram of Figure 6-39, and that at a lower temperature, (e), is like the butanol–water plot of Figure 6-44.

A similar *PTX* diagram for negative deviation has been given in Figure 6-38, but without the dome-shaped immiscibility feature it is much less interesting.

Liquid Immiscibility Without Strong Positive Deviations

It is possible to have limited miscibility in a solution in which the escaping tendencies of the components to the vapor differ only slightly from ideality. Then just as the butanol–water diagram, Figure 6-43, could be derived from Figures 6-22 and 6-40, so a combination of Figures 6-22 and 6-30 leads to Figure 6-47.

As a vapor of composition a is cooled, liquid phase β of composition c will begin to appear when the vapor is at point b. At the temperature of point d there will be vapor of composition e and solution β of composition f, in a

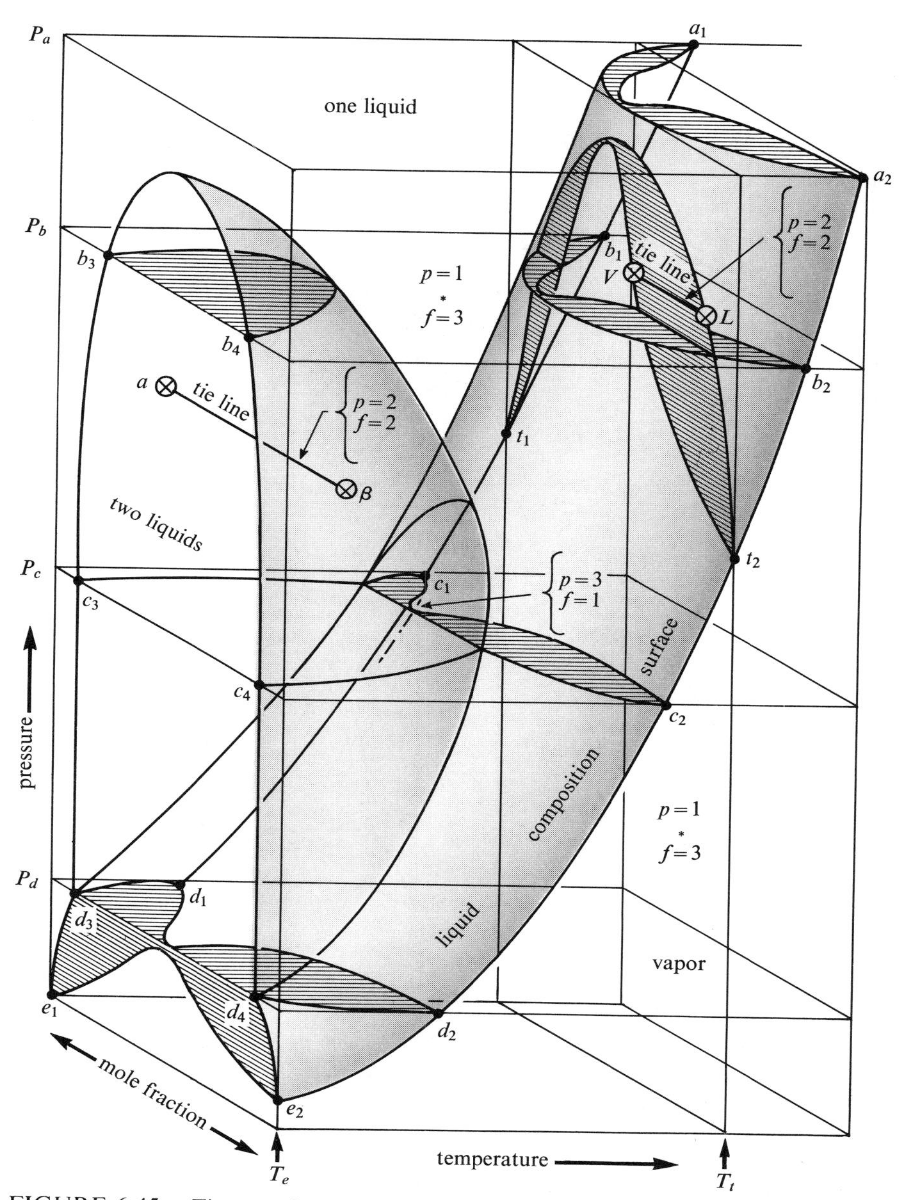

FIGURE 6-45. *The complete liquid–vapor phase diagram for solutions which show positive deviation from ideality. Figures 6-39 through 6-44 are sections through this PTX plot. The dome of liquid two-phase immiscibility at the left is also seen in Figure 6-26. The curved, convex double surface at the right is the analog for positive deviations of Figures 6-33 and 6-38. At the various points marked, p is the number of phases present and f is the number of degrees of freedom, from the phase rule for two components:* $f = 4 - p$. *The end faces of the plot, for* $X = 0$ *and* 1, *are the vapor pressure curves for the two pure components. Shading in the sections is for structural emphasis and does* not *represent tie lines.*

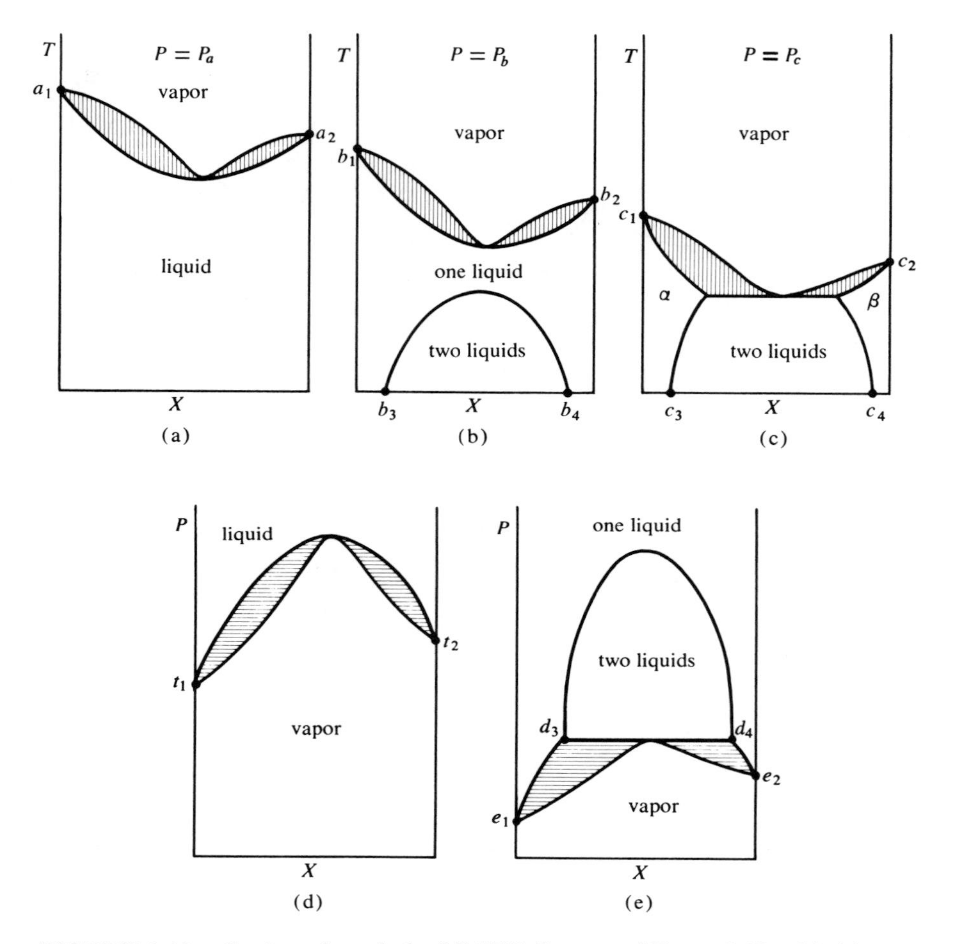

FIGURE 6-46. *Sections through the full PTX diagram of Figure 6-45.* (a)–(c) *are at constant pressure, and* (d) *and* (e) *are at constant temperature. Shading is for emphasis only, in conformity with Figure 6-45.*

vapor-to-liquid mole ratio of df/ed. Further removal of heat brings on the appearance of a second liquid phase, α. But unlike the butanol–water system, the new phase is not richer than the vapor in the minority component of phase β. Instead, it is intermediate in composition between vapor and β. The reaction which produces this new phase is not

$$\text{vapor} = \beta + \alpha \tag{6-88}$$

as in the butanol–water system, but is instead

$$\text{vapor} + \beta = \alpha \tag{6-89}$$

The new phase grows at the expense of *both* the old phases. If the butanol–water system is analogous to *eutectic* behavior in solid–liquid systems then this behavior is analogous to *peritectic* (see Figure 6-57). Only after all vapor has vanished can the temperature fall again for the two immiscible liquids.

Different behavior yet is observed starting from point h. Liquid phase β forms first at i, but at j phase β begins to dissociate according to the reaction

$$\beta + \text{vapor} = \alpha \tag{6-90}$$

until only vapor and α are left. At k the last vapor vanishes and only α remains.

Steam Distillation

In purifying organic compounds, it is often desirable to distill a substance which is volatile, but which decomposes before its normal boiling point. One method is to distill at reduced pressure, but an alternative method is steam distillation.

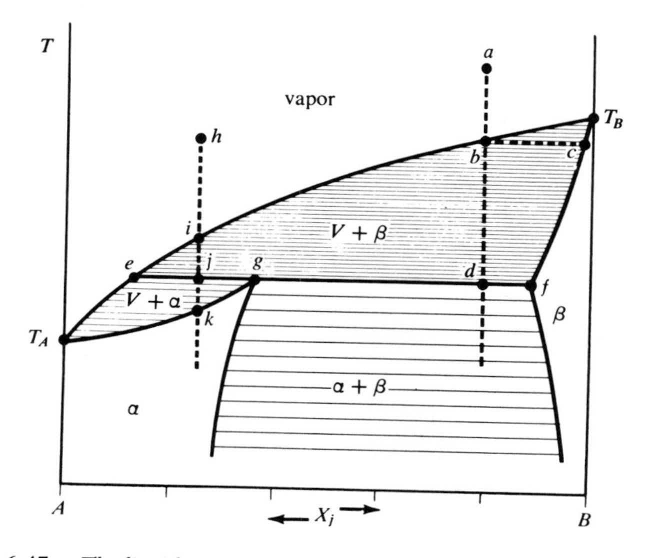

FIGURE 6-47. *The liquid–vapor analog of peritectic behavior in solid–liquid systems (Figure 6-57).*

The isobaric phase diagram for water and a typical organic liquid, nitrobenzene, is shown in Figure 6-48. Nitrobenzene and water form a constant boiling mixture of composition X_c at 99°C. If a steady supply of water as steam is provided, the temperature will remain at 99°C until all the nitrobenzene is used up. The mole ratio in the vapor will be that of the vapor pressures of the pure components at 99°C, which for water and nitrobenzene are 733 mm and 27 mm:

$$\frac{\text{moles water}}{\text{moles nitrobenzene}} = \frac{733}{27} = 27.1 \tag{6-91}$$

But with molecular weights of 18 and 123, the gram ratio of water to nitrobenzene is less:

$$\frac{\text{grams water}}{\text{grams nitrobenzene}} = \frac{733/18}{27/123} = 4 \tag{6-92}$$

Nitrobenzene can be distilled alone at 211°C, or can be distilled with water, with which it is immiscible, in a 1 to 4 weight ratio at 99°C. The ideal carrier

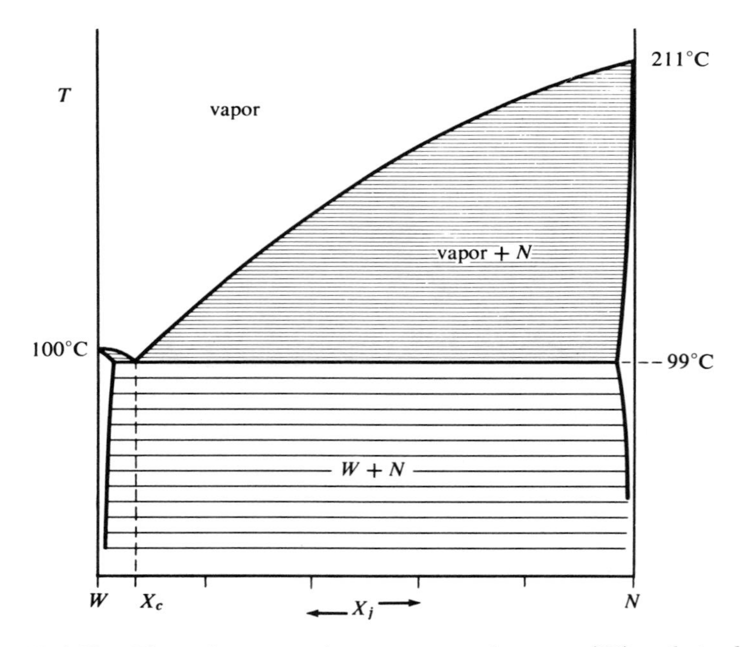

FIGURE 6-48. *Phase diagram at* 1 atm *pressure for water* (W) *and nitrobenzene* (N), *to illustrate the process of steam distillation.*

should be of low molecular weight and immiscible with the substance being distilled. Water serves admirably for many organic compounds.

Steam distillation can be used to determine rough molecular weights. As an example, terpinene steam distills at 95°C and 744 mm, in a distillate which is 55% terpinene by weight. At 95°C the vapor pressures of pure water and terpinene are 634 mm and 110 mm. The mole ratio is then

$$\frac{n_W}{n_T} = \frac{634}{110} = \frac{0.45/18}{0.55/M_T}$$

The approximate molecular weight of terpinene by this method is $M_T = 127$, to be compared with the true molecular weight of 136. Steam distillation can give a first order of magnitude figure for an unknown substance.

6-7 *TWO-COMPONENT SYSTEMS: LIQUID AND SOLID EQUILIBRIA*

Condensed phase diagrams are vitally important in studying and predicting the physical properties of solids, especially metals and alloys. But the most complicated such diagram can be broken down into a number of simpler types. What is more, all of these types have their analogs in the liquid–vapor diagrams which we have already seen. The chief difference is that changes in pressure have little effect upon equilibria between condensed phases, and all isobaric sections are virtually alike. The pattern of the solid state portion of Figure 6-45, if added to the left of the diagram, would extend up the P axis like the pattern in Christmas rock candy. In one isobaric slice, the phase rule for two components is $f' = 3 - p$.

Structure of Solid Solutions

How can we form a solution in the solid state? In a solid, the atoms or molecules are packed in a regular array to form a crystal, and an alien "solute" atom or molecule can enter only by fitting into the interstices in the original crystal lattice or by replacing an atom or molecule of "solvent." Interstitial solutions or alloys are possible only if the added atoms are small, and substitutional alloys are possible only if the solute and solvent are of similar size and crystallize in much the same way. Copper and zinc in roughly equimolar amounts crystallize in a body-centered cubic (bcc) lattice in which the ratio of Cu to Zn can vary over considerable limits. The other type, interstitial alloys, can be represented by steel, which is an interstitial solution of carbon in iron, and by the alloy Fe_4N. In this latter compound, the iron atoms occupy the four sites per cell of a face-centered cubic (fcc) lattice (Figure 6-50) while one nitrogen atom sits at the center of each cubic cell.

Body-centered, face-centered, and hexagonal close packing (hcp) are the three most efficient ways to pack structureless spheres, and are the most common crystal forms in metals and alloys. The most efficient packing of spheres in a plane is shown in Figure 6-49, where all neighboring spheres are touching, and the distance between centers d is the diameter of the sphere. A second equally well-packed layer can be fitted atop the first one so that its atom centers falls either at positions 2 or 3. If at 2, then a third layer can be placed at positions 3 or 1 with equal efficiency. If the sequence of layers is 1-2-1-2-1-2-1-2-, then the atoms are in hcp. If the stacking order is 1-2-3-1-2-3-1-2-3-1-2-3-, then the packing is called cubic close packing (ccp) or fcc. Both stacking sequences are illustrated in Figure 6-50, with the close-packed spheres replaced by much smaller ones for clarity. The numbers correspond to the layer numbering of Figure 6-49. Obviously many more layering schemes could be built. The metal americium, for example, crystallizes with a layer sequence 1-2-1-3-1-2-1-3-1-2-1-3-. But ccp and hcp are the most regular and by far the most common.

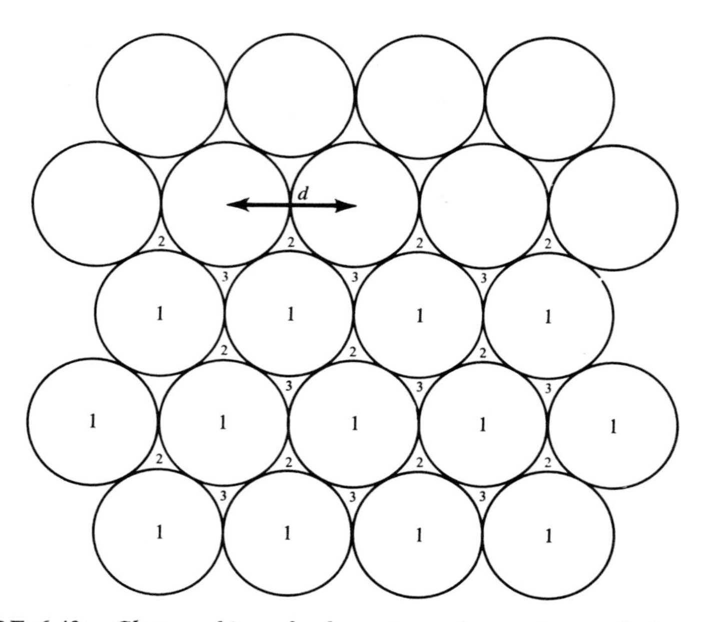

FIGURE 6-49. *Close packing of spheres in a plane. A second plane could be packed most closely to this one by having its sphere centers at the positions marked* 2 *or* 3. *If at* 2, *then the third plane could be centered at positions* 1 *like the first plane, or at* 3. *The first choice leads to hexagonal close packing* (*hcp*), *and the second to cubic close packing* (*ccp*).

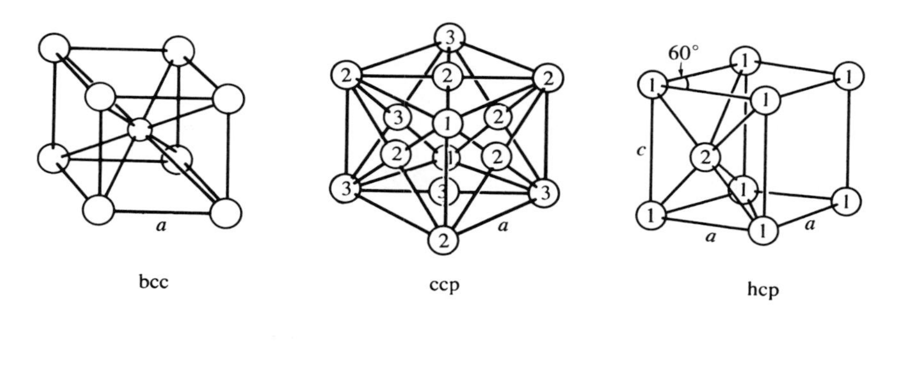

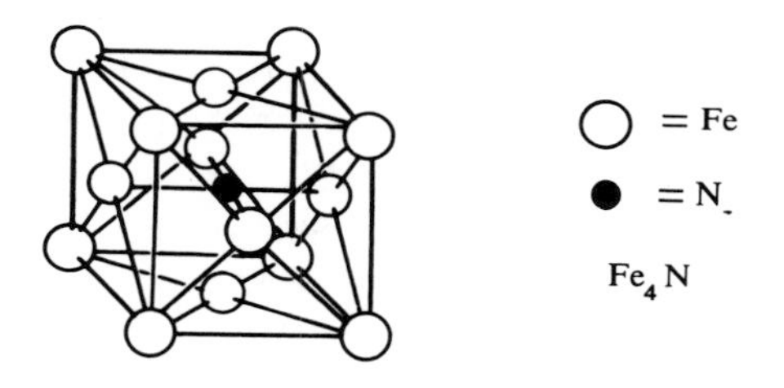

FIGURE 6-50. *The three most common crystal packing schemes in metals and alloys, and the atomic arrangement in* Fe_4N.

The basic unit cell edge in the ccp lattice, *a* in Figure 6-50, is $\sqrt{2}$ times the sphere contact diameter *d*, and there are four atoms per cell. (The atom count will come out right if you remember that the number of atoms in one kind of environment, such as the corners, must be divided by the number of adjacent unit cells with which this kind of atom is shared. Thus ccp has eight corner atoms, each shared between the eight cells which meet at a given corner, and six face atoms, each shared between two neighboring cells.) In the ideal hcp lattice, cell edge *a* equals the sphere contact diameter *d*, and the axial ratio of *c* to *a* is $2\sqrt{2}/\sqrt{3}$ or 1.633. There are two atoms per hcp cell. In both ccp and hcp, 26.0% of the solid is occupied by empty space between the packed atoms.

Body-centered cubic packing is only slightly more open, with 32% interstitial empty space. There are two atoms per cell, and the cell edge is $2/\sqrt{3}$ or 1.15 times the sphere contact diameter. In what is to follow on liquid–solid phase diagrams, changes in kind and dimensions of crystal packing are crucial in producing the observed phase behavior. Table 6-2 lists crystal packing data for some of the elements whose phase behavior we shall examine. Note that the packing energy differences are so small that Na can pack in either hcp or bcc, and Ni in either ccp or hcp. Although Cu when pure is ccp and

TABLE 6-2 CRYSTAL STRUCTURES OF SOME METALLIC ELEMENTS

ccp (four atoms/cell) (measurements are in angstroms)					
Element	Ni	Cu	Pd	Ag	La
Atomic radius	1.24	1.27	1.37	1.44	1.86
Cell dimension a	3.53	3.61	3.89	4.08	5.30
hcp (two atoms/cell)					
Element	Na	Ni	Zn	Cd	
Atomic radius	1.86	1.24	1.33	1.49	
a	3.77	2.65	2.66	2.98	
c	6.15	4.33	4.95	5.62	
bcc (two atoms/cell)					
Element	Na	K	Mo		
Atomic radius	1.86	2.31	1.36		
a	4.29	5.25	3.15		
Arsenic hexagonal cell (six atoms/cell)					
Element	As	Sb	Bi		
Atomic radius	1.26	1.44	1.55		
a	3.70	4.31	4.55		
c	10.55	11.27	11.86		

Zn is hcp, in the approximately equimolar β-brass alloy they share a bcc lattice. The hexagonal arsenic-type structure of Table 6-2 is a distorted packing which to a first approximation can be thought of as a packing of layers in the sequence 1-2-x-3-1-x-2-3-x-1-2-x-3-1-x-2-3-x-, where -x- indicates an unusually wide gap between layers and where the layers themselves are distorted from simple close packing.

"Ideal" Solid Solutions

If two substances crystallize in the same way when pure, and if their preferred crystal lattice dimensions are sufficiently close, then they can form a substitutional solid solution which satisfies the classical criterion for ideality—that all interatomic or intermolecular forces be *the same*. This is so for Cu and Ni (Figure 6-51), which when pure crystallize in ccp with cell dimensions 3.61 and 3.53 Å. Together they form a full range of ccp solid solutions with an ideal melting point diagram. This diagram should be compared to the benzene–toluene diagram of Figure 6-30. Au and Pt also form a nearly ideal solid solution, in the ccp lattice, and Sb and Bi do so in the hexagonal arsenic-type cell. It is possible to fractionate two such components by repeated melting and recrystallization, and the method is of some commercial importance.

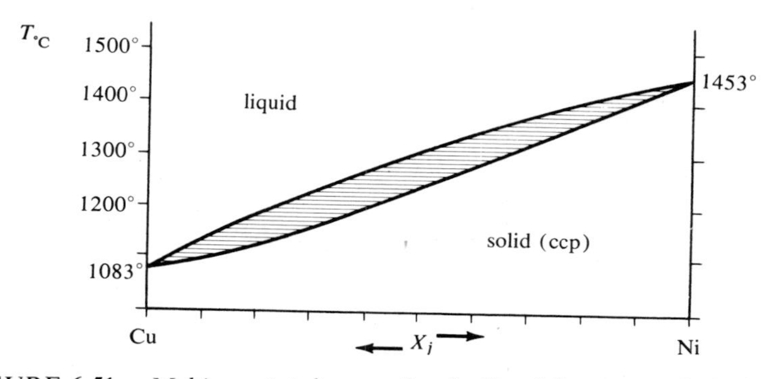

FIGURE 6-51. *Melting point diagram for the* Cu—Ni *system, which shows nearly ideal solid solution behavior. The lower curve is the solid composition and the upper curve is the liquid composition in equilibrium with it at a given temperature. The interpretation, including the use of the lever principle, is identical to that of Figure 6-30.* [*From M. Hansen,* The Constitution of Binary Alloys (*McGraw-Hill, New York, 1958*).]

If the atomic radii of two metals are sufficiently different, then it becomes difficult to accommodate one in the crystal lattice of the other, even though the lattices are of the same type. The ccp cell dimension of 3.89 Å for Pd is just enough greater than that of Ni so that too many Pd atoms stretch the Ni crystal lattice out of shape, and too many Ni atoms in Pd leave gaps which threaten the collapse of the lattice. There is therefore a greater escaping tendency of atoms into the melt from the strained lattice at intermediate compositions, and a lowering of the melting point. The melting point diagram, Figure 6-52, is a beautiful example of positive deviation, and should be compared with the ethanol–benzene boiling point diagram, Figure 6-40.

As and Sb form another example of positive deviation (Figure 6-53), and there have been reports of a region of immiscibility in the solid which would make the phase diagram resemble Figure 6-41 for the liquid–vapor system. Co and Cr, KNO_3 and $NaNO_3$ also form such solid solutions with positive deviation.

It is less easy to imagine a solid solution with negative deviation from ideality, or preferential *association* of unlike molecules. One example of this rare behavior is the mixture of *d*- and *l*-carvoxime, $C_{10}H_{14}NOH$. Each pure form melts at 72°C, but the 1 : 1 racemic mixture does not melt until 91°C (Figure 6-54). There are many examples known of association of unlike organic molecules in crystals, especially the stacking or pairing of ring structures, and the carvoxime behavior may really be quite common. But little is known about phase behavior of organic crystals, or will become known until organic semiconductors become as interesting to the industrial solid state physicists as metallic systems.

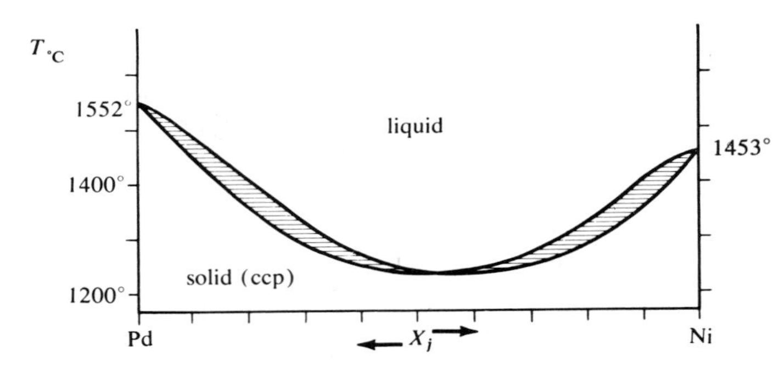

FIGURE 6-52. *Positive deviation from ideality in the* Pd–Ni *solid solution system.* (*From Hansen.*) *Compare Figure 6-40.*

Solid Phase Immiscibility

If the atomic or molecular dimensions of two substances which use the same lattice are too different, or if the substances prefer different lattices entirely, then the capacity of one to accept molecules of the other is finite, and there will be only a limited range of solubility. The Ag–Cu system (Figure 6-55)

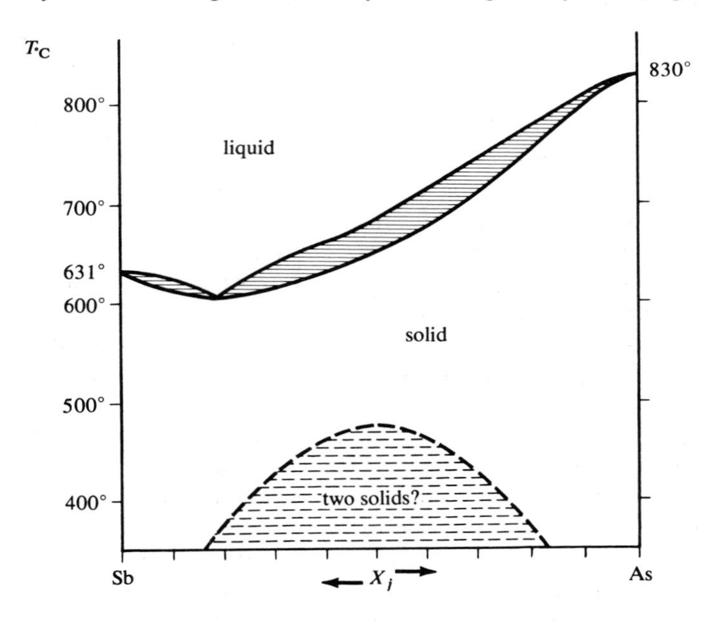

FIGURE 6-53. *Positive deviation in the* Sb–As *system, with a suggested region of two-phase solid immiscibility. The right half of the diagram, near the pure* As *ordinate, was measured at elevated pressures to keep the arsenic from subliming.* (*From Hansen.*) *Compare Figure 6-41.*

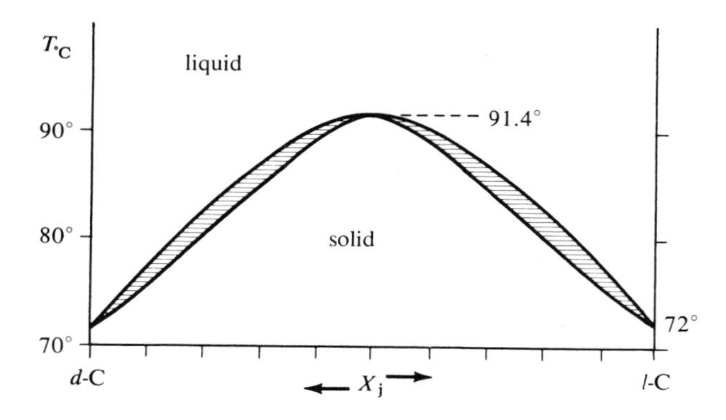

FIGURE 6-54. *Negative deviation from ideal solid solution behavior in the system of the d- and l-isomers of carvoxime. Compare Figure 6-37.*

shows positive deviation and regions of immiscibility of two solid phases, α and β. Both phases are ccp, but α is a distortion of the Ag lattice with $a = 4.08$ Å for pure Ag, and β is a distorted Cu lattice with $a = 3.61$ Å when pure. The analogy with the butanol–water liquid–vapor diagram is exact and the interpretation is the same. Line *u-w* is the saturation curve for the liquid in the α solid phase, and *u-v* gives the composition of the solid which

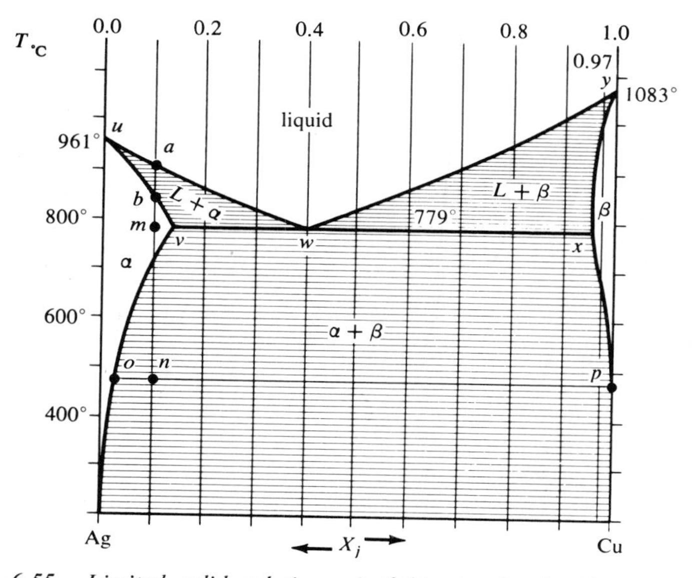

FIGURE 6-55. *Limited solid solution miscibility in the* Ag–Cu *system. The vertical lines are the concentrations for the cooling curves in Figure 6-56. Two-phase regions, solid–liquid or solid α–solid β, are shaded with tie lines. Compare Figure 6-43. (From Hansen.)*

crystallizes out at a given temperature. Line *u-w-y* can also be thought of as the melting point line for the solid, and *w* is the eutectic, or "easily melted" point.

One of the ways of preparing such a phase diagram is to make up melts of different proportions of the two components, allow them to cool, and follow the cooling curve of temperature versus time at a constant rate of heat removal. A family of such curves for the Ag–Cu system is shown in Figure 6-56. Pure liquid Ag ($X_{Cu} = 0.00$) cools above 961°C at a rate which is a function of its heat capacity. But at 961°C a second, crystalline phase appears and the temperature remains constant until all of the liquid is used up. The heat which diffuses away from the specimen is replaced by latent heat of fusion released upon crystallization. After the entire sample has crystallized, cooling again proceeds, but at a different rate reflecting the different heat capacity of the solid. The appearance of the solid will be that of large crystalline domains or grains of pure Ag, oriented at random relative to one another depending upon how the initial chance nucleations occurred.

A melt with 10 mole % Cu will have a different cooling curve. The slope will remain constant until approximately 905°C, when the first α phase begins

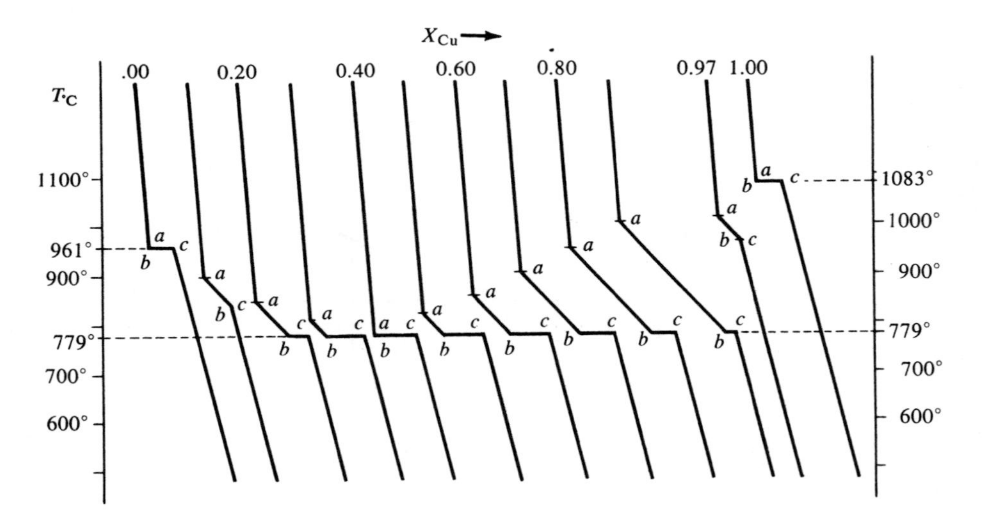

FIGURE 6-56. *A family of cooling curves for the* Ag–Cu *system of Figure 6-55. The region from the top of the curve to point a represents cooling of the liquid. From a to b, the liquid is continuing to cool, but at a slower rate since a solid phase is crystallizing out and giving off its heat of fusion. Section b to c is a eutectic halt until the last liquid crystallizes, longer or shorter depending on how much liquid was left when the liquid composition reached the eutectic point* (*w in Figure 6-55*). *From c to the bottom of the diagram, the solid phase or phases are cooling. It is from information such as these curves provide that phase diagrams like Figure 6-55 are constructed.*

to come down. The temperature will then fall more slowly, for latent heat of fusion is being released as phase α crystallizes. At 850°C there is a second break in the cooling curve and in the absence of any more liquid to crystallize, the temperature drop in the solid α phase is more rapid. These two cooling curve breaks define points *a* and *b* on the phase diagram. (Why, for pure Ag, does the onset of crystallization demand a temperature halt, while for the 10% mixture only a slowing of the temperature drop is observed? The facile answer is "The phase rule demands it." But do you really *understand* why?)

The 20 mole % cooling curve begins in a similar way; at 860°C there is a break in slope as phase α begins to come out. As the temperature falls and as the Ag-rich α phase crystallizes, the liquid becomes poorer in Ag and its composition falls along line *u-w*. At point *w* the liquid is saturated not only in the α phase, but in the β phase as well, and *both* begin to crystallize simultaneously. Relative amounts of the three phases are given by the lever principle, as before. With three phases there are *no* degrees of freedom, and the temperature is stabilized at 779°C. This stabilization, marked *b-c* on the cooling curve, is called a eutectic halt. After the liquid is gone, then the mixture of α and β phase crystals can be cooled at a new rate dependent upon their specific heats. The end product will be large crystals of α which grew between 860°C and 779°C, embedded in a matrix of tiny crystals of α and β which came down together during the eutectic halt.

The 30% Cu curve is similar, and provides another point on the *u-w* line on the phase diagram. The 40% line happens to go through the eutectic point *w* itself. There will be no crystallization at all until the eutectic temperature is reached, at which point both solid phases will come down simultaneously in a fine-grained eutectic matrix. The 50–90% cooling curves are like those at 20 and 30%, and help to map out curve *w-y*. But the solid now will be made up of large grains of β embedded in a eutectic matrix, rather than α. The 97% curve is analogous to the 10%, and the pure Cu curve to the pure Ag.

The physical properties of the alloy are strongly influenced by the size of the grains, or crystalline domains. Large-grained metals are softer and more ductile; for a slip dislocation, once started by mechanical stress, can travel a great distance before it is halted at a grain boundary. If the domains could be made smaller, the metal would be harder. The eutectic matrix is just such a small-grained form, and it is for this reason that the alloy is harder and stronger than either pure silver or pure copper.

If a 10% Cu alloy is cooled slowly to point *m* in Figure 6-55, the resulting phase will be soft and ductile. If it is cooled rapidly, or quenched, to point *n*, it will still be ductile, for the α phase, although now metastable relative to a mixture of α and β, will not have had time to equilibrate. Equilibrium in solid solutions is reached slowly if at all. Eventually, the original crystalline mass will separate into smaller grains of two phases, according to the reaction

$$\alpha_n = \alpha_o + \beta_p \tag{6-93}$$

where the subscripts refer to concentrations on Figure 6-55. This equilibrium mixture will now be much harder and less malleable. This is the process of age hardening, and can occasionally be seen as a crazed or crackle grain pattern on the surface of very old brass doorknobs. The other standard method of alloying to produce a hard metal was hinted at in passing when Fe_4N was mentioned. In an interstitial alloy, the slip of one layer of metal atoms over another is impeded by foreign atoms like C or N which act like sand in the bearings. Whether by grain boundaries or intercalated atoms, the goal in hardening metals is to make the free passage of one layer of atoms over another difficult.

Peritectic behavior, in contrast to the eutectic behavior just discussed, is seen in the $AgNO_3$–$NaNO_3$ system (Figure 6-57). The interpretation is analogous to that of Figure 6-47. As an α phase solid at point a is heated, it begins to melt at point b to a liquid of composition given by the other end of the horizontal tie line. At the temperature of point c, the liquid has composition d, and the solid phase α, e. But at this "peritectic temperature" a rather strange behavior is observed. As more α phase melts, liquid is formed,

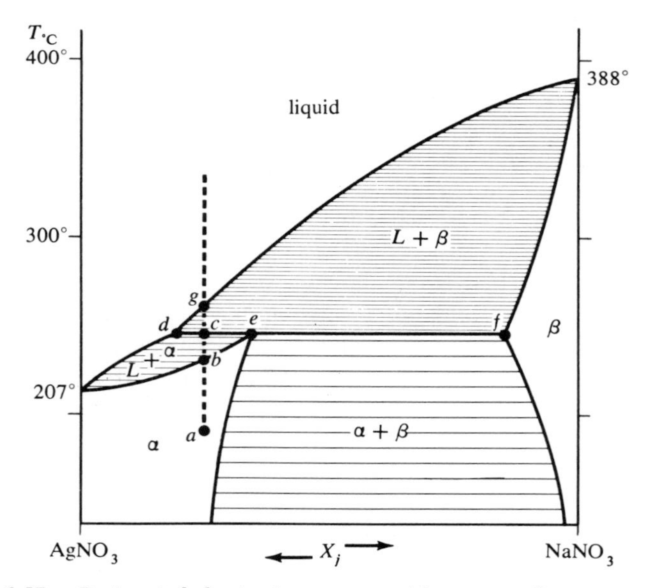

FIGURE 6-57. *Peritectic behavior in a system with a range of compositions possible in the solid phases.*

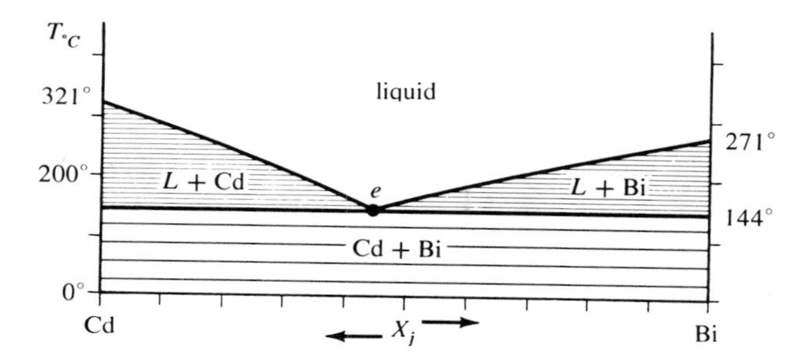

FIGURE 6-58. *Simple eutectic diagram of the* Cd–Bi *system, with complete immiscibility of the two solid phases. Figures 6-52, 6-55, and this one form a progression of increasing incompatibility of each pure component in the crystal lattice of the other. (From Hansen.)*

but so is a new solid phase β. The reaction is

$$\alpha = \text{liquid} + \beta \tag{6-94}$$

It is for this reason that this behavior is called peritectic, or "melting with change" behavior. What would a set of cooling curves for peritectic behavior look like, especially in the composition region between pure $AgNO_3$ and point e?

Cd and Bi crystallize in totally different forms, Cd in hcp and Bi in the hexagonal arsenic lattice (Table 6-2). They are totally insoluble in one another in the solid state, and their phase diagram is shown in Figure 6-58. This is just the limit of the Ag–Cu diagram with the α and β solubility zones flattened out against the vertical axes. The interpretation of the diagram is obvious. Point e is the eutectic point, where the solid melts to a liquid of *the same composition.* This behavior is also shown by KCl and AgCl (both the NaCl lattice, but unit cell distances of 6.29 and 5.55 Å, respectively), bromoform and benzene, and picric acid and TNT.

TABLE 6-3 TYPICAL EUTECTIC FREEZING MIXTURES

	T_e (°C)
H_2O and Na_2SO_4	− 1.1
H_2O and KCl	−10.7
H_2O and NH_4Cl	−15.4
H_2O and NaCl	−21.1
H_2O and NaBr	−28.0
H_2O and $CaCl_2$	−55

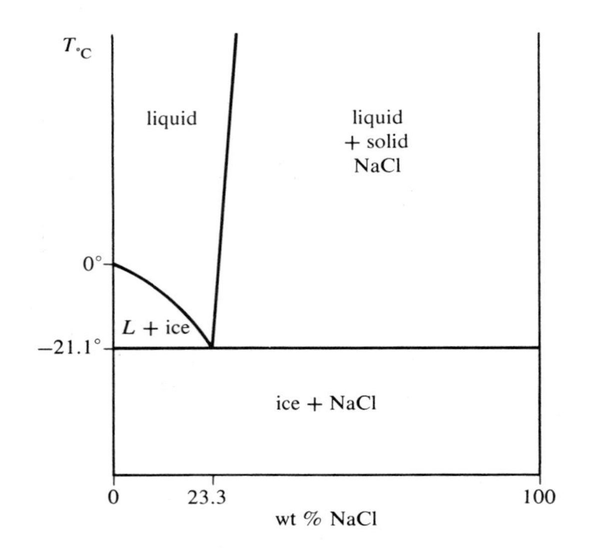

FIGURE 6-59. *Eutectic diagram for a salt water freezing bath.*

The eutectic halt is a useful means of maintaining a constant temperature. In the salt bath whose phase diagram is shown in Figure 6-59, as long as liquid, ice, and salt are all present, the temperature is fixed at the eutectic temperature of -21.1°C. If heat is added, ice melts to compensate for it; if heat is withdrawn, more water freezes out from the solution. Table 6-3 shows some other freezing mixtures using water, and others involving acetone or other organic solvents are common.

Complex Diagrams and Compound Formation

The most complex phase diagram can be understood if it is broken down into the basic types discussed so far. Consider two substances A and B which associate readily with one another in the solid solution. Then their phase diagram might appear as Figure 6-60a. In the extreme case of association, an intermediate compound might be formed, AB_2, and the phase diagram would appear like two Cu–Ni diagrams side by side. Now suppose that the compound AB_2 does not pack particularly well into the lattices of either A or B. Then the system would show positive deviation between A and AB_2, and AB_2 and B, as in Figure 6-60b. In the extreme case there might even be immiscibility of AB_2 with A and with B at lower temperatures, as in Figure 6-60c.

Only one step beyond this point brings us to the Ti_2O_3–TiO_2 diagram of Figure 6-61. This looks like two Ag–Cu diagrams side by side. There are three solid phases: α, β, and γ. The two-phase regions are marked. There

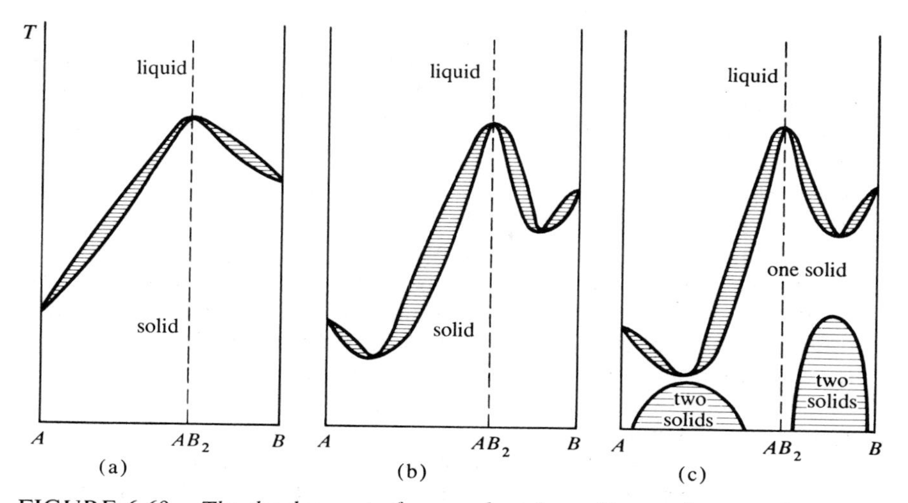

FIGURE 6-60. *The development of a complex phase diagram from an elementary one, with extreme negative deviation leading to intermediate compound formation (a), and solid phase immiscibility (c).*

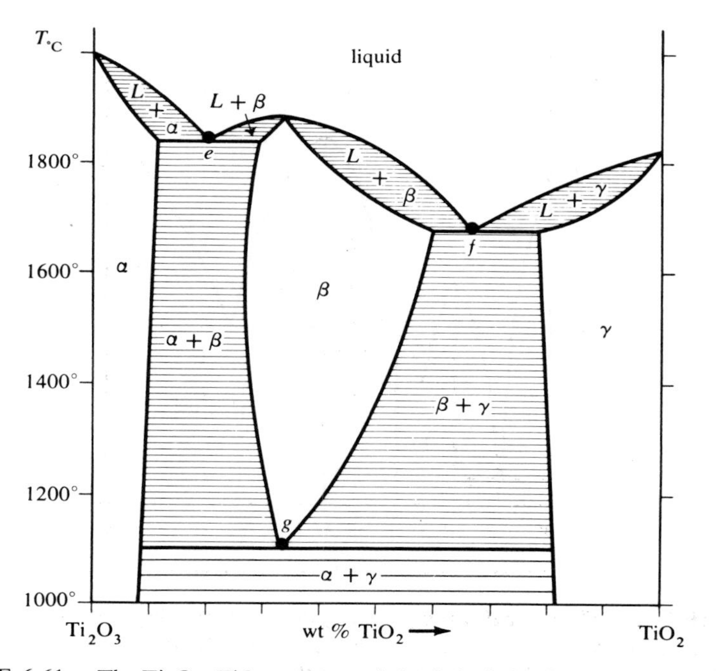

FIGURE 6-61. *The* Ti_2O_3–TiO_2 *portion of the full* Ti–O *phase diagram, showing two eutectic points e and f and a eutectic point g. This diagram can be thought of as the next logical step in the sequence begun in Figure 6-60.*

are two eutectic points e and f and a eutectoid point at g where, if equilibrium occurred quickly, one could observe the conversion of phase β to α and γ. Just as the type structure for the α phase is Ti_2O_3 and that of γ is TiO_2 although the exact compositions can vary widely, so the type formula for the β phase is Ti_3O_5.

Water and the various crystalline hydrates of sulfuric acid form the complex diagram of Figure 6-62. As sulfuric acid is added to water, the freezing point falls to a eutectic point at $-70°C$ with three phases present: ice, a 10% acid solution, and solid tetrahydrate. As more acid is added, the freezing point rises to $-25°C$, the melting point of pure crystalline tetrahydrate. This sort of behavior occurs with each of the other hydrated forms, which can be considered as pure compounds in a diagram like Figure 6-58. The eutectic point in aqueous solutions is often called the cryohydric point, and the eutectic mixture, a cryohydrate.

The Na–K system (Figure 6-63) shows an important variation, incongruent melting at a peritectic point p. This diagram is to Figure 6-57 as the Cd–Bi diagram was to the Ag–Cu. Na and K form an intermediate compound Na_2K which is insoluble in either the Na or the K lattice. Pure Na_2K melts at 6.6°C to form solid Na and a liquid poorer in Na than the original solid:

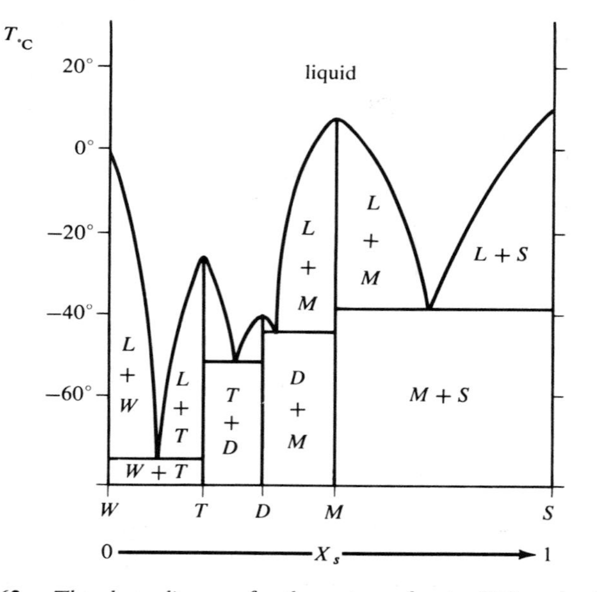

FIGURE 6-62. *The phase diagram for the system of water (W) and sulfuric acid (S). Intermediate crystal forms are the monohydrate (M), dihydrate (D), and trihydrate (T) of sulfuric acid. This diagram can be thought of as a fourfold repetition of Figure 6-21 or 6-58.*

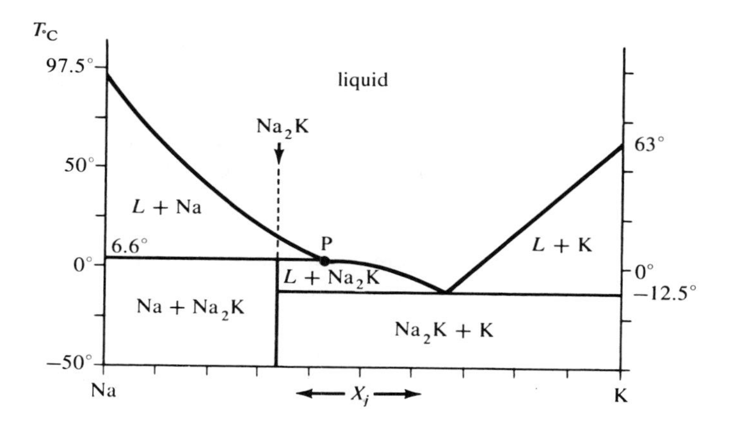

FIGURE 6-63. *The* Na–K *system, with an intermediate compound* Na_2K *which melts incongruently at its peritectic point p.* (*From Hansen.*)

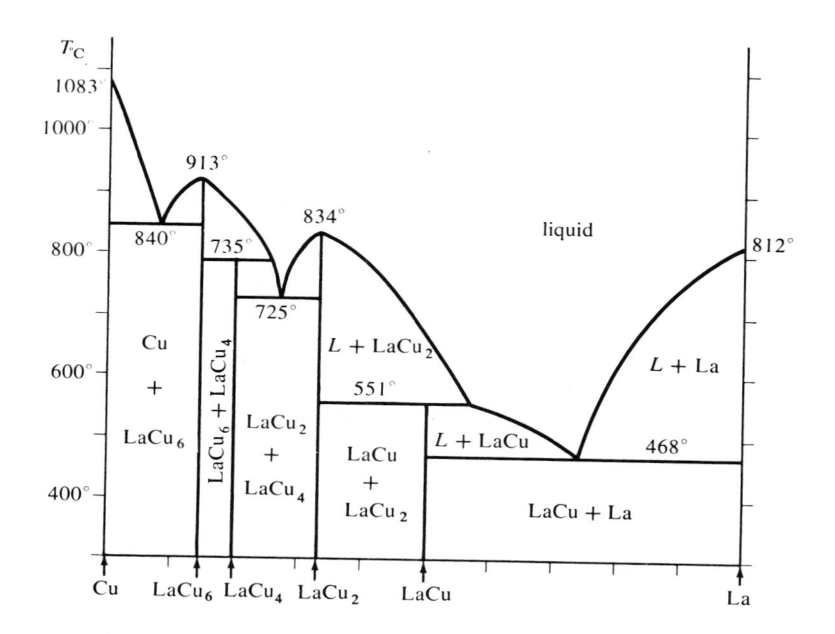

FIGURE 6-64. *A compound phase diagram of* Cu, La, *and four intermediate compounds, all mutually insoluble in the solid phase. This diagram can be considered as being built up from simple diagrams such as Figures 6-58 and 6-63.*

$$Na_2K_{(s)} = Na_{(s)} + \text{liquid} \tag{6-95}$$

With three phases, the system will remain at the peritectic temperature of 6.6°C until the last Na_2K is gone.

Copper and lanthanum form no less than four intermediate compounds, all insoluble with one another, two of them melting eutectically and two peritectically (Figure 6-64). The interpretation of this phase diagram should now be obvious.

Whatever the nature of the solid compatibility, the liquid or melt is usually completely miscible. But for the ice–phenol system (Figure 6-65) this is not so. If a one-phase solution at *a* is cooled to *b*, the solubility limit is reached and two liquid phases separate. The composition of one falls along the curve toward *d*, and the other toward point *e*. At this point, solid phenol begins crystallizing out, and continues to do so until the last of the phenol-rich liquid phase at *e* is gone. The phenol-poor liquid and solid phenol then coexist in a normal way down to a true eutectic at point *g*.

The postgraduate course in this discussion of complex diagrams is the important Cu–Zn system, which includes all of the brasses. Its phase

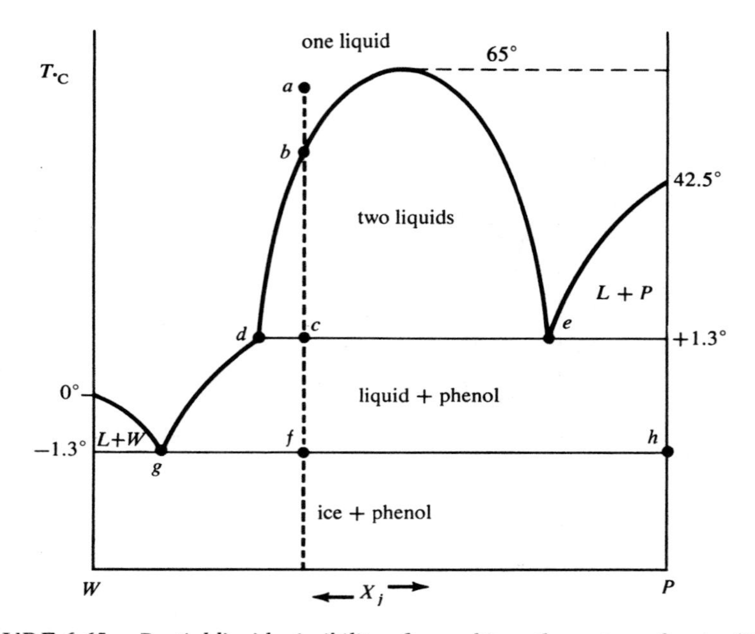

FIGURE 6-65. *Partial liquid miscibility after melting: the system of water (W) and phenol (P). With the relationship between this figure and Figure 6-25 as a guide, the solid liquid equilibrium surfaces could be added to the grand phase diagram of Figure 6-45. How would they change with pressure, the axis normal to the page?*

diagram, worked out patiently by a great many people over a great many years, is shown in Figure 6-66. There are six pure phases: α, β, γ, δ, ε, and η. α is the ccp lattice of pure Cu, and η is the hcp lattice of pure Zn. β-brass, the roughly equimolar form, is bcc with Cu and Zn atoms scattered at random throughout the lattice. The β' form below about 455°C is an ordered region in which all of the Cu atoms occupy the corners of the bcc lattice and the Zn atoms, the centers. This is the order–disorder transition discussed in connection with second order phase transitions, Section 5-6. Phase γ is a distorted bcc lattice with 52 atoms in a macrocell and the type composition Cu_5Zn_8. Phase δ is bcc again, and ε is hcp but with different lattice constants than the η form of nearly pure Zn.

Each of the solids except α decomposes peritectically to give liquid and the next form richer in Cu. There are no eutectic points, but there are two eutectoid points, one near 250° for the reaction

$$\beta' = \alpha + \gamma \tag{6-96}$$

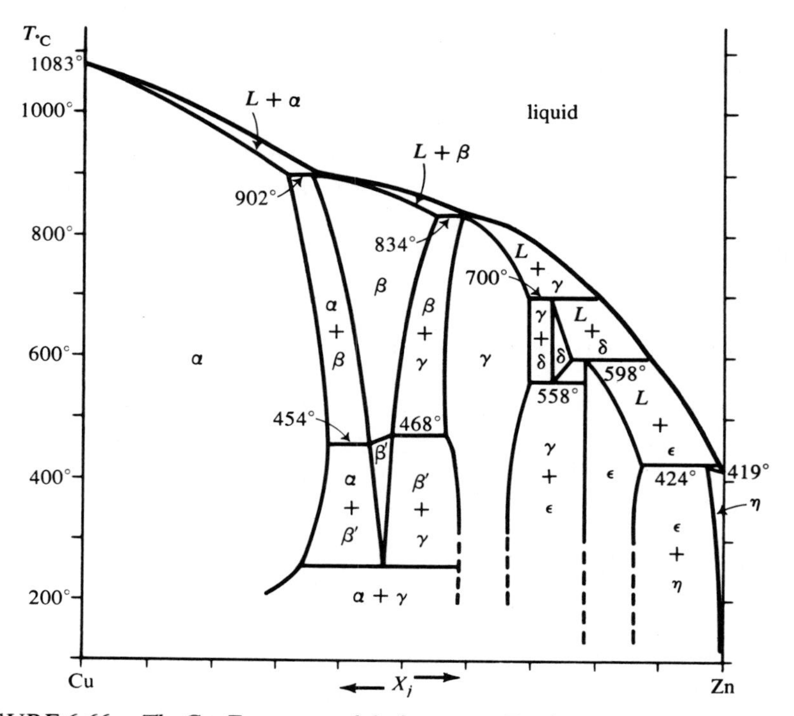

FIGURE 6-66. *The* Cu–Zn *system of the brasses. For discussion, see text.* (*From Hansen.*)

and another at 558° for

$$\delta = \gamma + \varepsilon \tag{6-97}$$

The commercially important brasses are 30–33% Zn, in the α phase near the α-β two-phase solubility boundary. Ordinary annealing at 600–650°C does not involve a phase change, but instead gives the grains a chance to grow and to make the metal more ductile. If a very high-Zn brass is then cooled to room temperature, it will enter a metastable two-phase zone, and very slow equilibration from α to a mixture of α and γ will take place.

Moly-B Disaster

An expensive example of the perils of ignoring phase diagrams was provided once in the aerospace industry when an engineer decided to use a new $500 molybdenum crucible (melting point, 2620°C) to melt powdered boron (melting point, 2037°C) into a one-pound ingot. The crucible with its charge of powdered boron was placed in a radio frequency induction furnace, the system was evacuated, and the crucible and charge were heated. Up to about 1900°C all went well, but shortly thereafter the boron appeared to be melting a good bit below its normal melting point. At about 1960°C the liquid began pouring through the bottom of what had been a new crucible, and back at 25°C the engineer was left with boron–molybdenum alloy all over the floor of his furnace and a very expensive piece of molybdenum pipe. The offending phase diagram is shown in Figure 6-67.

REFERENCES AND FURTHER READING

Many of the texts referred to in earlier chapters have material on solutions and on multicomponent phase diagrams. The Debye–Hückel theory is covered in Wall and in Lewis and Randall. In addition, the following four books are important for their treatment of phase diagrams and equilibria.

A. Findlay, *The Phase Rule* (Dover, New York, 1951).

S. T. Bowden, *The Phase Rule and Phase Reactions* (Macmillan, London, 1945).

J. E. Ricci, *The Phase Rule and Heterogeneous Equilibrium* (Van Nostrand, Princeton, N.J., 1951).

M. Hansen, *Constitution of Binary Alloys* (McGraw-Hill, New York, 1958), 2nd ed.

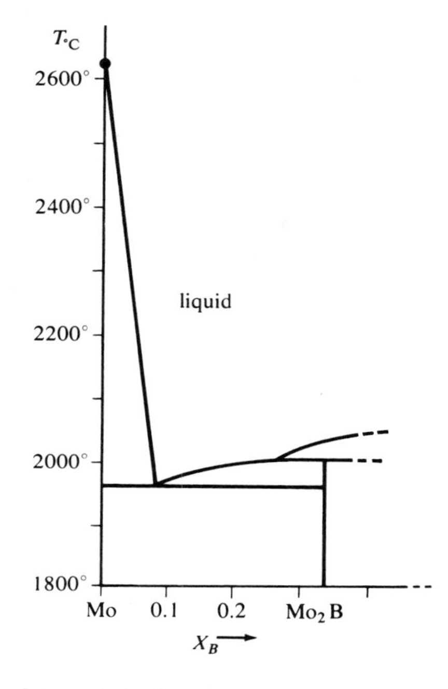

FIGURE 6-67. *The* Mo-*rich half of the* Mo–B *phase diagram.* (*From Hansen.*)

EXERCISES

6-1. Naphthalene, which melts at 80°C, forms an ideal solution in benzene. If the heat fusion of naphthalene is 35.1 cal/g, calculate the solubility of naphthalene in benzene at 25°C. At what temperature would the solubility of naphthalene be 85 g/mole of benzene?

6-2. If 1.346 g of the protein β-lactoglobulin are dissolved in enough water to make up 100 cm^3 of solution at 25°C, then the osmotic pressure is observed to be 9.91 cm of water. Find a rough molecular weight for the protein.

6-3. (a) A solution of a polymer of high molecular weight dissolved in benzene shows an osmotic rise of 1.15 cm of benzene at 25°C. If $K_f = 5.12$ for benzene, calculate the freezing point lowering of the solution and the vapor pressure lowering. The vapor pressure of pure benzene at this temperature is 94 mm of mercury. What do your results mean about the practicability of measuring freezing point lowering and vapor pressure lowerings of high polymer solutions?

(b) The polymer solution above contains 3.15 g of polymer per liter of solution. A solution half as concentrated shows an osmotic rise of 0.51 cm at 25°C. Estimate the molecular weight of the polymer.

6-4. The freezing point of a 0.02 mole fraction solution of acetic acid in benzene is 4.4°C. Acetic acid exists partly as a dimer in solution.

(a) Calculate the percent of monomer which has converted to the dimer form under these conditions.

(b) Calculate the standard free energy of dimerization in solution at 4.4°C. Under *standard* starting conditions, would the dimerization be spontaneous or not?

Find the other data which you need to solve this problem. Assume that both monomer and dimer form ideal solutions in benzene. The heat of fusion of benzene is 2.400 kcal/mole, and its melting point is 5.4°C.

6-5. How many independent components are there in each of the following systems:

(a) H_2O and NaCl
(b) graphite and diamond
(c) $NaCl + LiCl + CsCl + HCl + Cl_2$ at very high temperatures
(d) H_2O and C_2H_5OH
(e) H_2 and O_2 at high temperatures

If temperature range, time scale, or other variables matter in your answer, be sure to specify them.

6-6. Isobutyl alcohol (boiling point, 108°C) and isobutyl bromide (boiling point, 91°C) form an azeotropic mixture boiling at 89°C and containing 12% by weight of alcohol.

(a) Sketch the boiling point composition curve at constant pressure.

(b) Which component can be obtained pure by fractional distillation of a 50-50 mixture?

(c) If 150 g of alcohol are reacted to form the bromide in a 95% yield, and if the mixture is then fractionated, what weight of the pure isobutyl bromide will be obtained?

6-7. When a mixture of chlorobenzene (C_6H_5Cl) and water, with which it is immiscible, is heated, it boils at 91°C and 1 atm. The vapor pressures at 91°C of the constituents are 220 and 540 mm, respectively. Assuming that the vapors behave as ideal gases, calculate the molar composition of the distillate. Calculate its weight composition as well.

6-8. An unknown liquid, immiscible with water, is steam distilled and produces a distillate containing 28.6 cm^3 of the unknown per 100 cm^3 of distillate. The boiling point for the distillation is 98.2°C and the barometric pressure is 758 mm. The vapor pressure of water at 98.2°C is 712 mm. The density of the pure liquid is 1.83 g/cm^3. Calculate the molecular weight of the liquid.

6-9. The solubility of picric acid in benzene is

T (°C)	5	10	15	20	25	35
g/100 g benzene	3.70	5.37	7.29	9.56	12.66	21.38

The melting points of benzene and picric acid are 5.5 and 121.8°C. Calculate the heat of fusion of picric acid.

6-10. Two metals A and B form a simple eutectic system. Three samples are shown you consisting of the following:

(a) one of the pure components

(b) an alloy with the eutectic composition

(c) an alloy of another composition

What physical measurements or observations could you use to identify the samples?

6-11. In the system $NaCl \cdot H_2O$, a simple eutectic is observed at $-21°C$ for a solution containing 23.3% NaCl by weight, at which point $NaCl \cdot 2H_2O$ and ice crystallize out of the mixture. At $-9°C$, a peritectic point exists, and the dihydrate decomposes to form anhydrous NaCl and a solution 27% by weight NaCl. The temperature coefficient of solubility of anhydrous NaCl is very small and positive. Make a rough sketch for the system showing clearly the phases in equilibrium in the various areas and along the various curves of the diagram. (Express the horizontal axis in terms of weight percentage of NaCl.)

6-12. The following information is known about the cobalt–aluminum system:

(a) Pure Al melts at 658°C.

(b) The compound $CoAl_4$ decomposes at 945° to solid Co_2Al_5 and a liquid.

(c) Co_2Al_5 decomposes at 1170° to CoAl and liquid.

(d) CoAl melts congruently at 1630°.

(e) Pure cobalt melts at 1480°.

(f) Cooling curves for most melt concentrations between CoAl and pure Co show eutectic halts at 1375° of the type shown in illustration (a). Very nearly pure Co, or very nearly pure CoAl, however, show cooling curves like illustration (b), with no eutectic halt.

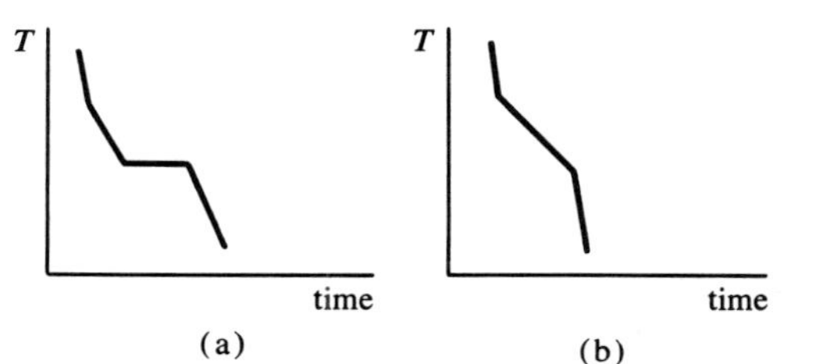

(a) (b)

(g) *All* melts of composition between $CoAl_4$ and pure Al show cooling curves with a eutectic halt at 600°.

Draw the phase diagram of T versus mole fraction for the Co–Al system. Label each region with the phases which are present at that temperature and over-all composition range.

6-13. Cooling curves for the binary alloy system of gold and antimony give the following information:

Wt. % Sb	0	10	20	30	40
Start of freezing (°C)	1063	730	470	400	445
Completion of freezing	1063	360	360	360	360

Wt. %	50	55	60	70	80	90	100
Start	455	460	495	545	580	610	631
Compl.	360	460	460	460	460	460	631

Plot the phase diagram for this system and identify all areas and phases. If a sample of over-all composition 40% by weight Sb were examined at 200°C, what components would be present and in roughly what proportions?

6-14. Phase diagrams such as the nicotine–water one of Figure 6-24 are sometimes a bit of a swindle because the pressure variable is inadequately controlled. Measurements of miscibility versus temperature by early workers were carried out in sealed capillaries, in which the pressure would rise with temperature. The only reason that such experiments now have any validity is that the dependence of miscibility on pressure is small within the pressure ranges involved. Data for Figure 6-24 were collected using 1-mm sealed glass capillaries made of special Jena glass; see C. S. Hudson, *Z. Phys. Chem.* **47**, 113 (1904). If the upper consolute temperature of 208°C occurs at a composition of 32% nicotine by volume, if the boiling point of nicotine at 1 atm pressure is 246°C, and if the vapor pressure of water at 208°C is 18.8 atm, estimate the true pressure within the capillary at the upper consolute point in Figure 6-24.

6-15. (a) A cup of coffee and a cup of tea are sitting side by side, each filled to the same point. One spoonful of coffee is transferred to the tea cup, thoroughly mixed in, and then one spoonful of the mixture is brought back to the coffee cup and mixed. At the end, which cup has more of the other cup's beverage in it? Note that there is a hard way and an easy way to solve this problem. Assume ideal solutions.

(b) Suppose that the liquids had been water and methyl ethyl ketone instead (Figure 6-25). What would your answer be? Is there now an easy shortcut? If not, what information would you need to solve the problem?

Chapter 7

THERMODYNAMICS AND LIVING SYSTEMS

IT IS possible to know thermodynamics without understanding it, or even to understand it without appreciating it. Thermodynamics should be more than a useful device for getting answers for a chemist; it should be a way of thought. For these three postulates and the definitions and ideas which go with them have a universality far beyond the tabulation of heats of reaction and predicting of equilibria. This chapter hopefully will give you some idea of the universality of thermodynamics, in applying it to that most interesting class of chemical systems, living organisms.

There is no intention here of providing a short course in biochemistry. Our only interest in specific reactions is in their illustration of a thermodynamic principle. We shall not hesitate where helpful to simplify or oversimplify to a point which would strike terror into the soul of a biologist. Nevertheless, the fact that thermodynamics deals with reality means that we will have to bring in a fair amount of biochemistry along the way. More background in possibly unfamiliar material which we shall gloss over quickly can be found in the references for further reading for this chapter. Lehninger and Klotz both have good treatments of free energy in biological reactions. Blum, although very dated in its descriptive parts, has in Chapters VII, XI, and XII a classic treatment of entropy, time, and evolution which has never been bettered. Krebs and Kornberg will be the standard source for thermodynamic data, and Oparin and Bernal provide a good background in the origin of life.

7-1 *SIMULTANEOUS OR COUPLED REACTIONS*

So far we have discussed individual chemical reactions as if they existed in isolation from one another. This of course is untrue. The proper if impractical field of observation in any chemical reaction is the entire material universe. It is safe to narrow down the field of study by some 27 orders of magnitude if we observe that the excluded parts of the universe neither change in any way related to what happens to the chosen chemicals nor themselves exert any influence on them. The smallest convenient system for study, unless it is quite artificial, will still have in it a great many reacting or potentially reacting chemical species. It is important to recognize that the chopping up of this system in our minds into a collection of individual chemical reactions is only a matter of convenience. The system is unaware of the dividing lines. The principles of thermodynamics apply to the system *as a whole.* Things which would be quite unlikely if left to themselves can happen with considerable frequency as a part of a larger process. Highly endoergonic reactions may be swept along by other reactions (our boundaries again) which liberate free energy. This is the principle of *coupled reactions.*

Free Energy Considerations

The general expression for the free energy change in a system in which component j changes in amount by dn_j, whether by chemical reaction or by physical addition or removal, has been derived previously:

$$dG = -SdT + VdP + \mu_1\,dn_1 + \mu_2\,dn_2 + \cdots$$

or

$$dG = -SdT + VdP + \sum_j \mu_j\,dn_j \tag{5-58}$$

Consider a system in which O_2 and H_2 are disappearing, H_2O is appearing, and CO and CO_2 are both present and either appearing or disappearing. This system can be thought of as being made up of two chemical reactions:

$$\begin{aligned} &(1) \qquad CO + \tfrac{1}{2}O_2 \rightarrow CO_2 \\ &(2) \qquad CO_2 + H_2 \rightarrow CO + H_2O \end{aligned} \tag{7-1}$$

The extents to which these two reactions proceed can be measured by means of reaction parameters ξ_1 and ξ_2. These coordinates measure the number of stoichiometric units of each reaction, as written, that have occurred. If reaction 1 proceeds in amount $d\xi_1$ and reaction 2 in amount $d\xi_2$, then the change in amount of each component can be expressed as

$$\begin{aligned} dn_{CO} &= -d\xi_1 + d\xi_2 \\ dn_{O_2} &= -\tfrac{1}{2}\, d\xi_1 \\ dn_{CO_2} &= d\xi_1 - d\xi_2 \\ dn_{H_2} &= -d\xi_2 \\ dn_{H_2O} &= +d\xi_2 \end{aligned} \tag{7-2}$$

These expressions can be substituted in Eq. 5-58 in order to obtain the over-all free energy change for the system:

$$dG = SdT + VdP + \widetilde{\Delta G}_1\, d\xi_1 + \widetilde{\Delta G}_2\, d\xi_2 \tag{7-3}$$

where

$$\begin{aligned} \widetilde{\Delta G}_1 &= \mu_{CO_2} - \tfrac{1}{2}\mu_{O_2} - \mu_{CO} \\ \widetilde{\Delta G}_2 &= \mu_{CO} + \mu_{H_2O} - \mu_{CO_2} - \mu_{H_2} \end{aligned} \tag{7-4}$$

Such a treatment can be extended to a system with any number of internal reactions. The free energy of the entire system can be given as a function of temperature, pressure, and the reaction coordinate for each individual reaction.

Treating the two reactions independently, it is evident from Eq. 7-3 that the reaction potential of each individual process is related to the over-all free energy change by

$$\begin{aligned} \widetilde{\Delta G}_1 &= \left(\frac{\partial G}{\partial \xi_1}\right)_{P,T,\xi_2} \\ \widetilde{\Delta G}_2 &= \left(\frac{\partial G}{\partial \xi_2}\right)_{P,T,\xi_1} \\ G &= G(T, P, \xi_1, \xi_2) \end{aligned} \tag{7-5}$$

For each of the reactions one can write a partial molar expression just as if each reaction were the only one present:

$$\left(\frac{\partial \widetilde{\Delta G}_1}{\partial T}\right)_{P,\xi_2} = -\widetilde{\Delta S}_1 \qquad \left(\frac{\partial \widetilde{\Delta G}_2}{\partial T}\right)_{P,\xi_1} = -\widetilde{\Delta S}_2 \tag{7-6}$$

$\widetilde{\Delta S}_1$ and $\widetilde{\Delta S}_2$ represent the partial molar entropy change of each reaction:

$$\begin{aligned} \widetilde{\Delta S}_1 &= \bar{S}_{CO_2} - \tfrac{1}{2}\bar{S}_{O_2} - \bar{S}_{CO} \\ \widetilde{\Delta S}_2 &= \bar{S}_{CO} + \bar{S}_{H_2O} - \bar{S}_{CO_2} - \bar{S}_{H_2} \end{aligned} \tag{7-7}$$

Similar expressions can be written for the partial molar volumes:

$$\left(\frac{\partial \widetilde{\Delta G}_1}{\partial P}\right)_{T,\,\xi_2} = \Delta V_1 \qquad \left(\frac{\partial \widetilde{\Delta G}_2}{\partial P}\right)_{T,\,\xi_1} = \widetilde{\Delta V}_2 \tag{7-8}$$

In general, all other thermodynamic functions applicable to the reaction potential for a single reaction also apply to each component reaction of a multireaction system with the proper alteration of notation. However, there are some new expressions which describe the relationship between the reaction potentials of different reactions occurring in the same system. Forming the Euler cross differential between extents of reaction in Eq. 7-3:

$$\left(\frac{\partial \widetilde{\Delta G}_1}{\partial \xi_2}\right)_{P,\,T} = \left(\frac{\partial \widetilde{\Delta G}_2}{\partial \xi_1}\right)_{P,\,T} \tag{7-9}$$

This tells us that the reaction potential of *each* reaction is equally sensitive to changes in the *other* reaction. If the two reactions have no chemical species in common, and are not coupled through any catalytic mechanism, then there will be no connection between the reaction potentials, and the Euler cross differentials above will be zero. On the other hand, if as in our example a product of reaction 1 is a reactant in reaction 2, then the extent of the first reaction will influence the reaction potential of the second reaction. Any time the two reactions have a common species, the differentials in Eq. 7-9 will be nonzero.

In a system consisting of several reactions the criterion for spontaneity is still that the over-all free energy of the system must decrease. From Eq. 7-3, at constant temperature and pressure, spontaneity of reaction requires that

$$dG = \widetilde{\Delta G}_1\, d\xi_1 + \widetilde{\Delta G}_2\, d\xi_2 < 0 \tag{7-10}$$

Each of the individual reaction potentials need not be negative as long as the over-all free energy change is negative. For example, if $\widetilde{\Delta G}_2$ is greater than zero, dG can still be negative if $d\xi_1$ is large, assuming that $\widetilde{\Delta G}_1$ is negative. One could say that the favorable reaction, liberating a lot of free energy, *drives* the unfavorable reaction. This is the principle of coupling. The mechanism of the coupling, strictly speaking, is outside the province of thermodynamics. But if, by a judicious sharing of chemical species or by a suitable enzymatically controlled mechanism, the coupling can be made to occur, then a thermodynamically unfavorable reaction can be induced to go by the presence of a second very exoergonic reaction. An analog of the coupling of chemical reactions is the coupling of weights by means of pulleys. With the proper system of pulleys, a lighter weight by traveling farther can be used to lift heavier weights. The purpose of elaborate structural and metabolic systems in the body is to couple reactions in much the same fashion. The "string" between the reactions is usually the product of one reaction, necessary as a reactant for the other. The "pulley system" ensuring efficient operation is the properly designed reaction apparatus *in vitro* or the spatial and temporal arrangement of enzymes and cofactors *in vivo*.

EXAMPLE 1. *Wet zinc and oxygen.* The system of wet zinc and atmospheric oxygen provides a nonbiological example of two coupled reactions. The over-all reaction is

$$Zn + H_2O + O_2 \rightarrow ZnO + H_2O_2 \tag{7-11}$$

For convenience this system can be thought of as two coupled reactions:

$$\begin{array}{lll} (1) & Zn + \frac{1}{2}O_2 \rightarrow ZnO & \widetilde{\Delta G}^0 \ll 0 \\ (2) & H_2O + \frac{1}{2}O_2 \rightarrow H_2O_2 & \widetilde{\Delta G}^0 > 0 \end{array} \tag{7-12}$$

In this case, reaction 1 provides the driving force to carry reaction 2, and the over-all reaction has a negative free energy. Unlike the previous example, this process does not involve a product of one reaction appearing as a reactant in the second. Instead, the coupling is a matter of reaction mechanism. When we divide an over-all reaction into coupled reactions, the division is based on energy considerations and not on mechanistic considerations. We are the ones who have done the division, not the system.

EXAMPLE 2. *Lactic dehydrogenase.* Coupling in biological reactions is controlled by enzyme molecules. Their primary function, as catalysts, is to accelerate the drive of a reaction toward equilibrium. In a closed system, this would only mean that the point of equilibrium where no further macroscopic change occurs would be reached more quickly. But in an open system, where reactants are constantly being supplied from the outside and products are being removed, the situation is different. Now *more* of a catalyzed reaction will occur in a given length of time than of an uncatalyzed one, and if the two reactions compete for the use of a common reactant, the catalyzed reaction will predominate. Enzymes can thus play a switching and controlling role in metabolism.

The other function of enzymes is more subtle, and is the one which concerns us here. Enzymes can link two reactions together so that when one reaction occurs, one particular second reaction will occur of all the many which could happen. If the first reaction is exoergonic and the second is endoergonic, then the free energy liberated by the first reaction is saved by the second, and the inherently difficult second reaction is made to take place by the first. How is this coupling brought about?

The coupling, on the atomic scale, appears to be purely mechanical (if Schrödinger's mechanics is included along with Newton's). Enzymes have catalytic sites on their surface which bind the reactants, hold them in place while reaction occurs, and then release the products. If two reactions are catalyzed simultaneously, then there will be two kinds of active site, and part of the function of the enzyme is to ensure that reaction at one site leads to reaction at the other. This phenomenon of the binding at one site on an enzyme leading to altered catalytic properties at another site is called *allostery*, and one of the prime goals of protein crystallographers now is to find out the detailed mechanics of the process.

The enzyme *lactate dehydrogenase* provides an example of coupling. It catalyzes the conversion of pyruvate to lactate in one of the steps just prior to the citric acid cycle, which we shall be examining in Section 7-3. At the same time, it causes a molecule of reduced nicotine adenine dinucleotide (NADH) to be oxidized (the structure of NADH is shown in Figure 7-4). The entire reaction is

$$\underset{\text{(pyruvate)}}{CH_3\text{—}CO\text{—}COO^-} + NADH + H^+ \rightleftharpoons \underset{\text{(lactate)}}{CH_3\text{—}CH(OH)\text{—}COO^-} + NAD^+ \quad (7\text{-}13)$$

with a standard free energy change of -6.0 kcal/mole. But there is no inevitable reason why the reduction of pyruvate should be accompanied by the oxidation of NADH; in fact, NADH is a universal oxidation–reduction carrier molecule which is linked to any number of metabolic breakdowns by the proper enzymes. The individual decoupled reactions are

$$CH_3\text{—}CO\text{—}COO^- + H_2O \rightarrow CH_3\text{—}CH(OH)\text{—}COO^- + \tfrac{1}{2}O_2 \qquad \widetilde{\Delta G}^0 = +46.4 \text{ kcal} \quad (7\text{-}14)$$

$$NADH + H^+ + \tfrac{1}{2}O_2 \rightarrow NAD^+ + H_2O \qquad \widetilde{\Delta G}^0 = -52.4 \text{ kcal} \quad (7\text{-}15)$$

(Here the use of O_2 as an oxidant and H_2O as a source of reducing hydrogen atoms is only a convenient formalism for recording standard free energies of reaction. Both O_2 and H_2O drop out when two such reactions are combined.) It is the task of lactate dehydrogenase to make sure that the second reaction occurs whenever the first one does, and in fact to make the second reaction provide the free energy to accomplish the first one.

Lactate dehydrogenase is a macromolecule made up of four apparently identical subunits of molecular weight 35,000 each. These subunits are shaped more or less like elongated footballs, and are packed into a square array in the molecule with their long axes parallel. Each subunit has one binding site for pyruvate–lactate, and another site for NADH–NAD^+. But the binding of NADH controls the binding of pyruvate. Pyruvate cannot bind to the enzyme until NADH has first bound, and, after the reaction, lactate must come off the enzyme before NAD^+. There is evidence that NADH or NAD^+ produces a change in the arrangement of subunits and perhaps in the folding of the subunits as well, and that pyruvate and lactate can bind only to this rearranged form. When pyruvate is reduced to lactate, it must obtain its reducing hydrogen atoms from a group on the surface of the enzyme. This group is then restored to its original state by the reducing hydrogens from the reaction (Eq. 7-15) above. It is the fact that NADH is present, bound to lactate dehydrogenase, and ready to donate protons that produces the coupling.

The exact mechanics of such coupling processes is just beginning to emerge, and as much as is known now is summarized in Dickerson and Geis. This aspect of the chemistry, strictly speaking, is in the domain of kinetics rather than thermodynamics, and is treated in more detail in Gardiner. Thermodynamics is concerned with what *can* happen from a free energy standpoint, and not with what *will* happen in a given time period in any given set of conditions. Nevertheless, it has been our viewpoint throughout that the molecular "why?" of thermodynamics makes the macroscopic "what?" much more meaningful.

7-2 *COUPLED REACTIONS AND METABOLISM*

The central idea of coupling in metabolic processes is that a reaction which is desirable for the organism but energetically unfavorable can be induced to go in a finite length of time by linking it through some molecular mechanism to other reactions which liberate a larger amount of free energy and which, if allowed to go unimpeded, would be wasteful.

One example of the coupling of an endoergonic and exoergonic reaction is the following:

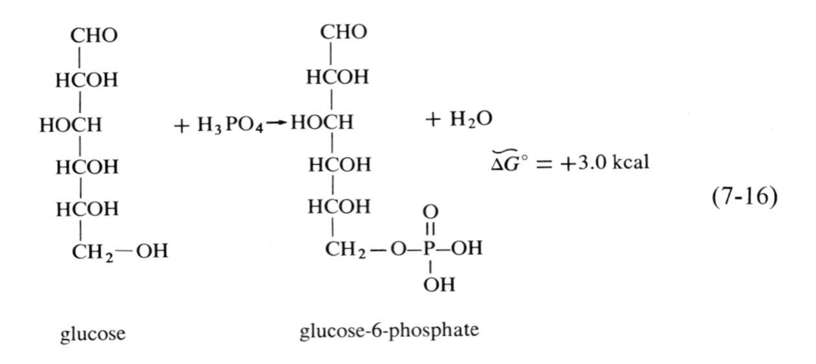

(7-16)

$$ATP \rightarrow ADP + H_3PO_4 \qquad \widetilde{\Delta G}^\circ = -8.1 \text{ kcal} \qquad (7\text{-}17)$$

Before glucose can be used in metabolism, it must be converted to the activated intermediate, glucose-6-phosphate (G-6-P). G-6-P is an activated compound in the sense that it has a higher free energy than glucose. But since this conversion process has a positive $\widetilde{\Delta G}^0$, this reaction would not be expected to occur by itself. Reaction 7-16 can be driven by coupling it with a second reaction (Eq. 7-17) which involves a central work horse of the entire chemical system of the body, adenosine triphosphate (ATP).[1] The structure of ATP is shown in Figure 7-1. The breakdown of ATP to adenosine diphosphate (ADP) proceeds with a large liberation of free energy. In the coupling of reactions 7-16 and 7-17 there is not only a sufficient impetus to drive the first reaction, but 5.1 kcal of wasted excess free energy as well:

$$\text{glucose} + \text{ATP} \rightarrow \text{G-6-P} + \text{ADP} \qquad \widetilde{\Delta G}^0 = -5.1 \text{ kcal} \qquad (7\text{-}18)$$

7-3 FREE ENERGY UTILIZATION IN METABOLISM CITRIC ACID CYCLE

The citric acid cycle is an appropriate system with which to illustrate the thermodynamics of life processes, for it is the central energy-storing mechanism for all oxygen-using organisms. The cycle is outlined in Figure 7-2, and the reactions and free energies are listed in Table 7-1.

Combustion of Acetic Acid

The net result of one complete trip around the citric acid cycle is the combustion of one mole of acetic acid. The over-all reaction is

[1] The enzyme which couples these two reactions is *hexokinase*. One molecule of yeast hexokinase is built up from four subunits of 24,000 molecular weight each. There seems to be a common pattern of two or four subunits of about this size in the various kinases and dehydrogenase enzymes which use ATP or NADH.

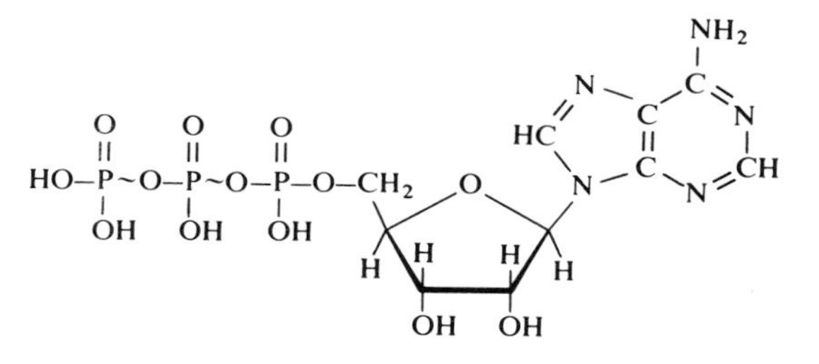

FIGURE 7-1. *The structure of adenosine triphosphate* (ATP). *It is built up from an adenine base* (*right*), *a five carbon sugar, ribose* (*center*), *and three phosphate groups. The last two phosphate bonds have an unusually high free energy of formation, making* ATP *an effective storage device in living organisms for chemical energy.*

$$\begin{aligned} &CH_3COOH_{(aq)} + 2O_{2(g)} \rightarrow 2H_2O_{(l)} + 2CO_{2(aq)} \\ &\widetilde{\Delta H}^0 = -217.3 \text{ kcal} \\ &\widetilde{\Delta G}^0 = -202.9 \text{ kcal} \\ &\widetilde{\Delta S}^0 = -49.6 \text{ e.u.} \end{aligned} \tag{7-19}$$

The entropy of this reaction is negative because of the disappearance of a gas from the reactant side of the equation. This entropy change reduces the total available heat of 217 kcal to only 203 kcal of useful free energy.

The worst possible utilization of this metabolic process would be the one-step combustion of acetic acid with the liberation of 217 kcal of heat, and the use of this heat to drive a heat engine. Since we are considering body processes, the appropriate upper temperature for a heat engine would be 98.6°F or 37°C, and the heat sink would be at room temperature or approximately 20°C. The efficiency of such a process operated between 37°C and 20°C would be

$$e = \frac{310 - 293}{310} \times 100\% = 5.5\% \tag{7-20}$$

And 5.5% of 217 kcal is a meager 12 kcal of heat. In fact, whenever the room temperature rose above 98.6°F such a heat engine would totally cease to function.

But we are not heat engines, and this one of the subtleties developed in the last three billion years of evolution. Instead of changing energy from chemical potential to heat to work, which is a very inefficient process, we transfer the chemical potential stored in one kind of molecule—that of the various foodstuffs—into the chemical potential of another type of molecule, ATP. Later on when muscular contraction or some other form of exertion is needed, we use this stored free energy of ATP. This stepwise process is considerably

TABLE 7-1 STANDARD FREE ENERGIES OF REACTIONS OF THE CITRIC ACID CYCLE[a]

	ΔG^0 (pH 7) (kcal)
1. acetate$^-$ + CoA + H^+ → acetyl-CoA + H_2O	+ 8.2
2. oxaloacetate^{2-} + acetyl-CoA + H_2O → citrate^{3-} + CoA + H^+	− 7.5
3. citrate^{3-} → isocitrate^{3-}	+ 1.6
4. isocitrate^{3-} + $\frac{1}{2}O_2$ + H^+ → α-ketoglutarate^{2-} + H_2O + CO_2	− 54.4
5. α-ketoglutarate^{2-} + $\frac{1}{2}O_2$ + CoA + H^+ → succinyl-CoA$^-$ + H_2O + CO_2	− 59.6
6. succinyl-CoA$^-$ + H_2O → succinate^{2-} + CoA + H^+	− 8.9
7. succinate^{2-} + $\frac{1}{2}O_2$ → fumarate^{2-} + H_2O	− 36.1
8. fumarate^{2-} + H_2O → malate^{2-}	− 0.9
9. malate^{2-} + $\frac{1}{2}O_2$ → oxaloacetate^{2-} + H_2O	− 45.3
1-9. acetate$^-$ + $2O_2$ + H^+ → $2H_2O$ + $2CO_2$	−202.9
10. ADP^{3-} + HPO_4^{2-} + H^+ → ATP^{4-}	+ 8.1
11. NAD^+ + H_2O → NADH + H^+ + $\frac{1}{2}O_2$	+ 52.4
12. $NADHP^+$ + H_2O → NADPH + H^+ + $\frac{1}{2}O_2$	+ 52.6
13. FAD + H_2O → $FADH_2$ + $\frac{1}{2}O_2$	+ 46.8

[a] All values from Krebs and Kornberg (1957) except No. 10, which is taken at pH 7 from Johnson (1960).

more efficient than the heat engine described above. The coupling process that helps the body to save the energy from the combustion of acetic acid involves the formation of ATP from ADP:

$$ADP + H_3PO_4 \rightarrow ATP \qquad \widetilde{\Delta G}^0 = +8.1 \text{ kcal (at pH 7)} \tag{7-21}$$

The free energy figures that appear in Table 7-1 and in the text of this chapter are taken from Krebs and Kornberg. The standard state used by biologists is slightly different from that used by physical chemists.

	Standard state
Physical chemists	H_2O—infinitely dilute solution. All other solution components—one molar concentration
Biochemists	Same as above, except for a standard hydrogen ion concentration of 10^{-7} molar (pH 7) instead of one molar (pH 0)

How does the difference in definition of the standard state affect the standard $\widetilde{\Delta G}^0$ values for reactions involving hydrogen ions? For reactions in which hydrogen is produced, the chemical potential at pH 7 is related to the physical chemists' standard state by

$$A + B \rightarrow C + x\mathrm{H}^+$$

$$\widetilde{\Delta G} = \widetilde{\Delta G}^0 + RT \ln \frac{[C](10^{-7})^x}{[A][B]} \tag{7-22}$$

$$\widetilde{\Delta G} - \widetilde{\Delta G}^0 = +xRT \ln 10^{-7} \cong -10 \text{ kcal per stoichiometric unit of } \mathrm{H}^+$$

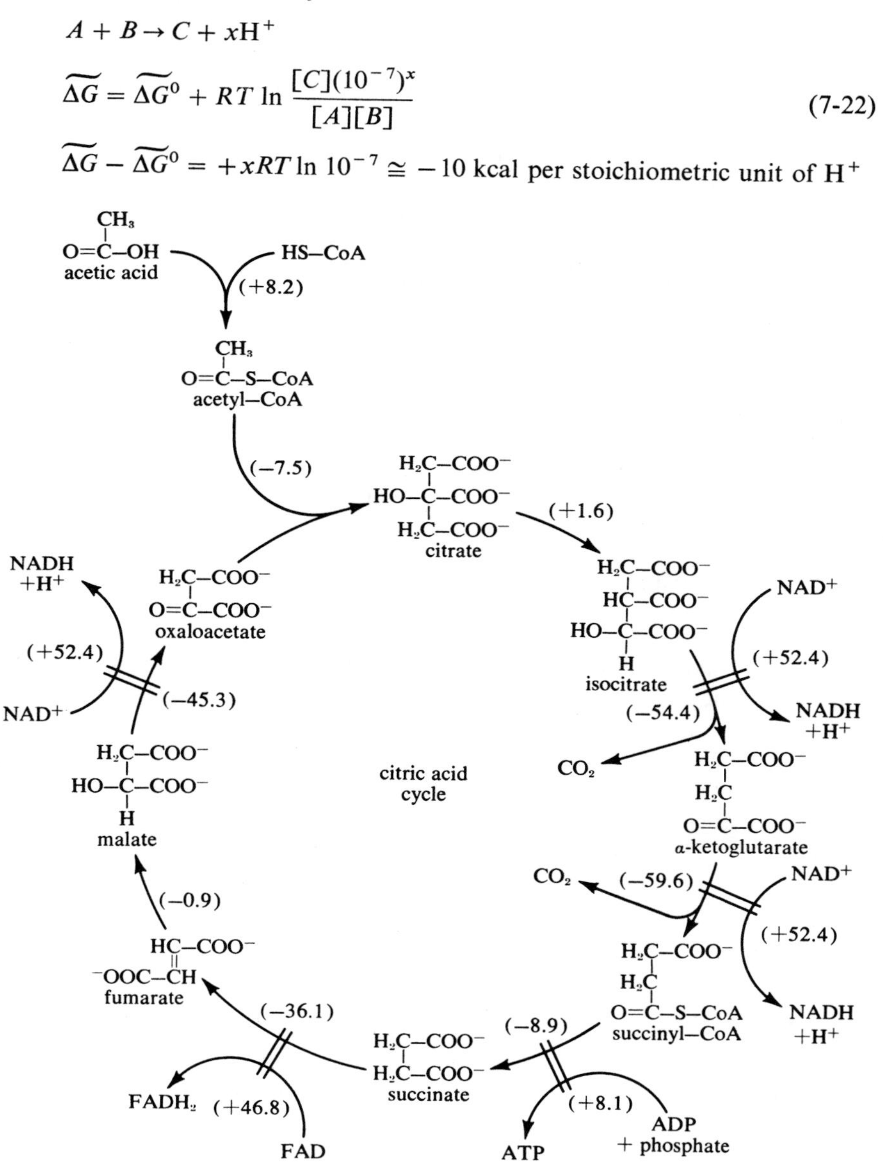

FIGURE 7-2. *The citric acid cycle, also called the Krebs cycle or the tricarboxylic acid cycle. The numbers in parentheses are the standard free energies or reaction potentials for the conversions as written in Table 7-1. A double bar indicates an enzymatic coupling of two reactions.*

Therefore, for reactions in which 1 mole of hydrogen ion is produced, the $\widetilde{\Delta G}^0$ at pH 7 is approximately 10 kcal less than the value used by physical chemists. The reaction is more spontaneous at pH 7 than at pH 0. The opposite is true for reactions in which hydrogen ion is consumed.

The combustion of acetic acid in biological systems is coupled with the formation of 12 molecules of ATP:

$$\begin{array}{ll} CH_3COOH_{(aq)} + 2O_{2(g)} \rightarrow 2H_2O_{(l)} + 2CO_{2(aq)} & \widetilde{\Delta G}^0 = -202.9 \text{ kcal} \\ 12ADP_{(aq)} + 12P_{i(aq)} \rightarrow 12ATP_{(aq)} & \widetilde{\Delta G}^0 = +\ 97.2 \text{ kcal} \\ \hline CH_3COOH_{(aq)} + 2O_{2(g)} + 12ADP_{(aq)} + 12P_{i(aq)} & \\ \rightarrow 2H_2O_{(l)} + 2CO_{2(aq)} + 12ATP_{(aq)} & \widetilde{\Delta G}^0 = -105.7 \text{ kcal} \end{array} \tag{7-23}$$

The symbol P_i is a shorthand notation for inorganic phosphate ion or acid, in whatever form it may be found, and illustrates the biochemist's typical disdain for balanced equations. With 12×8.1 kcal saved out of 203 kcal of total free energy, the efficiency of transfer of free energy is

$$e = \frac{97.2}{203} = 0.48 \tag{7-24}$$

Since it has taken about three billion years to develop such a living machine with 48% efficiency, it is not likely that we shall be able to improve upon it in the near future.

Any machine which has been three billion years in development might be expected to be sophisticated and somewhat complex. One of the components of this free energy transfer machine is the molecule adenosine triphosphate, ATP (Figure 7-1). ATP is sometimes written as A—P~P~P, where the wavy lines represent "high energy" bonds. In biochemical language, a high energy bond is one that liberates an unusually large amount of free energy when it is broken. For example,

$$\begin{array}{lll} ATP \rightarrow ADP + P_i & \widetilde{\Delta G}^0 = -8 \text{ kcal} & \multirow{2}{*}{\text{high energy}} \\ ADP \rightarrow AMP + P_i & \widetilde{\Delta G}^0 = -8 \text{ kcal} & \\ AMP \rightarrow \text{adenosine} + P_i & \widetilde{\Delta G}^0 = -2 \text{ kcal} & \text{normal} \end{array} \tag{7-25}$$

These first two phosphate bonds have quite large bond energies in comparison with those of most chemical systems. It is natural that they or some such bonds would have been selected in the process of chemical evolution as efficient ways of storing free energy.

Citric Acid Cycle

As Figure 7-2 shows, the combustion of acetic acid is not a one-step oxidation. One can imagine a one-step reaction that would liberate 203 kcal of free energy and form 12 ATP molecules. But such a reaction is absurd. Coupling the participants in such a reaction would require at least a 13-body reaction, and any good physicist can tell you the difficulties involved with even three-body reactions. It is impractical to couple the synthesis of 12 molecules of ATP to a one-step combustion. Living systems have evolved a more efficient way of performing this process, involving first the transfer of free energy from acetic acid to intermediates via the citric acid cycle, and then the final storage of the free energy in ATP by means of the terminal oxidation chain.

The various steps in the citric acid cycle are shown in Figure 7-2, the reactions involved appear in Table 7-1, and an energy level diagram of the successive steps is shown in Figure 7-3. One mole of acetic acid, after it has been initially activated by combination with a molecule called coenzyme A, or CoA, is combined with oxaloacetate to form 1 mole of citrate. This citrate is broken down in seven steps, each controlled by an enzyme system, releasing free energy in small packets and ultimately forming oxaloacetate which is ready once again to combine with more acetate. Along the way,

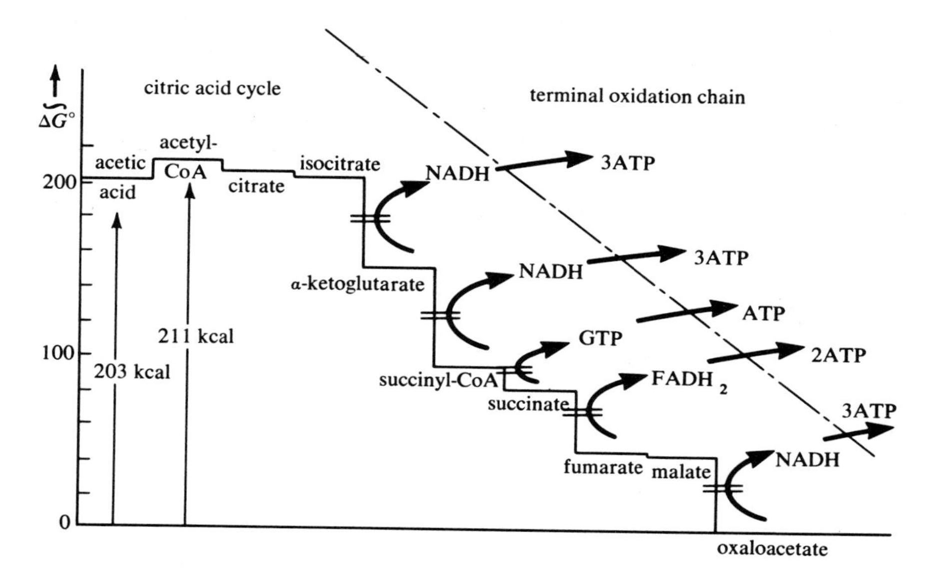

FIGURE 7-3. *The stepwise fall in free energy in one turn of the citric acid cycle. The products formed by coupled reactions, and their eventual role in forming* ATP, *are shown.*

five special molecules are reduced in coupled reactions which absorb free energy, and the energy liberated by the citric acid cycle is saved.

The entering substance for the cycle is the activated acetyl-CoA, not acetic acid itself. The use of acetic acid is actually an oversimplification to make the energy relationships clear. Acetyl-CoA may be produced by the prior breakdown of a great number of intermediate metabolites, all of which can feed into the funnel of the citric acid cycle. The formation of citrate and then of isocitrate are only molecular rearrangements which prepare for the first big free energy liberating step, the conversion of isocitrate to α-ketoglutarate (during which, incidentally, one of the two moles of CO_2 is released). This reaction is coupled via an enzyme system to the following endoergonic reaction:

$$NAD^+ + H_2O \rightarrow NADH + H^+ + \tfrac{1}{2}O_2 \qquad \widetilde{\Delta G}^0 = 52.4 \text{ kcal} \tag{7-26}$$

(A different enzyme system also couples α-ketoglutarate formation to a similar reaction involving $NADP^+$, but this is irrelevant here.) NAD^+, or nicotine adenine dinucleotide, is a complex molecule of the general class of ATP, and is shown in Figure 7-4a along with the reduced dinucleotide in Figure 7-4b. This entire molecule exists for the sole purpose of being reduced with the loss of a charge and addition of a hydrogen atom at the nicotine ring. Why the rest of the molecule is necessary is not yet known, and is a job for the structural physical chemists and enzyme chemists to figure out.

A second major energy transfer step follows immediately, as α-ketoglutarate is broken down to succinyl-CoA, again with the loss of a mole of CO_2. This reaction is also coupled to the formation of NADH. In the removal of CoA to form succinate, guanine diphosphate (GDP) is converted to GTP, and this compound then is used to make the first mole of ATP:

$$\begin{aligned} GDP + P_i &= GTP \\ GTP + ADP &= GDP + ATP \end{aligned} \tag{7-27}$$

When succinate is oxidized to fumarate, the reaction coupled to it is different:

$$FAD + H_2O = FADH_2 + \tfrac{1}{2}O_2 \tag{7-28}$$

This flavin adenine dinucleotide, again, is a complex molecule of the same general type as NAD^+ but with a flavin group in place of the nicotine ring. After a rearrangement step, a last coupled reaction converts malate to oxaloacetate and produces a third mole of NADH.

The over-all result from the operation of the citric acid cycle is the breakdown of acetic acid and the formation of three moles of NADH, one of $FADH_2$, and one of ATP. The free energy balance is shown in Table 7-2.

TABLE 7-2 ENERGY BALANCE FOR THE CITRIC ACID CYCLE

	ΔG^o (kcal)
acetyl-CoA + $2O_2 \rightarrow 2H_2O + 2CO_2$ + CoA	−211.1
3 [$NAD^+ + H_2O \rightarrow NADH + H^+ + \frac{1}{2}O_2$]	+157.2
$FAD + H_2O \rightarrow FADH_2 + \frac{1}{2}O_2$	+ 46.8
$ADP + P_i \rightarrow ATP$	+ 8.1
	+ 1.0

The discrepancy of 1 kcal is within the limits of accuracy of the free energy values used. Eventually, in the terminal oxidation chain, each NADH

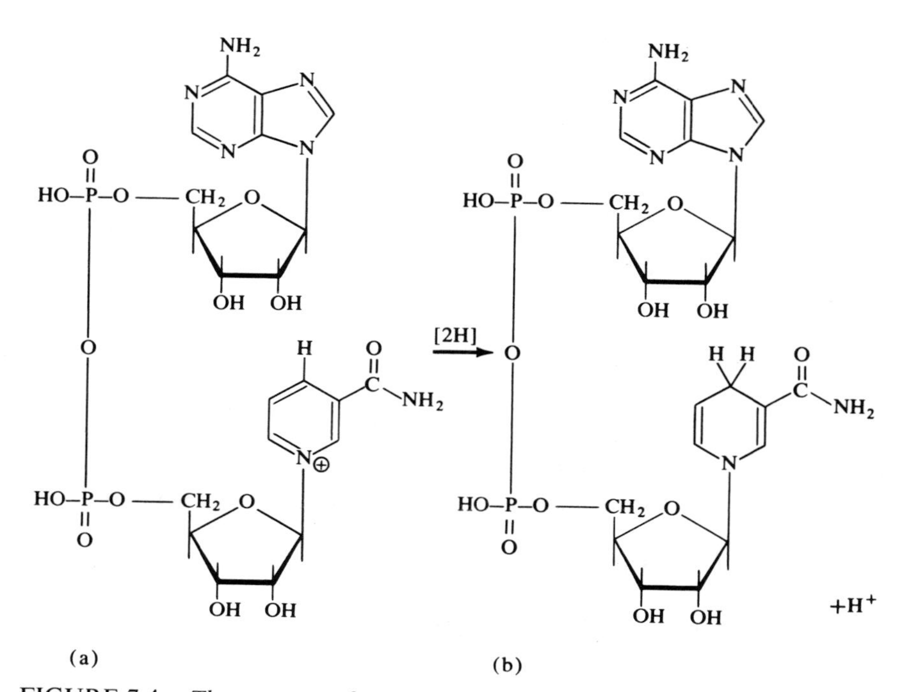

FIGURE 7-4. *The structure of nicotine adenine dinucleotide* (a) *oxidized,* NAD^+, *and* (b) *reduced,* NADH. *Note the similarity between the top half of the molecule as drawn here and* ATP. NAD^+ *is also called* DPN^+, *and* NADH, DPNH.

causes the formation of three moles of ATP, and each $FADH_2$, of two ATP, as shown in Figure 7-3.

The necessity for activating acetic acid by reacting it with CoA is now apparent. The citric acid cycle is carefully tailored to the energy liberated by the breakdown of acetyl-CoA, of no matter what source. Acetic acid itself does not possess enough free energy to drive the cycle.

Dependence of the Reaction Potential Upon Concentration

All of the reaction potentials given so far have been standard reaction potentials, at pH 7 and with all components except water and hydrogen ion at one molar concentration. This is obviously not always true under actual conditions. What effects do changes in reactant and product concentrations within the citric acid cycle have on the over-all process?

Let us take the malate to oxaloacetate conversion as an example:

$$\text{malate} + \tfrac{1}{2}O_2 \rightarrow \text{oxaloacetate} + H_2O \qquad \widetilde{\Delta G}^0 = -45.3 \text{ kcal}$$

At other than standard conditions the reaction potential is

$$\widetilde{\Delta G} = \widetilde{\Delta G}^0 + RT \ln \prod_j a_j^{\nu_j} \tag{7-29}$$

The activities can be approximated by

$$a_{\text{ox}} \cong c_{\text{ox}} \qquad a_{H_2O} \cong 1 \quad \text{(dilute solution)}$$

$$a_{\text{mal}} \cong c_{\text{mal}} \qquad a_{O_2} \cong P_{O_2}$$

If we assume a constant partial pressure of 1 atm for oxygen, then the reaction potential is

$$\widetilde{\Delta G} = \widetilde{\Delta G}^0 + RT \ln \frac{c_{\text{ox}}}{c_{\text{mal}}} \tag{7-30}$$

The ratio of concentrations of oxaloacetate to malate will determine how far the real $\widetilde{\Delta G}$ deviates from $\widetilde{\Delta G}^0$. At equilibrium $\widetilde{\Delta G} = 0$, and the ratio of concentrations represents the equilibrium constant for this reaction:

$$-\widetilde{\Delta G}^0 = RT \ln K_a$$

$$K_a = \exp\left(-\frac{\widetilde{\Delta G}^0}{RT}\right) = \frac{c_{\text{ox}}}{c_{\text{mal}}} = \exp\left(\frac{45{,}300}{600}\right) \tag{7-31}$$

$$K_a = e^{75.5} = 5 \times 10^{32}$$

At equilibrium there is a great excess of oxaloacetate over malate.

We know that this reaction is coupled[2] to another reaction which reduces NAD^+. Suppose that NAD^+ and NADH were present in equimolar amounts so that for this second half-reaction $\widetilde{\Delta G} = \widetilde{\Delta G}^0 = +52.4$ kcal. By adding the two half-reactions, we can determine the ratio of oxaloacetate to malate that would result in no free energy loss for the pair of reactions:

$$-52{,}400 = -45{,}300 + RT \ln \frac{c_{ox}}{c_{mal}}$$

$$\frac{c_{ox}}{c_{mal}} \cong 10^{-5} \qquad (7\text{-}32)$$

Thus, to establish the system so that the free energy released by the malate–oxaloacetate reaction is just enough to balance the NAD^+–NADH reaction, one must make the ratio of oxaloacetate to malate of the order of 10^{-5} rather than the equilibrium value of 5×10^{32}.

While it is true that a great excess of malate will balance the free energies in steps 9 and 11 of Table 7-1, it is also true that this large amount of malate will impede the fumarate–malate reaction of step 8 by exactly the same amount. In fact no matter what the intermediate concentration of malate, the free energy change in going from fumarate to oxaloacetate will be unchanged, other conditions being the same. That is, in any two-step process $A \rightarrow B$ and $B \rightarrow C$, the over-all free energy change is not affected by any change in the concentration of the intermediate B. This is easily demonstrated for the general case:

$$A \rightarrow B \qquad \widetilde{\Delta G}_{AB} = \widetilde{\Delta G}^0_{AB} + RT \ln \frac{a_B}{a_A}$$

$$B \rightarrow C \qquad \widetilde{\Delta G}_{BC} = \widetilde{\Delta G}^0_{BC} + RT \ln \frac{a_C}{a_B}$$

$$A \rightarrow C \qquad \widetilde{\Delta G}_{ABC} = \widetilde{\Delta G}^0_{AB} + \Delta G^0_{BC} + RT \ln\left(\frac{\not{a}_B\, a_C}{a_A\, \not{a}_B}\right)$$

The dependence upon concentration of B (more accurately the activity) conveniently cancels out the last equation. This means that the malate concentration can be adjusted in order to give efficient coupling between the

[2] The enzyme involved in this coupling step is malate dehydrogenase, one molecule of which is built from two subunits of molecular weight 32,500 each. Each subunit has one active site for malate–oxaloacetate and another for NAD^+–NADH. And, as with lactate dehydrogenase, interaction between sites is such that malate or oxaloacetate cannot bind unless NAD^+ or NADH has bound to its site first.

half-reactions in steps 9 and 11 *without* affecting the total free energy change for steps 8 and 9 combined.

These arguments can be applied to all of the other steps of the cycle. The *individual* free energies of Figure 7-2 and Table 7-1 are standard values, and the actual values will depend upon concentrations of the different intermediates. These concentrations will be maintained at suitable levels for efficient coupling at the critical energy-transfer steps. But these fine adjustments within the cycle have no effect upon the *over-all* free energy change in the breakdown of acetyl-CoA. For this reason standard reaction potentials are useful as guides to what is happening even when details of concentrations are not known.

7-4 *TERMINAL OXIDATION CHAIN*

In the terminal oxidation chain the free energy stored in the form of NADH and $FADH_2$ is released so that ATP can be synthesized. The oxidation of each of these free energy carriers releases approximately 50 kcal of free energy, still too large a quantity to be handled in a single step. Therefore, a stepwise process is used to control the release of this energy. This process is called the terminal oxidation chain and is a series of successive oxidation reductions as shown in Figure 7-5. The letters *b*, *c*, and *a* represent different iron-containing cytochrome proteins which participate in the chain with the iron atom of a heme group being oxidized and reduced. The first step in the chain, the oxidation of NADH is the reverse of the reduction shown in Figure 7-4.

Figure 7-5 may seem confusing, but thermodynamically it is quite sensible. The free energy staircase for this chain in Figure 7-6 illustrates the fact that

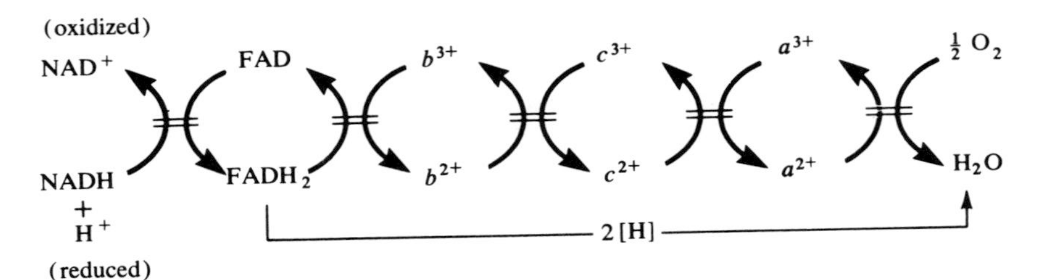

FIGURE 7-5. *The coupling of reactions in the terminal oxidation chain.* FAD *is flavin adenine dinucleotide, another oxidation–reduction carrier molecule of the same general class as* NAD^+. *The symbols b, c, and a represent cytochrome b, c, and a in their oxidized* (+3) *and reduced* (+2) *states. The cytochromes are heme proteins of various sizes, the smallest, cytochrome c, having a molecular weight of* 12,400. *In the final step, as cytochrome a is reoxidized, oxygen is reduced to water, using the protons that were lost by* $FADH_2$ *as it was reoxidized several steps earlier in the chain.*

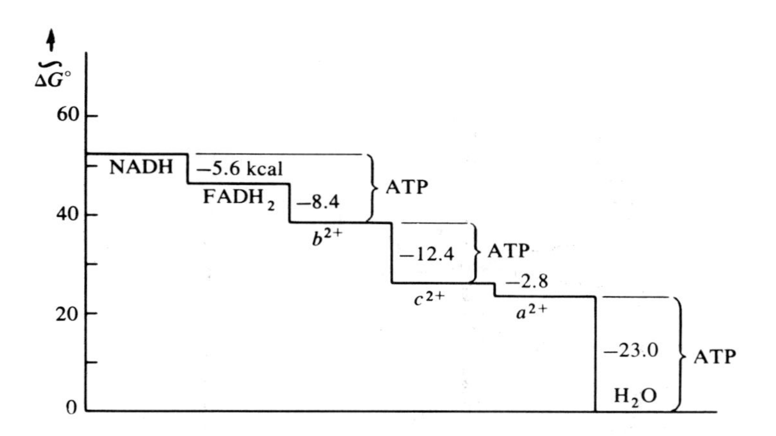

FIGURE 7-6. *The free energy staircase for the reactions in the terminal oxidation chain. The species at each step is that one which is reduced at that point. Coupled reactions produce three moles of* ATP *as shown.* $FADH_2$ *as a starting material enters the chain late and only produces two moles of* ATP.

the terminal oxidation chain is nothing more than a series of descending free energy steps. The component shown at each step in the staircase is the component which is then present in reduced form. One of the reduced nicotine adenine dinucleotide molecules from the citric acid cycle first reduces a flavin adenine dinucleotide molecule with a drop of 5.6 kcal of free energy. This in turn is reoxidized as it reduces a cytochrome *b* molecule and liberates 8.4 kcal more free energy. These two processes are coupled to the synthesis of one ATP molecule. Reduced cytochrome *b* then reduces cytochrome *c* and another ATP is formed. After a small adjustment of 2.8 kcal as cytochrome *a* is reduced, oxygen is finally reduced to water and the third ATP is formed. $FADH_2$ feeds into the chain below the first ATP and hence produces only two ATP molecules.

7-5 OVER-ALL METABOLIC PLAN

The complete metabolic plan of the energy-accumulating process is represented schematically in Figure 7-7. There is first an initial breakdown which causes a small drop in free energy. An intermediate process breaks down amino acids, hexose sugars, and fatty acids into molecular species which can be fed into the next stage, the citric acid cycle. The final stage is the reoxidation of NADH and $FADH_2$ in the terminal oxidation chain.

If glucose is the intermediate metabolite, then the free energy flow is as shown in Table 7-3. In a series of some ten steps, glucose is converted to two moles of pyruvate with a drop of 140 kcal of free energy. However, 65 kcal are saved by the net synthesis of 8 moles of ATP. The 2 moles of

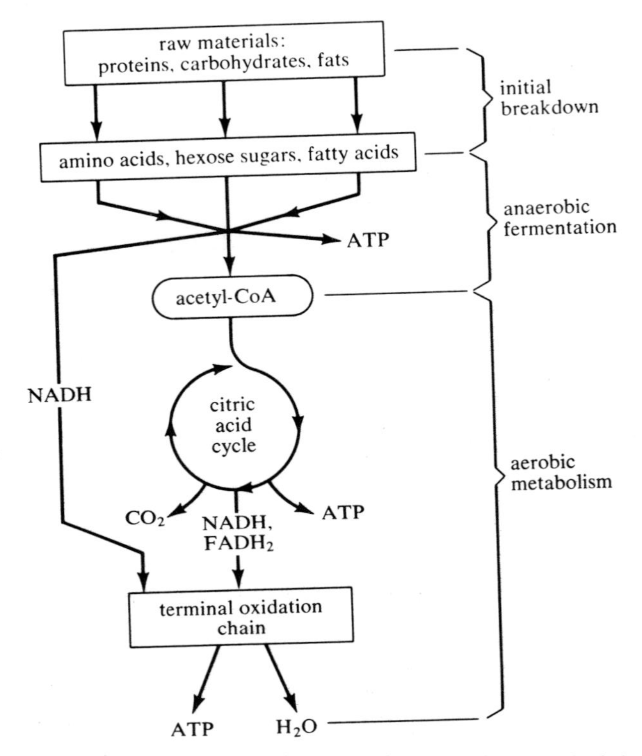

FIGURE 7-7. *A schematic outline of the metabolic machinery which breaks down foods and stores chemical free energy as the high energy bonds of* ATP.

TABLE 7-3 FREE ENERGY FLOW IN THE METABOLISM OF GLUCOSE

Reaction	ΔG^0 (kcal)	*Coupling*
1. glucose + $O_2 \rightarrow 2$ pyruvate$^-$ + $2H^+ + 2H_2O$	−140.1	8 ATP = + 64.8
2. 2 pyruvate$^-$ + $2H^+ + O_2$ + 2 coenzymeA $\rightarrow$ 2 acetyl-CoA + $2H_2O$ + $2CO_2$	−123.6	6 ATP = + 48.6
3. 2 acetyl-CoA + $4O_2 \rightarrow$ 2CoA + $2H_2O$ + $4CO_2$	−422.2	24 ATP = + 194.4
1 → 3. glucose + $6O_2 \rightarrow 6H_2O + 6CO_2$	−685.9	38 ATP = + 307.8

$$\text{efficiency of transfer} = \frac{307.8}{685.9} = 0.45$$

pyruvate are converted to 2 moles of acetyl-CoA, with a drop of 124 kcal in free energy and a synthesis of ATP. Finally, these two acetyl-CoA moles are degraded to CO_2 and water by the processes which we have been examining, dropping 422 kcal in free energy and synthesizing 24 moles of ATP. As Table 7-3 shows, the over-all efficiency of transfer of free energy from 1 mole of glucose to 38 moles of ATP is 45%.

Significance of $\widetilde{\Delta G}$

The reaction potential figures with which we have been working have all been standard reaction potentials. The true reaction potentials, as we have seen, depend upon the concentrations of reactants and products. If a large free energy output is to be obtained from a chemical reaction, then the concentrations of reactants must be kept high and the products, low. In an enclosed system, products will accumulate and reactants will be depleted, with a steady decrease in the free energy of the reaction until finally the point is reached at which $\widetilde{\Delta G}$ is zero. This, of course, is equilibrium.

The living machine is designed in such a way as to keep the $\widetilde{\Delta G}$ of an essential reaction as nearly constant as possible. This is accomplished by maintaining the input of reactants and by continuously removing products by diffusion, evaporation, excretion, and so on. A living organism is an open system, and, moreover, is one which maintains nearly a steady state, rather than a closed system which is eventually doomed to equilibrium.

There are two ways of looking at the $\widetilde{\Delta G}$ of a reaction. The first way is to consider $\widetilde{\Delta G}$ as a measure of the *available heat* (available for doing useful work) as opposed to the *total heat*. Diagrams showing the relationship between useful or available heat and total heat are given in Figure 7-8 for

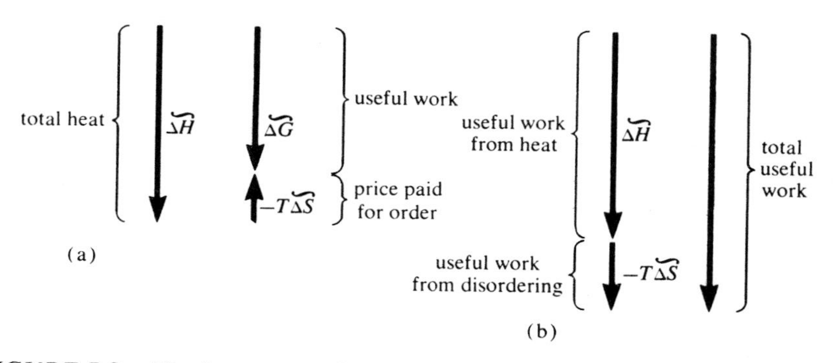

FIGURE 7-8. *The free energy of a reaction compared with the enthalpy change for* (a) *ordering, and* (b) *disordering during the course of the reaction.*

two different cases. Figure 7-8a shows a process in which the entropy decreases (order increases). Increasing the order of the system is not a something-for-nothing process. It requires work, and this work is manifested as nonfunctional $T\Delta S$ heat. The total heat emitted is still $\widetilde{\Delta H}$, but since $T\Delta S$ of it is tied up in the entropy process, only $\widetilde{\Delta G}$ of it is available to do useful work. The greater the ordering of the system, the more of the total heat is tied up, and the smaller the amount of heat left over to do useful work. On the other hand, for a reaction in which entropy rises and the system becomes more disordered, this disordering can be used to do useful work. Such a process is depicted in Figure 7-8b. The total amount of useful work in this process is then the sum of the $\widetilde{\Delta H}$ and $-T\Delta S$ terms.

Another way of looking at $\widetilde{\Delta G}$ is to consider it as a measure of the tendency of a system to undergo reaction. If there is a large excess of reactants, then $\widetilde{\Delta G}$ will be large in magnitude and negative, producing a large tendency for the reaction to go. Conversely, if products are in excess, the reaction will be reversed and $\widetilde{\Delta G} > 0$. In either case, $\widetilde{\Delta G}$ is a convenient measure of the drive of a system to reaction. The reaction potential in this view is analogous to the flow potential or hydrostatic head at a dam. If the dam were placed in a closed system and the water were let through to drive the turbines, the hydrostatic head would decrease as the water level below the dam rose, and that above the dam fell. Ultimately the water levels on both sides of the dam would become equal and a hydrostatic head would no longer exist. Such a situation corresponds to equilibrium in a closed chemical system, with $\widetilde{\Delta G} = 0$. However, in an open system, the reservoir above the dam would constantly be replenished and the water below the dam would be continuously drained away. The hydrostatic head would be maintained in much the same way as a large $-\widetilde{\Delta G}$ is maintained in an open biological system.

7-6 LIVING SYSTEMS: GENERAL THERMODYNAMIC CONSIDERATIONS

The second law, as we have learned it, is a death sentence to any isolated system. Any useful process that we want to carry out occurs between a high and a low free energy state. A heat engine must operate with a temperature gradient. Electrochemical cells must operate with chemical species whose reaction liberates free energy, or else with concentration gradients of a particular species. Other types of chemical engines are faced with the same limitations. In all cases, some sort of a gradient is needed to drive these processes, yet the processes themselves act to erode the gradients. Once these gradients are destroyed, the system is at equilibrium and can perform no more useful work.

Closed Systems

By Eq. 6-41, the free energy of any isolated reaction will approach zero as reactants are used up and products are formed, and equilibrium is approached. The usefulness of each successive unit of the reaction will be less than that of the unit before it. This will be equally true of each of the possible reactions between components in a more complex chemical system in isolation. Moreover, temperature gradients in the system will decrease with time. By the second law, every time that some heat is converted into work, some heat is also lost to a lower temperature. Only as long as there exist differences in temperature in the isolated system can heat-engine work be done, yet the operation of such processes leads to the elimination of temperature differences. Similarly, work can be obtained from concentration-cell processes only as long as there are concentration gradients across the system, yet operation of such processes spreads the components throughout the system and destroys the concentration gradients. As high free energy compounds are degraded to do work, the supply of such compounds decreases. Eventually, when the temperature everywhere is the same, when there is nothing left but a uniform mixture of low free energy compounds, then the system is thermodynamically dead.

Open Systems

A study of open systems as they exist in living organisms provides quite a contrast. These open systems are characterized by a steady state rather than an approach to equilibrium. Reactants are continuously added and products are continuously removed, so that $\widetilde{\Delta G}$ is kept large in magnitude and negative. This situation ensures that much useful work is obtained from every stoichiometric unit of chemical reaction. The efficiency of these chemical engines, as we have seen from previous examples, is usually high.

The particular types of open systems that we are considering—living systems—are not heat engines. The reason is obvious. Since we do not have significant temperature gradients within us, the efficiency of any heat-engine process would be very low. The body is essentially an isothermal system. Another characteristic of living systems, which again requires that they be open, is their high degree of organization. Entropy in these open systems is kept very low, and the systems are highly organized both in space and in time. The organization in space is immediately obvious, in the complex structures whose integrity is required if the organism is to function. But it is perhaps less often realized that the metabolism of a living organism is an intricate organization of chemical processes in time. Things must not only happen; they must happen at the right moment and in the right sequence if the elaborate chemical structure is to maintain itself. Such systems are by nature unstable, or at best metastable. As long as there is a constant influx of free energy and a constant drainoff of entropy, the system maintains its steady state. But as soon as this flow is ended, as soon as the free energy source is cut off and the entropy begins to rise, then the system begins to approach the equilibrium state which is conventionally known as death.

There exists a vitalist paradox which deals with open and closed systems. According to the second law, the entropy of any irreversible adiabatic process must always increase. On the other hand, it is a fact that the entropy of living systems remains small in spite of the irreversibility of their processes. Therefore, this argument runs, these living systems must violate the second law, which, as everybody knows, is inviolable. The conclusion is often drawn that this requires there to be some extra-material principle or agency which is responsible for maintaining living systems in a state of low entropy.

As with most paradoxes, a fundamental error of definition is at the root of the contradiction. Living systems are open systems, and the second law as employed above does not apply to open systems. In order to apply this entropy argument to a living organism, one must include with the organism all of its environment which exchanges matter, work, or energy with it. The system of organism-plus-total-environment is then a closed system, and, for such a system, entropy does indeed increase just as predicted by the second law.

Man in a Space Capsule—Example of a Closed System

A man living in a space capsule is a clear-cut example of a closed system. The entropy of the man is kept low as long as he is alive; but in order to accomplish this, the entropy of the space capsule has to increase. What had been food and oxygen now becomes carbon dioxide, water, and excrement. The over-all entropy of these waste products is sufficiently greater than the entropy of the foodstuffs and oxygen to counterbalance the low entropy of the man. Over-all entropy will increase with time; this fact places a limit on the length of time the man can remain in such a closed system. Eventually he will run out of high free energy components, the capsule will fill up with less organized components, and he will have to return to earth. By returning to earth and opening the capsule in order to remove waste materials and re-stock his supply of fuel and foodstuffs, he once again produces an open system.

The astronaut could avoid coming back for supplies by setting up some sort of a hydroponics farm on board. Rather than getting rid of waste products and restocking with new foodstuffs, he could feed the waste products to the algae in the hydroponics tanks and let the algae synthesize new high free energy compounds. The astronaut could live off these new foods, and the cycle could be maintained indefinitely in principle. But this would once again be an open system. Although matter would not be flowing across the boundaries of the system, energy would, in the form of the sunlight which drove the photosynthetic reactions of the algae. In this process, all of the carbon, nitrogen, oxygen, and other atoms involved in the photosynthetic and nutritional cycle could be regarded as the working substances in a machine designed to transfer free energy from the sun to the astronaut, or, alternatively, to absorb solar energy and use it to maintain a state of low entropy in the astronaut.

It is still true that any closed system works on borrowed time. In this respect, entropy has become popularized (by Eddington and Blum) as "time's arrow." On a macroscopic, statistical scale, there is a common direction in which all processes go spontaneously, and that is the direction of increasing entropy. Thus in a sense, entropy is the arrow which points out the way time is really flowing, in spite of the reversibility of simple microscopic mechanical processes.

Living organisms need a constant supply of free energy coupled with a constant elimination of entropy. As soon as this process is interrupted, the system breaks down. The most intricately structured parts of the system are the first to break down, and with humans this is the brain. A few minutes without oxygen and the brain suffers permanent damage, even though the rest of the body may recover. Exaggerating only slightly, one could say that for eighty years man struggles to maintain an island of low entropy in a sea of disorder; yet within minutes after his death, the universe is more disordered than if he had never existed. Trying to bring order into the universe as a whole through the existence of living systems is like trying to cool a large room by keeping the refrigerator door open.

7-7 THE PLANET AS AN OPEN THERMODYNAMIC SYSTEM—ORIGIN OF LIFE

Because we receive energy, if no significant amount of matter, from outside of our planet, the earth must be considered an open system. Living organisms all over the planet require high free energy compounds as sources of fuel. Where do they find these high free energy substances?

One of the first processes that developed when life was evolving was the process of anaerobic metabolism. Based on the best theories on the origin of life, there seem to be good natural reasons for high free energy compounds to have been synthesized in the primitive atmosphere and in the primitive seas. Out of this broth or soup, as people have referred to it, evolved the integrated chemical systems that we now call living organisms. These first organisms maintained their structure and metabolism by degrading high free energy compounds and ejecting the degradation products back into the soup.

One of the reasons that this theory is so widely accepted is that there are vestiges of this anaerobic metabolism in all living organisms. The first half of the pathway of glucose metabolism is an anaerobic process. In man, this anaerobic process degrades glucose to two molecules of lactic acid in a sequence that is coupled to the synthesis of two molecules of ATP. The ΔG of this process is -50 kcal, and its over-all efficiency is $e = 16 \text{ kcal}/50 \text{ kcal} = 0.32$. In other organisms, this process is slighly altered. In yeast, for example, two molecules of ethanol are formed instead of lactic acid. The $\widetilde{\Delta G}$ of this process is -56 kcal with an efficiency of 29%. One of the results of sudden, vigorous exercise is a buildup of lactic acid in the muscles and blood stream. This increase in lactic acid sometimes causes muscle cramps; and when the lactic acid is dissipated by a subsequent aerobic process, the discomfort disappears. Anaerobic metabolism of glucose is the same in humans and yeast except for the last step, when the process in yeast produces ethanol rather than lactic acid. The British biochemist Jevons has remarked that if only a few enzymes were changed in our metabolic system, how much more popular exercise would be!

The aerobic part of glucose metabolism is considerably more efficient. The breakdown no longer ends with lactic acid. Instead, another intermediate, pyruvic acid, is converted to acetyl-CoA which then enters the citric acid cycle and emerges as water and CO_2 at the end of the terminal oxidation chain. In this complete process, $\widetilde{\Delta G}$ is -686 kcal, and the over-all efficiency is about 45%. This metabolic process is coupled to the formation of 38 molecules of ATP per molecule of glucose, and not two. Not only is there an increase in the absolute *amount* of free energy retained per glucose molecule, but the efficiency of the conversion rises by 15%. The anaerobic metabolism just described is common to all living organisms on the earth. The aerobic process evolved at a later date and is not nearly so universal. The anaerobic part of the pathway is the older core of the process, inefficient as it may be.

If high free energy compounds were necessary for life to evolve, where did they come from? There are four likely energy sources for sythesizing these compounds abiologically: electrical discharge, ultraviolet light, natural radioactivity, and thermal energy. In 1953, Miller found that passing an electrical discharge through a replica of the postulated original atmosphere gave rise to simple amino acids and small organic compounds. Similar experiments using ultraviolet light as the energy source have had much the same results as the electrical discharge experiments. Recently, Calvin has found that the β decay of radioactive potassium is of the right energy to convert five molecules of HCN to one of adenine. If adenine was indeed formed in this manner during primitive times, then this discovery of Calvin's may hold the answer to why the adenine base has such a central position in biological systems rather than any of the other purine or pyrimidine bases.

Fox has discovered that thermal energy from hot springs and vulcanism can synthesize and polymerize amino acids in a way that may be relevant to the initial origin of proteins and of protozoan cell membranes. He has proposed that high free energy compounds could have been synthesized in the proper environments by use of the free energy available as heat.

The earth is thought to have had a reducing atmosphere in primitive times. It would have contained H_2, NH_3, HCN, N_2O, CO_2, CO, small hydrocarbons, but *no* free O_2. Without free O_2 or other oxidizing agents, high free energy compounds would have been relatively stable and could have accumulated in large quantities. The synthesis by electrical discharge, ultraviolet light, radioactivity, and heat would have been a slow but cumulative process. But once the enclosed packets of chemical reactions called living organisms had evolved from this primitive soup, their better-organized metabolisms could degrade high free energy compounds faster than they could be synthesized abiogenically. If no further sources of synthesis had been found, then the level of life would have been severely restricted by the limited availability of high free energy compounds.

Alternate Energy Sources—Photosynthesis

In the face of a dwindling supply of natural metabolites, the ability to use other sources of free energy would have been a tremendous advantage to survival. One source was inorganic matter, and even today, in deep petroleum deposits and similar out of the way corners removed from competition, there exist chemautotrophic bacteria which use inorganic sources of energy for synthesizing high free energy compounds. However, these anomalies are definitely an evolutionary backwater compared to another class which found a better source.

This better source was the sun, the largest energy source around, and the one which keeps our planet an open system. At an early stage of evolution some living systems developed the ability to use and later to synthesize molecules which could be electronically excited by photons from the sun. As these excited molecules returned to their ground states, the energy then released was coupled to reactions synthesizing the much needed high free energy compounds. Plants currently use the Mg porphyrin known as chlorophyll to trap solar energy and to use it in synthetic processes. Living systems which developed this photosynthetic process were henceforth no longer dependent upon their environment for high free energy compounds. They could make their own and store them until needed.

A third class of organisms failed to develop either energy source, or any other at all. An outside observer might have given these metabolic delinquents small chance of survival. But they did develop three outstanding properties: motion—the ability to go after food, sensory input—the ability to detect where the food is, and intelligence—the ability to use the sensory input to make the motions purposeful. Living things came to be divided into two classes: those which became efficient, the plants, and those which ate the efficient ones. The moral here seems to be that you do not have to be smart if you are ruthless.

This raises an interesting philosophical sideline. Most of us would regard animals, and particularly man, as being in some sense superior to plants.[3] It has been proposed rhetorically that this is one species' criterion, and that it is not the judgement that an oak tree would make. But oak trees do not make judgements at all, and this ability to judge is a property peculiar to animals, and to man in particular. But in what ways is man superior to an oak tree? We would say that the attributes which mark his superiority are mobility, better sensory data input, better sensory data analysis, independence of action based upon the analysis of these data, and, to a certain extent, intelligence and reason. But interestingly enough, all of these attributes are the talents of the predator. Animals, in lieu of their own abilities to make foods, had to develop mobility. They had to develop the ability to detect foods and to collect them by means of purposeful and efficient activity. Their initial goal was a simple one: to find their own dinner without becoming someone else's. But out of this effort evolved the nervous system, and out of the nervous system developed all that we would call human. So all of our "higher" attributes, all of our capacities for cultural, moral, and ethical refinements have their roots in talents which are the hallmarks of the predator.

Aerobic Metabolism

One of the by-products of photosynthesis was the liberation of oxygen into the atmosphere As a result, the atmosphere changed from a reducing one to an oxidizing one. This drastic change had several major consequences. First of all, previous conditions for the repeated evolution of life were destroyed. Compounds which had been stable in a reducing atmosphere were now subject to oxidation. Of course, this change was not fatal to those forms of life which by this time had developed alternate free energy sources.

A second result of the advent of aerobic metabolism was the development

[3] The question of the "superiority" of man is a touchy one to partisans on both sides of the philosophical fence. But, there is one criterion which arises directly from the discussion of this section, and which seems undeniable. Many organisms are exquisitely adapted for a particular environment, and as long as the environment does not change, or changes slowly compared with the generation time of an individual, the species can adapt and survive. But the higher vertebrates have evolved a general-purpose mechanism for coping with much more rapid fluctuations in the environment—a highly developed and efficient central nervous system. If mollusks, oak trees, and ants are adapted for a particular environment, then the higher mammals and man *par excellence* are adapted *for adaptation* to new environments. This is surely a higher order of behavior pattern.

of an ozone layer about the planet which screened out much of the ultraviolet light from the sun. This change was good in the sense that unfiltered ultraviolet light would be fatal to organisms living on the surface of the planet. Previously, life had been confined to the depths of the seas where ultraviolet light could not penetrate, but now life could spread over the entire planet.

The most important result of the switch to an oxidizing atmosphere was the development of the terminal oxidation process which allowed for more efficient use of high free energy compounds. As we have seen, aerobic metabolism is much more efficient than anaerobic metabolism. The two processes are compared in Figure 7-9, using glucose as a type metabolite.

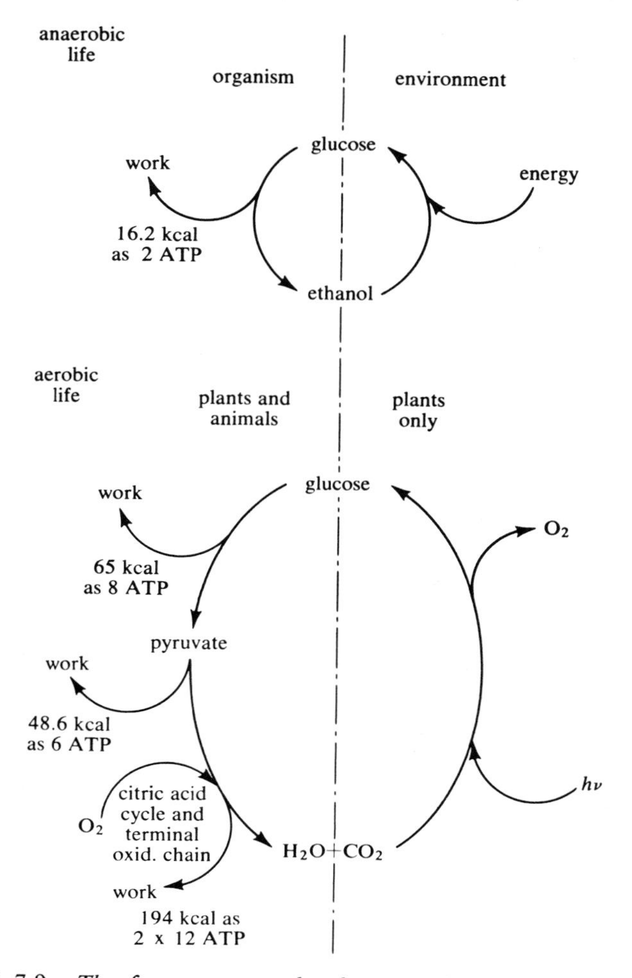

FIGURE 7-9. *The free energy cycles for anaerobic life* (*top*) *and aerobic life* (*bottom*).

With the evolution of aerobic metabolism, the basic pattern was set, and the subsequent history of life is one of expansion and diversification.

7-8 *LARGER THERMODYNAMIC SYSTEMS*

An over-all view of our planet would show it to be an open system nourished by energy from the sun. Life on its surface has been characterized by a competition for survival, first on the chemical level, next on the macromolecular level, and then on the biological level.[4] The most efficient and stable systems (in a dynamic sense) have the best chance of survival. Particularly stable are those isolated portions of space that are highly organized both in space and in time—living systems.

The more complex systems soon show themselves to have higher survival value, for the obvious reason that the right kind of complexity means a greater flexibility of response to stimulus and greater possibilities of adaptation to changing conditions. There is a steady increase of order in the course of evolution of the planet and a steady lowering of the entropy of the planet. One might be tempted to apply the vitalist paradox involving the second law to the planet as a whole. But this is as wrong as for individuals, and for the same reason. All of this process of ordering is being driven by the constant input of free energy from the sun. If entropy determines which way the hands of our planetary clockwork will move, then the sun is the mainspring.

The Solar System as a Closed System

It is reasonable to consider the solar system as a thermodynamically closed system. It is true that stars are used for navigation, and that, according to information theory, this influx of information to the solar system can be considered as an influx of negentropy (presumably the navigator would be slightly less confused, and slightly more ordered than before), but the total amount of energy, mass, or negentropy which flows across the boundaries of the solar system is negligible. Within these limits, the second law as proposed for isolated systems *does* apply. The solar system *is* running down. The total entropy of the sun plus all of its planets does increase daily. The increase of entropy on the sun as it burns down far surpasses the local fall of entropy on our particular planet. No matter how low the entropy might temporarily become on earth, when the sun burns out the game will be over.

This assumes that we cannot turn the solar system eventually into an open system. How might this be accomplished? The distances involved make it unlikely that we shall ever find it practical to bring in matter for fuel.

[4] We are now going through a critical phase in cultural or social evolution on this planet, which may show whether or not a highly developed and integrated central nervous system really does confer a higher long-range survival value on the species which possesses it.

Similarly, it is unlikely that we shall ever be able to tap an extra-solar energy source to any practical extent. The answer, if there is any, seems more likely to be that we shall open the system by absenting ourselves and going elsewhere. There is no hurry. The sun is approximately five billion years old, roughly half the age of the galaxy. The probable total lifetime of a star of the sun's class is around 50 billion years. But we might expect a metamorphosis to a form of red giant in around five billion more years, so we are roughly half way through the useful working life of this particular star. In view of the fact that this available time represents approximately five thousand times the lifetime of *Homo sapiens* so far, we can safely let the future take care of itself, and be more concerned with the disorder that we ourselves create on this particular planet.

The Universe

The solar system can be considered to be an open system by a considerable stretch of the imagination, but what about the universe in its entirety? If the material universe is considered to be a closed thermodynamic system, and if the laws of thermodynamics as we know them are assumed to hold throughout all time and all space, then the ultimate entropy death of the universe is inescapable. The entire history of the evolution of life on this planet, and on any other where it might happen to have appeared, is then only a ripple on a steadily descending stream. When the last star has burned down, when the last temperature differential has vanished, when the last high free energy compound has been degraded with no energy sources to reform it, then the history of the universe will be at an end.

Is there any alternative to the entropy death of the universe? There is none if the two assumptions above are valid, and it must be emphasized that there is not one shred of material evidence that they are wrong. Nevertheless, they leave some uncomfortable voids in the direction of beginnings as well as endings, and it is worth examining them both.

Violations of the first assumption are far too complex to be discussed here, and will be left for others to debate. But possible deviations from the laws of thermodynamics as we know them should be examined. These laws, as has been repeatedly stated in this book, are not universal and somehow ideal principles. They are *summaries of experience*. When one is forced to decide whether he believes that they will continue to hold at times and places with which he has had no experience, he must assess the relative probabilities of their holding or not holding, along the lines of Section 2-14. Virtually every rational person would assume that the laws of thermodynamics, like the laws of physics, will hold for all times and places which he is likely to experience. In fact, a propensity for this kind of assessment of probabilities is a good empirical definition of rationality. Santayana provided a beautiful practical rebuttal to solipsism with the remark that "Perhaps there *is* no real and objective material world, external to me and independent of me, which is responsible for the sensory impressions which I experience. Nevertheless, I have found by experience that it is to my advantage to act *as if there were*." Similarly, we can say "Perhaps the laws of thermodynamics, and the laws of physics in general, do not hold in new and untried circumstances. Nevertheless, people who assume that they do, and act accordingly, get on noticeably better than those who assume that they do not." Science makes progress only by proposing simple models and by holding to them until experimental facts force a change. One such simple model is the universality of the laws of thermodynamics.

But what about events and processes vastly removed from our own time and space; in particular, what about the question of origins and cosmology? May not the laws of conservation of mass-energy and of the steady increase of disorder be particular expressions of more general principles, peculiar to the present state of the universe, in the sense that nonquantized Newtonian mechanics is a limiting case of quantum mechanics for large dimensions and great masses? Such speculations are poor bases for planning scientific experiments in the here and now, but they are valid starting points for cosmology. There are three models which are serious contenders for an explanation of the origin of the universe

(1) The *expanding universe* model of Lemaitre, Hubble, and Eddington, which explains the observed red shift in stellar spectra by postulating that all of the matter in the universe was concentrated in a fantastically dense "super atom" some 12 billion years ago (assuming a uniform expansion rate), and has been expanding since. This is often called the "big bang" theory.

(2) The *pulsating universe* model, which avoids the Eddington "super atom" and likewise the question "What came just *before* the big bang?" by postulating a recurring cycle of alternating expansions (such as we are now experiencing) and contractions to some minimum radius.

(3) The *steady-state* model of Bondi, Gold, and Hoyle, which postulates an eternal yet expanding universe with continuous creation of new matter in interstellar space.

These three models are compared in Figure 7-10, which shows a measure of the size of the universe R and of the mean free energy per unit mass $\bar{G}$, plotted against time.

Each of these models has its difficulties with the laws of thermodynamics as we have learned them. The expanding universe model is in harmony with the first and second laws for all times *after* the primal event, but is silent about what happened, or what existed, just before. The steady-state model eliminates the problem of a moment of creation by denying its existence, but does this at the price of discarding the first law of thermodynamics as we know it. The pulsating model avoids the need for creation and retains the first law at the price of doing some strange things with entropy, or time, or both.

During an expansion period in the pulsating model, the galaxies move steadily apart as in the Eddington model. There is an irreversibility of real macroscopic phenomena, measured by a steady drive toward states of greater disorder. The entropy of the universe increases in all real processes. But what happens when the expansion has reached its crest and the galaxies begin to rush back to form the node before a new expansion? During the contraction period, there must be a spontaneous tendency toward the restoration of order, if the state of the universe at the end of the contraction is to be identical to that before expansion began. Entropy will *decrease* spontaneously, and the free energy which had been dissipated during the expansion process will reaccumulate. The second law will be reversed.

Is this a valid reason for rejecting the pulsating model out of hand? It is not, for it recasts the second law in a more general form which contains the old version as a special case applicable to all parts of space and all moments

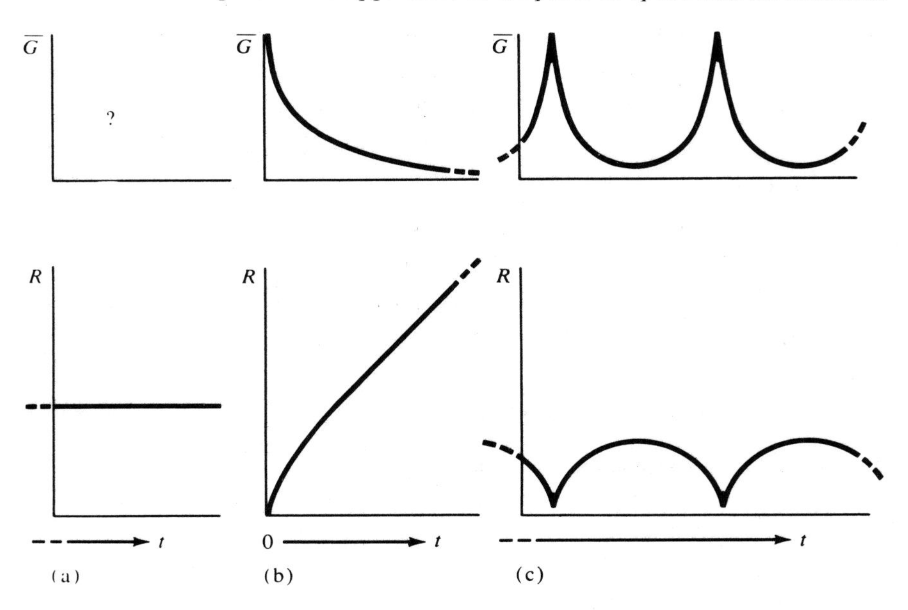

FIGURE 7-10. *A schematic plot of the size of the universe R and the mean free energy per unit of mass $\bar{G}$ as a function of time for* (a) *the steady state model,* (b) *the "big bang" model, and* (c) *the oscillating or pulsating universe model. The British astronomer Sandage and others have estimated that the period between nodes in model* (c) *might be something of the order of 80 billion years. Evidence from the red shift in stellar spectra suggests that we are presently something like 12 billion years to the right of a node, or of the ordinate in model* (b).

of time with which we shall ever have contact. Such a more general form can neither be proved nor disproved by experiments conducted in our own time and space.[5] There is another alternative, however. Boltzmann was disturbed about his entropy concept, and tried to define *time* itself as a measure of change whose positive direction is that in which entropy increases during a spontaneous process. By this definition the second law is automatically true, but one is left with the disturbing conclusion that *time itself* flows in opposite directions during the phases of expansion and contraction.

We have come a long way from cannon borings and primitive steam engines, to living system, cosmology, and time reversal. Few ideas in science have been so universal and all encompassing as those of thermodynamics, especially when coupled with molecular theory. What began as the most mundane of experiments with heat and work has grown into a quasi-philosophical framework of thought of great generality. And the fringes of the subject, where the applicability is open to argument, lead us to some of the most fundamental questions which can be asked.

Perhaps it is fitting to break up the solemnity of this discussion and to end this book with a parody of the end of Tennyson's "Locksley Hall."

> Yet I doubt not through the ages one increasing function runs,
> And the thoughts of men are numbered with the process of the suns.
>
> Massive constellations burning, mellow moons and golden skies,
> Then a star begins its cycle, and a solar system dies.
>
> I would rather hold it better men should perish one by one,
> Than that earth should stand forever till the universe is done.
>
> 'Mid the shadow of disorder, let our fleeting efforts grow,
> Better eighty years of striving than a cycle of Carnot.

REFERENCES AND FURTHER READING

Bioenergetics and Biological Mechanisms

H. F. Blum, *Time's Arrow and Evolution* (Princeton University Press, Princeton, 1969), Rev. ed. An influential classic that first appeared in 1951. One of the best treatments anywhere of the significance of entropy in the evolution of living organisms.

R. E. Dickerson and I. Geis. *The Structure and Action of Proteins* (Harper and Row, New York, 1969). The principles of folding and the structures of protein molecules, especially enzymes, and the way in which the structure

[5] It is possible, however, to choose between these models by looking for the *consequences* of the events involved, in our present universe. These, in the form of galaxy population counts, interstellar matter, and background radiation, are now being actively studied by astronomers.

of an enzyme molecule is responsible for its catalytic action. Extensively illustrated.

I. M. Klotz, *Energy Changes in Biochemical Reactions* (Academic Press, New York, 1967).

A. L. Lehninger, *Bioenergetics* (W. A. Benjamin, Inc., New York, 1965). Two good treatments of the subject at about the level of this chapter.

H. J. Morowitz, *Energy Flow in Biology* (Academic Press, New York, 1968). A more advanced treatment, including the inevitable thermodynamic evolution of order in open systems.

H. A. Krebs and H. L. Kornberg, *Energy Transformations in Living Matter* (Springer-Verlag, Berlin, 1957). A well-known and standard compilation of thermodynamics data for metabolic processes.

M. J. Johnson, " Enzymic Equilibria and Thermodynamics," in *The Enzymes*, P. D. Boyer, H. Lardy, and K. Myrbäck, Eds. (Academic Press, New York, 1960), 2nd ed., Vol. 3, p. 407. A similar compilation to that of Krebs and Kornberg.

Chemical Evolution and the Origin of Life

J. D. Bernal, *The Origin of Life* (World Publishing Company, Cleveland, 1967).

A. I. Oparin, *Life: Its Nature, Origin and Development* (Oliver and Boyd, Edinburgh, 1961).

Two books by pioneers in the field.

I. S. Shklovskii, and C. Sagan, *Intelligent Life in the Universe* (Holden-Day, San Francisco, 1966). A rambling but absorbing dialogue between a Russian and an American astronomer.

E. Schrödinger, *What is Life?* (Cambridge University Press, Cambridge, 1962) (1st ed. 1944). The inspiration for much of postwar molecular biology. Not as revolutionary as it once was, but interesting still because of its author.

Thermodynamics, Time, and Cosmology

H. Bondi, *Cosmology* (Cambridge University Press, Cambridge, 1952).

F. Hoyle, *Galaxies, Nuclei and Quasars* (Harper and Row, New York, 1965). Two excellent introductions to cosmology by the authors of one of the main models.

G. J. Whitrow, *The Structure of the Universe* (Hutchinson's University Library, London, 1950). An older but very readable introduction.

G. J. Whitrow, *The Natural Philosophy of Time* (Nelson, London, 1961). See especially the two chapters on the statistical theory of time, including Boltzmann's efforts, and Feynman's explanation of positrons as electrons moving backward in time. An extension of this idea would be the explanation that all of the electrons in the universe are indistinguishable because they are all the same electron.

EXERCISES

7-1. Compound A is broken down into E by the following four-step cyclic process:

$$
\begin{aligned}
A + B &\rightarrow C & \Delta G^{o} &= +\ 1 \text{ kcal} \\
C &\rightarrow D & \Delta G^{o} &= -55 \text{ kcal} \\
D &\rightarrow E + F & \Delta G^{o} &= -\ 3 \text{ kcal} \\
F &\rightarrow B & \Delta G^{o} &= -82 \text{ kcal}
\end{aligned}
$$

(a) What is the standard free energy of the reaction $A \rightarrow E$?
(b) Coupled to the second step is the reaction $G \rightarrow H$, for which $\Delta G^{o} = 60$ kcal. How is it that a reaction with a ΔG^{o} of 60 kcal can be made to go by one of only -55 kcal?
(c) Assuming equimolar concentrations of G and H, what would have to be the relative concentrations of C and D for perfect free energy coupling in this step?
(d) Demonstrate rigorously that the total free energy change in one turn of the cycle is independent of the concentrations of B, C, D, and F.
(e) If the same reaction $G \rightarrow H$ is also coupled to the fourth step, what is the over-all efficiency of transfer of free energy of the cycle under standard conditions?

7-2. (a) What is the thermodynamic advantage of coupled reactions in living systems?
(b) Why does the maintenance of low entropy in a living organism not violate sound thermodynamic principles?
(c) Why do living organisms not use heat-engine processes as energy sources?
(d) In one step of the citric acid cycle, the following reaction

$$\text{malate} \rightarrow \text{oxaloacetate} + 2[\text{H}] \qquad \widetilde{\Delta G}^{o} = -45.3 \text{ kcal}$$

is used to drive the following reaction

$$\text{NAD}^{+} + 2[\text{H}] \rightarrow \text{NADH} + \text{H}^{+} \qquad \widetilde{\Delta G}^{o} = 52.4 \text{ kcal}$$

In view of the reaction potentials as given, how is it that the pair of reactions is thermodynamically spontaneous? (The mechanical reason for the coupling of these two particular reactions involves the enzyme malate dehydrogenase, a macromolecule composed of two polypeptide chain subunits of 32,500 molecular weight each, which is unfortunately outside the scope of this chapter.)

7-3. (a) What is the difference between a closed and an open thermodynamic system?
(b) Is the city of Los Angeles a closed or an open system?
(c) Is the planet earth a closed or open system?
(d) Is the solar system a closed or open system?
(e) Is the entire material universe a closed or an open system? Justify your answer carefully.
(f) What does the second law have to say about closed systems, and in what way is this situation changed in an open system?
(g) What proof is there that the three great principles, or laws, of thermodynamics hold for all places in the universe and have held for all times past and to come?

APPENDIX 1 PHYSICAL CONSTANTS AND CONVERSION FACTORS

Avogadro's number	$N = 6.0232 \times 10^{23}$ molecules/mole
Electron charge	$e = 1.60206 \times 10^{-19}$ coulomb
Faraday's constant	$F = Ne = 9.6495 \times 10^{4}$ coulomb/equivalent
Electron rest mass	$m_e = 0.91083 \times 10^{-27}$ g
Proton rest mass	$m_p = 1672.4 \times 10^{-27}$ g
Speed of light	$c = 2.99793 \times 10^{10}$ cm/sec
Boltzmann's constant	$k = 1.38044 \times 10^{-16}$ erg/deg
Gas constant	$R = Nk = 8.31470 \times 10^{7}$ erg/deg mole
	$= 1.98726$ cal/deg mole
	$= 82.054$ cm^3 atm/deg mole
Planck's constant	$h = 6.62517 \times 10^{-27}$ erg sec
	$\hbar = h/2\pi = 1.05443 \times 10^{-27}$ erg sec
	$Nh = 3.99047 \times 10^{-3}$ erg sec
	$hc = 1.98618 \times 10^{-16}$ erg cm
	$hc/k = 1.43880$ cm deg
1 electron volt (eV)/molecule	$= 23.063$ kcal/mole
	$= 8066.04$ cm^{-1}
	$= 1.6021 \times 10^{-12}$ erg/molecule
1 kcal/mole	$= 349.73$ cm^{-1}
	$= 0.043360$ eV/molecule
	$= 6.9465 \times 10^{-14}$ erg/molecule

APPENDIX 2 SELECTED PROPERTIES OF DIATOMIC MOLECULES[a]

Molecule	Disso-ciation energy D_o (eV)	Reduced mass μ (at. units)	Funda-mental vibration frequency ω_e (cm^{-1})	First anhar-monicity constant $\omega_e x_e$ (cm^{-1})	Rotation constant B/c (cm^{-1})	Equilib-rium inter-nuclear separa-tion r_e (Å)
H_2	4.476	0.504	4395.2	117.9	60.800	0.742
HD	4.511	0.672	3817.1	95.0	45.655	0.742
D_2	4.554	1.007	3118.5	64.1	30.429	0.742
Li_2	1.03	3.509	351.4	2.6	0.6727	2.672
B_2	3.6	5.506	1051.3	9.4	1.212	1.589
C_2^{12}	3.6	6.002	1641.4	11.7	1.6326	1.312
N_2^{14}	7.37	7.004	2359.6	14.5	2.010	1.094
O_2^{16}	5.08	8.000	1580.4	12.1	1.446	1.207
F_2^{19}	<2.75	9.502	892.1	...	...	1.435
P_2^{31}	5.03	15.492	780.4	2.8	0.3033	1.894
S_2^{32}	≤ 4.4	15.991	725.7	2.9	0.2956	1.889
Cl_2^{35}	2.48	17.489	564.9	4.0	0.2438	1.988
Se_2^{80}	≤ 3.55	39.971	391.8	1.1	0.0907	2.157
$Br^{79}Br^{81}$	1.971	39.958	323.2	1.1	0.0809	2.284
$Cl^{35}I^{127}$	2.152	27.422	384.2	1.5	0.1142	2.321
I_2^{127}	1.542	63.467	214.6	0.6	0.0374	2.667
HCl^{35}	4.430	0.9799	2989.7	52.1	10.591	1.275
HBr^{79}	3.754	0.9956	2649.7	45.2	8.473	1.414
HI^{127}	3.056	1.0002	2309.5	39.7	6.551	1.604
$C^{12}N^{14}$	...	6.464	2068.7	13.1	1.899	1.172
$C^{12}O^{16}$	...	6.858	2170.2	13.5	1.931	1.128
$N^{14}O^{16}$	5.29	7.469	1904.0	14.0	1.705	1.151
Na_2^{23}	0.73	11.498	159.2	0.7	0.1547	3.078
$Na^{23}Cl^{35}$	3.58	13.951	380	1	...	2.51

[a] Data excerpted from a much longer compilation in G. Herzberg, *The Spectra of Diatomic Molecules*, (D. Van Nostrand, Princeton, 1950), 2nd ed. See also *Infrared and Raman Spectra of Polyatomic Molecules*, by the same author.

APPENDIX 3 THERMODYNAMIC FUNCTIONS FOR A HARMONIC OSCILLATOR[a]

x	$\frac{C_V}{R}$	$\frac{E-E_o}{RT}$	$-\frac{G-E_o}{RT}$	$\frac{S}{R}$ [b]
0.00	1.0000	1.0000	∞	∞
0.05	0.9998	0.9752	3.0206	3.9958
0.10	0.9992	0.9508	2.3522	3.3030
0.15	0.9981	0.9269	1.9712	2.8981
0.20	0.9967	0.9033	1.7078	2.6111
0.25	0.9948	0.8802	1.5087	2.3889
0.30	0.9925	0.8575	1.3502	2.2077
0.35	0.9898	0.8352	1.2197	2.0549
0.40	0.9868	0.8133	1.1096	1.9229
0.45	0.9833	0.7918	1.0151	1.8069
0.50	0.9794	0.7708	0.9328	1.7036
0.55	0.9752	0.7501	0.8603	1.6104
0.60	0.9705	0.7298	0.7959	1.5257
0.65	0.9655	0.7100	0.7382	1.4482
0.70	0.9602	0.6905	0.6863	1.3768
0.75	0.9544	0.6714	0.6394	1.3108
0.80	0.9483	0.6528	0.5966	1.2494
0.85	0.9419	0.6345	0.5576	1.1921
0.90	0.9352	0.6166	0.5218	1.1384
0.95	0.9281	0.5991	0.4890	1.0881
1.00	0.9207	0.5820	0.4587	1.0407
1.10	0.9050	0.5489	0.4048	0.9537
1.20	0.8882	0.5172	0.3584	0.8756
1.30	0.8703	0.4870	0.3182	0.8052
1.40	0.8515	0.4582	0.2832	0.7414
1.50	0.8318	0.4308	0.2525	0.6833
1.60	0.8114	0.4048	0.2255	0.6303
1.70	0.7904	0.3800	0.2017	0.5817
1.80	0.7687	0.3565	0.1807	0.5372
1.90	0.7466	0.3342	0.1620	0.4962
2.00	0.7241	0.3130	0.1454	0.4584
2.20	0.6783	0.2741	0.1174	0.3915
2.40	0.6320	0.2394	0.0951	0.3345
2.60	0.5859	0.2086	0.0772	0.2858
2.80	0.5405	0.1813	0.0627	0.2440
3.00	0.4963	0.1572	0.0511	0.2083
3.20	0.4536	0.1360	0.0416	0.1776
3.40	0.4129	0.1174	0.0340	0.1514
3.60	0.3743	0.1011	0.0277	0.1288
3.80	0.3380	0.0870	0.0226	0.1096
4.00	0.3041	0.0746	0.0185	0.0931
4.25	0.2652	0.0615	0.0144	0.0759
4.50	0.2300	0.0506	0.0112	0.0618
4.75	0.1986	0.0414	0.0087	0.0501
5.00	0.1707	0.0339	0.0068	0.0407
5.50	0.1264	0.0226	0.0041	0.0267
6.00	0.08968	0.01491	0.00248	0.01739
6.50	0.06371	0.00979	0.00151	0.01130
7.00	0.04476	0.00639	0.00091	0.00730
7.50	0.03115	0.00415	0.00055	0.00470
8.00	0.02148	0.00269	0.00034	0.00303
9.00	0.01000	0.00111	0.00012	0.00123

[a] From N. Davidson, *Statistical Mechanics* (McGraw-Hill, New York, 1962).

[b]
$$x = \frac{h\nu}{kT} = \frac{1.4388\ \omega}{T} \qquad \frac{C_V}{R} = \frac{x^2 e^x}{(e^x - 1)^2}$$

$$\frac{E - E_0}{RT} = \frac{x}{e^x - 1} \qquad -\frac{G - E_0}{RT} = -\ln(1 - e^{-x})$$

$$\frac{S}{R} = \frac{x}{e^x - 1} - \ln(1 - e^{-x})$$

APPENDIX 4 SELECTED VALUES OF THERMODYNAMIC PROPERTIES AT 298.15°K AND 1 atm

The following table has been modified from one in H. A. Bent, *The Second Law* (Oxford University Press, Oxford, 1965) but stems ultimately from the publications of the National Bureau of Standards by Frederick D. Rossini, entitled "Selected Values of Chemical Thermodynamic Properties." Other convenient tabulations are to be found in the *Chemical Rubber Company Handbook of Chemistry and Physics* and *Lange's Handbook of Chemistry*. The best tabulations of thermodynamic properties of molecules involved in biological processes are to be found in H. A. Krebs and H. L. Kornberg, *Energy Transformations in Living Matter* (Springer-Verlag, Berlin, 1957) and in the *Chemical Rubber Company Handbook of Biochemistry*.

The elements in this table are arranged in groups as in the periodic table, from Ia through VIIb. The standard heat (enthalpy) and free energy of formation, ΔH^0 and ΔG^0, are in kcal/mole. Third law molar entropies S^0 and heat capacities C_P are in cal/deg mole. (c) means crystalline solid, sometimes with the common name or symbol of the crystal form.

PERIODIC ARRANGEMENT OF THE ELEMENTS

Period	Group I a	Group I b	Group II a	Group II b	Group III a	Group III b	Group IV a	Group IV b	Group V a	Group V b	Group VI a	Group VI b	Group VII a	Group VII b	Group VIII	Group O
1	1 H															2 He
2	3 Li		4 Be			5 B		6 C		7 N		8 O		9 F		10 Ne
3	11 Na		12 Mg			13 Al		14 Si		15 P		16 S		17 Cl		18 A
4	19 K		20 Ca		21 Sc		22 Ti		23 V		24 Cr		25 Mn		26 Fe 27 Co 28 Ni	
		29 Cu		30 Zn		31 Ga		32 Ge		33 As		34 Se		35 Br		36 Kr
5	37 Rb		38 Sr		39 Y		40 Zr		41 Nb		42 Mo		43 Tc		44 Ru 45 Rh 46 Pd	
		47 Ag		48 Cd		49 In		50 Sn		51 Sb		52 Te		53 I		54 Xe
6	55 Cs		56 Ba		57–71[a]		72 Hf		73 Ta		74 W		75 Re		76 Os 77 Ir 78 Pt	
		79 Au		80 Hg		81 Tl		82 Pb		83 Bi		84 Po		85 At		86 Rn
7	87 Fr		88 Ra		89[b]											

[a] Rare earths: 57 La, 58 Ce, 59 Pr, 60 Nd, 61 Pm, 62 Sm, 63 Eu, 64 Gd, 65 Tb, 66 Dy, 67 Ho, 68 Er, 69 Tm, 70 Yb, 71 Lu.

[b] Actinide series: 89 Ac, 90 Th, 91 Pa, 92 U, 93 Np, 94 Pu, 95 Am, 96 Cm, 97 Bk (berkelium), 98 Cf (californium), 99 E (einsteinium), 100 Fm (fermium).

	Substance	ΔH^0_{298}	ΔG^0_{298}	S^0_{298}	C_{P298}
	Hydrogen				
	H(gas)	52.089	48.575	27.393	4.968
	H^+(aq)	0.0	0.0	0.0	0.0
	H_3O^+(aq)	−68.317	−56.690	16.716	17.996
	H_2(gas)	0.0	0.0	31.211	6.892
Ia	Lithium				
	Li(gas)	37.07	29.19	33.143	4.9680
	Li(c)	0.0	0.0	6.70	5.65
	Li^+(aq)	−66.554	−70.22	3.4	
	Li_2(gas)	47.6	37.6	47.06	8.52
	Li_2O(c)	−142.4	−133.9	9.06	
	LiH(gas)	30.7	25.2	40.77	7.06
	LiH(c)	−21.61	−16.72	5.9	8.3
	LiOH(c)	−116.45	−105.9	12	
	LiOH · H_2O(c)	−188.77	−163.3	17.07	
	LiF(c)	−146.3	−139.6	8.57	10.04
	LiCl(c)	−97.70	−91.7	(13.2)	
	LiBr(c)	−83.72	−81.2	(16.5)	
	LiI(c)	−64.79	−64		
	Li_2SO_4(c)	−342.83	−316.6	(27.0)	
	Li_2CO_3(c)	−290.54	−270.66	21.60	23.28
	$LiBH_4$(c)	−44.6			
	$LiAlH_4$(c)	−24.2			
	Sodium				
	Na(gas)	25.98	18.67	36.715	4.9680
	Na(c)	0.0	0.0	12.2	6.79
	Na^+(aq)	−57.279	−62.589	14.4	
	Na_2(gas)	33.97	24.85	55.02	
	NaO_2(c)	−61.9	−46.5		
	Na_2O(c)	−99.4	−90.0	17.4	16.3
	Na_2O_2(c)	−120.6	−102.8	(16.0)	
	NaH(gas)	29.88	24.78	44.93	7.002
	NaH(c)	−13.7			
	NaOH(c, II)	−101.99	−90.1	(12.5)	19.2
	NaOH · H_2O(c)	−175.17	−149.00	20.2	
	NaF(c)	−136.0	−129.3	14.0	11.0
	NaCl(c)	−98.232	−91.785	17.3	11.88
	$NaClO_4$(c)	−92.18	−61.4		
	NaBr(c)	−86.030	−83.1		
	NaI(c)	−68.84	−56.7		
	Na_2S(c)	−89.2	−86.6		
	Na_2SO_3(c)	−260.6	−239.5	34.9	28.7

Substance	ΔH^0_{298}	ΔG^0_{298}	S^0_{298}	C_{P298}
Na_2SO_4(c)	−330.90	−302.78	35.73	30.50
$Na_2SO_4 \cdot 10H_2O$(c)	−1033.48	−870.93	141.7	140.4
$NaNO_3$(c, II)	−111.54	−87.45	27.8	22.24
Na_2CO_3(c)	−270.3	−250.4	32.5	26.41
$Na_2C_2O_4$(c)	−314.3	−308		
$NaHCO_3$(c)	−226.5	−203.6	24.4	20.94
NaCN(c)	−21.46	−14.7		
Na_2SiO_3(c)	−363	−341	27.2	26.72
$NaBH_4$(c)	−43.82	−28.57	25.02	20.7
Na_2SiF_6(c)	−677	−610.4		
Potassium				
K(gas)	21.51	14.62	38.296	4.968
K(c)	0.0	0.0	15.2	6.97
K^+(aq)	−60.04	−67.46	24.5	
K_2(gas)	30.8	22.1	59.69	
K_2O(c)	−86.4	−76.2		
K_2O_2(c)	−118	−100.1		
KH(gas)	30.0	25.1	47.3	
KOH(c)	−101.78	−89.5		
KF(c)	−134.46	−127.42	15.91	11.73
$KF \cdot 2H_2O$(c)	−277.00	−242.7	36	
KHF_2(c)	−219.98	−203.73	24.92	18.37
KCl(c)	−104.175	−97.592	19.76	12.31
KCl(gas)	−51.6	−56.2	57.24	8.66
$KClO_3$(c)	−93.50	−69.29	34.17	23.96
$KclO_4$(c)	−103.6	−72.7	36.1	26.33
KBr(c)	−93.73	−90.63	23.05	12.82
$KBrO_3$(c)	−79.4	−58.2	35.65	25.07
KI(c)	−78.31	−77.03	24.94	13.16
KI_3(c)	−76.6	−73.5		
KIO_3(c)	−121.5	−101.7	36.20	25.42
K_2SO_4(c, II)	−342.66	−314.62	42.0	31.1
KNO_3(c)	−117.76	−93.96	31.77	23.01
KH_2PO_4(c)	−374.9	−339.2		
K_2CO_3(c)	−273.93	−255.5		
$K_2C_2O_4$(c)	−320.8	−296.7		
$KHCO_3$(c)	−229.3	−205.7		
KCN(c)	−26.90	−20		
$K_3Fe(CN)_6$(c)	−41.4	−3.3		
$K_4Fe(CN)_6$(c)	−125.1	−84.0		
$KMnO_4$(c)	−194.4	−170.6	41.04	28.5
$KCr(SO_4)_2$(c)	−562	−510		
$KCr(SO_4)_2 \cdot 12H_2O$(c)	−1383.1	−1164		

	Substance	ΔH^0_{298}	ΔG^0_{298}	S^0_{298}	C_{P298}
	$KAl(SO_4)_2$(c)	−589.24	−534.29	48.9	46.12
	$KAl(SO_4)_2 \cdot 12H_2O$(c)	−1447.74	−1227.8	164.3	155.6
	Rubidium				
	Rb(gas)	20.51	13.35	40.628	4.9680
	Rb(c)	0.0	0.0	16.6	7.27
	Rb^+(aq)	−58.9	−67.45	29.7	
	Rb_2O(c)	−78.9	−69.5		
	RbOH(c, II)	−98.9	−87.1		
	RbF(c)	−131.9	−124.3	27.2	
	RbCl(c)	−102.91	−96.8		12.3
	RbBr(c)	−93.03	−90.38	25.88	12.68
	RbI(c)	−78.5	−77.8	28.21	12.50
	Rb_2SO_4(c)	−340.50	−312.8		
	$RbNO_3$(c)	−117.04	−93.3		
	Cesium				
	Cs(gas)	18.83	12.24	41.944	4.9680
	Cs(c)	0.0	0.0	19.8	7.42
	Cs^+(aq)	−59.2	−67.41	31.8	
	Cs_2O(c)	−75.9	−65.6		
	CsOH(c)	−97.2	−84.9		
	CsF(c)	−126.9	−119.5		
	CsCl(c, II)	−103.5	−96.6		
	CsBr(c)	−94.3	−91.6	29	12.4
	CsI(c)	−80.5	−79.7	31	12.4
	Cs_2SO_4(c)	−339.38	−310.7		
	$CsNO_3$(c)	−118.11	−94.0		
	$CsAl(SO_4)_2 \cdot 12H_2O$(c)	−1449.5	−1281.5	164	148.1
IIa	Beryllium				
	Be(gas)	76.63	67.60	32.545	4.9680
	Be(c)	0.0	0.0	2.28	4.26
	Be^{2+}(aq)	−93	−85.2		
	BeO(c)	−146.0	−139.0	3.37	6.07
	Magnesium				
	Mg(gas)	35.9	27.6	35.504	4.9680
	Mg(c)	0.0	0.0	7.77	5.71
	Mg^{2+}(aq)	−110.41	−108.99	−28.2	
	MgO(c)	−143.84	−136.13	6.4	8.94
	$Mg(OH)_2$(c)	−221.00	−199.27	15.09	18.41
	MgF_2(c)	−263.5	−250.8	13.68	14.72

Substance	ΔH^0_{298}	ΔG^0_{298}	S^0_{298}	C_{P298}
$MgCl_2(c)$	−153.40	−141.57	21.4	17.04
$MgCl_2 \cdot 6H_2O(c)$	−597.42	−505.65	87.5	75.46
$Mg(ClO_4)_2(c)$	−140.6	−79.4		
$MgSO_4(c)$	−305.5	−280.5	21.9	23.01
$Mg(NO_3)_2(c)$	−188.72	−140.63	39.2	33.94
$MgCO_3(c)$	−266	−246	15.7	18.05
$MgSiO_3(c)$	−357.9	−337.2	16.2	19.56
$MgNH_4PO_4(c)$		−390		
Calcium				
Ca(gas)	46.04	37.98	36.99	4.968
Ca(c)	0.0	0.0	9.95	6.28
$Ca^{2+}(aq)$	−129.77	−132.18	−13.2	
CaO(c)	−151.9	−144.4	9.5	10.23
$CaH_2(c)$	−45.1	−35.8	10	
$Ca(OH)_2(c)$	−235.8	−214.33	18.2	20.2
$CaF_2(c)$	−290.3	−277.7	16.46	16.02
$CaBr_2(c)$	−161.3	−156.8	31	
$CaI_2(c)$	−127.8	−126.6	34	
$CaC_2(c)$	−15.0	−16.2	16.8	14.90
$CaCO_3$(c, calcite)	−288.45	−269.78	22.2	19.57
$CaCO_3$(c, aragonite)	−288.49	−269.53	21.2	19.42
$CaC_2O_4 \cdot H_2O$(c, precipt.)	−400.4	−361.9	37.28	36.40
$Ca(HCO_2)_2(c)$	−323.5	−300.8		
$Ca(CN)_2(c)$	−44.2	−33.1		
$CaSiO_3(c)$	−378.6	−358.2	19.6	20.38
$CaSO_4$(c, anhydrite)	−342.42	−315.56	25.5	23.8
$CaSO_4 \cdot \frac{1}{2}H_2O(c, \alpha)$	−376.47	−343.02	31.2	28.6
$CaSO_4 \cdot 2H_2O(c)$	−483.06	−429.19	46.36	44.5
$Ca_3N_2(c)$	−103.2	−88.1	25	22.5
$Ca(NO_3)_2(c)$	−224.00	−177.34	46.2	35.69
$Ca_3(PO_4)_2(c, \beta)$	−988.9	−932.0	56.4	54.45
$CaHPO_4(c)$	−435.2	−401.5	21	
$CaHPO_4 \cdot 2H_2O(c)$	−576.0	−514.6	40	
Strontium				
Sr(gas)	39.2	26.3	39.325	4.9680
Sr(c)	0.0	0.0	13.0	6.0
$Sr^{2+}(aq)$	−130.38	−133.2	−9.4	
$Sr(OH)_2(c)$	−229.3	−207.8	(21.0)	

	Substance	ΔH^0_{298}	ΔG^0_{298}	S^0_{298}	C_{P298}
	$SrO(c)$	−141.1	−133.8	13.0	10.67
	$SrF_2(c)$	−290.3	−277.8		
	$SrCl_2(c)$	−198.0	−186.7	28	18.9
	$SrBr_2(c)$	−171.1	−166.3		
	$SrI_2(c)$	−135.5	−135		
	$SrSO_4(c)$	−345.3	−318.9	29.1	
	$Sr(NO_3)_2(c)$	−233.25	−186		
	$SrCO_3(c)$	−291.9	−271.9	23.2	19.46
	Barium				
	Ba(gas)	41.96	34.60	40.699	4.9680
	Ba(c)	0.0	0.0	16	6.30
	$Ba^{2+}(aq)$	−128.67	−134.0	3	
	$BaO(c)$	−133.4	−126.3	16.8	11.34
	$BaO_2(c)$	−150.5	−135.8		
	$BaF_2(c)$	−286.9	−274.5	23.0	17.02
	$BaCl_2(c)$	−205.56	−193.8	30	18.0
	$BaCl_2 \cdot H_2O(c)$	−278.4	−253.1	40	28.2
	$BaCl_2 \cdot 2H_2O(c)$	−349.35	−309.8	48.5	37.10
IVa	Titanium				
	Ti(gas)	112	101	43.069	5.8385
	Ti(c)	0.0	0.0	7.24	6.010
	TiO_2(c, rutile III)	−218.0	−203.8	12.01	13.16
	$TiO^{2+}(aq)$		−138		
	$Ti_2O_3(c)$	−367	−346	18.83	23.27
	$Ti_3O_5(c)$	−584	−550	30.92	37.00
	$TiF_2(c)$	−198	−187.1		
	$TiF_3(c)$	−315	−290.9		
	$TiF_4(c)$	−370	−346.3		
	$TiCl_2(c)$	−114	−96		
	$TiCl_3(c)$	−165	−148		
	$TiCl_4(liq)$	−179.3	−161.2	60.4	37.5
Va	Vanadium				
	V(gas)	120	109	43.546	6.2166
	V(c)	0.0	0.0	7.05	5.85
	$V^{2+}(aq)$		−54.7		
	$V^{3+}(aq)$		−60.6		
	$VO_4^-(aq)$	−210.9	−203.9	(48.0)	
	$V_2O_3(c)$	−290	−271	23.58	24.83
	$V_2O_4(c)$	−344	−318	24.67	28.30
	$V_2O_5(c)$	−373	−344	31.3	31.00

	Substance	ΔH^0_{298}	ΔG^0_{298}	S^0_{298}	C_{P298}
VIa	Chromium				
	Cr(gas)	80.5	69.8	41.637	4.9680
	Cr(c)	0.0	0.0	5.68	5.58
	Cr^{2+}(aq)		−42.1		
	Cr^{3+}(aq)		−51.5	−73.5	
	Cr_2O_3(c)	−269.7	−250.2	19.4	28.38
	$Cr_2O_7^{2-}$(aq)	−349.1	−300.5	51.1	
	$Cr(OH)^{2+}$(aq)	−113.5	−103.0	−16.4	
	$HCrO_4^-$(aq)	−220.2	−184.9	16.5	
	CrO_4^{2-}(aq)	−213.75	−176.1	9.2	
	$Cr(OH)_2$(c)		−140.5		
	$Cr(OH)_3$(c)	−247.1	−215.3	(19.2)	
	CrF_3(c)	−265.2	−248.3	(22.2)	
	CrF_2(c)	−181.0	−170.7	(19.6)	
	$CrCl_2^+$(aq)	−130.0	−115.2		
	$CrCl_2$(c)	−94.56	−85.15	27.4	16.87
	$CrCl_3$(c)	−134.6	−118.0	30.0	21.53
	Molybdenum				
	Mo(gas)	155.5	144.2	43.462	4.9680
	Mo(c)	0.0	0.0	6.83	5.61
	MoO_4^{2-}(aq)	−254.3	−218.8		
	Tungsten (Wolfram)				
	W(gas)	201.6	191.6	41.552	5.0903
	W(c)	0.0	0.0	8.0	5.97
	WO_3(c, yellow)	−200.84	−182.47	19.90	19.48
VIIa	Manganese				
	Mn(gas)	68.34	58.23	41.493	4.9680
	Mn(c)	0.0	0.0	7.59	6.29
	Mn^{2+}(aq)	−52.3	−53.4	−20	
	MnO(c)	−92.0	−86.8	14.4	10.27
	MnO_2(c)	−124.2	−111.4	12.7	12.91
	MnO_4^-(aq)	−123.9	−101.6	45.4	
	MnO_4^{2-}(aq)		−120.4		
	Mn_2O_3(c)	−232.1	−212.3	(22.1)	25.8
	Mn_3O_4(c, I)	−331.4	−306.0	35.5	33.29
	$Mn(OH)_2$(c, precipt.)	−166.8	−146.9	21.1	
	$Mn(OH)_3$(c)	−212	−181	(23.8)	
	MnF_2(c)	−189	−179	22.2	16.24
	$MnCl_2 \cdot H_2O$(c)	−188.5	−164.5	(35.9)	
	MnS(c, green)	−48.8	−49.9	18.7	11.94
	MnS(c, precipt.)		−53.3		
	$MnSO_4$(c)	−254.24	−228.48	26.8	23.94
	$MnCO_3$(c)	−213.9	−195.4	20.5	19.48

	Substance	ΔH^0_{298}	ΔG^0_{298}	S^0_{298}	C_{P298}
VIII	Iron				
	Fe(gas)	96.68	85.76	43.11	6.13
	Fe(c)	0.0	0.0	6.49	6.03
	Fe^{2+}(aq)	−21.0	−20.30	−27.1	
	Fe^{3+}(aq)	−11.4	−2.53	−70.1	
	Fe_2O_3(c, hematite)	−196.5	−177.1	21.5	25.0
	Fe_3O_4(c, magnetite)	−267.9	−242.4	35.0	
	$Fe(OH)^{2+}$(aq)	−67.4	−55.91	−23.2	
	$Fe(OH)_2$(c)	−135.8	−115.57	19	
	$Fe(OH)_2{}^+$(aq)		−106.2		
	$Fe(OH)_3$(c)	−197.0	−166.0	(23.0)	
	$FeCl^{2+}$(aq)	−42.9	−35.9	−22	
	$FeCl_3$(c)	−96.8	−80.4	(31.1)	
	$FeBr^{2+}$(aq)	−34.2	−27.9		
	FeS(c)	−22.72	−23.32	16.1	
	FeS_2(c)	−42.52	−39.84	12.7	
	$FeSO_4$(c)	−220.5	−198.3	(27.6)	
	$FeSO_4 \cdot 7H_2O$(c)	−718.7	−597	(93.4)	
	$FeNO^{2+}$(aq)	−9.7	1.5	−10.6	
	Fe_3C(c, cementite)	5.0	3.5	25.7	25.3
	$Fe(CN)_6{}^{4-}$(aq)	126.7	170.4		
	Fe_2SiO_4(c)	−343.7	−319.8	35.4	31.75
	Cobalt				
	Co(gas)	105	94	42.881	5.5043
	Co(c)	0.0	0.0	6.8	6.11
	Co^{2+}(aq)	(−16.1)	−12.3	(−37.1)	
	Co^{3+}(aq)		29.6		
	$Co(OH)_2$(c)	−129.3	−109.0	(19.6)	
	$Co(OH)_3$(c)	−174.6	−142.6	(20.0)	
	CoF_2(c)	−157	−146.45	(20.0)	
	CoF_3(c)	−185	−168	(22.6)	
	$CoCl_2$(c)	−75.8	−65.5	25.4	18.8
	CoS(c, precipt.)	−19.3	−19.8	(16.1)	
	$CoSO_4$(c)	−205.5	−180.1	27.1	
	$Co(NO_3)_2$(c)	−100.9	−55.1	(46.0)	
	$Co(NH_3)_6{}^{2+}$(aq)		−57.7		
	$Co(NH_3)_6{}^{3+}$(aq)		−55.2		
	$Co(NH_3)_5Cl^{2+}$(aq)	−162.1	−86.2	96.1	

	Substance	ΔH^0_{298}	ΔG^0_{298}	S^0_{298}	C_{P298}
	Nickel				
	Ni(gas)	101.61	90.77	43.502	5.5986
	Ni(c)	0.0	0.0	7.20	6.21
	Ni^{2+}(aq)	(−15.3)	−11.53		
	$Ni(OH)_2$(c)	−128.6	−108.3	19	
	$Ni(OH)_3$(c)	−162.1	−129.5	(19.5)	
	$NiCl_2$(c)	−75.5	−65.1	25.6	18.6
	NiS(c, α)		−17.7		
	NiS(c, γ)		−27.3		
	NiO(c)	−58.4	−51.7	9.22	10.60
	$NiSO_4$(c)	−213.0	−184.9	18.6	33.4
	$NiSO_4 \cdot 6H_2O$(c, blue)	−642.5	−531.0	73.1	82
	$Ni(NH_3)^{2+}$(aq)		−60.1		
	$Ni(CN)_4^{2-}$(aq)	86.9	117.1	(33.0)	
Ib	Copper				
	Cu(gas)	81.52	72.04	39.744	4.968
	Cu(c)	0.0	0.0	7.96	5.848
	Cu^+(aq)	(12.4)	12.0	(−6.3)	
	Cu^{2+}(aq)	15.39	15.53	−23.6	
	CuO(c)	−37.1	−30.4	10.4	10.6
	Cu_2O(c)	−39.84	−34.98	24.1	16.7
	$Cu(OH)_2$(c)	−107.2	−85.3	(19.0)	
	$CuF_2 \cdot 2H_2O$(c)	−274.5	−235.2	36.2	
	CuCl(c)	−32.5	−28.2	20.2	
	$CuCl_2$(c)	−52.3	−42	(26.8)	
	$CuCl_2^-$(aq)	−66.1	−57.9	49.4	
	CuS(c)	−11.6	−11.7	15.9	11.43
	Cu_2S(c)	−19.0	−20.6	28.9	18.24
	$CuSO_4$(c)	−184.00	−158.2	27.1	24.1
	$CuSO_4 \cdot H_2O$(c)	−259.00	−219.2	35.8	31.3
	$CuSO_4 \cdot 3H_2O$(c)	−402.25	−334.6	53.8	49.0
	$CuSO_4 \cdot 5H_2O$(c)	−544.45	−449.3	73.0	67.2
	$Cu(NH_3)_4^{2+}$(aq)	(−79.9)	−61.2	192.8	
	$CuCO_3$(c)	−142.2	−123.8	21	
	Silver				
	Ag(gas)	69.12	59.84	41.3221	4.9680
	Ag(c)	0.0	0.0	10.206	6.092
	Ag^+(aq)	25.31	18.430	17.67	9
	Ag^{2+}(aq)		64.1		
	Ag_2O(c)	−7.306	−2.586	29.09	15.67
	AgCl(c)	−30.362	−26.224	22.97	12.14

	Substance	ΔH^0_{298}	ΔG^0_{298}	S^0_{298}	C_{P292}
	$AgClO_3$(c)	−5.73	16	(37.7)	
	$AgClO_4$(c)	−7.75	21	(38.8)	
	AgBr(c)	−23.78	−22.930	25.60	12.52
	$AgBrO_3$(c)		17.6		
	AgI(c)	−14.91	−15.85	27.3	13.01
	$AgIO_3$(c)	−41.7	−24.080	35.7	
	Ag_2S(c, rhombic)	−7.60	−9.62	34.8	
	Ag_2SO_4(c)	−170.50	−147.17	47.8	31.4
	$Ag(S_2O_3)_2{}^{3-}$(aq)	(−285.5)	−247.6		
	$Ag(SO_3)_2{}^{3-}$(aq)		−225.4		
	$AgNO_2$(c)	−10.605	4.744	30.62	18.8
	$AgNO_3$(c)	−29.43	−7.69	33.68	22.24
	$Ag(NH_3)_2{}^+$(aq)	−26.724	−4.16	57.8	
	Ag_2CO_3(c)	−120.97	−104.48	40.0	26.8
	$Ag_2C_2O_4$(c)	−159.1	−137.2	(48.0)	
	$Ag(CH_3CO_2)$(c)	−93.41	−74.2	(33.8)	
	Ag_2MoO_4(c)		−196.4		
	Ag_2CrO_4(c)	−176.2	−154.7	51.8	
	$Ag_4Fe(CN)_6$(c)		188.4		
	AgCN(c)	34.94	39.20	20.0	
	$Ag(CN)_2{}^-$(aq)	64.5	72.05	49.0	
	Gold				
	Au(gas)	89.29	72.83	43.12	4.9680
	Au(c)	0.0	0.0	11.4	6.03
	Au^+(aq)		39.0		
	Au^{3+}(aq)		103.6		
	$H_2AuO_3{}^-$(aq)		−45.8		
	$HAuO_3{}^{2-}$(aq)		−27.6		
	$AuO_3{}^{3-}$(aq)		−5.8		
	$Au(OH)_3$(c)	−100.0	−69.3	29.0	
	AuCl(c)	−8.4	−4.2	(24.0)	
	$AuCl_3$(c)	−28.3	−11.6	(35.0)	
	$AuCl_4{}^-$(aq)	−77.8	−56.2	61	
	$AuBr_4{}^-$(aq)	−45.5	−38.1	75	
	$Au(CN)_2{}^-$(aq)	58.4	51.5	99	
IIb	Zinc				
	Zn(gas)	31.19	22.69	38.45	4.968
	Zn(c)	0.0	0.0	9.95	5.99
	Zn^{2+}(aq)	−36.43	−35.184	−25.45	
	ZnO(c)	−83.17	−76.05	10.5	9.62
	$ZnO_2{}^{2-}$(aq)		−93.03		
	$Zn(NH_3)_4{}^{2+}$(aq)		−73.5		

Substance	ΔH^0_{298}	ΔG^0_{298}	S^0_{298}	C_{P298}
$Zn(OH)_2$(c)	−153.5	−132.6	(19.9)	
ZnS(c, sphalerite)	−48.5	−47.4	13.8	10.8
ZnS(c, wurtzite)	−45.3	−44.2	(13.8)	
ZnS(c, precipt.)	(−44.3)	(−43.2)		
$ZnCl_2$(c)	−99.4	−88.255	25.9	18.3
$ZnSO_4$(c)	−233.88	−208.31	29.8	28.0
$ZnSO_4 \cdot H_2O$(c)	−310.6	−269.9	34.9	34.7
$ZnSO_4 \cdot 6H_2O$(c)	−663.3	−555.0	86.8	80.8
$ZnSO_4 \cdot 7H_2O$(c)	−735.1	−611.9	92.4	93.7
$Zn(CN)_2$(c)	18.4	29	(22.9)	
$Zn(CN)_4^{2-}$(aq)	82.0	100.4		
$ZnCO_3$(c)	−194.2	−174.8	19.7	19.16
Cadmium				
Cd(gas)	26.97	18.69	40.067	4.968
Cd(c)	0.0	0.0	12.3	6.19
Cd^{2+}(aq)	−17.30	−18.58	−14.6	
CdO(c)	−60.86	−53.79	13.1	10.38
$Cd(OH)_2$(c)	−133.26	−112.46	22.8	
$CdCl_2$(c)	−93.00	−81.88	28.3	
$CdCl^+$(aq)		−51.8	5.6	
$CdCl_2$(aq, unionized)		−84.3	17	
$CdCl_3^-$(aq)		−115.9	50.7	
CdS(c)	−34.5	−33.60	17	
$CdSO_4$(c)	−221.36	−195.99	32.8	
$CdSO_4 \cdot H_2O$(c)	−294.37	−254.84	41.1	
$CdSO_4 \cdot \frac{8}{3}H_2O$(c)	−411.82	−349.63	57.9	
$Cd(CN)_2$(c)	39.0	49.7	(24.9)	
$Cd(CN)_4^{2-}$(aq)		111		
Mercury				
Hg(gas)	14.54	7.59	41.80	4.968
Hg(liq)	0.0	0.0	18.5	6.65
Hg^{2+}(aq)	41.59	39.38	−5.4	
Hg_2^{2+}(aq)		36.35		
HgO(c, red)	−21.68	−13.990	17.20	10.93
HgO(c, yellow)	−21.56	−13.959	17.5	
$HgCl_2$(c)	−55.0	−44.4	(34.5)	
Hg_2Cl_2(c)	−63.32	−50.35	46.8	24.3
$HgCl_4^{2-}$(aq)		−107.7		
Hg_2Br_2(c)	−49.42	−42.714	50.9	
HgI_2(c, red)	−25.2	−24.07	(42.6)	
HgI_2(c, yellow)	−24.55	−23.1	(42.6)	

	Substance	ΔH^0_{298}	ΔG^0_{298}	S^0_{298}	C_{P298}
	Hg_2I_2(c)	−28.91	−26.60	57.2	25.3
	HgS(c, red)	−13.90	−11.67	18.6	
	HgS(c, black)	−12.90	−11.05	19.9	
	Hg_2SO_4(c)	−177.34	−149.12	47.98	
	$Hg(CN)_2$(c)	62.5	74.3	(27.4)	
	$Hg(CN)_4^{2-}$(aq)	126.0	141.3		
	Hg_2CrO_4(c)		−155.75		
IIIb	Boron				
	B(gas)	97.2	86.7	36.649	4.971
	B(c)	0.0	0.0	1.56	2.86
	B_2O_3(c)	−302.0	−283.0	12.91	14.88
	B_2H_6(gas)	7.5	19.8	55.66	13.48
	B_5H_9(gas)	15.0	39.6	65.88	19
	HBO_2(c)	−186.9	−170.5	11	
	$H_2BO_3^-$(aq)	−251.8	−217.6	7.3	
	H_3BO_3(aq)	−255.2	−230.24	38.2	
	H_3BO_3(c)	−260.2	−230.2	21.41	19.61
	BF_3(gas)	−265.4	−261.3	60.70	12.06
	BF_4^-(aq)	−365	−343	40	
	BCl_3(gas)	−94.5	−90.9	69.29	14.97
	BCl_3(liq)	−100.0	−90.6	50.0	
	BBr_3(gas)	−44.6	−51.0	77.49	16.25
	BBr_3(liq)	−52.8	−52.4	54.7	
	Aluminum				
	Al(gas)	75.0	65.3	39.303	5.112
	Al(c)	0.0	0.0	6.769	5.817
	Al^{3+}(aq)	−125.4	−115.0	−74.9	
	AlO_2^-(aq)		−204.7		
	$H_2AlO_3^-$(aq)		−255.2		
	Al_2O_3(c)	−399.09	−376.77	12.186	18.88
	$Al(OH)_3$(amorphous)	−304.9	−271.9	(17.0)	
	AlF_3(c)	−311	−294	23	
	AlF_6^{3-}(aq)		−539.6		
	$AlBr_3$(c)	−125.8	−120.7	44	24.5
	$AlCl_3$(c)	−166.2	−152.2	40	21.3
	$(NH_4)Al(SO_4)_2 \cdot 12H_2O$(c)	−1419.40	−1179.02	166.6	163.3
	$Al_2(SO_4)_3$(c)	−820.98	−738.99	57.2	62.00
	$Al_2(SO_4)_3 \cdot 6H_2O$(c)	−1268.14	−1105.14	112.1	117.8

	Substance	ΔH^0_{298}	ΔG^0_{298}	S^0_{298}	C_{P298}
	Thallium				
	Tl(gas)	43.34	35.05	43.23	4.968
	Tl(c)	0.0	0.0	15.4	6.35
	Tl^+(aq)	1.38	−7.755	30.4	
	Tl^{3+}(aq)	27.7	50.0	−106	
	Tl(OH)(c)	−56.9	−45.5	17.3	
	$Tl(OH)_3$(c)	−156	−123.0	(24.4)	
	TlCl(gas)	−15	−21	61.1	8.66
	TlCl(c)	−48.99	−44.19	25.9	
	TlBr(gas)	−4	−13	63.8	8.81
	TlBr(c)	−41.2	−39.7	28.6	
	TlI(gas)	8	−3	65.6	8.86
	TlI(c)	−29.7	−29.7	29.4	
IVb	Carbon				
	C(gas)	171.698	160.845	37.761	4.9803
	C(c, diamond)	0.4532	0.6850	0.5829	1.449
	C(c, graphite)	0.0	0.0	1.3609	2.066
	CO(gas)	−26.4157	−32.8079	47.301	6.965
	CO_2(gas)	−94.0518	−94.2598	51.061	8.874
	CO_2(aq)	−98.69	−92.31	29.0	
	CH_4(gas)	−17.889	−12.140	44.50	8.536
	C_2H_2(gas)	54.194	50.0	47.997	10.499
	C_2H_4(gas)	12.496	16.282	52.45	10.41
	C_2H_6(gas)	−20.236	−7.860	54.85	12.585
	C_6H_6(gas)	19.820	30.989	64.34	19.52
	C_6H_6(liq)	11.718	41.30	29.756	
	$(HCOOH)_2$(gas)	−187.7	−163.8	83.1	
	HCOOH(gas)	−86.67	−80.24	60.0	
	HCOOH(liq)	−97.8	−82.7	30.82	23.67
	HCOOH(aq)	−98.0	−85.1	39.1	
	$HCOO^-$(aq)	−98.0	−80.0	21.9	
	H_2CO_3(aq)	−167.0	−149.00	45.7	
	HCO_3^-(aq)	−165.18	−140.31	22.7	
	CO_3^{2-}(aq)	−161.63	−126.22	−12.7	
	CH_3COOH(liq)	−116.4	−93.8	38.2	29.5
	CH_3COOH(aq)	−116.743	−95.51		
	CH_3COO^-(aq)	−116.843	−89.02		
	$(COOH)_2$(c)	−197.6	−166.8	28.7	26
	$(COOH)_2$(aq)	−195.57	−166.8		
	$HC_2O_4^-$(aq)	−195.7	−167.1	36.7	
	$C_2O_4^{2-}$(aq)	−197.0	−161.3	12.2	
	HCHO(gas)	−27.7	−26.2	52.26	8.45

Substance	ΔH^0_{298}	ΔG^0_{298}	S^0_{298}	C_{P298}
HCHO(aq)		−31.0		
CH_3OH(gas)	−48.10	−38.70	56.8	
CH_3OH(liq)	−57.036	−39.75	30.3	19.5
CH_3OH(aq)	−58.77	−41.88	31.63	
C_2H_5OH(gas)	−56.24	−40.30	67.4	
C_2H_5OH(liq)	−66.356	−41.77	38.4	26.64
CH_3CHO(gas)	−39.76	−31.96	63.5	15.0
CH_3CHO(aq)	−49.88			
$(CH_3)_2O$(gas)	−44.3	−27.3	63.72	15.76
C_2N_2(gas)	73.60	70.81	57.86	13.60
HCN(liq)	25.2	29.0	26.97	16.88
HCN(gas)	31.2	28.7	48.23	8.58
HCN(aq)	25.2	26.8	30.8	
CN^-(aq)	36.1	39.6	28.2	
HCNO(aq)	−35.1	−28.9	43.6	
CNO^-(aq)	−33.5	−23.6	31.1	
CH_3NH_2(gas)	−6.7	6.6	57.73	12.9
$CO(NH_2)_2$(c)	−79.634	−47.120	25.0	22.26
$CO(NH_2)_2$(aq)	−76.30	−48.72	41.55	
CH_3SH(gas)	−2.97	0.21	60.90	12.1
CS_2(gas)	27.55	15.55	56.84	10.91
CS_2(liq)	21.0	15.2	36.10	18.1
CNS^-(aq)	17.2	21.2	(36.0)	
CNCl(gas)	34.5	32.9	56.31	10.70
CF_4(gas)	−162.5	−151.8	62.7	
CCl_4(gas)	−25.50	−15.35	73.95	19.96
CCl_4(liq)	−33.34	−16.43	51.25	31.49
$COCl_2$(gas)	−53.30	−50.31	69.13	14.51
CH_3Cl(gas)	−19.58	−13.96	55.97	9.75
CH_3Br(gas)	−8.2	−5.9	58.74	10.18
$CHCl_3$(gas)	−24	−16	70.86	15.73
$CHCl_3$(liq)	−31.5	−17.1	48.5	27.8
Silicon				
Si(gas)	88.04	77.41	40.120	5.318
Si(c)	0.0	0.0	4.47	4.75
SiO(gas)	−26.72	−32.77	49.26	7.14
SiO_2(c, quartz)	−205.4	−192.4	10.00	10.62
SiH_4(gas)	−14.8	−9.4	48.7	10.24
SiF_4(gas)	−370	−360	68.0	18.2
SiF_6^{2-}(aq)	−558.5	−511		
$SiCl_4$(gas)	−145.7	−136.2	79.2	21.7
$SiCl_4$(liq)	−153.0	−136.9	57.2	34.7

	Substance	ΔH^0_{298}	ΔG^0_{298}	S^0_{298}	C_{P298}
	Germanium				
	Ge(gas)	78.44	69.50	40.106	7.346
	Ge(c)	0.0	0.0	10.14	6.24
	GeO(gas)	−22.8	−28.2	52.56	7.39
	Tin				
	Sn(gas)	72	64	40.245	5.081
	Sn(c, grey)	0.6	1.1	10.7	6.16
	Sn(c, white)	0.0	0.0	12.3	6.30
	SnO(c)	−68.4	−61.5	13.5	10.6
	SnO_2(c)	−138.8	−124.2	12.5	12.57
	$HSnO_2^-$(aq)		−98.0		
	$Sn(OH)_2$(c)	−138.3	−117.6	23.1	
	$Sn(OH)_6^{2-}$(aq)		−310.5		
	$SnCl_2$(c)	−83.6	−72.2	(29.3)	
	$SnCl_4$(liq)	−130.3	−113.3	61.8	39.5
	SnS(c)	−18.6	−19.7	23.6	
	Lead				
	Pb(gas)	46.34	38.47	41.890	4.968
	Pb(c)	0.0	0.0	15.51	6.41
	Pb^{2+}(aq)	0.39	−5.81	5.1	
	PbO(c, red)	−52.40	−45.25	16.2	
	PbO(c, yellow)	−52.07	−45.05	16.6	11.60
	$HPbO_2^-$(aq)		−81.0?		
	$Pb(OH)_2$(c)	−123.0	−100.6	21	
	PbO_2(c)	−66.12	−52.34	18.3	15.4
	Pb_3O_4(c)	−175.6	−147.6	50.5	35.14
	PbF_2(c)	−158.5	−148.1	29.0	
	$PbCl_2$(c)	−85.85	−75.04	32.6	18.4
	$PbBr_2$(c)	−66.21	−62.24	38.6	19.15
	PbI_2(c)	−41.85	−41.53	42.3	
	PbS(c)	−22.54	−22.15	21.8	11.83
	$PbSO_4$(c)	−219.50	−193.89	35.2	24.9
	$Pb(NO_3)_2$(c)	−107.35	−60.3	(50.9)	
	$PbCO_3$(c)	−167.3	−149.7	31.3	20.9
	PbC_2O_4(c)	−205.1	−180.3	(33.2)	
	$PbCrO_4$(c)	−225.2	−203.6	(36.5)	
	$PbMoO_4$(c)	−265.8	−231.7	(38.5)	
Vb	Nitrogen				
	N(gas)	112.965	108.870	36.6145	4.968
	N_2(gas)	0.0	0.0	45.767	6.960
	N_3^-(aq)	60.3	77.7	(32)	
	NO(gas)	21.600	20.719	50.339	7.137

Substance	ΔH^0_{298}	ΔG^0_{298}	S^0_{298}	C_{P298}
NO_2(gas)	8.091	12.390	57.47	9.06
NO_2^-(aq)	−25.4	−8.25	29.9	
NO_3^-(aq)	−49.372	−26.43	35.0	
N_2O(gas)	19.49	24.76	52.58	9.251
$N_2O_2^{2-}$(aq)	−2.59	33.0	6.6	
N_2O_4(gas)	2.309	23.491	72.73	18.90
N_2O_5(c)	−10.0	32	27.1	
NH_3(gas)	−11.04	−3.976	46.01	8.523
NH_3(aq)	−19.32	−6.36	26.3	
NH_4^+(aq)	−31.74	−19.00	26.97	
N_2H_4(aq)	8.16	30.56	(33)	
HN_3(gas)	70.3	78.5	56.74	10.02
HN_3(aq)	60.50	71.30	(48)	
HNO_2(aq)	−28.4	−12.82		
HNO_3(liq)	−41.404	−19.100	37.19	26.26
HNO_3(aq)	−49.372	−26.43	35.0	
NH_2OH(aq)	−21.7	−5.60	(40)	
$H_2N_2O_2$(aq)	−13.7	8.6	52	
NOCl(gas)	12.57	15.86	63.0	
NOBr(gas)	19.56	19.70	65.16	
NH_4Cl(c, II)	−75.38	−48.73	22.6	20.1
$(NH_4)_2SO_4$(c)	−281.86	−215.19	52.65	44.81
Phosphorus				
P(gas)	75.18	66.71	38.98	4.9680
P(c, white)	0.0	0.0	10.6	5.55
P(c, red)	−4.4	−3.3	(7.0)	
P_4(gas)	13.12	5.82	66.90	16.0
PO_4^{3-}(aq)	−306.9	−245.1	−52	
P_4O_{10}(c)	−720.0			
PH_3(gas)	2.21	4.36	50.2	
HPO_3^{2-}(aq)	−233.8	−194.0		
HPO_4^{2-}(aq)	−310.4	−261.5	−8.6	
$H_2PO_3^-$(aq)		−202.35	(19.0)	
$H_2PO_4^-$(aq)	−311.3	−271.3	21.3	
H_3PO_3(aq)	−232.2	−204.8	(40.0)	
H_3PO_4(aq)	−308.2	274.2	42.1	
PCl_3(gas)	−73.22	−68.42	74.49	
PCl_5(gas)	−95.35	−77.57	84.3	
$POCl_3$(gas)	−141.5	−130.3	77.59	
BPr_3(gas)	−35.9	−41.2	83.11	
Arsenic				
As(gas)	60.64	50.74	41.62	
As(c, grey metal)	0.0	0.0	8.4	5.97

	Substance	ΔH^0_{298}	ΔG^0_{298}	S^0_{298}	C_{P298}
	As_4(gas)	35.7	25.2	69	
	AsO^+(aq)		−39.1		
	AsO_4^{3-}(aq)	−208.0	−152	−34.6	
	$HAsO_4^{2-}$(aq)	−214.8	−169	0.9	
	$H_2AsO_3^-$(aq)	−170.3	−140.4		
	$H_2AsO_4^-$(aq)	−216.2	−178.9	28	
	H_3AsO_3(aq)	−177.3	−152.94	47.0	
	H_3AsO_4(aq)	−214.8	−183.8	49.3	
	AsF_3(gas)	−218.3	−214.7	69.08	30.3
	AsF_3(liq)	−226.8	−215.5	43.31	
	$AsCl_3$(gas)	−71.5	−68.5	78.2	
	$AsCl_3$(liq)	−80.2	−70.5	55.8	
	As_2S_3(c)	−35.0	−32.46	(26.8)	
	Antimony				
	Sb(gas)	60.8	51.1	43.06	4.968
	Sb(c, metal)	0.0	0.0	10.5	6.08
	SbO^+(aq)		−42.0		
	Sb_4O_6(c)	−336.8	−298.0	58.8	48.46
	Sb_2O_5(c)	−234.4	−200.5	29.9	28.1
	SbH_3(gas)	34	35.3	(53.0)	
	$SbCl_3$(gas)	−75.2	−72.3	80.8	18.5
	$SblC_3$(c)	−91.34	−77.62	(44.5)	
	$SbCl_5$(gas)	−93.9			
	Sb_2S_3(c)	−36.0	−32.0	(30.3)	
	Bismuth				
	Bi(gas)	49.7	40.4	44.67	4.968
	Bi(c)	0.0	0.0	13.6	6.1
	BiO^+(aq)		−34.54	25.317	
	Bi_2O_3(c)	−137.9	−118.7	36.2	27.2
	$BiCl_3$(gas)	−64.7	−62.2	85.3	19.0
	$BiCl_3$(c)	−90.61	−76.23	45.3	
	BiOCl(c)	−87.3	−77.0	20.6	
	Bi_2S_3(c)	−43.8	−39.4	35.3	
VIb	Oxygen				
	O(gas)	59.159	54.994	38.469	5.236
	O_2(gas)	0.0	0.0	49.003	7.017
	O_3(gas)	34.0	39.06	56.8	9.12
	OH(gas)	10.06	8.93	43.888	7.141
	OH^-(aq)	−54.957	−37.595	−2.52	−32.0
	H_2O(gas)	−57.798	−54.635	45.106	8.025
	H_2O(liq)	−68.317	−56.690	16.716	17.996

Substance	ΔH^0_{298}	ΔG^0_{298}	S^0_{298}	C_{P298}
H_2O_2(liq)	−44.84	−27.240	(22)	
H_2O_2(aq)	−45.68	−31.470		
Sulfur				
S(gas)	53.25	43.57	40.085	5.66
S(c, rhombic)	0.0	0.0	7.62	5.40
S(c, monoclinic)	0.071	0.023	7.78	5.65
S^{2-}(aq)	10.0	20.0		
SO(gas)	19.02	12.78	53.04	
SO_2(gas)	−70.76	−71.79	59.40	9.51
SO_3(gas)	−94.45	−88.52	61.24	12.10
SO_3^{2-}(aq)	−149.2	−118.8	10.4	
SO_4^{2-}(aq)	−216.90	−177.34	4.1	4.0
$S_2O_3^{2-}$(aq)	−154.0	−127.2	29	
HS^-(aq)	−4.22	3.01	14.6	
H_2S(gas)	−4.815	−7.892	49.15	8.12
H_2S(aq)	−9.4	−6.54	29.2	
HSO_3^-(aq)	−150.09	−126.03	31.64	
HSO_4^-(aq)	−211.70	−179.94	30.32	
H_2SO_3(aq)	−145.5	−128.59	56	
H_2SO_4(aq)	−216.90	−177.34	4.1	4.0
SF_6(gas)	−262	−237	69.5	
SO_2Cl_2(gas)		−73.6		
$(CH_3)_2SO$(liq)	−30.3	−18.9		
$(CH_3)_2SO_2$(liq)	−81.4	−64.7		
Selenium				
Se(gas)	48.37	38.77	42.21	4.968
Se(c, hex.)	0.0	0.0	10.0	5.95
Se^{2-}(aq)	31.6	37.2	20.0	
SeO_3^{2-}(aq)	−122.39	−89.33	3.9	
SeO_4^{2-}(aq)	−145.3	−105.42	5.7	
HSe^-(aq)	24.6	23.57	42.3	
H_2Se(gas)	20.5	17.0	52.9	
H_2Se(aq)	18.1	18.4	39.9	
$HSeO_3^-$(aq)	−123.5	−98.3	(30.4)	
$HSeO_4^-$(aq)	−143.1	−108.2	22.0	
H_2SeO_3(aq)	−122.39	−101.8	45.7	
H_2SeO_4(aq)	−145.3	−105.42	5.7	
SeF_6(gas)	−246.0	−222	75.10	
Tellurium				
Te(gas)	47.6	38.1	43.64	4.968
Te(c, II)	0.0	0.0	11.88	6.15
Te^{2-}(aq)		52.7		
HTe^-(aq)		37.7		

	Substance	ΔH^0_{298}	ΔG^0_{298}	S^0_{298}	C_{P298}
	H_2Te(gas)	36.9	33.1	56.0	
	H_2Te(aq)		34.1		
	TeO_2(c)	−77.69	−64.60	16.99	15.89
	H_2TeO_3(c)	−144.7	−115.7	47.7	
	TeF_6(gas)	−315	−292	80.67	
VIIb	Fluorine				
	F(gas)	18.3	14.2	37.917	5.436
	F^-(aq)	−78.66	−66.08	−2.3	−29.5
	F_2(gas)	0.0	0.0	48.6	7.52
	HF(gas)	−64.2	−64.7	41.47	6.95
	Chlorine				
	Cl(gas)	29.012	25.192	39.457	5.2203
	Cl^-(aq)	−40.023	−31.350	13.2	−30.0
	Cl_2(gas)	0.0	0.0	53.286	8.11
	ClO^-(aq)		−8.9	10.3	
	ClO_2(gas)	24.7	29.5	59.6	
	ClO_2^-(aq)	−16.5	−2.56	24.1	
	ClO_3^-(aq)	−23.50	−0.62	39	−18.0
	ClO_4^-(aq)	−31.41	−2	43.5	
	Cl_2O(gas)	18.20	22.40	63.70	
	HCl(gas)	−22.063	−22.769	44.617	6.96
	HCl(aq)	−40.023	−31.350	13.2	−30.0
	HClO(aq)	−27.83	−19.110	31.0	
	ClF_3(gas)	−37.0	−27.2	66.61	15.33
	Bromine				
	Br(gas)	26.71	19.69	41.8052	4.9680
	Br^-(aq)	−28.90	−24.574	19.29	−30.7
	Br_2(gas)	7.34	0.751	58.639	8.60
	Br_2(liq)	0.0	0.0	36.4	
	Br_3^-(aq)	−32.0	−25.3	(40)	
	BrO^-(aq)		−8		
	HBr(gas)	−8.66	−12.72	47.437	6.96
	BrO_3^-(aq)	−9.6	10.9	38.9	−19
	Iodine				
	I(gas)	25.482	16.766	43.184	4.9680
	I^-(aq)	−13.37	−12.35	26.14	−31.0
	I_2(gas)	14.876	4.63	62.280	8.81
	I_2(c)	0.0	0.0	27.9	13.14
	I_2(aq)	5.0	3.926		
	I_3^-(aq)	−12.4	−12.31	41.5	
	HI(gas)	6.2	0.31	49.314	6.97

	Substance	ΔH^0_{298}	ΔG^0_{298}	S^0_{298}	C_{P298}
	IO^-(aq)	(−34)	−8.5		
	HIO(aq)	(−38)	−23.5		
	IO_3^-(aq)		−32.250	28.0	
	ICl(gas)	4.2	−1.32	59.12	8.46
	ICl_2^-(aq)		−38.35		
	ICl_3(c)	−21.1	−5.40	41.1	
	IBr(gas)	9.75	0.91	61.8	
O	He(gas)	0.0	0.0	30.13	
	Ne(gas)	0.0	0.0	34.45	
	Ar(gas)	0.0	0.0	36.98	
	Kr(gas)	0.0	0.0	39.19	
	Xe(gas)	0.0	0.0	40.53	
	Rn(gas)	0.0	0.0	42.10	

APPENDIX 5A. *HEATS OF FORMATION OF MONATOMIC GASES FROM ELEMENTS IN THEIR STANDARD STATES AT 298°K (kcal/mole of monatomic gas)*

H						
52.09						
Li			C	N	O	F
37.07			171.70	112.96	59.16	18.3
Na	Mg	Al	Si	P	S	Cl
25.98	35.9	75.0	88.04	75.18	53.25	29.01
K	Ca		Ge	As	Se	Br
21.51	46.0		78.44	60.64	48.37	26.71
Rb			Sn	Sb	Te	I
20.51			72.	60.8	47.6	25.48
Cs			Pb	Bi		
18.83			46.34	49.7		

B.[a] *APPROXIMATE BOND ENERGIES*[b] *AT 298°K (kcal/mole)*

	C	N	O	F	Si	P	S	Cl	Br	I
H	98.8	93.4	110.6	134.6	70.4	76.4	81.1	103.2	87.5	71.4
C−	83.1	69.7	84.0	105.4	69.3		62.0	78.5	65.9	57.4
C=	147	147	174				114			
C≡	194	213								
N−	69.7	38.4		64.5				47.7		
N=	147	100								
N≡	213	226								
O−	84.0		33.2	44.2	88.2			48.5		
O=	174									

[a] From L. Pauling, *The Nature of the Chemical Bond* (Cornell University Press, Ithaca, N.Y., 1960), 3rd ed. See also T. L. Cottrell, *The Strengths of Chemical Bonds* (Butterworths, London, 1958) 2nd ed.

[b] This is an example of loose but conventional terminology. These are actually bond *enthalpies* at 298°K, in the sense that they were obtained from enthalpies of formation.

APPENDIX 6

n	$\int_{x=0}^{\infty} e^{-bx^2} x^n \, dx$
0	$\frac{1}{2}\left(\frac{\pi}{b}\right)^{\frac{1}{2}}$
1	$\frac{1}{2b}$
2	$\frac{1}{4b}\left(\frac{\pi}{b}\right)^{\frac{1}{2}}$
3	$\frac{1}{2b^2}$
4	$\frac{3}{8b^2}\left(\frac{\pi}{b}\right)^{\frac{1}{2}}$
n (odd)	$\frac{\left(\frac{n-1}{2}\right)!}{2b^{(n+1)/2}}$
n (even)	$\frac{(n-1)(n-3)\cdots 3\cdot 1}{2^{(n+2)/2}\, b^{n/2}}\left(\frac{\pi}{b}\right)^{\frac{1}{2}}$

n	$\int_0^{\pi} (\sin\theta)^n \, d\theta$
0	π
1	2
2	$\frac{\pi}{2}$
3	$\frac{3}{4}$
4	$\frac{3}{4}\frac{\pi}{2}$
5	$\frac{16}{15}$

Index